100 Years *of* Relativity

Space-Time Structure:
Einstein and Beyond

100 Years of Relativity

of

Relativity

Space-Time Structure:
Einstein and Beyond

editor

Abhay Ashtekar

Institute for Gravitational Physics and Geometry,
Pennsylvania State University, USA

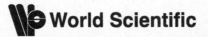

World Scientific

NEW JERSEY · LONDON · SINGAPORE · BEIJING · SHANGHAI · HONG KONG · TAIPEI · CHENNAI

Published by

World Scientific Publishing Co. Pte. Ltd.
5 Toh Tuck Link, Singapore 596224
USA office: 27 Warren Street, Suite 401-402, Hackensack, NJ 07601
UK office: 57 Shelton Street, Covent Garden, London WC2H 9HE

British Library Cataloguing-in-Publication Data
A catalogue record for this book is available from the British Library.

Cover Art by Dr. Cliff Pickover, Pickover.Com

First published 2005
Reprinted 2006

100 YEARS OF RELATIVITY
Space-Time Structure: Einstein and Beyond

ISBN-13 978-981-256-394-1
ISBN-10 981-256-394-6
ISBN-13 978-981-270-030-8 (pbk)
ISBN-10 981-270-030-7 (pbk)

Printed in Singapore

PREFACE

The goal of this volume is to describe how our understanding of space-time structure has evolved since Einstein's path-breaking 1905 paper on special relativity and how it might further evolve in the next century.

Preoccupation with notions of Space (the Heavens) and Time (the Beginning, the Change and the End) can be traced back to at least 2500 years ago. Early thinkers from Lao Tsu in China and Gautama Buddha in India to Aristotle in Greece discussed these issues at length. Over centuries, the essence of Aristotle's commentaries crystallized in the Western consciousness, providing us with mental images that we commonly use. We think of space as a three-dimensional continuum which envelops us. We think of time as flowing serenely, all by itself, unaffected by forces in the physical universe. Together, they provide a stage on which the drama of interactions unfolds. The actors are everything else — stars and planets, radiation and matter, you and me. In Newton's hands, these ideas evolved and acquired a mathematically precise form. In his masterpiece, the *Principia*, Newton spelled out properties of space and the absolute nature of time. The *Principia* proved to be an intellectual *tour de force* that advanced Science in an unprecedented fashion. Because of its magnificent success, the notions of space and time which lie at its foundations were soon taken to be obvious and self-evident. They constituted the pillars on which physics rested for over two centuries.

It was young Einstein who overturned those notions in his paper on special relativity, published on 26th September 1905 in *Annalen der Physik* (Leipzig). Lorentz and Poincaré had discovered many of the essential mathematical underpinnings of that theory. However, it was Einstein, and Einstein alone, who discovered the key *physical* fact — the Newtonian notion of absolute simultaneity is physically incorrect. Lorentz transformations of time intervals and spatial lengths are not just convenient mathematical ways of reconciling experimental findings. *They are physical facts.* Newton's assertion that the time interval between events is absolute and observer-independent fails in the real world. The Galilean formula for transformation of spatial distances between two events is physically incorrect. What seemed obvious and self-evident for over two centuries is neither. Thus, the model of space-time that emerged from special relativity was very different from that proposed in the *Principia.*

Space-time structure of special relativity has numerous radical ramifications — such as the twin "paradox" and the celebrated relation $E = Mc^2$ between mass and energy — that are now well known to all physicists. However, as in the *Principia*, space-time continues to remain a backdrop, an inert arena for dynamics, which cannot be acted upon. This view was toppled in 1915, again by Einstein, through his discovery of general relativity. The development of non-Euclidean geometries had led to the idea, expounded most eloquently by Bernhard Riemann in 1854, that the geometry of physical space may not obey Euclid's axioms — it may be curved due to the presence of matter in the universe. Karl Schwarzschild even tried to measure this curvature as early as 1900. But these ideas had remained speculative. General relativity provided both a comprehensive conceptual foundation and a precise mathematical framework for their realization.

In general relativity, gravitational force is not like any other force of Nature; it is encoded in the very geometry of space-time. Curvature of space-time provides a direct measure of its strength. Consequently, space-time is not an inert entity. It acts on matter and can be acted upon. As John Wheeler puts it: *Matter tells space-time how to bend* and *space-time tells matter how to move.* There are no longer any spectators in the cosmic dance, nor a backdrop on which things happen. The stage itself joins the troupe of actors. This is a profound paradigm shift. Since all physical systems reside in space and time, this shift shook the very foundations of natural philosophy. It has taken decades for physicists to come to grips with the numerous ramifications of this shift and philosophers to come to terms with the new vision of reality that grew out of it.

Einstein was motivated by two seemingly simple observations. First, as Galileo demonstrated through his famous experiments at the leaning tower of Pisa, the effect of gravity is universal: all bodies fall the same way if the only force acting on them is gravitational. This is a direct consequence of the equality of the inertial and gravitational mass. Second, gravity is *always* attractive. This is in striking contrast with, say, the electric force where unlike charges attract while like charges repel. As a result, while one can easily create regions in which the electric field vanishes, one cannot build gravity shields. Thus, gravity is omnipresent and nondiscriminating; it is everywhere and acts on everything the same way. These two facts make gravity unlike any other fundamental force and suggest that gravity is a manifestation of something deeper and universal. Since space-time is also omnipresent and the same for all physical systems, Einstein was led to regard gravity not as a force but a manifestation of space-time geometry. Space-time of general relativity is supple, depicted in the popular literature as a rubber sheet, bent by massive bodies. The sun, for example, bends space-time nontrivially. Planets like earth move in this curved geometry. In a precise mathematical sense, they follow the simplest trajectories called geodesics — generalizations of straight lines of the flat geometry of Euclid to the curved geometry of Riemann. Therefore, the space-time trajectory of earth is as "straight" as it could be. When projected into spatial sections defined by sun's rest frame, it appears elliptical from the flat space perspective of Euclid and Newton.

The magic of general relativity is that, through elegant mathematics, it transforms these conceptually simple ideas into concrete equations and uses them to make astonishing predictions about the nature of physical reality. It predicts that clocks should tick faster on Mont Blanc than in Nice. Galactic nuclei should act as giant gravitational lenses and provide spectacular, multiple images of distant quasars. Two neutron stars orbiting around each other must lose energy through ripples in the curvature of space-time caused by their motion and spiral inward in an ever tightening embrace. Over the last thirty years, precise measurements have been performed to test if these and other even more exotic predictions are correct. Each time, general relativity has triumphed. The accuracy of some of these observations exceeds that of the legendary tests of quantum electrodynamics. This combination of conceptual depth, mathematical elegance and observational success is unprecedented. This is why general relativity is widely regarded as the most sublime of all scientific creations.

Perhaps the most dramatic ramifications of the dynamical nature of geometry are gravitational waves, black holes and the big-bang.[a]

---

[a]History has a tinge of irony — Einstein had difficulty with all three! Like all his contemporaries, he believed in a static universe. Because he could not find a static solution to his original field equations, in 1917 he introduced a cosmological constant term whose repulsive effect could balance the gravitational attraction, leading to a large scale time independence. His belief in the static universe was so strong that when Alexander Friedmann discovered in 1922 that the original field equations admit a cosmological solution depicting an expanding universe, for a whole year, Einstein thought Friedmann had made an error. It was only in 1931, after Hubble's discovery that the universe is in fact dynamical and expanding that Einstein abandoned the extra term and fully embraced the expanding universe. The story of gravitational waves is even more surprising. In 1936 Einstein and Nathan Rosen submitted a paper to *Physical Review* in which they argued that Einstein's 1917 weak field analysis was misleading and in fact gravitational waves do not exist in full general relativity. It is now known that the paper was refereed by Percy

A. Ashtekar

Gravitational waves are ripples in the curvature of space-time. Just as an oscillating electric dipole produces electromagnetic waves, using the weak-field approximation of general relativity, Einstein showed already in 1917 that a time-changing mass quadrupole produces these ripples. In the 1960's Hermann Bondi, Rainer Sachs, Roger Penrose and others extended the theory to full nonlinear general relativity. They firmly established that these ripples are not "coordinate-gauge effects" but have direct physical significance. They carry energy; as Bondi famously put it, one could *boil water with them*. Through careful observations of a binary pulsar, now spanning three decades, Russell Hulse and Joseph Taylor showed that its orbit is changing precisely in the manner predicted by general relativity. Thus, the reality of gravitational waves is now firmly established. Gravitational wave observatories have been constructed to receive them directly on earth. It is widely expected that they will open a new window on the universe.

General relativity ushered-in the era of modern cosmology. At very large scales, the universe around us appears to be spatially homogeneous and isotropic. This is the grandest realization of the Copernican principle: our universe has no preferred place nor favored direction. Using Einstein's equations in 1922, Alexander Friedmann showed that such a universe cannot be static. It must expand or contract. In 1929, Edwin Hubble found that the universe is indeed expanding. This in turn implies that it must have had a beginning where the density of matter and curvature of space-time were infinite. This is the *big-bang*. Careful observations, particularly over the last decade, have shown that this event must have occurred some 14 billion years ago. Since then, galaxies are moving apart, the average matter content is becoming dilute. By combining our knowledge of general relativity with laboratory physics, we can make a number of detailed predictions. For instance, we can calculate the relative abundances of light chemical elements whose nuclei were created in the first three minutes after the big-bang; we can predict the existence and properties of a primal glow — the cosmic

---

Robertson, a renowned relativist, who pointed out the error in the argument. Einstein withdrew the paper saying that he had not authorized the Editors "to show the paper to specialists before it was printed" and "he saw no reason to address the — in any case erroneous — comments of your anonymous expert". Finally, he did not believe in the existence of black holes. As late as 1939, he published a paper in the *Annals of Mathematics* arguing that black holes could not be formed through the gravitational collapse of a star. The calculation is correct but the conclusion is an artifact of a nonrealistic assumption. Just a few months later, Robert Oppenheimer and Hartland Snyder — also members of the Princeton Institute of Advanced Study — published their now classic paper establishing that black holes do in fact result.

microwave background — that was emitted when the universe was some 400,000 years old; and we can deduce that the first galaxies formed when the universe was a billion years old. What an astonishing range of scales and variety of phenomena! In addition, general relativity also changed the philosophical paradigm to phrase questions about the Beginning. It is not just matter but the *space-time itself that is born at the big-bang*. In a precise sense, the big-bang is a boundary, a frontier, where space-time ends. General relativity declares that physics stops there; it does not permit us to look beyond.

Through black holes, general relativity opened up the third unforeseen vista. The first black hole solution to Einstein's equation was discovered already in 1916 by Schwarzschild while serving on the front lines during the First World War. However, acceptance of its physical meaning came very slowly. Black holes are regions in which the space-time curvature is so strong that even light cannot escape. Therefore, according to general relativity, they appear pitch black to outside observers. In the rubber sheet analogy, the bending of space-time is so extreme inside a black hole that space-time is *torn-apart*, forming a singularity. As at the big-bang, curvature becomes infinite. Space-time develops a final boundary and physics of general relativity simply stops.

And yet, black holes appear to be mundanely common in the universe. General relativity, combined with our knowledge of stellar evolution, predicts that there should be plenty of black holes with 10 to 50 solar masses, the end products of lives of large stars. Indeed, black holes are *prominent* players in modern astronomy. They provide the powerful engines for the most energetic phenomena in the universe such as the celebrated gamma ray bursts in which an explosion spews out, in a few blinding seconds, as much energy as from a 1000 suns in *their entire life time*. One such burst is seen every day. Centers of elliptical galaxies appear to contain a huge black hole of millions of solar masses. Our own galaxy, the Milky Way, has a black hole of some 3 million solar masses at its center.

General relativity is the best theory of gravitation and space-time structure we have today. It can account for a truly impressive array of phenomena ranging from the grand cosmic expansion to the functioning of the more mundane global positioning system on earth. But it is incomplete because it ignores quantum effects that govern the subatomic world. Moreover, the two theories are dramatically different. The world of general relativity has geometric precision, it is deterministic; the world of quantum physics is dictated by fundamental uncertainties, it is probabilistic. We maintain a happy, schizophrenic attitude, using general relativity to describe the large scale phenomena of astronomy and cosmology and quantum mechanics to account for properties of atoms and elementary particles. This is a viable strategy because the two worlds rarely meet. Nonetheless, from a conceptual standpoint, this is highly unsatisfactory. Everything in our experience as physicists tells us that there should be a grander, more complete theory from which general relativity and quantum physics arise as special, limiting cases. This would be the quantum theory of gravity. It would take us beyond Einstein.

Perhaps the deepest feature of general relativity is its encoding of gravity into the very geometry of space-time. It is often the case that the deepest features of a theory also bring out its limitations. General relativity is no exception. As we saw, it implies that space-time itself must end and physics must stop when curvature becomes infinite. However, when curvatures become large, comparable to the Planck scale, one can no longer ignore quantum physics. In particular, near the big-bang and black hole singularities the world of the very large and of the very small meet and predictions of classical general relativity cannot be trusted. One needs a quantum theory of gravity. Although they seem arcane at first, singularities are our gates to go beyond general relativity. Presumably, quantum space-time continues to exist and real physics cannot stop there. To describe what really happens, once again we must dramatically revise, our notions of space and time. We need a new syntax.

This volume is divided into three parts. The first discusses the conceptual transition from Newtonian notions of space-time to those of special and general relativity. The second part is devoted to the most striking features of general relativity, especially the three ramifications outlined above. These contributions cover well established theoretical results within general relativity as well as applications and experimental status of the theory. The third part presents various approaches to quantum gravity. Here the emphasis is on new conceptual elements underlying the emerging paradigms and predictions they lead to. Is the dimension of space-time more than four? What replaces the continuum at the Planck scale? Is there a specific type of underlying discrete structure? Must one replace space-time with another fundamental structure already at the classical level, without any reference to the Planck scale? Can a consistent quantum gravity theory emerge by itself or must it necessarily unify all forces of Nature? Must it simultaneously complete our understanding of the quantum measurement theory? Different approaches adopt different — and often diametrically opposite

— viewpoints to these fundamental questions and therefore the resulting theories have strikingly different features. This variety is essential at the present stage because we are still rather far from a fully satisfactory quantum gravity theory. Furthermore, as yet there is no hard experimental data to guide us. Therefore, as we celebrate the 100th anniversary of Einstein's *Annus Mirabilis* it is important that we maintain a long range perspective. Indeed, we would do well to avoid the traps that the celebrated biologist François Jacob warned all scientists about:

> *The danger for scientists is not to measure the limits of their science, and thus their knowledge. This leads to mix what they believe and what they know. Above all, it creates the certitude of being right [prematurely].*

The third part of this endeavors to maintain the necessary openness.

Abhay Ashtekar
September, 2005

ACKNOWLEDGMENTS

I am grateful to all contributors for graciously agreeing to work with somewhat pressing deadlines and for making a special effort to provide a long range perspective on the field to the broad community of physicists, not just experts in gravitation theory. I would also like to thank numerous referees who reviewed the articles and made valuable comments. Throughout this project, the authors and I received efficient help from the World Scientific staff. I would especially like to thank our contact Editor, Ms Lakshmi Narayan for her diligence, care, flexibility and willingness to cater to a number of last minute requests. Finally, the idea of commemorating the *Annus Mirabilis* with a special volume addressed to nonexperts came from Dr. K. K. Phua. My initial hesitation was overcome by his persuasive arguments and willingness to provide all the support we needed to complete this project.

This work was supported in part by NSF grants PHY 0090091, PHY 0354932 and PHY 0456931; Alexander von Humboldt Foundation of Germany; Sir C. V. Raman Chair of the Indian Academy of Science; and by the Eberly research funds of Penn State.

CONTENTS

Part I

From Newton to Einstein
Paradigm Shifts

It is as if a wall which separated us from the truth has collapsed. Wider expanses and greater depths are now exposed to the searching eye of knowledge, regions of which we had not even a pre-sentiment.
—Hermann Weyl, On the Discovery of General Relativity

CHAPTER 1

DEVELOPMENT OF THE CONCEPTS OF SPACE, TIME AND SPACE-TIME FROM NEWTON TO EINSTEIN

JOHN STACHEL

Dept. of Physics and Center for Einstein Studies, Boston University, 590 Commonwealth Avenue, Boston, MA 02215, USA

The concept of physical change brings together the concepts of space and time. The evolution of the latter two concepts, and of relation between them, in physical theories from Newtonian mechanics to general relativity is outlined, culminating in the development of the concept of space-time and its dynamization. The chrono-geometrical and inertio-gravitational space-time structures are defined, and the compatibility relations between the two are discussed. The philosophical debate between absolute and relational concepts of space and then of space-time is reviewed, as is the contrast between pre-general-relativistic theories with fixed background space-time structures and background-free general relativistic theories. Some implications of this contrast for the problem of quantum gravity are indicated.

1. Introduction: The Changing Nature of Change

This chapter discusses the development of the concepts of space and time as employed in the formulation of various physical theories. These two concepts in turn are intimately connected with the concept of *change*, so time, space, and change must be discussed together. When considered apart from the causes of change, they form the subject matter of *kinematics*.[a]

Traditionally, one distinguished between two types of change:[b]

(1) the *change of position in space* of some object in the course of *time*, i.e., *motion*; and

---

[a]Under what circumstances kinematics can and cannot be cleanly separated from dynamics will form an important topic for later discussion.

[b]Naturally, the two types can be combined to describe changes of the properties of an object as it changes its position in space.

(2) the *change over time* of some *property* of an object – quantitative or qualitative – at the same position in space.

Change of some property leads to a concept of time that I call *local*, since it applies to sequences of events at the *same place*; while motion leads to a concept of time that I call *global*, since it requires the comparison of times at *different places* in space.[c] When all physics was based on Newtonian mechanics, with its ultimately atomistic basis,[d] this distinction sufficed. The mathematical description of both types of change employed *total derivatives* (d/dt) with respect to the Newtonian *absolute time*. Even today, after abandonment of this concept, such descriptions are still useful when dealing with discrete *objects* (often called 'particles') and employing appropriate concepts of *local* and *global time* (see the detailed discussion later of the various concepts of time).

But the development, in the course of the 19th century, of the novel concept of (physical) *fields* filling all of space and changing over time, first in optics and then in electrodynamics, led to the need for a new type of mathematical description of change,[e] using *partial derivatives* with respect to a global temporal variable evaluated at some *fixed* point in space.[f] The failure of all attempts to bring the field concept into harmony with Newtonian mechanics ultimately led to the critical examination of such concepts as *at the same place at different times* and *at the same time at different places*, and hence to the special theory of relativity.

2. The Bronstein Cube

Starting from the concepts of time and space associated with Galilei-Newtonian physics, I shall then discuss the changes in these concepts necessitated by the special theory of relativity, and finally, emphasize the even

---

[c]The concepts of time and place each will be examined more critically in later sections.

[d]The mechanics of continuous media is based on the idea that one can follow the trajectory of each 'particle' of the medium, resulting in the so-called *Lagrangian* description of its dynamics.

[e]Attempts were made to incorporate the concept of field within Newtonian mechanics by regarding such fields as states of hypothetical mechanical media or *ether*. But all attempts to treat the ether as a mechanical system ultimately proved fruitless, and it came to be recognized that field is a new, non-mechanical concept. In spite of attempts to unify physics on the basis of either the field or particle concept, contemporary physics still has a dualistic basis.

[f]In point of fact, this first occurred in fluid dynamics, resulting in the so-called *Eulerian* description of fluid dynamics. But in this case, one can always revert to a Lagrangian description (see note 4).

BRONSTEIN CUBE

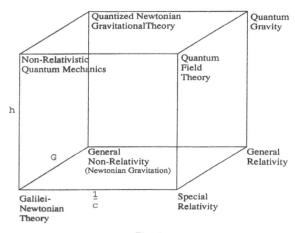

Fig. 1

more striking changes brought about by the advent of the general theory. My approach is not primarily historical, but what I call (in tribute to Ernst Mach) *historical-critical*. That is, while broadly following the historical order of development of the subject, I do not hesitate to violate that order by introducing more recent concepts and/or viewpoints whenever this facilitates the understanding of some current question.

A convenient way to orient oneself amidst the various physical theories to be discussed is to use what I call the *Bronstein cube* (see Fig. 1).[g] At the lower left-hand corner of the cube is the starting point of modern physics (not including gravitation), Galilei-Newtonian Theory. By adding the Newtonian gravitational constant G, the speed of light in vacuum c, and Planck's constant h in various combinations, one reaches each of the physical theories listed on the other corners of the cube. Most of these transitions are fairly clear, but some require further comment. The space-time structures of Galilei-Newtonian Theory and Special Relativity are *unique*; but when G is introduced (see Fig. 2), resulting in General Non-Relativity and General Relativity, these names denote *classes* of space-time structures — one for each solution to Newton's and Einstein's gravitational field equations

---

[g]For further discussion of the Bronstein cube, see Ref. 14.

BRONSTEIN SQUARE

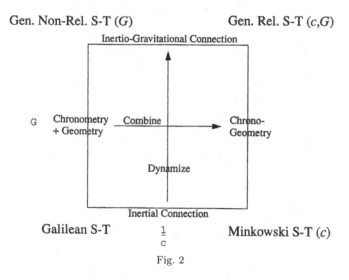

Gen. Non-Rel. S-T (*G*)   Gen. Rel. S-T (*c,G*)

Inertio-Gravitational Connection

G   Chronometry   Combine   Chrono-
     + Geometry   Geometry

                    Dynamize

Inertial Connection

Galilean S-T $\frac{1}{c}$ Minkowski S-T (*c*)

Fig. 2

respectively. Furthermore, Newtonian gravitation theory is considered in its modern, four-dimensional formulation, which takes into account the equivalence principle (see Sec. 11); I have adopted the name for this formulation of Newtonian theory introduced by Jürgen Ehlers – *General Non-Relativity* — which contrasts nicely with *General Relativity*. But the name should not be taken to imply that the theory has *no* relativity (i.e., no symmetry group of the space-time): indeed, the chronogeometry of this class of space-times is invariant under a wider group than that of Galilei-Newtonian space-time, as discussed in Secs. 5 and 8.

The upper right-hand corner of the cube, *Quantum Gravity*, also calls for further comment. It is not the name of an already-existing theory, but rather of a hope: The goal of much current research in theoretical physics is to find a theoretical framework wide enough to encompass (in some sense) both classical general relativity and (special-relativistic) quantum field theory. Unfortunately, in the current state of this subject, the edges of the cube do not commute (see Figs. 3, 4). Starting from quantum field theory on the road to quantum gravity and attempting to incorporate the Newtonian constant *G*, most particle physicists believe that one will arrive at some version

BRONSTEIN CUBE

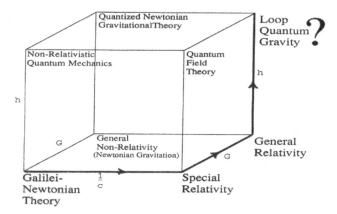

Fig. 3. General Relativists' Viewpoint.

BRONSTEIN CUBE

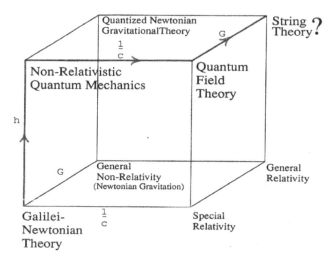

Fig. 4. Particle Physicists' Viewpoint.

of string theory, M-theory or what have you –opinion about the ultimate goal seems divided. Starting from general relativity and attempting to incorporate Planck's constant h, most general relativists believe that one must develop a diffeomorphism-invariant formulation of quantum theory – a goal of the loop quantum gravity program that has not yet been fully achieved. Only time, and successful incorporation of quantum-gravitational phenomena when they are discovered, will tell which path (if either) ultimately proves to be more fruitful.[h]

3. Demokritos versus Aristotle: 'Space' versus 'Place'

Before turning to the modern period, it is instructive to recall that ancient Greek natural philosophy was already the site of a conflict between two concepts of space: the *absolute* concept of the Greek atomists, according to which space, referred to as 'the void,' is a container, in which atoms of matter move about. Demokritos of Abdera asserted:

> By convention are sweet and bitter, hot and cold, by convention is color; in truth are atoms and void (Fragment 589 in Ref. 6, p. 422).

Aristotle criticized the atomists in these words:

> The believers in its [the void's] reality present it to us as if it were some kind of receptacle or vessel, which may be regarded as full when it contains the bulk of which it is capable, and empty when it does not (*Physics*, book VI).

He espoused the *relational* concept, according to which space has no independent existence, but is just a name for the collection of all spatial relations between material entities. A void cannot exist, motion is just the displacement of one portion of matter by another; thus, Aristotle's doctrine is really a theory of *place* rather than of space.

Aristotelianism triumphed together with Christianity, and atomism practically vanished from the western philosophical tradition for almost two millennia. With its revival in early modern times and subsequent adoption by Newton, the conflict between the absolute and relational concepts also revived as the 17th–18th century battle between Newtonianism and Cartesianism (the philosophy of René Descartes). Voltaire wittily observed:

---

[h]Elsewhere I have offered some arguments explaining why I favor the latter course, or some even more radical departure from background-dependent physics. See Ref. 15.

A Frenchman, who arrives in London, will find philosophy, like everything else, very much changed there. He had left the world a *plenum*, and he now finds it a *vacuum* (*Lettres philosophiques.* Letter XIV, On Descartes and Sir Isaac Newton).[i]

This time it was the Newtonian absolute concept of space that triumphed in spite of the cogent arguments of Leibniz and Huygens against it.

In fine, the better to resolve, if possible, every difficulty, he [Newton] proves, and even by experiments, that it is impossible there should be a *plenum*; and brings back the *vacuum*, which Aristotle and Descartes had banished from the world (Voltaire, *ibid.*, Letter XV, On Attraction).

As Leonhard Euler emphasized, absolute space seemed to be demanded by Newtonian dynamics.[j]

4. Absolute versus Relational Concepts of Space and Time

Einstein[4] (p. xiv) summarizes the conflict in these words:

Two concepts of space may be contrasted as follows: (a) space as positional quality of the world of material objects; (b) space as container of all material objects. In case (a), space without a material object is inconceivable. In case (b), a material object can only be conceived as existing in space; space then appears as a reality, which in a certain sense is superior to the material world.

A similar contrast can be made between two concepts of time, but now in my words:

(a) *time* as an ordering quality of the world of material events;
(b) *time* as a container of all material events.

On the basis of the first concept, time without a material event is inconceivable. On the basis of the second, a material event can only be conceived as existing in time; time then appears as a reality that in a certain sense is superior to the material world.

---

[i] Cited from the English translation.[20]
[j] See Ref. 5.

Logically, it would seem possible to combine either point of view about space with either point of view about time; but in fact most seventeenth-century natural philosophers either adopted the absolute viewpoint about both, like Newton; or the relational viewpoint, like Descartes, and following him Leibniz and Huygens.

5. Universal Temporal Order

I shall return later to the question of absolute versus relational concepts of time, but for the moment focus on a presupposition about temporal order common to both views before the advent of the special theory of relativity. Whether inherent in the nature of the events defining this order, or drawn from the immersion of these events in an independent temporal continuum, temporal order was presumed to be unique and universal. For a continuous sequence of events occurring 'at the same place,'[k] this order is indeed unique for each sequence of events, as is the associated concept of *local time*. The existence of a *universal* or *absolute*[l] time connecting events at different places depends on the existence of a *unique* temporal order common to *all* possible local times, i.e., all possible sequences of events at the 'same place', at all places.[m]

For events occurring at 'different places', a convention is still needed to define 'simultaneity' of events and allow introduction of a *global time*. Simultaneity must be an *equivalence relation*, i.e., reflexive (any event is simultaneous with itself), symmetric (if event a is simultaneous with event b, then b is simultaneous with a), and transitive (if a is simultaneous with b and b is simultaneous with c, then a is simultaneous with c). But an equivalence relation between the elements of any set divides the set into *equivalence classes*: Every element of the set belongs to one and only one class, all members of which are equivalent under the relation. If an absolute time exists, it provides a 'natural' equivalence relation between distant events (two events are temporally equivalent iff they occur at the same absolute time), and hence may be used to define *simultaneity* and a *global*

---

[k]For the moment, taking for granted the meaning of this expression.

[l]Note that the word 'absolute' is used here in a different sense than earlier in this chapter, where absolute and relational concepts of time and space were contrasted. It would be better to use the term 'universal' here, but absolute is so generally used in this context that I shall follow this usage in spite of the possibility of confusion. As will be seen below, in this context the contrast is between *absolute* or *universal* (i.e., independent of frame of reference) and *relative* (i.e., dependent on frame of reference).

[m]Mathematically, this is the requirement that the local time along any sequence of events be a perfect differential of the unique global time.

time. In the context of Newtonian kinematics, this is obvious, and usually is not even mentioned; but it is a convention nonetheless. We thereby transfer the local temporal ordering of events at one place (which place is chosen does not matter because of the existence of the absolute simultaneity relation) to a global temporal ordering of events at different places. Hence, in Galilei-Newtonian physics, the concepts of local and global time, although conceptually distinct, effectively coincide in the idea of a universal or absolute time. Operationally, this means that (ideal) clocks, once synchronized at one place in space, will always agree on their readings wherever they are compared again, and whatever their intermediate histories. This assertion has been tested, and found to be false — but only long after Newtonian kinematics had been replaced for other reasons.

6. Relative Space

Now I turn to the long-promised examination of the concept of 'at the same place'. Just as his dynamical theory presupposes the existence of a universal or absolute time, Newton believed that it also required the existence of a universal or absolute space. If this were so, it would be clear what 'at the same place' means: at the same place in absolute space. But Newton himself was aware of an 'operational' problem with the concept of absolute space: While it is rather easy to give an experimental prescription for deciding whether or not a body is in absolute rotation (Newton's bucket thought experiment, or Foucault's later actual pendulum) or (if we neglect gravitation) if it is absolutely accelerated linearly, there is no prescription enabling a decision on whether a body is at rest or in uniform (i.e., unaccelerated) linear motion with respect to a hypothetical absolute space. In other words, the laws of Newtonian mechanics do not allow the singling out, by means of the result of any mechanical experiment, of a state of absolute rest from the class of all uniform motions, which much late came to be called inertial systems.

If one believes, as did adherents of the mechanical world-view, that all physical phenomena could be explained as effects of mechanical interactions, this inability implied the abandonment of the autocratic concept of a single absolute space, and its replacement by a democracy of so-called *inertial frames* of reference, as was realized by the end of the 19th century.[n] An inertial frame can be defined as one in a state of un-accelerated motion, that is, one in which Newton's First Law is valid: An object acted

---

[n] In particular, Ludwig Lange introduced both the concept and the name 'inertial system' in 1885. See the discussion of Lange in Ref. 18, pp. 17ff.

upon by no (net) external forces moves in a straight line with constant velocity with respect to an inertial frame of reference.° This equivalence of all inertial frames has been called the *relativity principle* of Newtonian mechanics. There is a three-fold infinity of inertial frames, each of which is in a state of uniform motion with respect to all the others, and in each of which Newton's laws of mechanics are equally valid.

With the abandonment of absolute space, the concept 'at the same place' loses its absolute significance, and must be replaced by the concept 'at the same place relative to some inertial frame of reference.' P

In summary, whether or not adherents of the mechanical worldview were aware of it, Newtonian mechanics properly understood *does* require a universal or absolute concept of time, but leads to abandonment of the concept of absolute position in favor of a concept of *relative* position with respect to some reference frame, usually chosen to be inertial.

The concepts of absolute time and relative space may be given operational significance in terms of the measurement of temporal and spatial intervals with (ideal) measuring clocks and rods, respectively. For example, the temporal interval ('time') between two events, as measured by two clocks is independent of the inertial frame of reference in which the clocks are at rest (and therefore, they need not both be at rest in the same frame). On the other hand, the spatial interval (distance) between two non-simultaneous events, as measured by a rigid measuring rod for example, is not absolute, but depends on the inertial frame of reference in which the measuring rod is at rest. Since there is no preferred inertial frame of reference (no 'absolute space'), none of these relative distances can claim the title of 'the (absolute) distance'.

For simultaneous events, of course, the distance between them is independent of the inertial frame in which the measuring rod is at rest: It makes no difference whether the rod is at rest or in motion relative to the two events. When one speaks of 'the length' of an extended object, what is

---

°The phrase 'with constant velocity' implies that an appropriate definition of distant simultaneity must be adopted when, as in the special theory of relativity, the concept of absolute time is abandoned. When we come to consider gravitation, we shall have to examine the phrase: 'acted on by no (net) external forces' more critically.

POne might be tempted to generalize, and say: 'with respect to any arbitrary frame of reference'. But in both Galilei-Newtonian and special-relativistic theories, the inertial frames maintain their privileged role: The laws of physics take their simplest form when expressed relative to these frames – as long as gravitation is disregarded. As we shall see, when gravitation is taken into account, the inertial frames lose their privileged role even in Galilei-Newtonian theory.

meant is the distance between two simultaneous events, one at each end of the object, and in Newtonian physics this is an absolute concept.

The triumph of the wave theory of optics in the mid-19th century, and its subsequent union with electrodynamics in Maxwell's theory of electromagnetism, led to the introduction of a hypothetical medium, the *luminiferous ether*, as the seat of light waves and later of all electric and magnetic fields. After extensive debates in the course of the century about the nature of the interaction between the ether and ordinary matter[q], the viewpoint espoused by H. A. Lorentz prevailed: The ether fills all of space – even where ordinary matter is present – and remains immobile, even when matter moves through it. Ejected from mechanics, Newton's absolute space seemed to return in the form of this immobile ether, which provided a unique frame of reference for optical and electrodynamic phenomena. In particular, the speed of light (or any other electromagnetic wave) was supposed to have a fixed value with respect to the ether, but according to Newtonian kinematics (the Galileian law of addition of velocities) its value with respect to any other inertial frame of reference in motion relative to the ether should differ – in particular with respect to the inertial frames defined at each instant by the earth in its motion.

Yet all attempts to measure such a change in the velocity of light – or any other predicted effect of motion through the ether on the behavior of light or any other electrodynamic phenomenon – failed. Various efforts to resolve this dilemma within the framework of classical kinematics raised their own problems,[r] until Einstein[2] cut the Gordian knot, pointing out that:

> The theory to be developed - like every other electrodynamics - is based upon the kinematics of rigid bodies, since the assertions of any such theory concern relations between rigid bodies (coordinate systems), clocks, and electromagnetic processes. Insufficient consideration of this circumstance is at the root of the difficulties with which the electrodynamics of moving bodies currently has to contend.

7. Relative Time

A re-analysis of the foundations of Galilei-Newtonian kinematics led Einstein to pinpoint the concept of universal or absolute time as the source

---

[q]See Ref. 16.
[r]See *ibid.*

of the incompatibility of the relativity principle of Newtonian mechanics with Maxwell's electrodynamics of moving bodies. Once this concept was abandoned, he found that all reference to the ether could be eliminated, and the relativity principle — the full equivalence of all inertial frames — could be extended to *all* phenomena, electrodynamical and optical as well as mechanical. Elimination of the ether of course eliminated the need to find an explanation for the failure of all attempts to detect motion through it. Taken together with the relativity principle, the absence of a physically privileged ether frame implies that the speed of light must be the same in *all* inertial frames, a result that is clearly incompatible with the old kinematics. Einstein developed a new kinematics, in which temporal intervals — local or global — are no more universal or absolute than are spatial intervals: Both (global) time and space are relative to an inertial frame of reference. Since there is no universal or absolute time, clocks can no longer be synchronized absolutely, i.e., in a way that is independent of the method of synchronization; so the concept of distant simultaneity must be analyzed very carefully. As noted earlier, an element of arbitrariness (convention) enters into any definition of global time and hence of (relative) simultaneity. But, in each inertial frame, the choice of one convention is clearly superior since it results in the greatest simplicity in the expression of the laws of nature. This convention, applied in each inertial frame of reference, results in that frame having its own relative simultaneity. The convention can be defined operationally, for example, with the help of a system of equivalent clocks at rest in that frame. These clocks may be synchronized using any signal that has a speed in that frame that one may justifiably regard as independent of the direction of the signal. The speed of light in vacuum meets this criterion, and furthermore has the same value in all inertial frames if this synchronization convention is used in each. So, it provides a convenient, if not essential, signal for use in synchronization. The global time interval between two events taking place at different places now depends on ("is relative to") the inertial frame of reference, just as did the spatial interval between two (non-simultaneous) events in Galilei-Newtonian kinematics. Since simultaneity is now relative, the spatial interval between *any* two (space-like separated) events is also relative.

There is now, in many ways, much more symmetry between the properties of time intervals and of space intervals. For example, just as the proper length of an object is defined as the spatial interval between two simultaneous events, one at each end-point of the object, in an inertial frame of reference in which the object is at rest, the proper time read by a clock can be defined as the time interval between two ticks of the clock in a frame of reference in which the clock is at rest. A curious feature of the new concept of time is now apparent. According to the old concept, for any two non-simultaneous events, there is always a frame of reference in which both occur at the same place, because it is always possible to travel between the two events with sufficient speed. According to the new concept, the speed of light is a limiting velocity, which it is impossible to even reach, let alone exceed, by any material process. Hence, there are pairs of events that do not occur at the same place in any inertial frame of reference, but for which there is an inertial frame of reference with respect to which they occur simultaneously. Such pairs are said to be *space-like separated*. Similarly, there are pairs of events that do not occur at the same time with respect to any inertial frame of reference, but for which there is an inertial frame of reference with respect to which they occur at the same place. Such pairs are said to be *time-like separated*. Finally, there are pairs of events, neither space-like nor time-like separated, but which can be connected by a light signal. Such pairs are said to be *null* or *light-like separated*.

The local time is relative to the path in the special theory. For a continuous sequence of events taking place at some point that is not at rest (i.e., is in accelerated motion) in anyone inertial frame, we can define the *proper time* of the sequence as follows: Pick a finite sequence of events E_1, $E_2, \ldots, E_{(n-l)}$, E_n such that E_1 is the first and E_n the last. Calculate the proper time between the pairs of events $E_1 - E_2, \ldots, E_{(n-l)} - E_n$ in the sequence, and add them. Then take the limit of this sum while making the sequence of intermediate events more and more dense. The result is the *local time interval* of the sequence of events, usually called in relativity the *proper time interval*. This relativistic local or proper time between two events is quite distinct from the global or inertial-frame time. Most notably, it depends upon *the space-time path* between the two events, i.e., its elements are not perfect differentials. We are quite used to the idea that the (proper) distance travelled between any two points in space depends on the (space-like) path taken between them (e.g., the distance travelled between Boston and Beijing depends on whether you go by way of Paris or Cape Town), and that the straighter the path the shorter the distance. A similar situation now holds for time: The local (proper) time that elapses between two events depends on the (time-like) path taken between them. But there is an important difference:[s] the *straighter* the path, the *longer* the proper time elapsed.[t] This effect is the essence of the famed twin paradox.

We can summarize the contrast between the global and the local time as follows. The global time is *relative*, i.e., depends on the frame of reference. But no physical result can depend on this choice of global time used to describe a physical process. The local time is *absolute*, in the sense that it is independent of the frame of reference. But it is relative to a timelike path between two events.

8. Fixed Kinematical Structure

One could expatiate on many other curious features of the special-relativistic times (both local and global) as compared to the one New-tonian absolute time. Dramatic as are the differences between the two global concepts of time, however, they share an important common feature when compared with the general-relativistic concept. Both the Galilei-Newtonian and special-relativistic concepts of time (and of space as well) are based upon the existence of a *fixed* kinematic framework, the structure of which is independent of any dynamical processes taking place in space-time.

Whether one regards this kinematical framework as existing prior to and independently of these dynamical processes, or as defined by and dependent on them, depends on whether one adopts an absolute or a relational account of time and space.[u] But on either account, the kinematic structure is fixed once and for all by a ten-parameter Lie group, which has a representation as a group of symmetry transformations of the points of space-time. The dynamical laws governing any closed system are required to remain invariant under all transformations of this group; i.e., there is another representation of the group that acts on the basic dynamical variables of the system. There is a close connection between these two representations: to every space-time symmetry there corresponds a dynamical conservation law of a closed system.

---

[s]Note that, in spite of the closer analogy between space and time in special relativity, there is still a big difference between them!

[t]But note that, as a four-dimensional *extremum*, a time-like geodesic represents a saddle point among all possible paths between the two points. I am grateful to Roger Penrose for pointing this out to me.

[u]In special relativity, with its intimate commingling of space and time, it is hard to imagine how one could combine a relational account of one with an absolute account of the other.

In the case of Galilei-Newtonian space-time, this group is the inhomogeneous Galilei group. In the special-relativistic case, it is the inhomogeneous Lorentz (or Poincaré) group. Seven of the ten generators of the group take the same form in both groups: four spatio-temporal translations, expressing (respectively) the homogeneity of the relative space of each inertial frame and the uniformity of the time – absolute or relative – of each inertial frame; and three spatial rotations, expressing the isotropy of the relative space of each inertial frame. They correspond, respectively to the conservation of the linear momentum and energy, and of the angular momentum of the dynamical system.

The two groups differ in the form of the three so-called "boosts," which relate the spatial and temporal coordinates of any event with respect to two inertial frames in motion relative to each other. The Galilei-Newtonian boosts depend on the absolute time, which does not change from frame to frame; so they involve only a transformation of the spatial coordinates of an event with respect to the two frames. The special-relativistic boosts involve a transformation of the temporal as well as the spatial coordinates of an event. The boosts correspond to the center-of-mass conservation law for the dynamical system.

The well known spatial (Lorentz) contraction and time dilatation effects, which result from the different breakup of a spatio-temporal process (for that is what a ruler and a clock are) into spatial and temporal intervals with respect to two inertial frame in relative motion, can be deduced from the relativistic boost transformation formulae relating the global space and time coordinates of the two frames.

9. Fetishism of Mathematics

Before discussing the next topic, the four-dimensional formulation of space-time theories, I shall interject a word of caution about a possible pathology in theoretical physics that I call 'the fetishism of mathematics'. We invent physical theories to enable us to better comprehend and cope with the world, or rather with limited portions of the world. We employ mathematics as a vital tool in such attempts. Mathematical structures help us to correctly encode numerous, often extremely complicated, relations among physical concepts, relations of both a quantitative and a qualitative nature. If these mathematical structures have been judiciously chosen, they do more than encode the relations that led to their introduction: formal manipulation of the structures leads to the discovery of new relations between physical concepts that can then be tested successfully, and even to the development of new concepts, inherent int he theory but hitherto unrecognized. However, there are almost always redundant elements in the mathematical structure that have no obvious correspondents in the physical theory and perhaps indeed no relevance at all to the physical content of the theory. What is more, all-too-often there comes a point at which the mathematical structure leads to predictions that fail the test of experiment; or experiments yield relevant results that the theory seems unable to encompass. Then we may say that the mathematical structure and/or the physical theory, has reached its limits of validity. But before such a limit is reached, there is often a tendency to forget that the mathematical structure was introduced originally as a tool to help us encode known and discover new relations among physical concepts; that is, to forget about the constant dialectic between attempts to encompass new relations within the given structure and attempts to modify the structure itself in response to new relations that cannot be so encompassed. Instead, the mathematical structure is sometimes considered to be (or to represent) a more fundamental level of reality, the properties of which entail the concepts and relations of the physical theory and indeed those of the phenomenal world. Drawing on the language of Karl Marx, who speaks of 'the fetishism of commodities,'[v] I designate as 'the fetishism of mathematics' the tendency to endow the mathematical constructs of the human brain with an independent life and power of their own. Perhaps the most flagrant current examples of this fetishism are found in the realm of quantum mechanics (I need only mention the fetishism of Hilbert spaces; there are people who regard Hilbert spaces as real but tables and chairs as illusory!), but the fetishism of four-dimensional formalisms in relativity does not fall too far behind. In my exposition, I shall try not to fall into fetishistic language; but if I do, please regard it as no more than a momentary lapse.

The pioneers in the development of space-time diagrams were aware of the conceptual problems raised by this 'spatialization of time.'[w] While the representation of temporal intervals by spatial ones can be traced back to Aristotle, it was Nicholas Oresme who first plotted time and velocity in a two-dimensional diagram in the fourteenth century. He justified this step in these words:

---

[v]See Ref. 7.

[w]The phrase is due to Emile Myerson. For a historical review of the development of the concept of space-time, see Ref. 11; an enlarged English version, "Space-Time," containing full references for the historical citations in this chapter is available as a preprint.

And although a time and a line are [mutually] incomparable in quantity, still there is no ratio found as existing between time and time, which is not found among lines, and vice versa. (*A Treatise on the Configuration of Qualities and Motions*).

It was not until 1698 that Pierre Varignon combined spatial and temporal intervals in a single diagram. He commented:

Space and time being heterogeneous magnitudes, it is not really them that are compared together in the relation called velocity, but only the homogeneous magnitudes that express them, which are ... either two lines or two numbers, or any two other homogeneous magnitudes that one wishes (*Mémoire* of 6 July 1707).

Forty years later, Jean le Rond D'Alembert was even more explicit:

One cannot compare together two things of a different nature, such as space and time; but one can compare the relation of two portions of time with that of the parts of space traversed. ... One may imagine a curve, the abcissae of which represent the portions of time elapsed since the beginning of the motion, the corresponding ordinates designating the corresponding spaces traversed during these temporal portions: the equation of this curve expresses, not the relation between times and spaces, but if one may so put it, the relation of relation that the portions of time have to their unit to that which the portions of space traversed have to theirs (*Traité du dynamique*, 1743).

The only twentieth-century mathematician I have found who emphasizes the dimensional nature of physical quantities is J. A. Schouten, the eminent differential geometer. In Ref. 8, he distinguished between geometrical and physical quantities, pointing out (p. 126) that:

quantities in physics have a property that geometrical quantities do not. Their components change not only with transformations of coordinates but also with transformations of certain units.

Constantly bearing in mind this difference between mathematical and physical quantities can help avoid the fetishism of mathematics.

10. Four-Dimensional Formulation

Poincaré and Minkowski pioneered in showing how to represent special-relativistic transformations of space and time in a mathematically simple, elegant and fruitful form by the introduction of a four-dimensional formalism. With the use of this formalism, space and time coordinates can be combined in a four-dimensional diagram, representing what is now called *Minkowski space-time* or simply Minkowski space.[x] A point of Minkowski space is often called an *event*, even though strictly speaking that term should be reserved for some physical occurrence at this point of space-time. The history of a 'point of space' is represented by a so-called *world-line*: a one-dimensional curve in space-time representing the history of this point over proper (local) time, corresponding to a continuous sequence of events. The history of a three-dimensional region of space is represented in the diagram by a *world tube*; if the tube is filled with matter, it will consist of a congruence of (non-intersecting) world-lines that fills the entire tube (region of space-time). Such a congruence is often called a *timelike fibration* of the (region of) space-time.

An instant of (global) time is represented by a spacelike hypersurface; if there is also a timelike fibration in the region, the hypersurface will intersect each curve in the fibration once and only once. A (non-intersecting) family of such hypersurfaces filling all of (a region of) space-time represents a global time variable. Such a family is often called a *spacelike foliation* or more informally, a *slicing* of space-time.

To summarize: In the four-dimensional formalism, a (relative) space is represented by a particular timelike fibration, and a global time (absolute in Galilei-Newtonian space-time or relative in Minkowski space-time) by a particular spacelike foliation, of space-time. The proper time along any timelike world-line is the local time associated with a continuous sequence of events taking place along that world-line.

Of course, the introduction of time as a fourth dimension to facilitate the description of motion in space long predated relativity.[y] But in Galilei-Newtonian kinematics, due to the existence of the absolute time, four-dimensional space-time is foliated uniquely by a family of parallel

---

[x]Their original formulations were actually not as elegant and geometrically intuitive as they could have been. Concerned to make the analogy with Eulidean geometry as close as possible, both introduced an imaginary time coordinate and assimilated boosts to rotations in the resulting four-dimensional 'Euclidean' space, rather than to pseudo-rotations in what we now call Minkowski space.

[y]For some of this history, see Ref. 11.

hyperplanes of equal absolute time. In this sense, the four-dimensional unification of space and time remained somewhat formal.

Yet it did permit a simple geometrical representation of the relativity of space. An inertial frame of reference is represented by a timelike fibration consisting of a space-filling congruence of parallel time-like straight lines. In Galilei-Newtonian space-time, each line of such a fibration intersects each hyperplane of absolute time once and only once.

In Minkowski space-time, each inertial frame is represented by a *different* foliation of space-time into hyperplanes of equal global time (using the Poincaré-Einstein convention to define this global time) that are 'pseudo-orthogonal' (a concept soon to be explained) to the straight lines of the fibration defining the inertial frame. Since global space and time now are both relative, their special-relativistic unification is much less formal.[z] A Lorentz boost, which expresses the relation between two inertial frames (each represented by a different foliation and fibration), is represented by a so-called '*pseudo-rotation*' taking one foliation and fibration into the other.

If we examine the *kinematic structure* associated with four dimensional space-time more closely, we find that there are two distinct but interrelated structures, the *chronogeometrical* and the *inertial*. The chronogeometrical structure models the spatial geometry and the measure of local time, as these can be exemplified respectively, for example, by the behavior of a system of rigid rods and clocks all at rest relative to each other. The inertial structure models the behavior of physical bodies not acted on by external forces, i.e., behavior characterized by Newton's First Law (the law of inertia). If we can neglect the spatial dimensions (and possible multipole structure) of such a body, we refer to it as a (monopole, or) particle, and speak of the behavior of a freely-falling particle. Each of these structures is represented mathematically by a geometrical-object field: the chronogeometry is represented by a (*pseudo-*) *metric* (a symmetric second rank covariant tensor field); while the inertial structure is represented by a flat symmetric *affine connection* field. The two structures are not independent of each other, but must obey certain *compatibility conditions*, namely the covariant

---

[z]The presence in the special theory of an invariant fundamental speed c, usually referred to as the speed of light, enables both time and space coordinates to be expressed in commensurable units. The existence of such an invariant speed is a consequence of the existence of a group of space-time transformations treating all inertial frames on an equal footing (see, e.g. Ref. 9). From this point of view, Galilei-Newtonian kinematics represents the degenerate limiting case, in which this speed becomes infinite.

derivative of the metric tensor with respect to the affine connection must
vanish. Roughly speaking, this condition ensure that sets of freely falling
particles and light rays can be used to construct measuring rods and clocks;
or equivalently, that a set of freely falling measuring rods and clocks, chosen
so that they are at rest in some inertial frame of reference and suitably
synchronized, can be used to correctly measure distances and global times,
respectively, relative to that frame.

As mentioned above, in both Galilei-Newtonian and special-relativistic
theories, the kinematic structures are fixed and given *a priori*; these structures include the chronogeometrical and the inertial structures, and the
compatibility conditions between them. Stemming as it does from the law
of inertia common to both theories, the inertial structure is common to both
Galilei-Newtonian and special-relativistic kinematics: Mathematically, this
inertial structure is represented by a four-dimensional linear affine space;
which includes (among other things) the concepts of parallel straight lines,
parallel hyperplanes, and the equality of parallel vectors in space-time. A
fibration of space-time by a family of parallel, time-like straight lines represents a particular inertial frame.

The Galilei-Newtonian chrono-geometrical structure consists of two elements: a preferred *foliation* of space-time by a family of parallel hyperplanes transvected by each inertial fibration; each hyperplane represents
all possible events occurring at the same absolute time. Further, the geometry of each hyperplane is assumed to be Euclidean: it is spanned by a
triad of mutually orthogonal unit vectors at each point; parallel transport
of the triad at any point by the inertial connection produces the corresponding parallel triad at every other point.[aa] The triad plus the Euclidean
three-metric at each point enables one to construct a degenerate (rank-
three) four-dimensional contravariant metrical structure, invariant under
rotations of the tetrad, representing the Euclidean geometry that holds in
every inertial frame. In more detail, this Euclidean metric is constructed
using the dual triad of co-vectors in that frame. The compatibility between
the inertial and the chronometric and geometric structures is expressed by
the vanishing of the covariant derivatives, with respect to the flat affine connection field, of the triad vector fields and of the gradient of the absolute
time.

The special-relativistic (or Minkowskian) flat affine connection remains
the same; but the chronogeometry now is represented by a non-degenerate

---

[aa]Since the inertial connection is flat, parallel transport is independent of the path taken
between any two points.

four-dimensional pseudo-metric of signature two. A metric is a quadratic form used to compute the 'length-squared' of vectors, which is always a positive quantity. If the tensor does not assign a positive 'length-squared' to all vectors, i.e., if some of them have zero or negative 'length squared', the tensor is called a pseudo-metric. The Minkowski metric is such a pseudo-metric of signature two. This means that, when diagonalized, it has three plus terms and one minus term (or the opposite – at any rate three terms of one sign and one of the other resulting in a signature of two for the pseudo-metric tensor), which represent space and time respectively. If the 'length squared' of a vector computed with this pseudo-metric is positive, the vector is called *space-like*; if negative, the vector is called *time-like*; if zero, the vector is called *null* or *light-like*. Signature two for the four-dimensional Minkowski pseudo-metric implies that the null vectors at each point form a three-dimensional cone with vertex at that point, called the *null cone*. A time-like and a space-like vector are said to be *pseudo-orthogonal* if their scalar product vanishes when computed with the pseudo-metric. A fibration of parallel lines is time-like if it has time-like tangent vectors and such a fibration represents an inertial frame. A hyperplane pseudo-orthogonal to the fibration (that is, with all space-like vectors in the hyperplane pseudo-orthogonal to the tangents to the fibration) represents the set of all events of equal global (Poincaré-Einstein) time relative to that inertial frame; and the slicing (foliation) consisting of all hyperplanes parallel to this one represents the sequence of such global times relative to that inertial frame.

The compatibility conditions between pseudo-metric and flat affine structure are now expressed by the vanishing of the covariant derivative of the former with respect to the latter. In both the Newtonian and special-relativistic cases, the compatibility conditions require that the 'length-squared' of equal parallel vectors (space-like, time-like, or null) be the same.

11. Enter Gravitation: General Non-Relativity

As indicated above, in general relativity, both elements of the kinematic structure, the chronogeometrical and the inertial, lose their fixed, *a priori* character and become dynamical structures. There are two basic reasons for this:

(a) General relativity is a theory of gravitation and, even at the Newtonian level, gravitation transforms the fixed inertial structure into the dynamic inertio-gravitational structure.

(b) General relativity preserves the unique compatibility relation between the special-relativistic chrono-geometric and inertial structures; so that when the latter becomes dynamical, the former must follow suit.

It follows that, in general relativity, there is no kinematics prior to and independent of dynamics. That is, before a solution to the dynamical field equations is specified, there is neither an inertio-gravitational field nor a chrono-geometry of space-time. (Of course a similar comment will apply to any generally covariant field equations. For details, see Ref. 17.)

Let us look more closely at each of the two assertions above. First of all, an examination of the Newtonian theory of gravitation from the four-dimensional point of view shows that gravitation *always* dynamizes the inertial structure of space-time.[bb] As we have seen, the concept of inertial structure is based on the behavior of freely falling bodies, i.e., the trajectories of bodies (structureless particles[cc]) not acted upon by any (net) external force. If one neglects gravitation, force-free motions can be readily identified (in principle); but because of the equivalence principle, the presence of gravitation effectively nullifies the distinction between forced and force-free motions. How could we realize a force-free motion in principle (that is, ignoring purely practical difficulties)? First of all, the effect of non-gravitational forces (electrical, magnetic, etc.) on a particle can be either *neutralized* or shielded from. But gravitation is universal: it cannot be neutralized or shielded. This still would not constitute a fatal difficulty if we could correct for the effect of gravitation on the motion of an otherwise force-free body (as we often do for non-gravitational forces when we cannot shield from them). But the effect of gravitation is universal in a second. sense: As Galileo supposedly demonstrated using the leaning tower of Pisa, it has the same effect on the motion of all bodies (this is often called the weak principle of equivalence). This would still not constitute a fatal difficulty if it were possible to single out the class of inertial frames of reference in a way that is independent of the concept of force-free motion. This possibility is tacitly assumed when gravitation is described, in the tradition of Newton, as a force pulling objects off their inertial paths. But inertial frames cannot be defined independently of inertial motions, which are in turn defined as force-free motions! So our attempt to distinguish between

---

[bb]See, e.g., Ref. 13.
[cc]That is, bodies, for which any internal structure beyond their monopole mass may be neglected.

inertial and gravitational effects ends up in a vicious circle. This is true of Newtonian gravitation discussed here as it is true of gravitation in general relativity, soon to be discussed.

The only alternative left is to admit that one cannot distinguish in any absolute sense between the effects of inertia and gravitation on the motion of a body. There is a single *inertio-gravitational field*, and 'free falls' are the motions of bodies subject only to this field. (As noted above, in principle one can eliminate, or correct for, the effects on a body's motion of all non-gravitational fields.) Gravitation is no longer treated as a force pulling a body off a '*free*' inertial paths in a flat space-time, but rather as a factor entering into the determination of the 'free' inertio-gravitational paths in a non-flat space-time.

As a consequence of this new outlook on gravitation, the class of preferred frames of reference must be enlarged: It is now impossible in principle to distinguish between an inertial frame of reference with a (possibly time-dependent but) spatially homogeneous gravitational field in some fixed direction, and a (possibly time-dependent but) linearly accelerated frame of reference with no gravitational field. So, when Newtonian gravitation is taken into account, all linearly accelerated frames of reference are equally valid: linear acceleration is relative. Note that rotating frames of reference can still be distinguished from non-rotating ones, as Newton's bucket experiment demonstrates; so rotational acceleration is still absolute. Within this enlarged class of preferred frames, the Galilei-Newtonian chronogeometric structure, consisting of the absolute time plus Euclidean geometry in each relative space, can still be postulated in a way that is independent of all dynamics including that of the inertio-gravitational field but still remain compatible with the latter.

12. Dynamizing the Inertial Structure

As noted above, the Galilei-Newtonian inertial structure must be *dynamized* in order to include Newtonian gravitation as demanded by the equivalence principle. The resulting *inertio-gravitational structure* is still modeled mathematically by an affine connection. But now, this connection is no longer fixed *a priori* as a flat connection, describing a linear affine space-time. Rather it becomes a *non-flat connection*, subject to dynamical field equations specifying just how the presence of matter produces a curvature of the resulting space-time. These equations describe the relation of the Newtonian inertio-gravitational field to its material sources, i.e., they are the four-dimensional analogue of:

$$\mathrm{Div}\, \boldsymbol{a} = 4\pi G\rho\,,$$

where $\boldsymbol{a}$ is the Newtonian gravitational acceleration, G is the Newtonian gravitational constant, and ρ is the mass density. In their four-dimensional form, these equations look remarkably like Einstein's equations: They relate the Ricci tensor of the Newtonian field to the Newtonian stress-energy tensor for matter.[dd]

What happens to the compatibility conditions between the kinematic chronogeometrical structure, which is still fixed and non-dynamical, and the now-dynamized inertio-gravitational structure? These conditions remain valid, but now do *not* fix uniquely the non-flat inertio-gravitational structure. They allow just enough leeway in the choice of the latter to impose the four-dimensional equivalent of the law relating the Newtonian gravitational acceleration $\boldsymbol{a}$ to the Newtonian gravitational potential φ:[ee]

$$\boldsymbol{a} = -\mathrm{Grad}\,\varphi\,.$$

Thus, even at the Galilei-Newtonian level, gravitation, when interpreted as part of a four-dimensional inertio-gravitational structure, dynamizes the affine structure making it non-flat. A breakup of the resulting affine structure into inertial and gravitational parts is not absolute (i.e., frame-independent), but depends on (is relative to) the state of motion –in particular, the state of acceleration of the frame of reference being considered. The preferred group of transformations, under which the chrono-geometry is invariant, is enlarged to include transformations between all linearly accelerated frames. When going from one such frame of reference to another that is accelerated with respect to the first, the Newtonian gravitational potential φ must now be interpreted as a *connection potential*. It does not remain invariant, as it did under transformations from one inertial frame to another; but is subject to a more complicated law of transformation that follows from the four-dimensional transformation law for the inertio-gravitational structure.

What is the exact nature of this new inertio-gravitational structure? It is not described by a tensor field, but by a more general geometrical object called an *affine connection*.[ff] Before gravitation is introduced, the

---

[dd]See Refs. 1 and 12.
[ee]Combining the two equations, one gets the four-dimensional analogue of Poisson's equation for the gravitational potential.
[ff]For an introduction to these concepts, see, e.g., Ref. 13.

linear affine space modeling the inertial structure has a property described mathematically as flatness or lack of curvature, which corresponds to the vanishing of a tensor formed from the affine connection field and its first derivatives and called the *Riemann tensor*.[gg] In the presence of a Newtonian gravitational field this tensor no longer vanishes, but (as discussed above) is subject to a set of field equations, which translate Poisson's equation for the Newtonian gravitational potential into a form that describes just how matter modifies the inertio-gravitational field.

The Riemann tensor of a connection can be interpreted physically with the help of the equation of *geodesic deviation*. In a linear affine space-time (affine-flat connection), the paths of freely-falling test particles are modeled by time-like straight lines in the space-time. The spatial distance between any pair of such particles at the same absolute time varies linearly with that time. In other words, the two particles have *no* relative acceleration. In the non-flat affine space-time of a Newtonian inertio-gravitational field, the paths of freely-falling particles are time-like geodesics of the affine connection of that space-time – the straightest possible paths in such a space.[hh] A neighboring pair of freely-falling particles (modeled by a pair of time-like geodesics) now generally has a non-vanishing relative acceleration (also called the *tidal force* between them), the magnitude of which in various directions is proportional to corresponding components of the affine curvature tensor of the space-time. Measurement of such relative accelerations (tidal forces) thus constitutes a direct measurement of components of the affine curvature tensor. This curvature tensor, as mentioned above, also enters in a very simple way into the four-dimensional formulation of Newton's field law of gravitation.

13. General Relativity

Up to now we have been discussing the Newtonian theory of gravitation, which is still based on the concept of absolute time. Now we must ap-

---

[gg]Although the names are often used interchangeably, I shall distinguish between the *Riemann tensor* associated with a connection and the *curvature tensor* associated with a (pseudo-)metrical tensor field. Even though the components of each are the same in the case of general relativity, their geometrical interpretation is quite different. The curvature tensor at a point can be interpreted in terms of the Gaussian or sectional curvatures of all two-sections through that point, and it is hard to see what this interpretation has to do with gravitation. The geometric interpretation of the Riemann tensor of the connection and its relation to gravitation are discussed in the text.

[hh]Here 'straight' means that the curve is *autoparallel*, i.e., it never pulls away from the direction of its tangent vector.

ply the lessons learned about gravitation from this theory to the task of
formulating a relativistic theory of gravitation without an absolute time,
i.e., a theory based on the lessons learned from the special theory of relativity. The most important lesson of the four-dimensional formulation
of Newtonian theory is that gravitational phenomena are best incorporated into an inertio-gravitational structure, described mathematically by
a non-flat affine connection subject to certain dynamical (gravitational)
field equations involving the Riemann tensor of that connection. But an
important feature distinguishes special-relativistic theories from Galilei-
Newtonian theories: In special relativity, the compatibility relations between the chrono-geometrical and inertial structures restrict the latter completely – the chrono-geometrical structure uniquely determines the inertial
structure — while in General Non-Relativity this is not the case. Hence,
when attempting to incorporate gravitation by converting the inertial structure into an inertio-gravitational structure and dynamizing it, we are con-
fronted with a choice. Either:

(a) give up the unique compatibility relation between chrono-geometrical
    and inertial structures, which is the road taken by the various special-
    relativistic theories of gravitation;[ii] or
(b) preserve the unique relation, hence dynamizing the chrono-geometry
    along with the inertio-gravitational field, which is the path chosen in
    general relativity.[jj]

The pseudo-metric tensor field both represents the chrono-geometrical
structure and, via the compatibility conditions between metric and connection, provides the "potentials" for the affine connection field that represents the inertio-gravitational structure. The latter is subject to a set of
field equations relating its Riemann tensor to a tensor describing all non-
gravitational sources of the gravitational interactions, the so-called stress-
energy tensor.[kk] Formally, these field equations for the affine curvature

---

[ii]This includes all attempts to give a "special-relativistic" interpretation of the general-
relativistic field equations.

[jj]Of course, one might *both* dynamize the chronogeometry *and* give up the unique relation.
Various tensor-scalar theories of gravitation, for example, have attempted to do this.
But so far the minimal assumption, that it suffices to dynamize the chronogeometry
without changing its unique relation to the inertio-gravitational field, has survived all
experimental challenges by such theories.

[kk]In most cases, construction of the stress-energy tensor involves the metric tensor field
(but not its derivatives — the so called "minimal coupling" assumption), so that it is
not simply a case of solving the gravitational equations for a given source. One must
solve the coupled set of equations for gravitational and source fields.

tensor look almost exactly the same as the four-dimensional Newtonian equations. Both involve a "trace" of the Riemann tensor called the *Ricci tensor*. But in the case of general relativity, as emphasized above, the affine structure, including its curvature tensor, is completely determined by the metric tensor field.

14. Differentiable Manifolds, Fiber Bundles[ll]

Up to now, I have not discussed the nature of the mathematical space, with which the metric and affine fields of general relativity are associated. In the case of Galilei-Newtonian and special-relativistic space-times, the unique structure of these spaces is determined by the respective kinematical symmetry groups of these theories: The inhomogeneous Galilei group leads to Galilei-Newtonian space-time, while the Poincaré group leads to Minkowski space-time. Both are linear affine spaces with flat affine structure. In general relativity, *no* kinematic space-time structure is given *a priori*, so there is no preferred kinematic symmetry group singling out a subgroup of the class of all allowed one-one point transformations, or *automorphisms*, of the underlying four-dimensional. mathematical space.[mm]

It might seem that any four-dimensional topological space would do as a mathematical starting point: its symmetry group consists of all the *homeomorphisms* that is all bicontinuous automorphisms of the space. But to do physics, we need tensors and other geometric-object fields on the space, on which one may carry out various differential operations that require the introduction of coordinates. But we want to carry out such operations in a way that is independent of the particular choice of coordinate system. So we need to impose a *differentiable structure* on the underlying topological spaces, which are chosen to be four-dimensional *differentiable manifolds*; such manifolds have the group of *diffeomorphisms* (that is, differentiable homeomorphisms) as their symmetry group.[40]

This differentiable manifold constitutes the *base space* of a *fiber bundle*, which consists of such a base manifold B, a total manifold E, and a projection operator π from E into B, which turns E into a fibered manifold. (The inverse operation π^{-1} acting on the points of B produces the fibering of E.) Geometric object fields of a certain type, such as the metric tensor and the affine connection used in general relativity, constitute the fibered

---

[ll]For a more detailed discussion of the topic of this section, see, e.g. Ref. 19 or 17.
[mm]By 'allowed' I mean with the conditions of differentiability appropriate to the theory under consideration. I shall not discuss this question any further here.

manifold appropriate for each particular theory. A *cross section* σ of the fibered manifold (i.e., a suitably smooth choice of one point on each fiber) constitutes a particular field of the chosen type, and if selected according to the appropriate rule, this cross section will be a solution to the field equations of the theory of that type.

For fibered manifolds, it is natural to consider only *fiber-preserving diffeomorphisms*, i.e., diffeomorphisms of E that take fibers into fibers. By the coordinate-free definition of a geometric object[nn], to each diffeomorphism of the base manifold there corresponds a unique fiber-preserving diffeomorphism of the fibered manifold. Given any cross section σ of E that is a solution to the field equations, if we carry out any diffeomorphism of B and the corresponding fiber-preserving diffeomorphism of E, the effect on is to produce an essentially equivalent physical solution.

But diffeomorphisms of B and fiber-preserving diffeomorphisms of E many be carried out *independently*. In the absence of any fixed kinematic symmetry group, one is led to require the *covariance* of any set of field equations: Given any cross section σ of the fibered manifold that satisfies the field equations, any other cross section σ' generated from the first by a fiber-preserving diffeomorphism, called the *carried-along* or *dragged-along* cross-section, is also a solution. Note that this does *not* imply that σ and σ' describe the *same* physical solution. If the points of B are individuated in some way that is independent of the cross-section chosen, they will indeed be different solutions. But, in the absence of any independent individuation, all the carried-along solutions will correspond to the same physical solution, and I call such a theory *generally covariant* (see below). By this definition, general relativity is a generally covariant theory.

15. The General-Relativistic Revolution

It must be emphasized just how revolutionary are the steps involved in the development of the general theory of relativity: identification of the distinction between chronogeometrical and inertio-gravitational structures and the compatibility conditions between them; dynamization of both structures; and the associated requirement of general covariance. In many ways, these steps involve a much greater break with traditional physics than the steps leading from Galilei-Newtonian physics to the special theory of relativity. I shall discuss four radical differences between the general-relativistic and all pre-general-relativistic concepts of space and time.

[nn]See Ref. 17.

First of all, there is no longer such a thing as an 'empty' region of space-time. Wherever there is space-time there is at the very least an inertio-gravitational field. Returning for a moment to the controversy between absolute and relational concepts of time and space, it seems difficult to sustain the absolute position with regard to general relativity. As Einstein[3] (p. 155) puts it:

> [The metric tensor components] describe not only the field, but at the same time also the topological and metrical structural properties of the manifold. ... There is no such thing as empty space, i.e., a space without a field. Space-time does not claim existence on its own, but only as a structural quality of the field.

Secondly, the space-time structures no longer form a fixed stage, on which the drama of matter and fields is enacted; space and time have become actors in the drama. Curious possibilities thereby arise, such is the possibility of solutions to the field equations that represent space-times containing closed time-like world lines. A time-like world line represents the possible history of a particle. So in such a space-time, the history of some particles (with open world lines) includes a doubly-infinite (past and future) amount of proper time, while for others (with closed' world lines) only a finite amount of proper time elapses before the particles 'loops back upon themselves', so to speak.

Thirdly, not only is the local structure (local, in the sense of a finite but limited region) of space-time dynamized; the global structure (global in the sense of the entire topology) is no longer given *a priori*. For each solution to the gravitational field equations given locally (i.e., on a patch of space-time), one must work out the global topology of the maximally extended manifold(s) – criteria must be given for the selection of such an extension (or extensions if one is not uniquely selected) – compatible with the local space-time structure of that solution. Solutions are possible that are spatially and/or temporally finite but unbounded. In the former case, someone always marching straight ahead into the universe could end up back where s/he started. In the latter case, the entire history of the universe would repeat itself after a finite time had elapsed.

Fourthly, until the choice of a cross section of the fibered manifold on which both the chronogeometrical and inertio-gravitational structures live, the points of the base space have no individuating properties. Many textbooks on general relativity refer to these points as 'events,' thereby suggesting that they are physically individuated apart from the choice of a metric, which merely defines the chronogeometrical relations between events given *priori*. This is incorrect for at least two reasons. As noted earlier, without the specification of a particular metric tensor field, we cannot specify the global topology of the base space differentiable manifold. But, even locally, there is no means of physically individuating *a priori* the points of a region of a bare manifold. If particles or other, non-gravitational fields are present in the region, then their properties may suffice to individuate the points of the region. But even then, without a metric, it is generally impossible to fully interpret the physical properties of such non-gravitational entities.

If we confine ourselves to otherwise empty regions, in which only the chronogeometrical (and corresponding inertio-gravitational) field is present, then the points of such a region cannot be physically individuated by anything but the properties of this field.[oo]

16. The Second Relativization of Time

Among the properties of any physical event are its position in space-time: the 'here' and 'now' of the event. One cannot give meaning to the concepts 'here' and 'now' of an event in otherwise empty regions of space-time without use of the chronogeometry (metric tensor). This leads to a second relativization of the concept of time in general relativity that is even more drastic than that required by the special theory of relativity.

To recall what has been said earlier: In Galilei-Newtonian kinematics, the absolute time function as a global time, the same for all rigid reference frames, regardless of their states of motion, even when gravitation is taken into account; and it also serves as a local time, the time elapsed between two events is the same for all time-like paths between them.

In special-relativistic kinematics, this unique concept of absolute time splits into two different concepts, both of which are relativized. Time in the global sense (the time used to compare events at different places) is relative to the inertial frame chosen, i.e., to one member of a preferred class of reference frames. Time in the local sense (the proper time that elapses between events along a time-like path) is relative to the path and differs for different paths between the same two events.

In both pre-relativistic and special-relativistic kinematics, the use of different frames of reference leads to different descriptions of some process undergone by a dynamical system: The expression for the initial state of

---

[oo]For a more detailed discussion of this point, see Ref. 10.

the system at some time, and for the changes in that state as it evolves in accord with the dynamical laws governing the system, depend on (are relative to) the choice of inertial frame. These different descriptions are all concordant with one another, leading, for example, to the same prediction for the outcome of any experiment performed on the system.[pp] Mathematically, this is because different descriptions amount to no more than different slicings (foliations) and fibrations of the background space-time, in which a unique four-dimensional description of the dynamical process can be given.

The Galilei-Newtonian absolute time and the special-relativistic global (inertial-frame relative) time have this in common: They are defined kinematically in a way that is independent of any dynamical process.[qq]

In general relativity, no time – either global (frame-dependent) or local (path-dependent) – can be defined in a way that is independent of the dynamics of the inertio-gravitational field. In general relativity, that is, time is relative in another sense of the word: It is relative to dynamics; in particular, it is relative to the choice of a solution to the gravitational field equations.

The proper (local) time is of course relative to a path in the manifold, but also depends on the particular solution in a new way: Whether a path in the base space manifold is time-like or not cannot be specified *a priori*. Similarly, introduction of a frame-relative global time depends upon a slicing (foliation) of the space-time into three-dimensional space-like slices, and whether such a slicing of the base manifold is even space-like cannot be specified *a priori*. Indeed, while a space-like slicing is always possible locally, a global space-like foliation may not even exist in a particular space-time. Even if such global space-like slicings are possible for a particular solution, there is generally no preferred family of slicings (such as the spacelike hyperplanes of Minkowski space) because in general the metric tensor of a space-time has no symmetries. The symmetries of a particular metric tensor field may entail preferred slicings. For example, there is a class of preferred slicings of a static metric such that the spatial geometry of the slices does not change from slice to slice. Hence, there is a preferred global frame relative to any static solution. As in the pre-general-relativistic case, different choices of slicings (frames) may lead to different descriptions of the same

---

[pp]This is the case in classical physics. Quantum mechanically, the different descriptions must all lead to predictions of the same probability for a given process.
[qq]Of course, the *measurement* of such time intervals depends on certain dynamical processes, but that is a different question.

dynamical process as a temporal succession of states; but these descriptions will be concordant with one another just because they are different slicings of the same four-dimensional space-time process.

Of course, in general relativity, different dynamical processes cannot take place in the same space-time.[rr] The gravitational and the non-gravitational dynamical equations must be solved as a single, coupled system. A space-time that is a solution to the gravitational equations is associated with a dynamical system that is a solution to the non-gravitational dynamical equations. Hence no separation between space-time kinematics and the dynamics of the gravitational and other fields is possible.

17. What is the Question?

As a result of these changes, there is a basic difference between the type of questions that one can ask in pre-general-relativistic physical theories and in general-relativistic theories. A typical pre-general-relativistic question takes the form: *Here* is a point in space and *now* is a moment in time (or, if you prefer, *here-and-now* is a point in space-time): What is (are) the value(s) of some physical quantity (quantities) 'here' and 'now'?[ss] We answer such questions theoretically by solving some set of dynamical equations for the quantity (quantities) in question and evaluating the solution 'here' and 'now'.

In general relativity, as we have seen, 'here' and 'now' cannot be defined *before* we have a chronogeometry, which presupposes that we have already solved the dynamical equations for the metric tensor field. So 'here' and 'now' cannot be part of the initial question. In general relativity, we must start by specifying some solution to these field equations and any other dynamical field equations that may be involved in the questions we want to ask. Then either:

(1) we don't ask any questions that depend on specification of a 'here' and 'now': e.g., we limit ourselves to questions that depend only on globally defined quantities, such as the total mass or charge of a system; or

(2) we construct a 'here' and 'now' from the given solution to the field equations; then they will be part of the answer, not part of the question.

[rr]Barring exceptional cases, in which different dynamical processes happen to have the same stress-energy tensor.

[ss]Of course in practice, and in quantum field theory even in principle, such questions will always involve finite regions of space-time; but this issue is easily handled by the introduction of test functions with support in such regions.

Such a construction could involve just the pure gravitational field, i.e., be based upon just the metric tensor field; or, it could involve the other non-gravitational field variables if they are present in the region of space-time under consideration.

I shall not go into further detail here, but end by noting that this problem reappears in all attempts to construct a quantum theory that incorporates general relativity — a quantum gravity. A typical quantum question concerns a process, and takes the form: If one physical quantity has been found to have a certain value 'here' and 'now', what is the probability that some (possibly the same) physical quantity will be found to have some other value 'there' and 'then'? Again, a quantum gravity question cannot take this form because 'here' and 'now', and 'there' and 'then' somehow have to be constructed as part of the answer, not part of the question.

Acknowledgement

I thank Dr. Mihaela Iftime for a careful reading of the manuscript and many valuable suggestions for its improvement.

References

1. J. Ehlers, Über den Newtonschen Grenzwert der Einsteinschen Gravitationstheorie, in *Grundlagenprobleme der modernen Physik*, eds. J. Nitsch *et al.* (Bibliographisches Institut, Mannheim, 1981), pp. 65–84.
2. A. Einstein, Zur Elektrodynamik bewegter Korper, *Annalen der Physik* **17**, 891–921 (1905); Translation from John Stachel *et al.* (eds.), *The Collected Papers of Albert Einstein, Vol. 2, The Swiss Years; Writings, 1902–1909* (Princeton University Press, 1989), p. xxiii.
3. A. Einstein, Relativity and the Problem of Space, *Appendix V* to *Relativity: The Special and the General Theory*, 15th edn. (Crown, New York, 1952).
4. A. Einstein, 'Foreword' to Max Jammer, *Concepts of Space* (Harvard University Press, Cambridge, 1954), pp. xi–xvi.
5. L. Euler, *Reflexions sur l'espace et le temps. Memoir de l'Académie des Sciences de Berlin 4*: 324–333 (1748); Reprinted in Leonhard Euler, *Opera Omnia*, Series 3, Vol. 2, pp. 376–383; English translation by Soshichi Uchii, Euler on Space and Time, Physics and Metaphysics <www.bun.kyoto-u.ac.jp/phisci/Newsletters/newslet_41.html>.
6. G. S. Kirk and J. E. Raven, *The Presocratic Philosophers* (Cambridge University Press, 1957).
7. K. Marx, *Capital A Critique of Political Economy*, Vol. 1 (Penguin, Harmondsworth, 1976), Chapter I, Section 4 entitled: "The fetishism of the Commodity and Its Secret," pp. 163–177.
8. J. A. Schouten, *Tensor Analysis for Physicists*, 2nd edn. (Dover Publications, Mineola, New York, 1989).

9. J. Stachel, Special relativity from measuring rods, in *Physics, Philosophy and Psychoanalysis: Essays in Honor of Adolf Gruebaum*, eds. Robert S. Cohen and Larry Laudan, (Reidel, Boston, 1983), pp. 255–272.

10. J. Stachel, The meaning of general covariance: The hole story, in *Philosophical Problems of the Internal and External Worlds/Essays on the Philosophy of Adolf Grünbaum*, eds. John Earman *et al.* (University of Pittsburgh Press/Universitätsverlag Konstanz, 1993), pp. 129–160.

11. J. Stachel, Espace-temps, in *Dictionnaire d'histoire et philosophie des sciences*, ed. Dominique Lecourt (Presses Universitaires de France, Paris, 1999), pp. 376–379.

12. J. Stachel, Einstein's Intuition and the Post-Newtonian Approximation, To appear in the Proccedings of the Conference "Topics in Mathematical Physics, General Relativity and Cosmology on the Occasion of the 75th Birthday of Professor Jerzy F. Plebanski" Mexico City, August 2002 (World Scientific, Singapore).

13. J. Stachel, The Story of Newstein, or Is Gravity Just Another Pretty Force? To appear in Jürgen Renn and Matthias Schimmel (eds.), *The Genesis of General Relativity: Documents and Interpretations, Vol. 3, Alternative Approaches to General Relativity* (Kluwer, Dordrecht, 2004).

14. J. Stachel, A Brief History of Space-Time, in Ignazio Ciuffolini, Daniele Dominic and Luca Lusanna (eds.) *2001: A Relativistic Spacetime Odyssey* (World Scientific, Singapore, 2003), pp. 15–34.

15. J. Stachel, Structure, Individuality and Quantum Gravity, in *Structural Foundation of Quantum Gravity*, eds. D. P. Rickles, S. R. D. French and J. Saatsi (Oxford University Press, 2005).

16. J. Stachel, Fresnel's Formula as a Challenge to 19th Century Optics and Electrodynamics of Moving Media, in *The Universe of General Relativity*, eds. Jean Eisenstaedt and Anne Kox (Birkhäuser, Boston, 2005).

17. J. Stachel and M. Iftime, Fibered Manifolds, Natural Bundles, Structured Sets, G-Sets and all that. The Hole Story from Space-Time to Elementary Particles 2005, <arXiv.gr-qc/0505138v2 28 May 2005>.

18. R. Torretti, *Relativity and Geometry* (Pergamon Press, Oxford, 1983).

19. A. Trautman, Fiber bundles, gauge fields, and gravitation, *General Relativity and Gravitation* **1**, 287–308 (1980).

20. Voltaire, Modern History Sourcebook: "Voltaire (1694–1778) Letters on Newton from the Letters on the English or *Lettres Philosophiques*, c. 1778" (2005), <www.fordham.edu/halsall/mod/1778voltaire-newton.html>.

Part II

Einstein's Universe
Ramifications of General Relativity

When Henry Moore visited the University of Chicago ... I had the occasion to ask him how one should view sculpture: from afar or from nearby. Moore's response was that the greatest sculpture can be viewed —indeed should be viewed— from all distances since new aspect of beauty will be revealed at every scale. In the same way, the general theory of relativity reveals strangeness in the proportion at any level in which one may explore its consequences.
—S. Chandrasekhar, (Truth and Beauty)

CHAPTER 2

GRAVITATIONAL BILLIARDS, DUALITIES AND HIDDEN SYMMETRIES

H. NICOLAI

*Max-Planck-Institut für Gravitationsphysik, Albert-Einstein-Institut,
Am Mühlenberg 1, D-14476 Golm, Germany*

The purpose of this article is to highlight the fascinating, but only very incompletely understood relation between Einstein's theory and its generalizations on the one hand, and the theory of indefinite, and in particular hyperbolic, Kac Moody algebras on the other. The elucidation of this link could lead to yet another revolution in our understanding of Einstein's theory and attempts to quantize it.

1. Introduction

As we look back 90 years to take stock of what has been achieved since Einstein explained gravity in terms of spacetime geometry and its curvature, the progress is impressive. Even more impressive is the wealth of structure contained in Einstein's equations which has been revealed by these developments. Major progress has been made concerning

- Exact solutions (Schwarzschild, Reissner-Nordström, Kerr, axisymmetric stationary solutions,...)
- Cosmological applications (standard FRW model of cosmology, inflationary universe,...)
- Mathematical developments (singularity theorems, black hole uniqueness theorems, studies of the initial value problem, global stability results,...)
- Conceptual developments (global structure and properties of spacetimes, horizons, black hole entropy, quantum theory in the context of cosmology, holography,...)
- Canonical formulations (Dirac's theory of constrained systems, ADM formalism, Ashtekar's variables,...)

- Higher dimensions (Kaluza Klein theories, brane worlds,...)
- Unified theories 'beyond' Einstein (supergravity, superstrings, supermembranes and M(atrix) theory,...)
- Quantizing gravity (perturbative and canonical quantization, path integral approaches, dynamical triangulations, spin networks and spin foams,...)

All these subjects continue to flourish and are full of promise for further and exciting developments (hinted at by the dots in the above list). No doubt many of them will be discussed and elaborated in other contributions to this volume. In this article, we will concentrate on yet another line of research that evolved out of the study of Einstein's equations and its locally supersymmetric extensions, and which points to another deep, and still mostly unexplored property of Einstein's theory. It may well be that the main discoveries in this direction still remain to be made, and that, once they have been made, they will also have a profound impact on attempts to quantize Einstein's theory (or some bigger theory containing it). This is the subject of

- Hidden symmetries and dualities

The first hint of these symmetries appeared in Ref. 1, where a transformation between two static solutions of Einstein's equations was given, which in modern parlance is nothing but a T-duality transformation. A decisive step was Ehlers' discovery in 1957 of a solution generating symmetry,[2] nowadays known as the 'Ehlers $SL(2,\mathbb{R})$ symmetry' which acts on certain classes of solutions admitting one Killing vector. In 1970, R. Geroch demonstrated the existence of an *infinite dimensional* extension of the Ehlers group acting on solutions of Einstein's equations with two commuting Killing vectors (axisymmetric stationary solutions).[3] In the years that followed, the Geroch group was extensively studied by general relativists with the aim of developing 'solution generating techniques' (see[4,5] and references therein for an *entrée* into the literature). The field received new impetus with the discovery of 'hidden symmetries' in supergravities, most notably the exceptional $E_{7(7)}$ symmetry of maximal $N = 8$ supergravity.[6] These results showed that the Ehlers and Geroch groups were but special examples of a more general phenomenon.[7,8,9,10,11] With the shift of emphasis from solution generating techniques to the Lie algebra and the group theoretical structures underlying these symmetries, it became clear that the Geroch symmetry is a non-linear and non-local realization of an affine Lie group (a loop group with a central extension), with Lie algebra $A_1^{(1)} = \widehat{\mathfrak{sl}(2, \mathbb{R})}_{ce}$. This completed earlier results by general relativists who had previously realized 'half' of this affine symmetry in terms of 'dual potentials'.[12] Likewise, generalizations of Einstein's theory, and in particular its locally supersymmetric extensions were shown to possess similar infinite dimensional symmetries upon reduction to two dimensions. These results also provided a direct link to the integrability of these theories in the reduction to two dimensions, i.e. the existence of Lax pairs for the corresponding equations of motion.[13,14,9]

All these duality invariances of Einstein's theory and its extensions apply only to certain truncations, but do not correspond to properties of the *full* theory, or some extension thereof. Our main point here will be to review and highlight the evidence for even larger symmetries which would *not* require any truncations, and whose associated Lie algebras belong to a special class of infinite dimensional Lie algebras, namely the so-called *indefinite Kac Moody Algebras*.[15,16,17] We will discuss two examples of such Lie algebras, namely the rank three algebra AE_3,[18] which is related to Einstein's theory in four dimensions, and secondly (but only very briefly), the maximal rank 10 algebra E_{10}, which is singled out from several points of view, and which is related to maximal $D = 11$ supergravity.[19] We can thus phrase the central open question as follows:

Is it possible to extend the known duality symmetries of Einstein's equations to the full non-linear theory without any symmetry reductions?

A perhaps more provocative, way to pose the question is

Is Einstein's theory integrable?

In this form, the question may indeed sound preposterous to anyone with even only a passing familiarity with the complexities of Einstein's equations, which are not only the most beautiful, but also the most complicated partial differential equations in all of known mathematical physics. What is meant here, however, is not the usual notion of integrability in the sense that one should be able to write down the most general solution in closed form. Rather, it is the 'mappability' of the initial value problem for Einstein's theory, or some M theoretic extension thereof, onto a group theoretical structure that itself is equally intricate, and so infinite in extent that we may never be able to work it out completely, although we know that it exists. Even a partial answer to the above question would constitute a great advance, and possibly clarify other unsolved problems of general relativity. To name but one: the 'conserved charges' associated with these Lie algebras would almost certainly be linked to the so far elusive 'observables' of pure gravity (which might better be called 'perennials'[20]) – which we believe should exist, though no one has ever been able to write down a single one explicitly.

Last but not least, duality symmetries have come to play a prominent role in modern string theory in the guise of T, S and U dualities, where they may provide a bridge to the non-perturbative sector of the theory (see[21,22,23] and references therein). Here, we will not dwell too much on this side of the story, however, because the duality groups considered in string theory so far are descendants of the *finite dimensional* Lie groups occurring in $D \geq 4$ supergravity, whereas here we will be mostly concerned with the *infinite dimensional* symmetries that emerge upon reduction to $D \leq 2$ dimensions, and whose role and significance in the context of string theory are not understood. Still, it seems clear that infinite dimensional symmetries may play a key role in answering the question, what M Theory – the conjectural and elusive non-perturbative and background independent formulation of superstring theory – really is, because that question may well be closely related (or even equivalent) to another one, namely

What is the symmetry underlying M Theory?

There has been much discussion lately about the maximally extended hyperbolic Kac Moody algebra E_{10} as a candidate symmetry underlying M Theory, *i.e.* $D = 11$ supergravity and the other maximally supersymmetric theories related to IIA and IIB superstring theory, see [24,25,26,27,28,29], and [30,31]. A conceptually different proposal was made in [32], and further elaborated in [33,34,35], according to which it is the 'very extended' indefinite KM algebra E_{11} that should be viewed as the fundamental symmetry (E_{11} contains E_{10}, but is no longer hyperbolic, but see [36] for a discussion of such 'very extended algebras'). A 'hybrid' approach for uncovering the symmetries of M-theory combining [25] and [32] has been adopted in [37,38]. Although our focus here is mostly on pure gravity in four space-time dimensions and its associated algebra AE_3, we will very briefly mention these developments in the last section.

Whatever the outcome of these ideas and developments will be, the very existence of a previously unsuspected link between two of the most beautiful concepts and theories of modern physics and mathematics, respectively — Einstein's theory of gravity on the one hand, and the theory of indefinite and hyperbolic Kac Moody algebras on the other — is most remarkable and surely has some deep significance.

2. Known Duality Symmetries

We first review the two types of duality symmetries of Einstein's theory that have been known for a long time. The first concerns the *linearized* version of Einstein's equations and works in any space-time dimension. The second is an example of a *non-linear* duality, which works only for the special class of solutions admitting two commuting Killing vectors (axisymmetric stationary and colliding plane wave solutions). This second duality is more subtle, not only in that it is non-linear, but in that it is linked to the appearance of an *infinite dimensional symmetry*.

2.1. *Linearized duality*

The duality invariance of the linearized Einstein equations generalizes the well known duality invariance of electromagnetism in four spacetime dimensions. Recall that Maxwell's equations *in vacuo*

$$\partial^\mu F_{\mu\nu} = 0 \qquad , \qquad \partial_{[\mu}F_{\nu\rho]} = 0 \tag{1}$$

are invariant under $U(1)$ rotations of the complex field strength

$$\mathcal{F}_{\mu\nu} := F_{\mu\nu} + i\tilde{F}_{\mu\nu} \tag{2}$$

with the dual ('magnetic') field strength

$$\tilde{F}_{\mu\nu} := \frac{1}{2}\epsilon_{\mu\nu\rho\sigma}F^{\rho\sigma} \tag{3}$$

The action of this symmetry can be extended to the combined electromagnetic charge $q = e + ig$, where e is the electric, and g is the magnetic charge. The partner of the one-form electric potential A_μ is a *dual magnetic* one-form potential $\tilde{A}_\mu$, obeying

$$\tilde{F}_{\mu\nu} := \partial_\mu\tilde{A}_\nu - \partial_\nu\tilde{A}_\mu \tag{4}$$

Observe that this dual potential can only be defined *on-shell*, when $F_{\mu\nu}$ obeys its equation of motion, which is equivalent to the Bianchi identity for $\tilde{F}_{\mu\nu}$. Consequently, the $U(1)$ duality transformations constitute an *on-shell symmetry* because they are valid only at the level of the equations of motion. The two potentials A_μ and $\tilde{A}_\mu$ are obviously *non-local* functions of one another. Under their exchange, the equations of motion and the

Bianchi identities are interchanged. Moreover, the equations of motion and the Bianchi identities can be combined into a single equation

$$\partial^\mu \mathcal{F}_{\mu\nu} = 0 \tag{5}$$

Analogous duality transformations to the electromagnetic ones exhibited above exist for p-form gauge theories in arbitrary spacetime dimensions D (these theories are always *abelian* for $p > 1$). More precisely, an 'electric' p-form potential $A_{\mu_1...\mu_p}$ is dual to a 'magnetic' $(D-p-2)$ potential $\tilde{A}_{\mu_1...\mu_{D-p-2}}$. A prominent example is the 3-form potential of $D = 11$ supergravity,[19] with a dual 6-form magnetic potential. Upon quantization, the duality becomes a symmetry relating the weak and strong coupling regimes by virtue of the Dirac quantization condition $eg = 2\pi i\hbar$. This is one of the reasons why such dualities have recently acquired such an importance in string theory, and why they are thought to provide an inroad into the non-perturbative structure of the theory.

Does there exist a similar symmetry for Einstein's equations? Remarkably, for *linearized* Einstein's equations in arbitrary space-time dimension D the answer is *yes*. [39,40,41,32,25,42,43] However, this answer will already illustrate the difficulties one encounters when one tries to extend this symmetry to the full theory. To exhibit it, let us expand the metric as $g_{\mu\nu} = \eta_{\mu\nu} + h_{\mu\nu}$, where $\eta_{\mu\nu}$ is the Minkowski metric[a], and the linearized fluctuations $h_{\mu\nu}$ are assumed to be small so we can neglect higher order terms. The linearized Riemann tensor is

$$R^L_{\mu\nu\rho\sigma}(h) = \partial_\mu\partial_\rho h_{\nu\sigma} - \partial_\nu\partial_\rho h_{\mu\sigma} - \partial_\mu\partial_\sigma h_{\nu\rho} + \partial_\nu\partial_\sigma h_{\mu\rho} \tag{6}$$

The linearized Einstein equations therefore read

$$R^L_{\mu\nu}(h) = \partial^\rho\partial_\rho h_{\mu\nu} - \partial_\mu\partial^\rho h_{\rho\nu} - \partial_\nu\partial^\rho h_{\rho\mu} + \partial_\mu\partial_\nu h^\rho{}_\rho = 0 \tag{7}$$

where indices are raised and lowered by means of the flat background metric $\eta^{\mu\nu}$. To reformulate thes equations in analogy with the Maxwell equations in such a way that $R^L_{\mu\nu} = 0$ gets interchanged with a Bianchi identity, we define

$$C_{\mu\nu|\rho} := \partial_\mu h_{\nu\rho} - \partial_\nu h_{\mu\rho} \tag{8}$$

This 'field strength' is of first order in the derivatives like the Maxwell field strength above, but it is *not* invariant under the linearized coordinate

---

[a]It is noteworthy that the construction given below appears to work only for a flat Minkowskian background.

transformations

$$\delta h_{\mu\nu} = \partial_\mu \xi_\nu + \partial_\nu \xi_\mu \quad \Longrightarrow \quad \delta C_{\mu\nu|\rho} = \partial_\rho(\partial_\mu \xi_\nu - \partial_\nu \xi_\mu) \neq 0 \qquad (9)$$

This is a first difficulty: unlike ordinary gauge theories, Einstein's theory needs *two* derivatives for gauge covariance.

The 'Bianchi identity' now reads

$$\partial_{[\mu} C_{\nu\rho]|\sigma} = 0 \quad ; \qquad (10)$$

and is obviously different from the usual Bianchi identity on the Riemann tensor $R^L_{[\mu\nu\ \rho]\sigma} = 0$. The linearized Einstein equations are now recovered from the equation of motion

$$\partial^\mu C_{\mu\nu|\rho} = 0 \qquad (11)$$

if we impose the gauge condition

$$C_{\mu\nu}{}^\nu = 0 \qquad (12)$$

(imposing this condition is possible precisely because $C_{\mu\nu|\rho}$ is *not* gauge invariant). Noticing that (10) and (11) are completely analogous to Maxwell's equations, we now introduce the 'dual field strength'

$$\tilde{C}_{\mu_1...\mu_{D-2}|\nu} = \varepsilon_{\mu_1...\mu_{D-2}}{}^{\rho\sigma} C_{\rho\sigma|\nu} \qquad (13)$$

It is then easy to see that vanishing divergence for one of the field strengths implies vanishing curl for the other, and vice versa. Furthermore,

$$\tilde{C}_{[\mu_1...\mu_{D-2}|\nu]} = 0 \quad \Longleftrightarrow \quad C_{\mu\nu}{}^\nu = 0 \qquad (14)$$

On shell, where $\partial_{[\mu_1} \tilde{C}_{\mu_2...\mu_{D-2}]|\nu} = 0$, we can therefore introduce a 'dual graviton field' $\tilde{h}_{\mu_1...\mu_{D-3}|\nu}$, analogous to the dual 'magnetic' potential $\tilde{A}_\mu$, with associated 'field strength'

$$\tilde{C}_{\mu_1...\mu_{D-2}|\nu} := \partial_{[\mu_1} \tilde{h}_{\mu_2...\mu_{D-2}]|\nu} \qquad (15)$$

Let us stress that this dual 'field strength' exists only *on-shell*, i.e. when the linearized Einstein equations are satisfied. The tracelessness condition (14) requires

$$\tilde{h}_{[\mu_1...\mu_{D-3}|\nu]} = 0 \qquad (16)$$

This is a second new feature *vis-à-vis* Maxwell and p-form gauge theories: for $D \geq 5$, the dual graviton field transforms in a *mixed Young tableau representation*. The associated gauge transformations are also more involved, as the gauge parameters may likewise belong to non-trivial representations.[42]

It does not appear possible to extend this duality invariance to the full non-linear theory in any obvious way. A generalization does not appear to exist even at first non-trivial order beyond the linear approximation, at least not in a way that would be compatible order by order with the background Lorentz invariance of the free theory. More succinctly, the No-Go Theorem of [42] asserts that there exists no continuous deformation of the free theory with these properties. On the other hand, experience has taught us that there is no No-Go Theorem without a loophole! So we simply interpret this result as evidence that one must search in a different direction, giving up one or more of the seemingly 'natural' assumptions that went into its proof. An example how one might possibly evade these assumptions is the one-dimensional 'geodesic' σ-model over infinite dimensional cosets which will be introduced in section 6, and which renounces manifest space-time Lorentz invariance.

2.2. A nonlinear duality: the Geroch group

Unlike for the free spin-2 theory discussed in the foregoing section, there does exist a version of Einstein's theory possessing a non-linear and non-local duality symmetry, but it suffers from a different limitation: it works only when Einstein's theory is dimensionally reduced to two space or space-time dimensions, i.e. in the presence of two commuting Killing vectors. For definiteness, we will take the two Killing vectors to be spacelike, and choose coordinates such that they are (locally) given by $\partial/\partial y$ and $\partial/\partial z$: this means that the symmetry acts on solutions depending on two of the four spacetime coordinates, namely (t, x). In a suitable gauge we can then write the line element as[5]

$$ds^2 = \Delta^{-1}\lambda^2(-dt^2 + dx^2) + (\rho^2\Delta^{-1} + \Delta\tilde{B}^2)dy^2 + 2\Delta\tilde{B}dy\,dz + \Delta dz^2 \ , \quad (17)$$

where the metric coefficients depend only on (t, x). The metric coefficient $\tilde{B}$ is the third component of the Kaluza Klein vector field $(B_\mu, B_2) \equiv (0, 0, \tilde{B})$ that would arise in the reduction of Einstein's theory to three dimensions. The metric ansatz (17) can be further simplified by switching to Weyl canonical coordinates where ρ is identified with the time coordinate

$$\rho(t, x) = t \Longrightarrow \quad \tilde{\rho} = x \qquad\qquad (18)$$

This particular choice is adapted to cosmological solutions, where $t \geq 0$ with the singularity 'occurring' on the spacelike hypersurface at time $t = 0$.

This is also the physical context in which we will consider the gravitational billiards in the following section.[b]

Here we will not write out the complete Einstein equations for the metric ansatz (17) (see, however,[5,9,11]) but simply note that upon dimensional reduction, the fields $(\Delta, \tilde{B})$ with $\Delta \geq 0$ coordinatize a homogeneous σ-model manifold $SL(2, \mathbb{R})/SO(2)$.[44] The equation for $\tilde{B}$ reads

$$\partial_\mu(t^{-1}\Delta^2\partial^\mu \tilde{B}) = 0 \tag{19}$$

with the convention, in this subsection only, that $\mu, \nu = 0, 1$. Because in two dimensions, every divergence-free vector field can be (locally) rewritten as a curl, we can introduce the dual 'Ehlers potential' $B(t, x)$ by means of

$$t\Delta^{-2}\partial_\mu B = \epsilon_{\mu\nu}\partial^\nu \tilde{B} \tag{20}$$

The Ehlers potential obeys the equation of motion

$$\partial_\mu(t\Delta^{-2}\partial^\mu B) = 0 \tag{21}$$

The combined equations of motion for Δ and B can be compactly assembled into the so-called Ernst equation[5]

$$\Delta\partial_\mu(t\partial^\mu \mathcal{E}) = t\partial_\mu\mathcal{E}\partial^\mu\mathcal{E} \tag{22}$$

for the complex Ernst potential $\mathcal{E} := \Delta + iB$. The pair (Δ, B) again parametrizes a coset space $SL(2, \mathbb{R})/SO(2)$, but different from the previous one.

To write out the non-linear action of the two $SL(2, \mathbb{R})$ symmetries, one of which is the Ehlers symmetry, we use a notation that is already adapted to the Kac Moody theory in the following chapters. The relation to the more familiar 'physicist's notation' for the $SL(2, \mathbb{R})$ generators is given below:

$$e \sim J^+ \quad, \quad f \sim J^- \quad, \quad h \sim J^3 \tag{23}$$

In writing the variations of the fields, we will omit the infinitesimal parameter that accompanies each transformation. The Ehlers group is generated by [9,45]

$$\begin{aligned}
e_3(\Delta) &= 0 &, \quad e_3(B) &= -1 \\
h_3(\Delta) &= -2\Delta &, \quad h_3(B) &= -2B \\
f_3(\Delta) &= 2\Delta B &, \quad f_3(B) &= B^2 - \Delta^2
\end{aligned} \tag{24}$$

---

[b]If the Weyl coordinate ρ is taken to be spacelike, we would be dealing with a generalization of the so-called Einstein-Rosen waves.

The other $SL(2,\mathbb{R})$, often referred to as the Matzner Misner group, is generated by

$$e_2(\Delta) = 0 \quad , \quad e_2(\tilde{B}) = -1$$
$$h_2(\Delta) = 2\Delta \quad , \quad h_2(\tilde{B}) = -2\tilde{B}$$
$$f_2(\Delta) = -2\Delta\tilde{B} \quad , \quad f_2(\tilde{B}) = \tilde{B}^2 - \frac{\rho^2}{\Delta^2} \tag{25}$$

(the numbering of the generators has been introduced in accordance with the numbering that will be used later in section 5). The Geroch group is now obtained by intertwining the two $SL(2,\mathbb{R})$ groups, that is by letting the Ehlers group act on $\tilde{B}$, and the Matzner Misner group on B, and by iterating this procedure on the resulting 'dual potentials'. It is not difficult to see that, in this process, one 'never comes back' to the original fields, and an infinite tower of dual potentials is generated.[3] The Geroch group is then realized on this infinite tower; when projecting down this action onto the original fields, one ends up with a non-linear and non-local realization of this group.

The mathematical proof that the Lie algebra underlying the Geroch group is indeed $A_1^{(1)} \equiv \widehat{\mathfrak{sl}(2,\mathbb{R})}_{ce}$ proceeds by verification of the bilinear relations (no summation on j)[45,46]

$$[e_i, f_j] = \delta_{ij} h_j \ , \quad [h_i, e_j] = A_{ij} e_j \ , \quad [h_i, f_j] = -A_{ij} f_j \tag{26}$$

for $i, j = 2, 3$, with the (Cartan) matrix

$$A_{ij} = \begin{pmatrix} 2 & -2 \\ -2 & 2 \end{pmatrix} \tag{27}$$

The subscript 'ce' on $\widehat{\mathfrak{sl}(2,\mathbb{R})}_{ce}$ is explained by the existence of a *central extension* of the loop algebra, with the *central charge* generator

$$c := h_2 + h_3 \tag{28}$$

This charge acts on the conformal factor λ as a scaling operator, but leaves all other fields inert [8,9,11]. Finally, the trilinear Serre relations

$$[f_2, [f_2, [f_2, f_3]]] = [f_3, [f_3, [f_3, f_2]]] = 0 \tag{29}$$

are satisfied on all fields (the corresponding relations for the e generators are trivially fulfilled). Together, (26) and (29) are just the *defining relations* (Chevalley Serre presentation [15,16,17]) for the affine Lie algebra $A_1^{(1)}$.

Evidently, the relation (20) between $\tilde{B}$ and the Ehlers potential B is a *nonlinear* extension of the duality

$$\partial_\mu \varphi = \epsilon_{\mu\nu} \partial^\nu \tilde{\varphi} \tag{30}$$

valid for free scalar fields in two dimensions. The main difference is that, whereas in the free field case, and more generally for p-form gauge theories in higher dimensions, a second dualization brings us back to the field from which we started (modulo integration constants), iterating the duality transformations (24) and (25) does not, as we pointed out already. *It is therefore the intrinsic non-linearity of Einstein's theory that explains the emergence of an infinite chain of dualizations, and consequently of an infinite dimensional symmetry.*

3. Gravitational Billiards and Kac-Moody Algebras

The duality transformations reviewed in the previous section are invariances of mutilated versions of Einstein's theory only. On the other hand, what we are really after, are symmetries that would not require any such truncations. The symmetries we are about to discuss next considerably extend the ones discussed so far, but have not actually been shown to be symmetries of Einstein's theory, or some extension thereof. There are two reasons for this. First, the full gravitational field equations are far more complicated than the truncations discussed in the foregoing section — as evidenced by the circumstance that no exact solutions appear to be known that would not make use of some kind of symmetry reduction in one way or another (in the appropriate coordinates). Consequently, any extension of the known symmetries to the full theory, which by necessity would be very non-local, will not be easy to identify. The second difficulty is that the Lie algebras that are conjectured to arise in this symmetry extension belong to the class of *indefinite Kac Moody algebras*. However, after more than 35 years of research in the theory of Kac Moody algebras, we still do not know much more about these algebras beyond their mere existence — despite the fact that they can be characterized by means of a simple set of generators and relations! [c] The main encouragement therefore derives from the fact that there exists this link between these two seemingly unrelated areas, which provides more than just a hint of an as yet undiscovered symmetry of Einstein's theory. A key role in deriving these results was played by an analysis of Einstein's equations near a spacelike singularity in terms of *gravitational billiards*, to which we turn next.

---

[c]See remarks after Table 1 to appreciate the challenge.

3.1. BKL dynamics and gravitational billiards

A remarkable and most important development in theoretical cosmology was the analysis of spacelike (cosmological) singularities in Einstein's theory by Belinskii, Khalatnikov and Lifshitz (abbreviated as 'BKL' in the remainder), and their discovery of chaotic oscillations of the spacetime metric near the initial singular hypersurface;[47] see also [48,49,50,51,52]. There is a large body of work on BKL cosmology, see [53,54,55] for recent reviews and extensions of the original BKL results. In particular, there is now convincing evidence for the correctness of the basic BKL picture both from numerical analyses (see e.g. [56,57]) as well as from more rigorous work [58,59,60,61]. It has also been known for a long time that the chaotic oscillations of the metric near the singularity can be understood in terms of *gravitational billiards*, although there exist several different realizations of this description, cf. [51,54,55] and references therein. The one which we will adopt here, grew out of an attempt to extend the original BKL results to more general matter coupled systems, in particular those arising in superstring and M theory[62,63,24,64,65]. It is particularly well suited for describing the relation between the BKL analysis and the theory of indefinite Kac Moody algebras, which is our main focus here, and which we will explain in the following section. See also [66,67] for an alternative approach.

We first summarize the basic picture, see [65] for a more detailed exposition. Our discussion will be mostly heuristic, and we shall make no attempt at rigorous proofs here (in fact, the BKL hypothesis has been rigorously proven only with very restrictive assumptions [59,57,60,61,68], but there is so far no proof of it in the general case). Quite generally, one considers a big-bang-like space-time with an initial singular spacelike hypersurface 'located' at time $t = 0$. It is then convenient to adopt a pseudo-Gaussian gauge for the metric (we will leave the number of spatial dimensions d arbitrary for the moment)

$$ds^2 = -N^2 dt^2 + g_{ij} dx^i dx^j \tag{31}$$

and to parametrize the spatial metric g_{ij} in terms of a frame field, or dreibein, θ^a (a one form) [d]

$$g_{ij} dx^i \otimes dx^j = \sum_{a=1}^{d} \theta^a \otimes \theta^a \tag{32}$$

---

[d]The summation convention is in force for the coordinate indices $i, j, \ldots$, but suspended for frame indices $a, b, \ldots$.

For this frame field we adopt the so-called Iwasawa decomposition

$$\theta^a = e^{-\beta^a} \mathcal{N}^a{}_i dx^i \qquad (33)$$

by splitting off the (logarithmic) scale factors β^a from the off-diagonal frame (and metric) degrees of freedom $\mathcal{N}^a{}_i$, which are represented by an upper triangular matrix with 1's on the diagonal. The spatial metric then becomes

$$g_{ij} = \sum_{a=1}^{d} e^{-2\beta^a} \mathcal{N}^a{}_i \mathcal{N}^a{}_j \qquad (34)$$

The main advantage of the Iwasawa decomposition is that it matches precisely with the triangular decomposition (48) below, which is valid for any Kac Moody algebra. Furthermore, it turns out that, in the limit $t \to 0$ all the interesting action takes place in the scale factors β^a, whereas the $\mathcal{N}$ as well as the matter degrees of freedom asymptotically 'come to rest' in this limit. Similarly, the metric and other degrees of freedom at *different spatial points* should decouple in this limit, as the spatial distance between them exceeds their horizon[e]. The basic hypothesis underlying the BKL analysis is therefore that spatial gradients should become less and less important in comparison with time derivatives as $t \to 0$, such that the resulting theory should be effectively describable in terms of a one dimensional reduction, in which the complicated partial differential equations of Einstein's theory are effectively replaced by a continuous infinity of ordinary differential equations.

To spell out this idea in more detail, let us insert the above metric ansatz into the Einstein-Hilbert action, and drop all spatial derivatives (gradients), so that this action is approximated by *a continuous superposition of one-dimensional systems*. One then obtains (still in d spatial dimensions)

$$S[g_{ij}] = \frac{1}{4} \int d^d x \int dx^0 \tilde{N}^{-1} \left[\left(\mathrm{tr} \, (\mathrm{g}^{-1} \dot{\mathrm{g}})^2 - (\mathrm{tr} \, \mathrm{g}^{-1} \dot{\mathrm{g}})^2 \right) \right] \qquad (35)$$

in a matrix notation where $\mathrm{g}(t) \in GL(d, \mathbb{R})$ stands for the matrix (g_{ij}) representing the spatial components of the metric at each spatial point, and $\tilde{N} \equiv N\sqrt{g}$ is a rescaled lapse function. Neglecting the off-diagonal

---

[e]One might even view this decoupling as a direct consequence of the *spacelike* nature of the singularity.

degrees of freedom, this action is further simplified to

$$S[\beta^a] = \frac{1}{4} \int d^d x \int dx^0 \tilde{N}^{-1} \left[\sum_{a=1}^{d} (\dot{\beta}^a)^2 - \left(\sum_{a=1}^{d} \dot{\beta}^a \right)^2 \right]$$

$$\equiv \frac{1}{4} \int d^d x \int dx^0 \tilde{N}^{-1} G_{ab} \dot{\beta}^a \dot{\beta}^b \qquad (36)$$

where G_{ab} is the restriction of the superspace metric (*à la* Wheeler-DeWitt) to the space of scale factors. A remarkable, and well known property of this metric is its *indefinite signature* $(- + \cdots +)$, with the negative sign corresponding to variations of the conformal factor. This indefiniteness will be crucial here, because it directly relates to the indefiniteness of the generalized Cartan-Killing metric on the associated Kac Moody algebra. In the Hamiltonian description the velocities $\dot{\beta}^a$ are replaced by their associated momenta π_a; variation of the lapse $\tilde{N}$ yields the *Hamiltonian constraint*

$$\mathcal{H} = \sum_a \pi_a^2 - \frac{1}{d-1} \left(\sum_a \pi_a \right)^2 \equiv G^{ab} \pi_a \pi_b \approx 0 \qquad (37)$$

Here G^{ab} is the inverse of the superspace metric, i.e. $G^{ac} G_{bc} = \delta_c^a$. The constraint (37) is supposed to hold at each spatial point, but let us concentrate at one particular spatial point for the moment. It is easy to check that (37) is solved by the well known conditions on the Kasner exponents. In this approximation, one thus has a Kasner-like metric at each spatial point, with the Kasner exponents depending on the spatial coordinate. In terms of the β-space description, we thus have the following picture of the dynamics of the scale factors at each spatial point. The solution to the constraint (37) corresponds to the motion of a relativistic massless particle (often referred to as the 'billiard ball' in the remainder) moving in the forward lightcone in β-space along a lightlike line w.r.t. the 'superspace metric' G_{ab}. The Hamiltonian constraint (37) is then re-interpreted as a relativistic dispersion relation for the 'billiard ball'.

Of course, the above approximation does not solve the Einstein equations, unless the Kasner exponents are taken to be constant (yielding the well known Kasner solution). Therefore, in a second step one must now take into account the spatial dependence and the effects of non-vanishing spatial curvature, and, eventually, the effect of matter couplings. At first sight this would seem to bring back the full complications of Einstein's equations. Surprisingly, this is not the case. Namely, one can show (at least heuristically) that[65]

(1) except for a finite number of them, the infinite number of degrees of freedom encoded in the spatially inhomogeneous metric, and in other fields, *freeze* in that they tend to finite limits as $t \to 0$; and

(2) the dynamics of the remaining 'active' diagonal metric degrees of freedom can be asymptotically described in terms of a simple billiard dynamics taking place in the β-space of (logarithmic) scale factors.

This result can be expressed more mathematically as follows. In the limit $t \to 0$, the effect of the remaining degrees of freedom consists simply in modifying the gravitational Hamiltonian (37) at a given spatial point by the addition of an *effective potential* that may be pictured as arising from 'integrating out' all but the diagonal degrees of freedom. Accordingly, the free Hamiltonian constraint (37) is now replaced by an *effective Hamiltonian constraint*

$$\mathcal{H}(\beta^a, \pi_a, Q, P) = G^{ab}\pi_a\pi_b + \sum_A c_A(Q, P)\exp\left(-2w_A(\beta)\right) \qquad (38)$$

where β^a, π_a are the canonical variables corresponding to the diagonal metric degrees of freedom, and Q, P denote the remaining canonical degrees of freedom. The quantities w_A appearing in the exponential are generically *linear forms* in β,

$$w_A(\beta) = \sum_a (w_A)_a \beta^a \qquad (39)$$

and are usually referred to as *'wall forms'*. It is crucial that the precise form of the coefficient functions $c_A(Q, P)$ — which is very complicated — does not matter in the BKL limit, which is furthermore dominated by a finite number of leading contributions for which $c_{A'}(Q, P) \geq 0$. The detailed analysis[65] reveals various different kinds of walls: gravitational walls due to the effect of spatial curvature, symmetry (or centrifugal) walls resulting from the elimination of off-diagonal metric components, electric and magnetic p-form walls, and dilaton walls. It is another non-trivial result that all these walls are *timelike* in β-space, that is, they have spacelike normal vectors.

The emergence of dominant walls is a consequence of the fact that, in the limit $t \to 0$, when $\beta \to \infty$, most of the walls 'disappear from sight', as the 'soft' exponential walls become steeper and steeper, eventually rising to infinity. Perhaps a useful analogy here is to think of a mountainscape, defined by the sum of the exponential potentials $C_A e^{-2w_A}$; when the mountaintops rise into the sky, only the nearest mountains remain visible to the observer in the valley. The Hamiltonian constraint (38) then takes the limiting form

$$H_\infty(\beta^a, \pi_a) = G^{ab}\pi_a\pi_b + \sum_{A'} \Theta_\infty\left(-2w_{A'}(\beta)\right) \qquad (40)$$

where the sum is only over the dominant walls (indexed by A'), and Θ_∞ denotes the infinite step function

$$\Theta_\infty(x) := \begin{cases} 0 & \text{if } x < 0 \\ +\infty & \text{if } x > 0 \end{cases} \qquad (41)$$

In conclusion, the original Hamiltonian simplifies dramatically in the BKL limit $t \to 0$. The dynamics of (40) is still that of a massless relativistic particle in β-space, but one that is confined in a 'box'. Hence, this particle undergoes occasional collisions with the 'sharp' walls: when the argument of the Θ_∞ function is negative, i.e. between the walls, this particle follows a free relativistic motion characterized by the appropriate Kasner coefficients; when the particle hits a walls (where Θ_∞ jumps by an infinite amount), it gets reflected with a corresponding change in the Kasner exponents (these reflections are also referred to as *Kasner bounces*). Because the walls are timelike, the Kasner exponents get rotated by an element of the orthochronous Lorentz group in β-space at each collision.

In summary, we are indeed dealing with a *relativistic billiard* evolving in the forward lightcone in β-space. The billiard walls ('cushions') are the hyperplanes in β-space determined by the zeros of the wall forms, i.e. $w_{A'}(\beta) = 0$. The chamber, in which the motion takes place, is therefore the wedge-like region defined by the inequalities [f].

$$w_{A'}(\beta) \geq 0 \qquad (42)$$

As for the long term (large β) behavior of the billiard, there are two possibilities:[64]

(1) The chamber characterized by (42) is entirely contained in the forward lightcone in β-space (usually with at least one edge on the lightcone). In this case, the billiard ball will undergo infinitely many collisions because, moving at the speed of light, it will always catch up with one of the walls. The corresponding metric will then exhibit infinitely

---

[f]Note that this is a *space-time picture* in β-space: the walls *recede* as t tends to 0, and $\beta \to \infty$. The actual 'billiard table' can be defined as the projection of this wedge onto the unit hyperboloid $G_{ab}\beta^a\beta^b = -1$ in β-space.[65] See also [53,54,55] for previous work and alternative descriptions of the billiard.

many Kasner bounces between $0 < t < \epsilon$ for any $\epsilon > 0$, hence *chaotic oscillations*.[g]

(2) The chamber extends beyond the lightcone, because some walls intersect outside the lightcone. In this case the billiard ball undergoes finitely many oscillations until its motion is directed towards a region that lies outside the lightcone; it then never catches up with any wall anymore because no 'cushion' impedes its motion. The corresponding metrics therefore exhibit a *monotonic Kasner-like behavior* for $0 < t < \epsilon$ for sufficiently small $\epsilon > 0$.

The question of chaotic vs. regular behavior of the metric near the singularity is thereby reduced to determining whether the billiard chamber realizes case 1 or case 2, and this is now a matter of a simple algebraic computation. In the case of monotonic Kasner-like behavior one can exploit these results and prove rigorous theorems about the behavior of the solution near the singularity.[60,68]

3.2. *Emergence of Kac Moody symmetries*

The billiard description holds not only for gravity itself, but generalizes to various kinds of matter couplings extending the Einstein-Hilbert action. However, these billiards have no special regularity properties in general. In particular, the dihedral angles between the 'walls' bounding the billiard might depend on continuous couplings, and need not be integer submultiples of π. In some instances, however, the billiard can be identified with the fundamental Weyl chamber of a symmetrizable Kac Moody algebra of indefinite type[h], with Lorentzian signature metric.[24,64,69] Such billiards are also called 'Kac Moody billiards'. Examples are pure gravity in any number of spacetime dimensions, for which the relevant KM algebra is AE_d, and superstring models[24] for which one obtains the rank 10 algebras E_{10} and BE_{10}, in line with earlier conjectures made in.[70] Furthermore, it was understood that chaos (finite volume of the billiard) is equivalent to hyperbolicity of the underlying Kac Moody algebra.[64] Further examples of the emergence of Lorentzian Kac Moody algebras can be found in.[69]

---

[g]Although we utilise this term in a somewhat cavalier manner here, readers can be assured that this system is indeed chaotic in the rigorous sense. For instance, projection onto the unit hyperboloid in β-space leads to a finite volume billiard on a hyperbolic manifold of constant negative curvature, which is known to be strongly chaotic.

[h]From now on we abbreviate 'Kac Moody' by 'KM', and 'Cartan subalgebra' by 'CSA'.

The main feature of the gravitational billiards that can be associated with KM algebras is that *there exists a group theoretical interpretation of the billiard motion:* the asymptotic BKL dynamics is equivalent, at each spatial point, to the asymptotic dynamics of a one-dimensional nonlinear σ-model based on a certain infinite dimensional coset space $G/K(G)$, where the KM group G and its maximal compact subgroup $K(G)$ depend on the specific model. In particular, *the β-space of logarithmic scale factors, in which the billard motion takes place, can be identified with the Cartan subalgebra (CSA) of the underlying indefinite Kac-Moody algebra.* The dominant walls that determine the billiards asymptotically are associated with the simple roots of the KM algebra. We emphasize that it is precisely the presence of gravity, which comes with an indefinite (Lorentzian) metric in the β-superspace, hence a Cartan-Killing metric of indefinite signature, which forces us to consider *infinite dimensional* KM groups. By contrast, the finite dimensional simple Lie algebras, which can also be considered as KM algebras, but which were already classified long ago by Cartan, are characterized by a positive definite Cartan-Killing metric.

The σ-model formulation to be introduced and elaborated in section 6 enables one to go beyond the BKL limit, and to see the beginnings of a possible identification of the dynamics of the scale factors *and* of all the remaining variables with that of a non-linear σ-model defined on the cosets of the KM group divided by its maximal compact subgroup.[25,27] In that formulation, the various types of walls can thus be understood directly as arising from the large field limit of the corresponding σ-models. So far, only two examples have been considered in this context, namely pure gravity, in which case the relevant KM algebra is AE_3,[65] and the bosonic sector of $D = 11$ supergravity, for which the relevant algebra is the maximal rank 10 hyperbolic KM algebra E_{10}; we will return to the latter model in the final section. Following Ref. 25, 27 one can introduce for both models a precise identification between the purely t-dependent σ-model quantities obtained from the geodesic action on the $G/K(G)$ coset space on the one hand, and the fields and their spatial gradients evaluated at a given, but arbitrarily chosen spatial point on the other.

3.3. *The main conjecture*

To sum up, it has been established that

(1) in many physical theories of interest (and all the models arising in supergravity and superstring theory), the billiard region in which the

dynamics of the active degrees of freedom takes place can be identified with the Weyl chamber of some Lorentzian KM algebra; and

(2) the concept of a nonlinear σ-model on a coset space $G/K(G)$ can be generalized to the case where G is a Lorentzian KM group, and $K(G)$ its 'maximal compact subgroup'; furthermore, these (one-dimensional) σ-models are asymptotically equivalent to the billiard dynamics describing the active degrees of freedom as $t \to 0$.

So far, these correspondences between gravity or supergravity models on the one hand, and certain KM coset space σ-models on the other, work only for truncated versions of both models. Namely, on the gravity side one has to restrict the dependence on the spatial coordinates, whereas the KM models must be analyzed in terms of a 'level expansion', in which only the lowest orders are retained, and the remaining vast expanse of the KM Lie algebra remains to be understood and explored. There are, however, indications that, at least as far as the higher order spatial gradients on the (super)gravity side are concerned, the correspondence can be further extended: the level expansions of AE_3, and other hyperbolic KM algebras contain all the requisite representations needed for the higher order spatial gradients[25] (as well as an exponentially increasing number of representations for which a physical interpretation remains to be found [71]). This observation gave rise to the key conjecture[25] for the correspondence between $D = 11$ supergravity and the $E_{10}/K(E_{10})$ coset model, which we here reformulate in a somewhat more general manner:

The time evolution of the geometric data at each spatial point, i.e. the values of all the fields and their spatial gradients, can be mapped onto some constrained null geodesic motion on the infinite dimensional $G/K(G)$ coset space.

If true, this conjecture would provide us with an entirely new way of describing and analyzing a set of (non-linear) partial differential equations in terms of an ordinary differential equation in infinitely many variables, by 'spreading' the spatial dependence over an infinite dimensional Lie algebra, and thereby mapping the cosmological evolution onto a single trajectory in the corresponding coset space. In the remainder of this article we will therefore spell out some of the technical details that lead up to this conjecture.

4. Basics of Kac Moody Theory

We here summarize some basic results from the theory of KM algebras, referring the reader to[15,16,17] for comprehensive treatments. Every KM algebra $\mathfrak{g} \equiv \mathfrak{g}(A)$ can be defined by means of an integer-valued Cartan matrix A and a set of generators and relations. We shall assume that the Cartan matrix is symmetrizable since this is the case encountered for cosmological billiards. The Cartan matrix can then be written as ($i, j = 1, \ldots r$, with r denoting the *rank* of $\mathfrak{g}(A)$)

$$A_{ij} = \frac{2\langle \alpha_i | \alpha_j \rangle}{\langle \alpha_i | \alpha_i \rangle} \tag{43}$$

where $\{\alpha_i\}$ is a set of r simple roots, and where the angular brackets denote the invariant symmetric bilinear form of $\mathfrak{g}(A)$.[15] Recall that the roots can be abstractly defined as linear forms on the Cartan subalgebra (CSA) $\mathfrak{h} \subset \mathfrak{g}(A)$. The generators, which are also referred to as Chevalley-Serre generators, consist of triples $\{h_i, e_i, f_i\}$ with $i = 1, \ldots, r$, and for each i form an $\mathfrak{sl}(2, \mathbb{R})$ subalgebra. The CSA $\mathfrak{h}$ is then spanned by the elements h_i, so that

$$[h_i, h_j] = 0 \tag{44}$$

The remaining relations generalize the ones we already encountered in Eqs. (26) and (29): Furthermore,

$$[e_i, f_j] = \delta_{ij} h_j \tag{45}$$

and

$$[h_i, e_j] = A_{ij} e_j , \quad [h_i, f_j] = -A_{ij} f_j \tag{46}$$

so that the value of the linear form α_j, corresponding to the raising operator e_j, on the element h_i of the preferred basis $\{h_i\}$ of $\mathfrak{h}$ is $\alpha_j(h_i) = A_{ij}$. More abstractly, and independently of the choice of any basis in the CSA, the roots appear as eigenvalues of the adjoint action of any element h of the CSA on the raising (e_i) or lowering (f_i) generators: $[h, e_i] = +\alpha_i(h)e_i$, $[h, f_i] = -\alpha_i(h)f_i$. Last but not least we have the so-called Serre relations

$$\mathrm{ad}\,(e_i)^{1-A_{ij}}(e_j) = 0 , \quad \mathrm{ad}\,(f_i)^{1-A_{ij}}(f_j) = 0 \tag{47}$$

A key property of every KM algebra is the triangular decomposition

$$\mathfrak{g}(A) = \mathfrak{n}^- \oplus \mathfrak{h} \oplus \mathfrak{n}^+ \tag{48}$$

where $\mathfrak{n}^+$ and $\mathfrak{n}^-$, respectively, are spanned by the multiple commutators of the e_i and f_i which do not vanish on account of the Serre relations or the

Jacobi identity. To be completely precise, $\mathfrak{n}^+$ is the quotient of the free Lie algebra generated by the e_i's by the ideal generated by the Serre relations (*idem* for $\mathfrak{n}^-$ and f_i). In more mundane terms, when the algebra is realized, in a suitable basis, by infinite dimensional matrices, $\mathfrak{n}^+$ and $\mathfrak{n}^-$ simply consist of the 'nilpotent' matrices with nonzero entries only above or below the diagonal. Exponentiating them formally, one obtains infinite dimensional matrices again with nonzero entries above or below the diagonal.

A main result of the general theory is that, for positive definite A, one just recovers from these relations Cartan's list of finite dimensional Lie algebras (see e.g.[72] for an introduction). For non positive-definite A, on the other hand, the associated KM algebras are infinite dimensional. If A has only one zero eigenvalue, with all other eigenvalues strictly positive, the associated algebra is called *affine*. The simplest example of such an algebra is the the $A_1^{(1)}$ algebra underlying the Geroch group, which we already encountered and discussed in section 2.2, with Cartan matrix (27). While the structure and properties of affine algebras are reasonably well understood,[15,17] this is not so for *indefinite* A, when at least one eigenvalue of A is negative. In this case, very little is known, and it remains an outstanding problem to find a manageable representation for them.[15,16] In particular, there is not a single example of an indefinite KM algebra for which the root multiplicities, *i.e.* the number of Lie algebra elements associated with a given root, are known in closed form. The scarcity of results is even more acute for the 'Kac-Moody groups' obtained by formal exponentiation of the associated Lie algebras. As a special, and important case, the class of Lorentzian KM algebras includes *hyperbolic* KM algebras whose Cartan matrices are such that the deletion of any node from the Dynkin diagram leaves either a finite or an affine subalgebra, or a disjoint union of them.

The 'maximal compact' subalgebra $\mathfrak{k}$ is defined as the invariant subalgebra of $\mathfrak{g}(A)$ under the standard Chevalley involution, *i.e.*

$$\theta(x) = x \quad \text{for all } x \in \mathfrak{k} \tag{49}$$

with

$$\theta(h_i) = -h_i\,, \quad \theta(e_i) = -f_i\,, \quad \theta(f_i) = -e_i \tag{50}$$

More explicitly, it is the subalgebra generated by multiple commutators of $(e_i - f_i)$. For finite dimensional $\mathfrak{g}(A)$, the inner product induced on the maximal compact subalgebra $\mathfrak{k}$ is negative-definite, and the orthogonal complement to $\mathfrak{k}$ has a positive definite inner product. This is not true, however, for indefinite A. It is sometimes convenient to introduce the operation of *transposition* acting on any Lie algebra element E as

$$E^T := -\theta(E) \tag{51}$$

The subalgebra $\mathfrak{k}$ is thus generated by the 'anti-symmetric' elements satisfying $E^T = -E$; after exponentiation, the elements of the maximally compact subgroup K formally appear as 'orthogonal matrices' obeying $k^T = k^{-1}$.

Often one uses a so-called Cartan-Weyl basis for $\mathfrak{g}(A)$. Using Greek indices $\mu, \nu, \ldots$ to label the root components corresponding to an arbitrary basis H_μ in the CSA, with the usual summation convention and a Lorentzian metric $G_{\mu\nu}$ for an indefinite $\mathfrak{g}$, we have $h_i := \alpha_i^\mu H_\mu$, where α_i^μ are the 'contravariant components', $G_{\mu\nu}\alpha_i^\nu \equiv \alpha_{i\,\mu}$, of the simple roots α_i ($i = 1, \ldots r$), which are linear forms on the CSA, with 'covariant components' defined as $\alpha_{i\,\mu} \equiv \alpha_i(H_\mu)$. To an arbitrary root α there corresponds a set of Lie-algebra generators $E_{\alpha,s}$, where $s = 1, \ldots, \text{mult}\,(\alpha)$ labels the (in general) multiple Lie-algebra elements associated with α. The root multiplicity $\text{mult}\,(\alpha)$ is always one for finite dimensional Lie algebras, and also for the real (= positive norm) roots, but generically grows exponentiallly with $-\alpha^2$ for indefinite A. In this notation, the remaining Chevalley-Serre generators are given by $e_i := E_{\alpha_i}$ and $f_i := E_{-\alpha_i}$. Then,

$$[H_\mu, E_{\alpha,s}] = \alpha_\mu E_{\alpha,s} \tag{52}$$

and

$$[E_{\alpha,s}, E_{\alpha',t}] = \sum_u c_{\alpha\alpha'}^{s,t,u} E_{\alpha+\alpha',u} \tag{53}$$

The elements of the Cartan-Weyl basis are normalized such that

$$\langle H_\mu | H_\nu \rangle = G_{\mu\nu} , \quad \langle E_{\alpha,s} | E_{\beta,t} \rangle = \delta_{st}\delta_{\alpha+\beta,0} \tag{54}$$

where we have assumed that the basis satisfies $E_{\alpha,s}^T = E_{-\alpha,s}$. Let us finally recall that the Weyl group of a KM algebra is the discrete group generated by reflections in the hyperplanes orthogonal to the simple roots.

5. The Hyperbolic Kac Moody Algebra AE_3

As we explained, the known symmetries of Einstein's theory for special types of solutions include the Ehlers and Matzner Misner $SL(2, \mathbb{R})$ symmetries, which can be combined into the Geroch group $\widehat{SL(2, \mathbb{R})}_{ce}$. Furthermore, in the reduction to one time dimension, Einstein's theory is invariant under a rigid $SL(3, \mathbb{R})$ symmetry acting on the spatial dreibein. Hence, any conjectured symmetry of Einstein's theory should therefore contain these symmetries as subgroups. Remarkably, there is a hyperbolic KM group with precisely these properties, whose Lie algebra is furthermore the simplest hyperbolic KM algebra containing an affine subalgebra.[46] This is the algebra AE_3, with Cartan matrix

$$A_{ij} = \begin{pmatrix} 2 & -1 & 0 \\ -1 & 2 & -2 \\ 0 & -2 & 2 \end{pmatrix} \tag{55}$$

The $\mathfrak{sl}(2,\mathbb{R})$ subalgebra corresponding to the third diagonal entry of A_{ij} is associated with the Ehlers group. The affine subgroup corresponding to the submatrix (27) is the Geroch group[3] already discussed in section 2.2. The $SL(3,\mathbb{R})$ subgroup containing the the the Matzner-Misner $SL(2,\mathbb{R})$ group, is generated by (e_1, f_1, h_1) and (e_2, f_2, h_2), corresponding to the submatrix

$$\begin{pmatrix} 2 & -1 \\ -1 & 2 \end{pmatrix} \tag{56}$$

As we said, not much is known about AE_3; in particular, there is no 'list' of its (infinitely many) generators, nor of its structure constants (which are certainly too numerous to fit in any list, see below!). Nevertheless, in order to gain some 'feeling' for this algebra, we will now work out the beginnings of its decomposition into irreducible representations of its $SL(3,\mathbb{R})$ subgroup. Of course, this decomposition refers to the *adjoint action* of the $\mathfrak{sl}(3,\mathbb{R})$ subalgebra embedded in AE_3. More specifically, we will analyze the lowest terms of the nilpotent subalgebra $\mathfrak{n}^+$. To do so, we first define, for any given root α, its $\mathfrak{sl}(3,\mathbb{R})$ level ℓ to be the number of times the root α_3 appears in it, to wit, $\alpha = m\alpha_1 + n\alpha_2 + \ell\alpha_3$. The algebra AE_3 thereby decomposes into an infinite irreducible representations of its $\mathfrak{sl}(3,\mathbb{R})$ subalgebra[i]. As is well known,[72] the irreducible representations of $\mathfrak{sl}(3,\mathbb{R})$ can be conveniently characterized by their Dynkin labels $[p_1, p_2]$. In terms of the Young tableau description of $\mathfrak{sl}(3,\mathbb{R})$ representations, the first Dynkin label p_1 counts the number of columns having two boxes, while p_2 counts the number of columns having only one box. For instance, $[p_1, p_2] = [1, 0]$ labels an antisymmetric two-index tensor, while $[p_1, p_2] = [0, 2]$ denotes a symmetric two-index tensor. The dimension of the representation $[p_1, p_2]$ is $(p_1 + 1)(p_2 + 1)(p_1 + p_2 + 2)/2$.

---

[i]A different decomposition would be one in terms of the affine subalgebra $A_1^{(1)} \subset AE_3$;[18] however, the representation theory of $A_1^{(1)}$ is far more complicated and much less developed than that of $\mathfrak{sl}(3,\mathbb{R})$.

The level $\ell = 0$ sector, which includes the third Cartan generator h_3, is the $\mathfrak{gl}(3, \mathbb{R})$ subalgebra with generators $K^i{}_j$ (where $i, j = 1, 2, 3$) and commutation relations

$$[K^i{}_j, K^k{}_l] = \delta^k_j K^i{}_l - \delta^i_l K^k{}_j \tag{57}$$

corresponding to the $GL(3, \mathbb{R})$ group acting on the spatial components of the vierbein. The restriction of the AE_3-invariant bilinear form to the level-0 sector is

$$\langle K^i{}_j | K^k{}_l \rangle = \delta^i_l \delta^k_j - \delta^i_j \delta^k_l \tag{58}$$

The identification with the Chevalley-Serre generators is

$$e_1 = K^1{}_2 , \quad f_1 = K^2{}_1 , \quad h_1 = K^1{}_1 - K^2{}_2$$
$$e_2 = K^2{}_3 , \quad f_2 = K^3{}_2 , \quad h_2 = K^2{}_2 - K^3{}_3$$
$$h_3 = -K^1{}_1 - K^2{}_2 + K^3{}_3 \tag{59}$$

showing how the over-extended CSA generator h_3 enlarges the original $\mathfrak{sl}(3, \mathbb{R})$ generated by (e_1, f_1, h_1) and (e_2, f_2, h_2) to the Lie algebra $\mathfrak{gl}(3, \mathbb{R})$. The CSA generators are related to the 'central charge' generator c by

$$c = h_2 + h_3 = -K^1{}_1 \tag{60}$$

which acts as a scaling on the conformal factor[8,9,11] (here realized as the 1-1 component of the vierbein).

To determine the representations of $\mathfrak{sl}(3, \mathbb{R})$ appearing at levels $\ell = \pm 1$, we observe that, under the adjoint action of $\mathfrak{sl}(3, \mathbb{R})$, i.e. of (e_1, f_1, h_1) and (e_2, f_2, h_2), the extra Chevalley-Serre generator f_3 is a highest weight vector:

$$e_1(f_3) \equiv [e_1, f_3] = 0$$
$$e_2(f_3) \equiv [e_2, f_3] = 0 \tag{61}$$

The Dynkin labels of the representation built on this highest weight vector f_3 are $(p_1, p_2) = (0, 2)$, since

$$h_1(f_3) \equiv [h_1, f_3] = 0$$
$$h_2(f_3) \equiv [h_2, f_3] = 2f_3 \tag{62}$$

As mentioned above, the representation $(p_1, p_2) = (0, 2)$ corresponds to a symmetric (two-index) tensor. Hence, at the levels ± 1 we have AE_3 generators which can be represented as symmetric tensors $E^{ij} = E^{ji}$

and $F_{ij} = F_{ji}$. One verifies that all algebra relations are satisfied with $(a_{(ij)} \equiv (a_{ij} + a_{ji})/2)$

$$[K^i{}_j, E^{kl}] = \delta^k_j E^{il} + \delta^l_j E^{ki}$$
$$[K^i{}_j, F_{kl}] = -\delta^i_k F_{jl} - \delta^i_l F_{kj}$$
$$[E^{ij}, F_{kl}] = 2\delta^{(i}_{(k} K^{j)}{}_{l)} - \delta^{(i}_k \delta^{j)}_l \left(K^1{}_1 + K^2{}_2 + K^3{}_3\right)$$
$$\langle F_{ij} | E^{kl} \rangle = \delta^{(k}_i \delta^{l)}_j \tag{63}$$

and the identifications

$$e_3 = E^{33} \ , \quad f_3 = F_{33} \tag{64}$$

As one proceeds to higher levels, the classification of $\mathfrak{sl}(3, \mathbb{R})$ representations becomes rapidly more involved due to the exponential increase in the number of representations with level ℓ. Generally, the representations that can occur at level $\ell + 1$ must be contained in the product of the level-ℓ representations with the level-one representation $(0, 2)$. Working out these products is elementary, but cumbersome. For instance, the level-two generator $E^{ab|jk} \equiv \varepsilon^{abi} E_i{}^{jk}$, with labels $(1, 2)$, is straightforwardly obtained by commuting two level-one elements

$$[E^{ij}, E^{kl}] = \varepsilon^{mk(i} E_m{}^{j)l} + \varepsilon^{ml(i} E_m{}^{j)k} \tag{65}$$

A more economical way to identify the relevant representations is to work out the relation between Dynkin labels and the associated highest weights, using the fact that the highest weights of the adjoint representation are the roots. More precisely, the highest weight vectors being (as exemplified above at level 1) of the 'lowering type', the corresponding highest weights are *negative* roots, say $\Lambda = -\alpha$. Working out the associated Dynkin labels one obtains

$$p_1 \equiv p = n - 2m \ , \quad p_2 \equiv q = 2\ell + m - 2n \tag{66}$$

As indicated, we shall henceforth use the notation $[p_1, p_2] \equiv [p, q]$ for the Dynkin labels. This formula is restrictive because all the integers entering it must be non-negative. Inverting this relation we get

$$m = \tfrac{2}{3}\ell - \tfrac{2}{3}p - \tfrac{1}{3}q$$
$$n = \tfrac{4}{3}\ell - \tfrac{1}{3}p - \tfrac{2}{3}q \tag{67}$$

with $n \geq 2m \geq 0$. A further restriction derives from the fact that the highest weight must be a root of AE_3, viz. its square must be smaller or equal to 2:

$$\Lambda^2 = \tfrac{2}{3}\left(p^2 + q^2 + pq - \ell^2\right) \leq 2 \tag{68}$$

Consequently, the representations occurring at level ℓ must belong to the list of all the solutions of (67) which are such that the labels m, n, p, q are non-negative integers and the highest weight Λ is a root, i.e. $\Lambda^2 \leq 2$. These simple diophantine equations/inequalities can be easily evaluated by hand up to rather high levels.

Although the above procedure substantially reduces the number of possibilities, it does not tell us how often a given representation appears, i.e. its *outer multiplicity* μ. For this purpose we have to make use of more detailed information about AE_3, namely the root multiplicities computed in.[18,15] Matching the combined weight diagrams with the root multiplicities listed in table H_3 on page 215 of,[15] one obtains the following representations in the decomposition of AE_3 w.r.t. its $\mathfrak{sl}(3, \mathbb{R})$ subalgebra up to level $\ell \leq 5$, where we also indicate the root coefficients (m_1, m_2, ℓ), the norm and multiplicity of the root α, and the outer multiplicity of the representation $[p, q]$:

Table 1. Decomposition of AE_3 under $\mathfrak{sl}(3, \mathbb{R})$ for $\ell \leq 5$.

ℓ	$[p,q]$	α	α^2	mult α	μ
1	[0,2]	(0,0,1)	2	1	1
2	[1,2]	(0,1,2)	2	1	1
3	[2,2]	(0,2,3)	2	1	1
	[1,1]	(1,3,3)	-4	3	1
4	[3,2]	(0,3,4)	2	1	1
	[2,1]	(1,4,4)	-6	5	2
	[1,0]	(2,5,4)	-10	11	1
	[0,2]	(2,4,4)	-8	7	1
	[1,3]	(1,3,4)	-2	2	1
5	[4,2]	(0,4,5)	2	1	1
	[3,1]	(1,5,5)	-8	7	3
	[2,0]	(2,6,5)	-14	22	3
	[0,1]	(3,6,5)	-16	30	2
	[0,4]	(2,4,5)	-6	5	2
	[1,2]	(2,5,5)	-12	15	4
	[2,3]	(1,4,5)	-4	3	2

The above table does not look too bad, but appearances are deceptive, because the number of representations grows exponentially with the level! For AE_3, the list of representations with their outer multiplicities is meanwhile available up to $\ell \leq 56$ [71]; the total number of representations up to that level is 20 994 472 770 550 672 476 591 949 725 720 [j], larger than 10^{31}! This number should suffice to convince readers of the 'explosion' that takes place in these algebras as one increases the level. Similar decompositions can be worked out for the indefinite Kac-Moody algebras E_{10} and E_{11} [71], and for E_{10} under its D_9 and $A_8 \times A_1$ subalgebras.[26,28]. The real problem, however, is not so much the large number of representations, but rather the absence of any discernible structure in these tables, at least up until now.

6. Nonlinear σ-Models in One Dimension

Notwithstanding the fact that we know even less about the groups associated with indefinite KM algebras, it is possible to formulate nonlinear σ-models in one time dimension and thereby provide an effective and unified description of the asymptotic BKL dynamics for several physically important models. The basic object of interest is a one-parameter dependent KM group element $\mathcal{V} = \mathcal{V}(t)$, assumed to be an element of the coset space $G/K(G)$, where G is the group obtained by formal exponentiation of the KM algebra $\mathfrak{g}$, and $K(G)$ its maximal compact subgroup, obtained by formal exponentiation of the associated maximal compact subalgebra $\mathfrak{k}$ defined above. For finite dimensional $\mathfrak{g}(A)$ our definitions reduce to the usual ones, whereas for indefinite KM algebras they are formal constructs to begin with. In order to ensure that our definitions are meaningful operationally, we must make sure at every step that any finite truncation of the model is well defined and can be worked out explicitly in a finite number of steps.

In physical terms, $\mathcal{V}$ can be thought of as an extension of the vielbein of general relativity, with G and $K(G)$ as generalizations of the $GL(d, \mathbb{R})$ and local Lorentz symmetries of general relativity. For infinite dimensional G, the object $\mathcal{V}$ thus is a kind of '∞-bein', that can be associated with the 'metric'

$$\mathcal{M} := \mathcal{V}^T \mathcal{V} \tag{69}$$

which is invariant under the left action ($\mathcal{V} \to k\mathcal{V}$) of the 'Lorentz group' $K(G)$. Exploiting this invariance, we can formally bring $\mathcal{V}$ into a triangular gauge

$$\mathcal{V} = \mathcal{A} \cdot \mathcal{N} \implies \mathcal{M} = \mathcal{N}^T \mathcal{A}^2 \mathcal{N} \tag{70}$$

---

[j]T. Fischbacher, private communication.

where the abelian part $\mathcal{A}$ belongs to the exponentiation of the CSA, and the nilpotent part $\mathcal{N}$ to the exponentiation of $\mathfrak{n}^+$. This formal Iwasawa decomposition, which is the infinite dimensional analog of (33), can be made fully explicit by decomposing $\mathcal{A}$ and $\mathcal{N}$ in terms of bases of $\mathfrak{h}$ and $\mathfrak{n}^+$ (using the Cartan Weyl basis)

$$\mathcal{A}(t) = \exp\left(\beta^\mu(t)\, H_\mu\right),$$

$$\mathcal{N}(t) = \exp\left(\sum_{\alpha \in \Delta_+} \sum_{s=1}^{\text{mult}(\alpha)} \nu_{\alpha,s}(t)\, E_{\alpha,s}\right) \tag{71}$$

where Δ_+ denotes the set of positive roots. The components β^μ, parametrizing a generic element in the CSA $\mathfrak{h}$, will turn out to be in direct correspondence with the metric scale factors β^a in (34). The main technical difference with the kind of Iwasawa decompositions used in section 3.1 is that now the matrix $\mathcal{V}(t)$ is infinite dimensional for indefinite $\mathfrak{g}(A)$, in which case the decomposition (71) is, in fact, the only sensible parametrization available! Consequently, there are now infinitely many ν's, whence $\mathcal{N}$ contains an infinite tower of new degrees of freedom. Next we define

$$\dot{\mathcal{N}}\mathcal{N}^{-1} = \sum_{\alpha \in \Delta_+} \sum_{s=1}^{\text{mult}(\alpha)} j_{\alpha,s} E_{\alpha,s} \quad \in \mathfrak{n}^+ \tag{72}$$

with

$$j_{\alpha,s} = \dot{\nu}_{\alpha,s} + \text{``}\nu\dot{\nu} + \nu\nu\dot{\nu} + \cdots\text{''} \tag{73}$$

(we put quotation marks to avoid having to write out the indices). To define a Lagrangian we consider the quantity

$$\dot{\mathcal{V}}\mathcal{V}^{-1} = \dot{\beta}^\mu H_\mu + \sum_{\alpha \in \Delta_+} \sum_{s=1}^{\text{mult}(\alpha)} \exp\left(\alpha(\beta)\right) j_{\alpha,s} E_{\alpha,s} \tag{74}$$

which has values in the Lie algebra $\mathfrak{g}(A)$. Here we have set

$$\alpha(\beta) \equiv \alpha_\mu \beta^\mu \tag{75}$$

for the value of the root α ($\equiv$ linear form) on the CSA element $\beta = \beta^\mu H_\mu$. Next we define

$$P := \frac{1}{2}\left(\dot{\mathcal{V}}\mathcal{V}^{-1} + (\dot{\mathcal{V}}\mathcal{V}^{-1})^T\right)$$

$$= \dot{\beta}^\mu H_\mu + \frac{1}{2} \sum_{\alpha \in \Delta_+} \sum_{s=1}^{\text{mult}(\alpha)} j_{\alpha,s} \exp\left(\alpha(\beta)\right)(E_{\alpha,s} + E_{-\alpha,s}) \tag{76}$$

where we arranged the basis so that $E^T_{\alpha,s} = E_{-\alpha,s}$. The KM-invariant σ-model Lagrangian is defined by means of the KM-invariant bilinear form

$$\mathcal{L} = \frac{1}{2} n^{-1} \langle P|P \rangle$$

$$= n^{-1} \left(\frac{1}{2} G_{\mu\nu} \dot{\beta}^\mu \dot{\beta}^\nu + \frac{1}{4} \sum_{\alpha \in \Delta_+} \sum_{s=1}^{\text{mult}(\alpha)} \exp\left(2\alpha(\beta)\right) j_{\alpha,s} j_{\alpha,s} \right) \quad (77)$$

Here the Lorentzian metric $G_{\mu\nu}$ is the restriction of the invariant bilinear form to the CSA, cf. (54). The 'lapse function' n ensures that our formalism is invariant under reparametrizations of the time variable. Remarkably, this action defined by the above Lagrangian is *essentially unique* because there are no higher order polynomial invariants for indefinite KM algebras.[15]

After these preparations we are now ready to specialize to the algebra AE_3. In this case this Lagrangian (77) contains the Kasner Lagrangian (35) as a special truncation. More specifically, retaining only the level zero fields (corresponding to the 'sub-coset' $GL(3,\mathbb{R})/O(3)$)

$$\mathcal{V}(t)\Big|_{\ell=0} = \exp(h^a{}_b(t) K^b{}_a) \quad (78)$$

and defining from $h^a{}_b$ a vielbein by matrix exponentiation $e^a{}_b \equiv (\exp h)^a{}_b$, and a corresponding contravariant metric $g^{ab} = e^a{}_c e^b{}_c$, it turns out that the bilinear form (58) reproduces the Lagrangian (35) (for the special case of three spatial dimensions). This means that *we can identify the restriction $G_{\mu\nu}$ of the Cartan-Killing metric to the CSA with the superspace metric G_{ab} in the superspace of scale factors β in (35).*

At level $\ell = 1$, we have the fields ϕ_{ij} associated with the level-one generators E^{ij}. Observe that for $D = 4$, these are precisely the spatial components of the dual graviton introduced in (15) — in other words, we have rederived the result of section 2.2 by a purely group theoretical argument! (This argument works likewise for $D > 4$.) This leads to a slightly less restricted truncation of our KM-invariant σ-model

$$\mathcal{V}(t)\Big|_{\ell=0,1} = \exp(h^a{}_b(t) K^b{}_a) \exp(\phi_{ab} E^{ab}) \quad (79)$$

In the gauge $n = 1$, the Lagrangian now has the form $\mathcal{L} \sim (g^{-1}\dot{g})^2 + g^{-1}g^{-1}\dot{\phi}\dot{\phi}$, where g denotes the covariant metric g_{ij}. As the ϕ_{ij}'s enter only through their time derivatives, their conjugate momenta Π^{ij} are constants of the motion in this $|\ell| \leq 1$ truncation. Eliminating the ϕ's in terms of the constant momenta Π yields

$$V_\phi(g) \propto +g_{ij}g_{kl}\Pi^{ik}\Pi^{jl} \tag{80}$$

This potential can be identified with the leading (weight-2) gravitational potential, if we identify the structure constants $C^i{}_{jk}$ defined by $d\theta^i = C^i{}_{jk}\theta^j \wedge \theta^k$, with the momenta conjugate to ϕ_{ij} as

$$\Pi^{ij} = \varepsilon^{kl(i}C^{j)}{}_{kl} \tag{81}$$

Consequently, the BKL dynamics at each spatial point is equivalent to the $|\ell| \leq 1$ truncation of the AE_3-invariant dynamics defined by (77). The fields $\phi_{ij}(t)$ parametrizing the components of the AE_3 coset element along the $\ell = 1$ generators are canonically conjugate to the structure constants $C^i{}_{jk}$. The proper physical interpretation of the higher level fields remains yet to be found.

Varying (77) w.r.t. the lapse function n gives rise to the constraint that the coset Lagrangian vanish. Defining the canonical momenta

$$\pi_a := \frac{\delta \mathcal{L}}{\delta \dot{\beta}^a} = n^{-1}G_{ab}\dot{\beta}^b \tag{82}$$

and the (non-canonical) momentum-like variables

$$\Pi_{\alpha,s} := \frac{\delta \mathcal{L}}{\delta \dot{j}_{\alpha,s}} = \frac{1}{2}n^{-1}\exp\left(2\alpha(\beta)\right)j_{\alpha,s} \tag{83}$$

and recalling the equivalence of the Cartan Killing and superspace metrics noted above, we are led to the Hamiltonian constraint of the σ-model, which is given by

$$\mathcal{H}(\beta^a, \pi_a, ...) = \frac{1}{2}G^{ab}\pi_a\pi_b + \sum_{\alpha \in \Delta_+}\sum_{s=1}^{\text{mult}(\alpha)}\exp\left(-2\alpha(\beta)\right)\Pi_{\alpha,s}\Pi_{\alpha,s} \tag{84}$$

where β^a, π_a are now the diagonal CSA degrees of freedom, and the dots stand for infinitely many off-diagonal (Iwasawa-type) canonical variables, on which the $\Pi_{\alpha,s}$ depend.

The evident similarity of (38) and (84) is quite striking, but at this point we can only assert that the two expressions coincide asymptotically, when they both reduce to a relativistic billiard. Namely, because the coefficients of the exponentials in (84) are non-negative, we can apply exactly the same reasoning as for the gravitational billiards in section 3.1. One then finds that the off-diagonal components $\nu_{\alpha,s}$ and the momentum-like variables $\Pi_{\alpha,s}$ get frozen asymptotically (again, we may invoke the imagery of a mountainscape, now defined by exponential potentials for all roots). In the present KM setup, all the walls enter on the same footing; there is nothing left of the distinctions between different types of walls (symmetry, gravitational, electric, and so on). The only important characteristic of a wall is its height ht $\alpha \equiv n_1 + n_2 + \cdots$ for a root decomposed along simple roots as $\alpha = n_1 \alpha_1 + n_2 \alpha_2 + \cdots$. The asymptotic Hamiltonian hence is dominated by the walls associated to the *simple roots*:

$$\mathcal{H}_\infty(\beta, \pi) = \frac{1}{2}\pi^a \pi_a + \sum_{i=1}^r \Theta_\infty\big(-2\alpha_i(\beta)\big) \tag{85}$$

where the sum is over the simple roots only, and the motion of the β^a is confined to the fundamental Weyl chamber $\alpha_i(\beta) \geq 0$.

The billiard picture for pure gravity in four dimensions is now readily understood in terms of the Weyl group of AE_3,[64] which is just the modular group $PGL(2, \mathbb{Z})$,[18] and the simple roots of AE_3. For the $\mathfrak{sl}(3, \mathbb{R})$ subalgebra, which has two simple roots, the Weyl group is the permutation group on three objects. The two hyperplanes orthogonal to these simple roots can be identified with the symmetry (centrifugal) walls. The third simple root extending (56) to the full rank 3 algebra (55) can be identified the dominant curvature (gravitational) wall.

To conclude: *in the limit where one goes to infinity in the Cartan directions, the dynamics of the Cartan degrees of freedom of the coset model become equivalent to a billiard motion within the Weyl chamber, subject to the zero-energy constraint* $\mathcal{H}_\infty(\beta, \pi) = 0$. Therefore, in those cases where the gravitational billiards of section 3.1 are of KM-type, they are asymptotically equivalent to the KM σ-models over $G/K(G)$.

7. Finale: E_{10} – The Ultimate Symmetry?

There can be little doubt that the algebra, which from many points is the most intriguing and most beautiful, is the maximal rank hyperbolic KM algebra E_{10}, which is an infinite dimensional extension of the better known finite dimensional exceptional Lie algebras E_6, E_7 and E_8. [72] There are two other rank-10 hyperbolic KM algebras DE_{10} and BE_{10} (respectively related to type I supergravity, and Einstein Maxwell supergravity in ten dimensions), but they appear to be less distinguished. The emergence of E_{10} in the reduction of $D = 11$ supergravity to one dimension had first been conjectured in.[70] A crucial new feature of the scheme proposed here, which is based on a hyperbolic σ-model defined by means of the geodesic action (77) is that it retains a residual spatial dependence, which on the

σ-model side is supposed 'to be spread' over the whole E_{10} Lie algebra. Thereby all degrees of freedom of the original theory should still be there, unlike for a *bona fide* reduction to one dimension.

Just like AE_3 the KM algebra E_{10} algebra is recursively defined via its Chevalley-Serre presentation in terms of generators and relations and its Dynkin diagram which we give below.

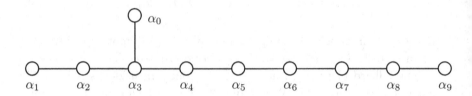

The nine simple roots $\alpha_1, \ldots, \alpha_9$ along the horizontal line generate an $A_9 \equiv \mathfrak{sl}(10, \mathbb{R})$ subalgebra. One of the reasons why E_{10} is distinguished is that its root lattice is the unique even self-dual Lorentzian lattice $\mathrm{II}_{1,9}$ (such lattices exist only in dimensions $d = 2 + 8n$.[73])

For the corresponding σ-model a precise identification can be made between the purely t-dependent σ-model quantities obtained from the geodesic action on the $E_{10}/K(E_{10})$ coset space on the one hand, and certain fields of $D = 11$ supergravity and their spatial gradients evaluated at a given, but arbitrarily chosen spatial point on the other.[25,27] The simple and essentially unique geodesic Lagrangian describing a null world line in the infinite-dimensional coset manifold $E_{10}/K(E_{10})$ thus reproduces the dynamics of the bosonic sector of eleven-dimensional supergravity in the vicinity of a space-like singularity. This result can be extended to *massive* IIA supergravity,[26] where also parts of the fermionic sector were treated for the first time, and to IIB supergravity in.[28] Related results had been previously obtained in the framework of E_{11}.[74,75,76]

A main ingredient in the derivation of these results is the level decomposition of E_{10} w.r.t. the A_9, D_9, and $A_8 \times A_1$ subalgebras of E_{10}, respectively, which generalizes the $\mathfrak{sl}(3, \mathbb{R})$ decomposition of AE_3 made in section 5. In all cases, one obtains precisely the field representation content of the corresponding supergravity theories at the lowest levels, and for all these decompositions, the bosonic supergravity equations of motion, when restricted to zeroth and first order spatial gradients, match with the corresponding σ-model equations of motion at the lowest levels. In particular, the self-duality of the five-form field strength in type IIB supergravity is implied by the dynamical matching between the $E_{10}/K(E_{10})$ σ-model and the supergravity equations of motion, and does not require local supersymmetry or some other extraneous argument for its explanation.

Combining the known results, we can summarize the correspondence between the maximally supersymmetric theories and the maximal rank regular subalgebras of E_{10} as follows

$$A_9 \subset E_{10} \quad \Longleftrightarrow \quad D = 11 \text{ supergravity}$$
$$D_9 \subset E_{10} \quad \Longleftrightarrow \quad \text{massive IIA supergravity}$$
$$A_8 \times A_1 \subset E_{10} \quad \Longleftrightarrow \quad \text{IIB supergravity}$$

The decompositions of E_{10} w.r.t. its other rank 9 regular subalgebras $A_{D-2} \times E_{11-D}$ (for $D = 3, \ldots, 9$) will similarly reproduce the representation content of maximal supergravities in D space-time dimensions as the lowest level representations.

We conclude by repeating the main challenge that remains: one must extend these correspondences to higher levels and spatial gradients, and find a physical interpretation for the higher level representations, whose number exhibits an exponential growth similar to the growth in the number of excited string states (see, however, [29] for recent progress concerning the link between higher order M Theory corrections and the E_{10} root lattice). Because this will inevitably require (or entail) a detailed understanding of indefinite and hyperbolic KM algebras, it might also help in solving the core problem of the theory of Kac Moody algebras, a problem that has vexed almost a generation of researchers.

Acknowledgments

This work was supported in part by the European Network HPRN-CT-2000-00122 and by the the German Israeli Foundation (GIF) Project Nr. I. 645 130-14-1999. It is a great pleasure to thank T. Damour, T. Fischbacher, M. Henneaux and A. Kleinschmidt for enjoyable collaborations and innumerable discussions that have shaped my understanding of the results reported here, and I.H.E.S., Bures-sur-Yvette, for continued support during several visits there. I am also very grateful to F. Englert, A. Feingold, A. Giveon, L. Houart and E. Rabinovici for enlightening discussions at various earlier stages of this work.

References

1. H.A. Buchdahl, "Reciprocal static solutions of the equation $G_{\mu\gamma} = 0$", Quart. J. Math. Oxford **5**, 116 (1954)
2. J. Ehlers, Dissertation Hamburg University (1957)
3. R. Geroch, "A method for generating solutions of Einstein's equations", J. Math. Phys. **12**, 918 (1971);
   "A method for generating solutions of Einstein's equations. 2", J. Math. Phys. **13**, 394 (1972).
4. W. Dietz and C. Hoenselars (eds.), Solutions of Einstein's equations: techniques and results (Springer, 1984)
5. D. Kramer et al.: "Exact Solutions of Einstein's Field Equations", Cambridge University Press
6. E. Cremmer and B. Julia, "The SO(8) supergravity", Nucl. Phys. **B159** (1979) 141
7. B. Julia, LPTENS 80/16, Invited paper presented at Nuffield Gravity Workshop, Cambridge, Eng., Jun 22 - Jul 12, 1980.
8. B. Julia, in the Proceedings of the Johns Hopkins Workshop on Current Problems in Particle Physics "Unified Theories and Beyond" (Johns Hopkins University, Baltimore, 1984).
9. P. Breitenlohner and D. Maison, "On the Geroch group", Ann. Inst. Henri Poincaré **46**, 215 (1986).
10. P. Breitenlohner, D. Maison and G. W. Gibbons, "Four-Dimensional Black Holes From Kaluza-Klein Theories," Commun. Math. Phys. **120**, 295 (1988).
11. H. Nicolai, in "Recent Aspects of Quantum Fields", Proceedings Schladming 1991, Lecture Notes in Physics (Springer Verlag, 1991)
12. W. Kinnersley and D. Chitre, J. Math. Phys. **18** (1977) 1583; **19** (1978) 1926 and 2037
13. D. Maison, Phys. Rev. Lett. **41** (1978) 521
14. V. Belinskii and V. Zakharov, "Integration of Einstein's equations by means of inverse scattering problem technique and construction of exact soliton solutions, Sov. Phys. JETP **48** (1978) 985
15. V.G. Kac, Infinite Dimensional Lie Algebras, 3rd edn., Cambridge University Press, 1990
16. R.V. Moody and A. Pianzola, Lie Algebras with Triangular Decomposition, Wiley, New York, 1995
17. P. Goddard and D.I. Olive, "Kac-Moody and Virasoro algebras in relation to quantum physics", Int. J. Mod . Phys. **A1**, 303 (1986)
18. A.J. Feingold and I.B. Frenkel, "A hyperbolic Kac-Moody algebra and the theory of Siegel modular form of genus 2", Math. Ann. **263**, 87 (1983)
19. E. Cremmer, B. Julia and J. Scherk, "Supergravity Theory In 11 Dimensions," Phys. Lett. B **76**, 409 (1978).
20. K. Kuchar, "Canonical quantum gravity", `gr-qc/9304012`
21. A. Giveon, M. Porrati and E. Rabinovici, "Target space duality in string theory", Phys. Rep. **244** (1994) 77, `hep-th/9401139`

22. C.M. Hull and P.K. Townsend, "Unity of superstring dualities", Nucl. Phys. **B438** (1995) 109, `hep-th/9410167`
23. N.A. Obers and B. Pioline, "U Duality and M Theory", Phys. Rep. **318** (1999) 113, `hep-th/9809039`
24. T. Damour and M. Henneaux, "E(10), BE(10) and arithmetical chaos in superstring cosmology," Phys. Rev. Lett. **86**, 4749 (2001) [arXiv:hep-th/0012172].
25. T. Damour, M. Henneaux and H. Nicolai, E_{10} and a small tension expansion of M Theory, Phys. Rev. Lett. **89** (2002): 221601, `hep-th/0207267`
26. A. Kleinschmidt and H. Nicolai, JHEP **0407**:041 (2004) `hep-th/0407101`
27. T. Damour and H. Nicolai, "Eleven dimensional supergravity and the $E_{10}/K(E_{10})$ sigma-model at low A_9 levels", `hep-th/0410245`
28. A. Kleinschmidt and H. Nicolai, "IIB Supergravity and E_{10}", Phys. Lett. **B606** (2005) 391, `hep-th/0411225`
29. T. Damour and H. Nicolai, Higher order M theory corrections and the Kac Moody algebra E_{10}, `hep-th/0504153`
30. J. Brown, O. Ganor and C. Helfgott, M theory and E_{10}: Billiards, Branes, and Imaginary Roots, JHEP **0408**:063 (2004)`hep-th/0401053`;
31. J. Brown, S. Ganguli, O. Ganor and C. Helfgott, $E(10)$ orbifolds, `hep-th/0409037`
32. P. C. West, "E_{11} and M Theory", Class. Quant. Grav. **18**, 4443–4460 (2001), [arXiv:hep-th/0104081]
33. P.C. West, "E_{11} and central charges", Phys. Lett. **B575** (2003) 333, `hep-th/0307098`; "Brane dynamics, central charges and $E(11)$", `hep-th/0412336`
34. P.C. West, "Some simple predictions for $E(11)$ symmetry", `hep-th/0407088`
35. A. Kleinschmidt and P.C. West, "Representations of G^{+++} and the role of spacetime", JHEP **0402** (2004) 033, `hep-th/0312247`
36. M. Gaberdiel, D.I. Olive and P.C. West, "A class of Lorentzian Kac Moody algebras", Nucl. Phys. **B645** (2002) 403, `hep-th/0205068`
37. F. Englert, L. Houart, A. Taormina and P.C. West, "The symmetry of M theories", JHEP **0309** (2003) 020, `hep-th/0304206`
38. F. Englert and L. Houart, G^{+++} invariant formulation of gravity and M-theories: Exact BPS solutions, JHEP **0401** (2004) 002 [arXiv:hep-th/0311255]; G^{+++} invariant formulation of gravity and M-theories: Exact intersecting brane solutions, JHEP **0405** (2004) 059 [arXiv:hep-th/0405082].
39. T. Curtright, Phys. Lett. **B165** (1985) 304
40. N.A. Obers, B. Pioline and E. Rabinovici, "M-theory and U-duality on T^d with gauge backgrounds, Nucl. Phys. **B 525** (1998) 163, `hep-th/9712084`
41. C.M. Hull, "Strongly coupled gravity and duality", Nucl. Phys. **B 583** (2000) 237, `hep-th/0004195`
42. X. Bekaert, N. Boulanger and M. Henneaux, "Consistent deformations of dual formulations of linearized gravity: a no go result", Phys. Rev. **D67** (2003) 044010, `hep-th/0210278`
43. M. Henneaux and C. Teitelboim, "Duality in linearized gravity", Phys. Rev. **D71**: 024018 (2005), `gr-qc/0408101`

44. G. Neugebauer and D. Kramer, Annalen der Physik **24** (1969) 62

45. H. Nicolai and A. Nagar, "Infinite dimensional symmetries in gravity", in *Gravitational Waves*, eds. I. Ciufolini, V. Gorini, U. Moschella and F. Fré, IoP, Bristol, 2001

46. H. Nicolai, "A Hyperbolic Lie Algebra From Supergravity," Phys. Lett. **B276**, 333 (1992).

47. V.A. Belinskii, I.M. Khalatnikov and E.M. Lifshitz, "Oscillatory approach to a singular point in the relativistic cosmology," Adv. Phys. **19**, 525 (1970).

48. V.A. Belinskii, I.M. Khalatnikov and E.M. Lifshitz, "Construction of a general cosmological solution of the Einstein equation with a time singularity", Sov. Phys. JETP **35**, 838-841 (1972).

49. V.A. Belinskii, I.M. Khalatnikov and E.M. Lifshitz, "A general solution of the Einstein equations with a time singularity," Adv. Phys. **31**, 639 (1982).

50. C.W. Misner, "Mixmaster universe," Phys. Rev. Lett. **22**, 1071-1074 (1969).

51. C. W. Misner, "Quantum Cosmology. 1," Phys. Rev. **186**, 1319 (1969); "Minisuperspace," In *J R Klauder, Magic Without Magic*, San Francisco 1972, 441-473.

52. D.M. Chitre, Ph. D. thesis, University of Maryland, 1972.

53. M. P. Ryan, "Hamiltonian cosmology," *Springer-Verlag, Heidelberg (1972)*.

54. M. P. Ryan and L. C. Shepley, "Homogeneous Relativistic Cosmologies," *Princeton, USA: Univ. Pr. (1975) 320 P. (Princeton Series In Physics)*.

55. R. T. Jantzen, "Spatially homogeneous dynamics: A unified picture," arXiv:gr-qc/0102035.

56. B. K. Berger, "Numerical Approaches to Spacetime Singularities," arXiv:gr-qc/0201056.

57. B. K. Berger and V. Moncrief, "Signature for local mixmaster dynamics in $U(1)$ symmetric cosmologies", Phys. Rev. **D62**: 123501 (2000) arXiv:gr-qc/0006071

58. N. J. Cornish and J. J. Levin, "The mixmaster universe is chaotic," Phys. Rev. Lett. **78**, 998 (1997) [arXiv:gr-qc/9605029]; "The mixmaster universe: A chaotic Farey tale," Phys. Rev. D **55**, 7489 (1997) [arXiv:gr-qc/9612066].

59. B. K. Berger, D. Garfinkle, J. Isenberg, V. Moncrief and M. Weaver, "The singularity in generic gravitational collapse is spacelike, local, and oscillatory," Mod. Phys. Lett. A **13**, 1565 (1998) [arXiv:gr-qc/9805063].

60. L. Andersson and A.D. Rendall, "Quiescent cosmological singularities," Commun. Math. Phys. **218**, 479-511 (2001) [arXiv:gr-qc/0001047].

61. E. Anderson, "Strong-coupled relativity without relativity," arXiv:gr-qc/0205118.

62. T. Damour and M. Henneaux, "Chaos in superstring cosmology," Phys. Rev. Lett. **85**, 920 (2000) [arXiv:hep-th/0003139]; [See also short version in Gen. Rel. Grav. **32**, 2339 (2000).]

63. T. Damour and M. Henneaux, "Oscillatory behaviour in homogeneous string cosmology models," Phys. Lett. B **488**, 108 (2000) [arXiv:hep-th/0006171].

64. T. Damour, M. Henneaux, B. Julia and H. Nicolai, "Hyperbolic Kac-Moody algebras and chaos in Kaluza-Klein models," Phys. Lett. B **509**, 323 (2001) [arXiv:hep-th/0103094].

65. T. Damour, M. Henneaux and H. Nicolai, Class. Quant. Grav. **20** (2003) R145-R200, `hep-th/0212256`
66. V.D. Ivashchuk and V.N. Melnikov, "Billiard Representation For Multidimensional Cosmology With Multicomponent Perfect Fluid Near The Singularity," Class. Quantum Grav. **12**, 809 (1995).
67. V. D. Ivashchuk and V. N. Melnikov, "Billiard representation for multidimensional cosmology with intersecting p-branes near the singularity," J. Math. Phys. **41**, 6341 (2000) [arXiv:hep-th/9904077].
68. T. Damour, M. Henneaux, A. D. Rendall and M. Weaver, "Kasner-like behaviour for subcritical Einstein-matter systems," arXiv:gr-qc/0202069, to appear in Ann. Inst. H. Poincaré.
69. T. Damour, S. de Buyl, M. Henneaux and C. Schomblond, "Einstein billiards and overextensions of finite-dimensional simple Lie algebras," arXiv:hep-th/0206125.
70. B. Julia, in *Lectures in Applied Mathematics*, AMS-SIAM, vol 21 (1985), p.355.
71. H. Nicolai and T. Fischbacher, in: *Kac Moody Lie Algebras and Related Topics*, eds. N. Sthanumoorthy and K.C. Misra, Contemporary Mathematics 343, American Mathematical Society, 2004, `hep-th/0301017`
72. J. E. Humphreys, "Introduction to Lie Algebras and Representation Theory", Graduate Texts in Mathematics 9, Springer Verlag, 1980
73. J.H. Conway and N.J.A. Sloane, "Sphere Packings, Lattices and Groups", Grundlehren der mathematischen Wissenschaften 290, 2nd edition, Springer Verlag, 1991
74. I. Schnakenburg and P. C. West, "Kac-Moody symmetries of IIB supergravity," Phys. Lett. B **517**, 421 (2001) [arXiv:hep-th/0107181].
75. I. Schnakenburg and P. C. West, "Massive IIA supergravity as a non-linear realisation," Phys. Lett. B **540**, 137 (2002) [arXiv:hep-th/0204207].
76. A. Kleinschmidt, I. Schnakenburg and P. West, *Very-extended Kac-Moody algebras and their interpretation at low levels*, Class. Quant. Grav. **21** (2004) 2493 [arXiv:hep-th/0309198].

CHAPTER 3

THE NATURE OF SPACETIME SINGULARITIES

ALAN D. RENDALL

Max-Planck-Institut für Gravitationsphysik
Albert-Einstein-Institut
Am Mühlenberg 1, D14476 Golm, Germany

Present knowledge about the nature of spacetime singularities in the context of classical general relativity is surveyed. The status of the BKL picture of cosmological singularities and its relevance to the cosmic censorship hypothesis are discussed. It is shown how insights on cosmic censorship also arise in connection with the idea of weak null singularities inside black holes. Other topics covered include matter singularities and critical collapse. Remarks are made on possible future directions in research on spacetime singularities.

1. Introduction

The issue of spacetime singularities arose very early in the history of general relativity and it seems that Einstein himself had an ambiguous relationship to singularities. A useful source of information on the confusion surrounding the subject in the first half century of general relativity is Ref. 1. The present article is a survey of the understanding we have of spacetime singularities today.

Before concentrating on general relativity, it is useful to think more generally about the concept of a singularity in a physical theory. In the following the emphasis is on classical field theories although some of the discussion may be of relevance to quantum theory as well. When a physical system is modelled within a classical field theory, solutions of the field equations are considered. If it happens that at some time physically relevant quantities become infinite at some point of space then we say that there is a singularity. Since the physical theory ceases to make sense when basic quantities become infinite a singularity is a sign that the theory has been applied beyond its domain of validity. To get a better description a theory of wider applicability should be used. Note that the occurrence of singularities does not say that a theory is bad - it only sets limits on the domain of physical phenomena where it can be applied.

In fact almost any field theory allows solutions with singularities if attention is not restricted to those solutions which are likely to be physically relevant. In this context a useful criterion is provided by the specification of solutions by initial data. This means that we only consider solutions which have the property that there is some time at which they contain no singularities. Then any singularities which occur must be the result of a dynamical evolution. With this motivation, singularities will be discussed in the following in the context of the initial value problem. Only those singularities are considered which develop from regular initial configurations. This has the consequence that linear field theories, such as source-free Maxwell theory, are free of singularities.

In the case of the Einstein equations, the basic equations of general relativity, the notion of singularity becomes more complicated due to the following fact. A solution of the Einstein equations consists not just of the spacetime metric, which describes the gravitational field and the geometry of spacetime, but also the spacetime manifold on which the metric is defined. In the case of a field theory in Newtonian physics or special relativity we can say that a solution becomes singular at certain points of spacetime, where the basic physical quantities are not defined. Each of these points can be called a singularity. On the other hand, a singularity in general relativity cannot be a point of spacetime, since by definition the spacetime structure would not be defined there.

In general relativity the wordline of a free particle is described by a curve in spacetime which is a timelike or null geodesic, for a massive or massless particle respectively. There is also a natural class of time parameters along such a geodesic which, in the timelike case, coincide up to a choice of origin and a rescaling with the proper time in the rest frame of the particle. If the worldline of a particle only exists for a finite time then clearly something has gone seriously wrong. Mathematically this is called geodesic incompleteness. A spacetime which is a solution of the Einstein equations is said to be *singular* if it is timelike or null geodesically incomplete. Informally we say in this case that the spacetime 'contains a singularity' but the definition does not include a description of what a 'singularity' or 'singular point' is. There have been attempts to define ideal points which could be added to spacetime to define a mathematical boundary representing singularities but these have had limited success.

When working practically with solutions of the Einstein equations it is necessary to choose coordinates or other similar auxiliary objects in order to have a concrete description. In general relativity we are free to use any coordinate system and this leads to a problem when considering singularities. Suppose that a metric written in coordinates is such that the components of the metric become infinite as certain values of the coordinates are approached. This could be a sign that there is a spacetime singularity but it could also simply mean that those coordinates break down at some points of a perfectly regular solution. This might be confirmed by transforming to new coordinates where the metric components have a regular extension through the apparent singularities. A way of detecting singularities within a coordinate system is to find that curvature invariants become infinite. These are scalar quantities which measure the curvature of spacetime and if they become infinite this is a sure sign that a region of spacetime cannot be extended. It is still not completely clear what is happening since the singular values of the coordinates might correspond to singular behaviour in the sense of geodesic incompleteness or they might be infinitely far away.

A breakthrough in the understanding of spacetime singularities was the singularity theorem of Penrose[2] which identified general conditions under which a spacetime must be geodesically incomplete. This was then generalized to other situations by Hawking and others. The singularity theorems are proved by contradiction. Their strength is that the hypotheses required are very general and their weakness is that they give very little information about what actually happens dynamically. If the wordline of a particle ceases to exist after finite proper time then it is reasonable to ask for an explanation, why the particle ceased to exist. It is to be expected that some extreme physical conditions play a role. For instance, the matter density or the curvature, representing tidal forces acting on the particle, becomes unboundedly large. From this point of view one would like to know that curvature invariants become unbounded along the incomplete timelike or null geodesics. The singularity theorems give no information on this question which is that of the nature of spacetime singularities. The purpose of the following is to explain what is known about this difficult question.

The hypotheses of the singularity theorems do not include very stringent assumptions about the matter content of spacetime. All that is needed are certain inequalities on the energy-momentum tensor $T_{\alpha\beta}$, the energy conditions[3]. Let V^α and W^α be arbitrary future pointing timelike vectors.

The *dominant energy condition* is that $T_{\alpha\beta}V^\alpha W^\beta \geq 0$. The *strong energy condition* is, provided the cosmological constant is zero, equivalent to the condition that $R_{\alpha\beta}V^\alpha V^\beta \geq 0$ where $R_{\alpha\beta}$ is the Ricci tensor. The *weak energy condition* is that $T_{\alpha\beta}V^\alpha V^\beta \geq 0$. The weak energy condition has the simple physical interpretation that the energy density of matter is nonnegative in any frame of reference. The vector V^α is the four-velocity of an observer at rest in that frame of reference. It is not reasonable to expect that the nature of spacetime singularities can be determined on the basis of energy conditions alone - more detailed assumptions on the matter content are necessary.

It follows from the above discussion that spacetime singularities should be associated with reaching the limits of the physical validity of general relativity. Quantum effects can be expected to come in. If this is so then to go further the theory should be replaced by some kind of theory of quantum gravity. Up to now we have no definitive theory of this kind and so it is not clear how to proceed. The strategy to be discussed in the following is to work entirely within classical general relativity and see what can be discovered. It is to be hoped that this will provide useful input for the future investigation of singularities within a more general context. The existing attempts to study the question of singularities within different approaches to quantum gravity, including the popular idea that quantum gravity should eliminate the singularities of classical general relativity, will not be discussed here. For a discussion of one direction where progress is being made, see the articles of A. Ashtekar and M. Bojowald in this volume.

A key question about singularities in general relativity is whether they are a disaster for the theory. If a singularity can be formed and then influence the evolution of spacetime then this means a breakdown of predictability for the theory. For we cannot (at least within the classical theory) predict anything about the influence a singularity will have. A singularity which can causally influence parts of spacetime is called a naked singularity. It is important for the predictive power of general relativity that naked singularities be ruled out. This has been formulated more precisely by Penrose as the cosmic censorship hypothesis[4,5]. In fact there are two variants of this, weak and strong cosmic censorhip. Despite the names neither of these implies the other[6]. Proving the cosmic censorship hypothesis is one of the central mathematical problems of general relativity. In fact the task of finding the right formulation of the conjecture is already a delicate one. It is necessary to make a genericity assumption and to restrict the matter fields allowed. More details on this are given in later sections.

One of the most important kinds of singularity in general relativity is the initial cosmological singularity, the big bang. The structure of cosmological singularities is the subject of section 2. Another important kind of singularity is that inside black holes. The recent evolution of ideas about the internal structure of black holes is discussed in section 3. An important complication in the study of singularities resulting from the properties of gravity is that they may be obscured by singularities due to the description of matter. This is the theme of section 4. In section 5 singularities are discussed which arise at the threshhold of black hole formation and which are still quite mysterious. Section 6 takes a cautious look at the future of research on spacetime singularities.

2. Cosmological Singularities

The simplest cosmological models are those which are homogeneous and isotropic, the FLRW (Friedmann-Lemaître-Robertson-Walker) models with some suitable choice of matter model such as a perfect fluid. In this context it is seen that the energy density blows up at some time in the past. An early question was whether this singularity might be an artefact of the high symmetry. The intuitive idea is that if matter collapses in such a way that particles are aimed so as to all end up at the same place at the same time there will be a singularity. On the other hand if this situation is perturbed so that the particles miss each other the singularity might be removed. On the basis of heuristic arguments, Lifshitz and Khalatnikov[7] suggested that for a generic perturbation of a FLRW model there would be no singularity. We now know this to be incorrect. This work nevertheless led to a very valuable development of ideas in the work of Belinskii, Khalatnikov and Lifshitz[8,9] which is one of the main sources for our present picture of cosmological singularities.

What was the problem with the original analysis? An ansatz was made for the form of the metric near the singularity and it was investigated how many free functions can be accomodated in a certain formal expansion. It was found that there was one function less than there is in the general solution of the Einstein equations. It was concluded that the most general solution could not have a singularity. This shows us something about the strengths and weaknesses of heuristic arguments. These are limited by the range of possibilities that have occurred to those producing the heuristics. Nevertheless they may, in expert hands, be the most efficient way of getting nearer to the truth.

It was the singularity theorems, particularly the Hawking singularity theorem, which provided convincing evidence that cosmological singularities do occur for very large classes of initial data. In particular they showed that the presence of a singularity (in the sense of geodesic incompleteness) is a stable property under small perturbations of the FLRW model. Thus a rigorous mathematical theorem led to progress in our understanding of physics. The use of mathematical theorems is very appropriate because the phenomena being discussed are very far from most of our experience of the physical world and so relying on physical intuition alone is dangerous.

The singularity theorems give almost no information on the nature of the singularities. In order to go further it makes sense to attempt to combine rigorous mathematics, heuristic arguments and numerical calculations and this has led to considerable progress.

The picture developed by Belinskii, Khalatnikov and Lifshitz (BKL) has several important elements. These are:

- Near the singularity the evolution of the geometry at different spatial points decouples so that the solutions of the partial differential equations can be approximated by solutions of ordinary differential equations.
- For most types of matter the effect of the matter fields on the dynamics of the geometry becomes negligible near the singularity
- The ordinary differential equations which describe the asymptotics are those which come from a class of spatially homogeneous solutions which constitute the mixmaster model. General solutions of these equations show complicated oscillatory behaviour near the singularity.

The first point is very surprising but a variety of analytical and numerical studies appear to support its validity. The extent to which the above points have been confirmed will now be discussed.

The mixmaster model is described by ordinary differential equations and so it is a huge simplification compared to the full problem. Nevertheless even ordinary differential equations can be very difficult to analyse. The solutions show complicated behaviour in the approach to the singularity and this is often called chaotic. This description is somewhat problematic since many of the usual concepts for defining chaos are not applicable. This point will not be discussed further here. For many years the oscillations in solutions of the mixmaster model were studied by heuristic and numerical techniques. This led to a consistent picture but turning this picture into mathematical theorems was an elusive goal. Finally this was achieved in the work of Ringström[10] so that the fundamental properties of the mixmaster model are now mathematically established.

With the mixmaster model under control, the next obvious step in confirming the BKL picture would be to show that it serves as a template for the behaviour of general solutions near the singularity. The work of BKL did this on a heuristic level. Attempts to recover their conclusions in numerical calculations culminated in the work of Garfinkle[11]. Previously numerical investigations of the question had been done under various symmetry assumptions. Solutions without symmetry were handled for the first time in Ref. 11. On the analytical side things do not look so good. There is not a single case with both inhomogeneity and mixmaster oscillations which has been analysed rigorously and this represents an outstanding challenge. One possible reason why it is so difficult will be described below.

One of the parts of the BKL picture contains the qualification 'for most types of matter'. There are exceptional types of matter where things are different. A simple example is a massless linear scalar field. It was already shown in Ref. 12 that in the presence of a scalar field the BKL analysis leads to different conclusions. It is still true that the dynamics at different spatial points decouples but the evolution is such that important physical quantities are ultimately monotone instead of being oscillatory as the singularity is approached. In this case it has been possible to obtain a mathematical confirmation of the BKL picture. In Ref. 13 it was shown that there are solutions of the Einstein equations coupled to a scalar field which depend on the maximal number of free functions and which have the asymptotic behaviour near the singularity predicted by the BKL picture.

As a side remark, note that in many string theory models there is a scalar field, the dilaton, which might kill mixmaster oscillations. Also, a BKL analysis of the vacuum Einstein equations in higher dimensions shows that the oscillations of generic solutions vanish when the spacetime dimension is at least eleven[14] and string theory leads to the consideration of models of dimension greater than four. So could mixmaster oscillations be eliminated in low energy string theory? An investigation in Ref. 15 shows that they are not. The simplifying effect of the dilaton and the high dimension is prevented by other form fields occurring in string theory. With certain values of the coupling constants in field theories of the type coming up in low energy string theory there is monotone behaviour near the singularity and theorems can be proved[16]. However the work of Ref. 15 shows that these do not include the values of the coupling constants coming from the string theories which are now standard.

A feature which makes oscillations so difficult to handle is that they are in general accompanied by large spatial gradients. Consider some physical quantity $f(t, x)$ in the BKL picture in a case without oscillations. Then it is typical that quantities like $\partial_i f / f$, where the derivatives are spatial derivatives, remain bounded near the singularity. However it can happen that this is only true for most spatial points and that there are exceptional spatial points where it fails. In a situation of mixmaster type where there are infinitely many oscillations as the singularity is approached the BKL picture predicts that there will be more and more exceptional points without limit as the singularity is approached. It has even been suggested by Belinskii that this shows that the original BKL assumptions are not self-consistent[17]. In any case, it seems that the question, in what sense the BKL picture provides a description of cosmological singularities, is a subtle one.

Large spatial gradients can also occur in solutions where the evolution is monotone near the singularity. It can happen that before the monotone stage is reached there are finitely many oscillations and that these produce a finite number of exceptional points. In the context of Gowdy spacetimes this has been shown rather explicitly. The features with large spatial gradients (spikes) were discovered in numerical work[18] and later captured analytically[19]. This allowed the behaviour of the curvature near the singularity to be determined.

An important issue to be investigated concerning cosmological singularities is that of cosmic censorship. In this context it is strong cosmic censorship which is of relevance and a convenient mathematical formulation in terms of the initial value problem has been given by Eardley and Moncrief[20]. To any initial data set for the Einstein equations there exists a corresponding maximal Cauchy development. (For background on the initial value problem for the Einstein equations see Ref. 21.) The condition that a spacetime is uniquely determined by initial data is global hyperbolicity. The maximal Cauchy development is in a well-defined sense the largest globally hyperbolic spacetime with the chosen initial data. It may happen that the maximal Cauchy development can be extended to a larger spacetime, which is then of course no longer globally hyperbolic. The boundary of the initial spacetime in the extension is called the Cauchy horizon. The extended spacetime can no longer be uniquely specified by initial data and this corresponds physically to a breakdown of predictability. A famous example where this happens is the Taub-NUT spacetime[3]. This is a highly symmetric solution of the Einstein vacuum equations. The extension which is no longer globally hyperbolic contains closed timelike curves.

How can the existence of the Taub-NUT and similar spacetimes be reconciled with strong cosmic censorship? A way to do this would be to show that this behaviour only occurs for exceptional initial data and that for generic data the maximal globally hyperbolic development is inextendible. This has up to now only been achieved in the simplified context of classes of spacetimes with symmetry. These classes of spacetimes are not generic and so they do not directly say anything about cosmic censorship. However they provide model problems where more can be learned about the conceptual and technical issues which arise in trying to prove cosmic censorship. This kind of restricted cosmic censorship has been shown for many spatially homogeneous spacetimes in Ref. 22 and Ref. 23 and for plane symmetric solutions of the Einstein equations coupled to a massless scalar field[24]. The most general, and most remarkable, result of this kind up to now is the proof by Ringström[25] of strong cosmic censorship restricted to the class of Gowdy spacetimes. He shows that all the solutions in this class of inhomogeneous vacuum spacetimes with symmetry are geodesically complete in the future[26] and that for generic initial data the Kretschmann scalar $R^{\alpha\beta\gamma\delta}R_{\alpha\beta\gamma\delta}$ tends to infinity uniformly as the singularity is approached. Major difficulties in doing this are the fact that there do exist spacetimes in this class where the maximal Cauchy development is extendible and that spikes lead to great technical complications. Roughly speaking, Ringström shows under a genericity assumption that the most complicated thing that can happen in the approach to the singularity is that there are finitely many spikes of the kind constructed in Ref. 19.

Another kind of partial result is to show that an expanding cosmological spacetime is future geodesically complete. This can be interpreted as saying that any singularities must lie in the past. There is up to now just one example of this for spacetimes not required to satisfy any symmetry assumptions. This is the work of Andersson and Moncrief[27] where they show that any small but finite vacuum perturbation of the initial data for the Milne model has a maximal Cauchy development which is future geodesically complete.

Already in the class of homogeneous and isotropic spacetimes there are models with an initial singularity which recollapse and have a second singularity in the future. Not much is known about general criteria for recollapse. The closed universe recollapse conjecture[28] says that any spacetime with a certain type of topology (admitting a metric of positive scalar curvature) and satisfying the dominant and strong energy conditions must recollapse.

No counterexample is known but the conjecture has only been proved in cases with high symmetry[29,30].

3. Black Hole Singularities

One of the most famous singular solutions of the Einstein equations is the Schwarzschild solution representing a spherical black hole. There is a singularity inside the black hole where the Kretschmann scalar diverges uniformly. It looks very much like a cosmological singularity. The singularity is not visible to far away observers. The points of spacetime from which no future-directed causal geodesic can escape to infinity constitute by definition the black hole region and its boundary is the event horizon. The situation in the Schwarzschild solution can be described informally by saying that the singularity is covered by an event horizon. The idea of weak cosmic censorship, a concept which will not be precisely defined here, is that any singularity which arises in gravitational collapse is covered by an event horizon. For more details see Ref. 6 and Ref. 31.

The central question which is to be answered is what properties of the Schwarzschild solution are preserved under perturbations of the initial data. Christodoulou has studied the spherical gravitational collapse of a scalar field in great detail[32]. Among his results are the following. There are initial data leading to the formation of naked singularities but for generic initial data this does not happen. The structure of the singularity has been analysed and it shows strong similarities to the Schwarzschild case.

A key concept in the Penrose singularity theorem is that of a *trapped surface*. It has been shown by Dafermos[33] that some of the results of Christodoulou can be extended to much more general spherically symmetric spacetimes under the assumptions that there exists at least one trapped surface and that the matter fields present are well-behaved in a certain sense. They should not form singularities outside the black hole region. This condition on the matter fields was verified for collisionless matter in Ref. 34. The fact that it is satisfied for certain non-linear scalar fields led to valuable insights in the discussion of the formation of naked singularities in a class of models motivated by string theory[35,36].

When the Schwarzschild solution is generalized to include charge or rotation the picture changes dramatically. In the relevant solutions, the Reissner-Nordström and Kerr solutions, the Schwarzschild singularity is replaced by a Cauchy horizon. At one time it was hoped that this was an artefact of high symmetry and that a further perturbation would turn it back into a curvature singularity. There was also a suggested mechanism for this, namely that radiation coming from the outside would undergo an unlimited blue shift as it approached the potential Cauchy horizon. Things turned out to be more complicated, as discovered by Poisson and Israel[37].

The new picture in Ref. 37 for a perturbed charged black hole was that the Cauchy horizon, where the metric is smooth, would be replaced by a null hypersurface where, although the metric remains continuous and non-degenerate, the curvature blows up. They called this a weak null singularity. The heuristic work of Ref. 37 was followed up by numerical work[38] and was finally turned into rigorous mathematics by Dafermos[39]. Perhaps the greatest significance of this work on charged black holes is its role as a model for rotating black holes. For the more difficult case of rotation much less is known although there is some heuristic analysis[40]. At this point it is appropriate to make a comment on heuristic work which follows on from remarks in the last section. For several years it was believed, on the basis of a heuristic analysis in Ref. 41, that a positive cosmological constant would stabilize the Reissner-Nordström Cauchy horizon. This turned out, however, to be another case where not all relevant mechanisms had been thought of. In a later heuristic analysis[42] it was pointed out that there is another instability mechanism at work which reverses the conclusion.

The case of weak null singularities draws attention to an ambiguity in the definition of strong cosmic censorship. The formulation uses the concept of extension of a spacetime. To have a precise statement is must be specified how smooth a geometry must be in order to count as an extension. This may seem at first sight like hair splitting but in the case of weak null singularities the answer to the question of strong cosmic censorship is quite different depending on whether the extension is required to be merely continuous or continuously differentiable. A related question is whether the extension should be required to satisfy the Einstein equations in some sense.

A question which does not seem to have been investigated is that of the consistency of weak null singularities with the BKL picture. It is typical to study black holes in the context of isolated systems. In reality we expect that black holes form in cosmological models which expand for ever. Do such 'cosmological black holes' show the same features in their interior as asymptotically flat ones? If so then this would indicate the existence of large classes of cosmological models whose singularities do not fit into the BKL picture. (It was never claimed that this picture must apply to all cosmological singularities.) A major difficulty in investigating this issue is that the class of solutions of the Einstein equations of interest does not seem to be consistent with any symmetry assumptions. A related question is that of the relationship between weak cosmic censorship, which is formulated in asymptotically flat spacetimes, and strong cosmic censorship, which makes sense in a cosmological context.

There are important results showing that no black holes form under certain circumstances. In the fundamental work of Christodoulou and Klainerman[43] it was shown that small asymptotically flat data for the Einstein vacuum equations lead to geodesically complete spacetimes. See also Ref. 44.

4. Shells and Shocks

A serious obstacle to determining the structure of spacetime singularites is that many common matter models develop singularities in flat space. This is in particular the case for matter models which are phenomenological rather than coming directly from fundamental physics. These matter models, when coupled to the Einstein equations, must be expected to lead to singularities which have little to do with gravitation which we may call matter singularities. These singularities are just a nuisance when we want to study spacetime singularities as fundamental properties of Einstein gravity.

There has been much study of the Einstein equations coupled to dust. It is not clear that they teach us much. In flat space dust forms shell-crossing singularities where a finite mass of dust particles end up at the same place at the same time. The density blows up there. In curved space this leads to naked singularities[45]. These occur away from the centre in spherical symmetry. Finite time breakdown of self-gravitating dust can also be observed in cosmological spacetimes[46]. This shows the need for restricting the class of matter considered if a correct formulation of cosmic censorship is to be found. In a more realistic perfect fluid the pressure would be expected to eliminate these singularities. On the other hand it is to be expected that shocks form, as is well-known in flat space. The breakdown of smoothness in self-gravitating fluids with pressure was proved in Ref. 47. At this point we must once again confront the question of what is a valid extension. In some cases solutions with fluid may be extended beyond the time when the classical solution breaks down[48]. The extended solution is such that the basic fluid variables are bounded but their first derivatives are not. The uniqueness of these solutions in terms of their initial data is not known but uniqueness results have recently been obtained in the flat space case[49].

A matter model which is better behaved than a fluid is collisionless matter described by the Vlasov equation. It forms no singularities in flat space and there are various cases known where self-gravitating collisionless matter can be proved to form no singularities. For instance this is the case for small spherically symmetric asymptotically flat initial data[50]. There is no case known where collisionless matter does form a matter singularity. Also in spherical symmetry it never forms a singularity away from the centre so that the analogue of shell-crossing singularities is ruled out[51]. In view of the investigations up to now collisionless matter seems to be as well-behaved as vacuum with respect to the formation of singularities.

5. Critical Collapse

Evidence for a new kind of singularity in gravitational collapse was discovered by Choptuik[52]. His original work concerned the spherically symmetric collapse of a massless scalar field but it has been extended in many directions since then. The basic idea is as follows. For small initial data the corresponding solution disperses leaving behind flat space. For very large data a black hole is formed. If a one-parameter family of data is taken interpolating between these two extreme cases what happens to the evolutions for intermediate values of the parameter? It is found that there is a unique parameter value (the critical value) separating the two regimes and that near the critical value the solutions show interesting, more or less universal, behaviour. The study of these phenomena is now known under the name of critical collapse.

Most of the work which has been done on critical collapse is numerical. There is a heuristic picture involving dynamical systems which is useful in predicting certain features of the results of numerical calculations. Up to now there are no rigorous results on critical phenomena. It is interesting to note that at least some of the features of critical collapse are not unique to gravity and may be seen in many systems of partial differential equations[53].

The results on critical collapse indicate the occurrence of a class of naked singularities arising from non-generic initial data which are qualitatively different from those discussed above. They represent an additional technical hurdle in any attempt to prove cosmic censorship in general.

6. Conclusion

In recent years it has been possible to go beyond the classical results on spacetime singularities contained in the singularity theorems of Penrose and Hawking and close in on the question of the nature of these singularities in various ways. In the case of cosmological singularities a key influence has been exerted by the picture of Belinskii, Khalatnikov and Lifshitz (BKL). In the case of black hole singularities the old idea that they should be similar to cosmological singularities has been replaced by the new paradigm of weak null singularities due to Poisson and Israel. A new kind of singularity has emerged in the work of Choptuik on critical collapse. It remains to be seen whether the Einstein equations have further types of singularities in store for us.

New things can happen if we go beyond the usual framework of the singularity theorems. The cosmological acceleration which is now well-established by astronomical observations corresponds on the theoretical level to a violation of the strong energy condition and suggests that a reworking of the singularity theorems in a more general context is necessary. Exotic types of matter which have been introduced to model accelerated cosmological expansion go even further and violate the dominant energy condition. This can lead to a cosmological model running into a singularity when still expanding[54]. This is known as a 'big rip' singularity[55] since physical systems are ripped apart in finite time as the singularity is approached. The study of these matters is still in a state of flux.

Returning to the more conventional setting where the dominant energy condition is satisfied, we can ask what the future holds for the study of spacetime singularities in classical general relativity. A fundamental fact is that our understanding is still very incomplete. Two developments promise improvements. The first is that the steady increase in computing power and improvement of numerical techniques means that numerical relativity should have big contributions to make. The second is that advances in the theory of hyperbolic partial differential equations are providing the tools needed to make further progress with the mathematical theory of solutions of the Einstein equations. As illustrated by the examples of past successes surveyed in this paper the numerical and mathematical approaches can complement each other very effectively.

References

1. Earman, J. 1995 Bangs, crunches, whimpers and shrieks. Singularities and acausalities in relativistic spacetimes. Oxford University Press, Oxford.
2. Penrose, R. 1965 Gravitational collapse and spacetime singularities. Phys. Rev. Lett. 14, 57-59.
3. Hawking, S. W. and Ellis, G. F. R. 1973 The large-scale structure of spacetime. Cambridge University Press, Cambridge.
4. Penrose, R. 1969 Gravitational collapse: the role of general relativity. Riv. del Nuovo Cimento 1, 252-276.

5. Penrose, R. 1979 Singularities and time asymmetry. In: Hawking, S. W. and Israel, W. General relativity: an Einstein centenary survey. Cambridge University Press, Cambridge.

6. Wald, R. M. 1997 Gravitational collapse and cosmic censorship. Preprint gr-qc/9710068.

7. Lifshitz, E. M. and Khalatnikov, I. M. 1963 Investigations in relativistic cosmology. Adv. Phys. 12, 185-249.

8. Belinskii, V. A., Khalatnikov, I. M. and Lifshitz, E. M. 1970 Oscillatory approach to a singular point in the relativistic cosmology. Adv. Phys. 19, 525-573

9. Belinskii, V. A., Khalatnikov, I. M. and Lifshitz, E. M. 1982 A general solution of the Einstein equations with a time singularity. Adv. Phys., 31, 639-667.

10. Ringström, H. 2001 The Bianchi IX attractor. Ann. H. Poincaré 2, 405-500.

11. Garfinkle, D. 2004 Numerical simulations of generic singularities. Phys.Rev.Lett. 93, 161101.

12. Belinskii, V. A. and Khalatnikov, I. M. 1973 Effect of scalar and vector fields on the nature of cosmological singularities. Sov. Phys. JETP 36, 591-597.

13. Andersson, L. and Rendall, A. D. 2001 Quiescent cosmological singularities. Commun. Math. Phys. 218, 479-511.

14. Demaret, J., Henneaux, M. and Spindel, P. 1985 Nonoscillatory behaviour in vacuum Kaluza-Klein cosmologies. Phys. Lett. 164B, 27-30.

15. Damour, T. and Henneaux, M. 2000 Chaos in superstring cosmology. Phys. Rev. Lett. 85, 920-923.

16. Damour, T., Henneaux, M., Rendall, A. D. and Weaver, M. 2002 Kasner-like behaviour for subcritical Einstein-matter systems. Ann. H. Poincare 3, 1049-1111.

17. Belinskii, V. A. 1992 Turbulence of a gravitational field near a spacetime singularity. JETP Lett. 56, 421-425.

18. Berger, B. K. and Moncrief, V. 1994 Numerical investigation of spacetime singularities. Phys. Rev. D48, 4676-4687.

19. Rendall, A. D. and Weaver, M. 2001 Manufacture of Gowdy spacetimes with spikes. Class. Quantum Grav. 18, 2959-2975

20. Moncrief, V. and Eardley, D. M. 1981 The global existence problem and cosmic censorship in general relativity. Gen. Rel.Grav. 13, 887-892.

21. Friedrich, H. and Rendall, A. D. 2000 The Cauchy problem for the Einstein equations. In B. G. Schmidt (ed) Einstein's Field Equations and Their Physical Implications. Lecture Notes in Physics 540. Springer, Berlin.

22. Chruściel, P. T. and Rendall, A. D. 1995 Strong cosmic censorship in vacuum space-times with compact locally homogeneous Cauchy surfaces. Ann. Phys. 242, 349-385.

23. Rendall, A. D. 1995 Global properties of locally spatially homogeneous cosmological models with matter. Math. Proc. Camb. Phil. Soc. 118, 511-526.

24. Tegankong, D. 2005 Global existence and future asymptotic behaviour for solutions of the Einstein-Vlasov-scalar field system with surface symmetry. Preprint gr-qc/0501062.

25. Ringström, H. 2004 Strong cosmic censorship in T^3 Gowdy spacetimes. Preprint.

26. Ringström, H. 2004 On a wave map equation arising in general relativity. Commun. Pure Appl. Math. 57, 657-703.

27. Andersson, L. and Moncrief, V. 2004 Future complete vacuum spacetimes. In: Chruściel, P. T. and Friedrich, H. The Einstein equations and the large scale behaviour of gravitational fields. Birkhäuser, Basel.

28. Barrow, J. D., Galloway, G. J. and Tipler, F. J. 1986 The closed-universe recollapse conjecture. Mon. Not. R. Astron. Soc. 223, 835-844.

29. Burnett, G. 1995 Lifetimes of spherically symmetric closed universes. Phys. Rev. D51, 1621-1631.

30. Burnett, G. and Rendall, A. D. 1996 Existence of maximal hypersurfaces in some spherically symmetric spacetimes. Class.Quant.Grav. 13, 111-124

31. Christodoulou, D. 1999 On the global initial value problem and the issue of singularities. Class. Quantum Grav. A23-A35,

32. Christodoulou, D. 1999 The instability of naked singularities in the gravitational collapse of a scalar field. Ann. Math. 149, 183-217.

33. Dafermos, M. 2004 Spherically symmetric spacetimes with a trapped surface. Preprint gr-qc/0403032.

34. Dafermos, M. and Rendall, A. D. 2004 An extension principle for the Einstein-Vlasov system in spherical symmetry. Preprint gr-qc/0411075.

35. Hertog, T., Horowitz, G. T. and Maeda, K. 2004 Generic cosmic censorship violation in anti de Sitter space. Phys. Rev. Lett. 92, 131101.

36. Dafermos, M. 2004 On naked singularities and the collapse of self-gravitating Higgs fields. Preprint gr-qc/0403033.

37. Poisson, E. and Israel, W. 1990 Internal structure of black holes. Phys. Rev. D41, 1796-1809.

38. Hod, S. and Piran, T. 1998 Mass inflation in dynamical collapse of a charged scalar field. Phys. Rev. Lett. 81, 1554-1557.

39. Dafermos, M. 2003 Stability and instability of the Cauchy horizon for the spherically symmetric Einstein-Maxwell-scalar field equations. Ann. Math. 158, 875-928.

40. Ori, A. 1999 Oscillatory null singularity inside realistic spinning black holes. Phys. Rev. Lett. 83, 5423-5426.

41. Mellor, F., Moss, I. G. 1990 Stability of black holes in de Sitter space. Phys. Rev. D41, 403-409.

42. Brady, P. R., Moss, I. G. and Myers, R. C. 1998 Cosmic censorship: as strong as ever. Phys. Rev. Lett. 80, 3432-3435.

43. Christodoulou, D. and Klainerman, S. 1993 The global nonlinear stability of the Minkowski space. Princeton University Press, Princeton.

44. Lindblad, H. and Rodnianski, I. 2004 The global stability of Minkowski spacetime in harmonic gauge. Preprint math.AP/0411109.

45. Yodzis, P., Seifert, H.-J. and Müller zum Hagen, H. 1973 On the occurrence of naked singularities in general relativity. Commun. Math. Phys. 34, 135-148.

46. Rendall, A. D. 1997 Existence and non-existence results for global constant

mean curvature foliations. Nonlinear Analysis, Theory, Methods and Applications 30, 3589-3598.

47. Rendall, A. D. and Ståhl, F. 2005 Shock waves in plane symmetric spacetimes. Unpublished.

48. Barnes, A. P., LeFloch, P. G., Schmidt, B. G. and Stewart, J. M. 2004 The Glimm scheme for perfect fluids on plane-symmetric Gowdy spacetimes. Class. Quantum Grav. 21, 5043-5074.

49. Bressan, A. 2000 Hyperbolic systems of conservation laws. The one-dimensional Cauchy problem. Oxford University Press, Oxford.

50. Rein, G. and Rendall, A. D. 1992 Global existence of solutions of the spherically symmetric Vlasov-Einstein system with small initial data. Commun. Math. Phys. 150, 561-583.

51. Rein, G., Rendall, A. D. and Schaeffer, J. 1995 A regularity theorem for solutions of the spherically symmetric Vlasov-Einstein system. Commun. Math. Phys. 168, 467-478.

52. Choptuik, M. W. 1993 Universality and scaling in the gravitational collapse of a scalar field. Phys. Rev. Lett. 70, 9-12.

53. Bizoń, P. and Tabor, Z. 2001 On blowup for Yang-Mills fields. Phys. Rev. D64, 121701.

54. Starobinsky, A. A. 2000 Future and origin of our universe: modern view. Grav. Cosmol. 6, 157-163.

55. Caldwell. R. R., Kamionkowski, M. and Weinberg, N. N. 2003 Phantom energy and cosmic doomsday. Phys. Rev. Lett. 91, 071301.

CHAPTER 4

BLACK HOLES – AN INTRODUCTION

PIOTR T. CHRUŚCIEL

Université François-Rabelais de Tours
UMR CNRS 6083 "Mathématiques et Physique Théorique"
Parc de Grandmont, F37200 Tours, France
piotr.chrusciel@lmpt.univ-tours.fr

This chapter is an introduction to the mathematical aspects of the theory of black holes, solutions of vacuum Einstein equations, possibly with a cosmological constant, in arbitrary dimensions.

1. Stationary Black Holes

Stationary solutions are of interest for a variety of reasons. As models for compact objects at rest, or in steady rotation, they play a key role in astrophysics. They are easier to study than non-stationary systems because stationary solutions are governed by elliptic rather than hyperbolic equations. Further, like in any field theory, one expects that large classes of dynamical solutions approach a stationary state in the final stages of their evolution. Last but not least, explicit stationary solutions are easier to come by than dynamical ones.

1.1. *Asymptotically flat examples*

The simplest stationary solutions describing compact isolated objects are the spherically symmetric ones. A theorem due to Birkhoff shows that in the vacuum region any spherically symmetric metric, even without assuming stationarity, belongs to the family of Schwarzschild metrics, parameterized by a positive mass parameter m:

$$g = -V^2 dt^2 + V^{-2} dr^2 + r^2 d\Omega^2 , \tag{1}$$

$$V^2 = 1 - \tfrac{2m}{r} , \quad t \in \mathbb{R} , \ r \in (2m, \infty) . \tag{2}$$

Here $d\Omega^2$ denotes the metric of the standard 2-sphere. Since the metric (1) seems to be singular as $r = 2m$ is approached, there arises the need to understand the geometry of the metric (1) there. The simplest way to do that, for metrics of the form (1) is to replace t by a new coordinate v defined as

$$v = t + f(r) \,, \quad f' = \frac{1}{V^2} \,, \tag{3}$$

leading to

$$v = t + r + 2m \ln(r - 2m) \,.$$

This brings g to the form

$$g = -(1 - \frac{2m}{r})dv^2 + 2dvdr + r^2 d\Omega^2 \,. \tag{4}$$

We have $\det g = -r^4 \sin^2 \theta$, with all coefficients of g smooth, which shows that g is a well defined Lorentzian metric on the set

$$v \in \mathbb{R} \,, \quad r \in (0, \infty) \,. \tag{5}$$

More precisely, (4)-(5) is an analytic extension of the original space-time[a] (1).

It is easily seen that the region $\{r \leq 2m\}$ for the metric (4) is a *black hole region*, in the sense that

observers, or signals, can enter this region, but can never leave it. (6)

In order to see that, recall that observers in general relativity always move on *future directed timelike curves*, that is, curves with timelike future directed tangent vector. For signals the curves are *causal future directed*, these are curves with timelike or null future directed tangent vector. Let, then, $\gamma(s) = (v(s), r(s), \theta(s), \varphi(s))$ be such a timelike curve, for the metric (4) the timelikeness condition $g(\dot{\gamma}, \dot{\gamma}) < 0$ reads

$$-(1 - \frac{2m}{r})\dot{v}^2 + 2\dot{v}\dot{r} + r^2(\dot{\theta}^2 + \sin^2\theta\dot{\varphi}^2) < 0 \,.$$

This implies

$$\dot{v}\left(-(1 - \frac{2m}{r})\dot{v} + 2\dot{r} \right) < 0 \,.$$

---

[a]The term *space–time* denotes a smooth, paracompact, connected, orientable and time–orientable Lorentzian manifold.

It follows that $\dot{v}$ does not change sign on a timelike curve. The usual choice of time orientation corresponds to $\dot{v} > 0$ on future directed curves, leading to

$$-(1 - \frac{2m}{r})\dot{v} + 2\dot{r} < 0 \,.$$

For $r \leq 2m$ the first term is non-negative, which enforces $\dot{r} < 0$ on all future directed timelike curves in that region. Thus, r is a strictly decreasing function along such curves, which implies that future directed timelike curves can cross the *event horizon* $\{r = 2m\}$ only if coming from the region $\{r > 2m\}$. The same conclusion applies for causal curves, by approximation.

Note that we could have chosen a time orientation in which future directed causal curves satisfy $\dot{v} < 0$. The resulting space-time is then called a *white hole* space-time, with $\{r = 2m\}$ being a *white hole event horizon*, which can only be crossed by those future directed causal curves which originate in the region $\{r < 2m\}$.

The transition from (1) to (4) is not the end of the story, as further extensions are possible. For the metric (1) a maximal analytic extension has been found independently by Kruskal, Szekeres, and Fronsdal, see Ref. 73 for details. This extension is visualised[b] in Figure 1. The region I there corresponds to the space-time (1), while the extension just constructed corresponds to the regions I and II.

A discussion of causal geodesics in the Schwarzschild geometry can be found in R. Price's contribution to this volume.

Higher dimensional counterparts of metrics (1) have been found by Tangherlini. In space-time dimension $n + 1$, the metrics take the form (1) with

$$V^2 = 1 - \frac{2m}{r^{n-2}} \,, \tag{7}$$

and with $d\Omega^2$ — the unit round metric on S^{n-1}. The parameter m is the *Arnowitt-Deser-Misner mass* in space-time dimension four, and is proportional to that mass in higher dimensions. Assuming again $m > 0$, a maximal analytic extension can be constructed using a method of Walker[92] (which applies to all spherically symmetric space-times),[c] leading to a space-time with global structure identical to that of Figure 1 (except for the replacement $2M \to (2M)^{1/(n-2)}$ there). Global coordinate systems for the stan-

---

[b]I am grateful to J.-P. Nicolas for allowing me to use his electronic figure.[78]
[c]A generalisation of the Walker extension technique to arbitrary Killing horizons can be found in Ref. 85.

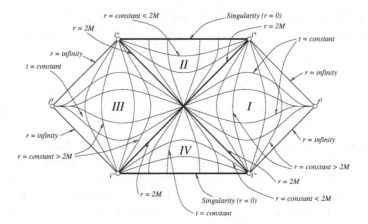

Fig. 1. The Carter-Penrose diagram for the Kruskal-Szekeres space-time with mass M. There are actually two asymptotically flat regions, with corresponding event horizons defined with respect to the second region. Each point in this diagram represents a two-dimensional sphere, and coordinates are chosen so that light-cones have slopes plus minus one.

dard maximal analytic extensions can be found in Ref. 67. The isometric embedding, into six-dimensional Euclidean space, of the $t = 0$ slice in a $(5 + 1)$–dimensional Tangherlini solution is visualised in Figure 2.

One of the features of the metric (1) is its *stationarity*, with Killing vector field $X = \partial_t$. A Killing field, by definition, is a vector field the local flow of which preserves the metric. A space–time is called *stationary* if there exists a Killing vector field X which approaches ∂_t in the asymptotically flat region (where r goes to ∞, see below for precise definitions) *and* generates a one parameter groups of isometries. A space–time is called *static* if it is stationary and if the stationary Killing vector X is hypersurface-orthogonal, i.e. $X^\flat \wedge dX^\flat = 0$, where

$$X^\flat = X_\mu dx^\mu = g_{\mu\nu} X^\nu dx^\mu \ .$$

A space–time is called *axisymmetric* if there exists a Killing vector field Y, which generates a one parameter group of isometries, and which behaves like a *rotation*: this property is captured by requiring that all orbits 2π periodic, and that the set $\{Y = 0\}$, called the *axis of rotation*, is non-empty. Killing vector fields which are a non-trivial linear combination of a time translation and of a rotation in the asymptotically flat region are called *stationary-rotating*, or *helical*. Note that those definitions require completeness of orbits of all Killing vector fields (this means that the equation $\dot{x} = X$ has a global

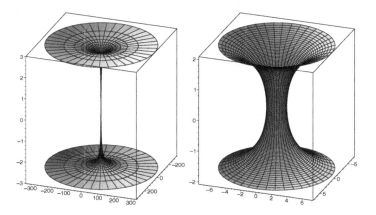

Fig. 2. Isometric embedding of the *space-geometry* of a $(5 + 1)$–dimensional Schwarzschild black hole into six-dimensional Euclidean space, near the throat of the Einstein-Rosen bridge $r = (2m)^{1/3}$, with $2m = 2$. The variable along the vertical axis asymptotes to $\approx \pm 3.06$ as r tends to infinity. The right picture is a zoom to the centre of the throat. The corresponding embedding in $(3 + 1)$–dimensions is known as the *Flamm paraboloid*.

solution for all initial values), see Refs. 22 and 51 for some results concerning this question.

In the extended Schwarzschild space-time the set $\{r = 2m\}$ is a null hypersurface $\mathscr{E}$, the Schwarzschild event horizon. The stationary Killing vector $X = \partial_t$ extends to a Killing vector $\hat{X}$ in the extended spacetime which becomes tangent to and null on $\mathscr{E}$, except at the "bifurcation sphere" right in the middle of Figure 1, where $\hat{X}$ vanishes. The global properties of the Kruskal–Szekeres extension of the exterior Schwarzschild[d] spacetime, make this space-time a natural model for a non-rotating black hole.

There is a rotating generalisation of the Schwarzschild metric, also discussed in the chapter by R. Price in this volume, namely the two parameter family of *exterior Kerr metrics*, which in Boyer-Lindquist coordinates take

---

[d]The exterior Schwarzschild space-time (1) admits an infinite number of non-isometric vacuum extensions, even in the class of maximal, analytic, simply connected ones. The Kruskal-Szekeres extension is singled out by the properties that it is maximal, vacuum, analytic, simply connected, with all maximally extended geodesics γ either complete, or with the curvature scalar $R_{\alpha\beta\gamma\delta}R^{\alpha\beta\gamma\delta}$ diverging along γ in finite affine time.

the form

$$g = -\frac{\Delta - a^2 \sin^2\theta}{\Sigma}dt^2 - \frac{2a\sin^2\theta(r^2 + a^2 - \Delta)}{\Sigma}dtd\varphi +$$
$$+ \frac{(r^2 + a^2)^2 - \Delta a^2 \sin^2\theta}{\Sigma}\sin^2\theta d\varphi^2 + \frac{\Sigma}{\Delta}dr^2 + \Sigma d\theta^2 . \tag{8}$$

Here

$$\Sigma = r^2 + a^2\cos^2\theta , \qquad \Delta = r^2 + a^2 - 2mr = (r - r_+)(r - r_-) ,$$

and $r_+ < r < \infty$, where

$$r_\pm = m \pm (m^2 - a^2)^{\frac{1}{2}} .$$

The metric satisfies the vacuum Einstein equations for any real values of the parameters a and m, but we will only discuss the range $0 \le a < m$. When $a = 0$, the Kerr metric reduces to the Schwarzschild metric. The Kerr metric is again a vacuum solution, and it is stationary with $X = \partial_t$ the asymptotic time translation, as well as axisymmetric with $Y = \partial_\varphi$ the generator of rotations. Similarly to the Schwarzschild case, it turns out that the metric can be smoothly extended across $r = r_+$, with $\{r = r_+\}$ being a smooth null hypersurface $\mathscr{E}$ in the extension. The simplest extension is obtained when t is replaced by a new coordinate

$$v = t + \int_{r_+}^r \frac{r^2 + a^2}{\Delta}dr , \tag{9}$$

with a further replacement of φ by

$$\phi = \varphi + \int_{r_+}^r \frac{a}{\Delta}dr . \tag{10}$$

It is convenient to use the symbol $\hat{g}$ for the metric g in the new coordinate system, obtaining

$$\hat{g} = -\left(1 - \frac{2mr}{\Sigma}\right)dv^2 + 2drdv + \Sigma d\theta^2 - 2a\sin^2\theta d\phi dr$$
$$+ \frac{(r^2 + a^2)^2 - a^2\sin^2\theta\Delta}{\Sigma}\sin^2\theta d\phi^2 - \frac{4amr\sin^2\theta}{\Sigma}d\phi dv . \tag{11}$$

In order to see that (11) provides a smooth Lorentzian metric for $v \in \mathbb{R}$ and $r \in (0, \infty)$, note first that the coordinate transformation (9)-(10) has been tailored to remove the $1/\Delta$ singularity in (8), so that all coefficients are now analytic functions on $\mathbb{R} \times (0, \infty) \times S^2$. A direct calculation of the determinant of $\hat{g}$ is somewhat painful, a simpler way is to proceed as follows: first, the calculation of the determinant of the metric (8) reduces to that of a two-by-two determinant in the (t, φ) variables, leading to $\det g = -\sin^2 \theta \Sigma^2$. Next, it is very easy to check that the determinant of the Jacobi matrix $\partial(v, r, \theta, \phi)/\partial(t, r, \theta, \varphi)$ is one. It follows that $\det \hat{g} = -\sin^2 \theta \Sigma^2$ for $r > r_+$. Analyticity implies that this equation holds globally, which (since Σ has no zeros) establishes the Lorentzian signature of $\hat{g}$ for all positive r.

Let us show that the region $r < r_+$ is a black hole region, in the sense of (6). We start by noting that ∇r is a causal vector for $r_- \leq r \leq r_+$, where $r_- = m - \sqrt{m^2 + a^2}$. A direct calculation using (11) is again somewhat lengthy, instead we use (8) in the region $r > r_+$ to obtain there

$$\hat{g}(\nabla r, \nabla r) = g(\nabla r, \nabla r) = g^{rr} = \frac{1}{g_{rr}} = \frac{\Delta}{\Sigma} = \frac{(r - r_+)(r - r_-)}{r^2 + a^2 \cos^2 \theta} . \tag{12}$$

But the left-hand-side of this equation is an analytic function throughout the extended manifold $\mathbb{R} \times (0, \infty) \times S^2$, and uniqueness of analytic extensions implies that $\hat{g}(\nabla r, \nabla r)$ equals the expression at the extreme right of (12). (The intermediate equalities are of course valid only for $r > r_+$.) Thus ∇r is spacelike if $r < r_-$ or $r > r_+$, null on the "Killing horizons" $\{r = r_\pm\}$, and timelike in the region $\{r_- < r < r_+\}$. We choose a time orientation so that ∇r is future pointing there.

Consider, now, a future directed causal curve $\gamma(s)$. Along γ we have

$$\frac{dr}{ds} = \dot{\gamma}^i \nabla_i r = g_{ij} \dot{\gamma}^i \nabla^j r = g(\dot{\gamma}, \nabla r) < 0$$

in the region $\{r_- < r < r_+\}$, because the scalar product of two future directed causal vectors is always negative. This implies that r is strictly decreasing along future directed causal curves in the region $\{r_- < r < r_+\}$, so that such curves can only leave this region through the set $\{r = r_-\}$. In other words, no causal communication is possible from the region $\{r < r_+\}$ to the "exterior world" $\{r > r_+\}$.

The Schwarzschild metric has the property that the set $g(X, X) = 0$, where X is the "static Killing vector" ∂_t, coincides with the event horizon $r = 2m$. This is not the case any more for the Kerr metric, where we have

$$g(\partial_t, \partial_t) = \hat{g}(\partial_v, \partial_v) = \hat{g}_{vv} = -\left(1 - \frac{2mr}{r^2 + a^2 \cos^2 \theta}\right) ,$$

and the equation $\hat{g}(\partial_v, \partial_v) = 0$ defines a set called the *ergosphere*:

$$\mathring{r}_\pm = m \pm \sqrt{m^2 - a^2 \cos^2 \theta} ,$$

see Figures 3 and 4. The ergosphere touches the horizons at the axes of

Fig. 3.   A coordinate representation[81] of the outer ergosphere $r = \mathring{r}_+$, the event horizon $r = r_+$, the Cauchy horizon $r = r_-$, and the inner ergosphere $r = \mathring{r}_-$ with the singular ring in Kerr space-time. Computer graphics by Kayll Lake.[66]

symmetry $\cos\theta = \pm 1$. Note that $\partial\mathring{r}_\pm/\partial\theta \neq 0$ at those axes, so the ergosphere has a cusp there. The region bounded by the outermost horizon $r = r_+$ and the outermost ergosphere $r = \mathring{r}_+$ is called the *ergoregion*, with X spacelike in its interior. We refer the reader to Refs. 15 and 79 for an exhaustive analysis of the geometry of the Kerr space-time.

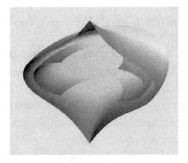

Fig. 4.   Isometric embedding in Euclidean three space of the ergosphere (the outer hull), and part of the event horizon, for a rapidly rotating Kerr solution. The hole arises due to the fact that there is no global isometric embedding possible for the event horizon when $a/m > \sqrt{3}/2$.[81] Somewhat surprisingly, the embedding fails to represent accurately the fact that the cusps at the rotation axis are pointing inwards, and not outwards. Computer graphics by Kayll Lake.[66]

The hypersurfaces $\{r = r_\pm\}$ provide examples of *null acausal boundaries*. Causality theory shows that such hypersurfaces are threaded by a family of null geodesics, called *generators*. One checks that the stationary-rotating Killing field $X + \omega Y$, where $\omega = \frac{a}{2mr_+}$, is null on $\{r > r_+\}$, and hence *tangent* to the generators of the horizon. Thus, the generators are rotating with respect to the frame defined by the stationary Killing vector field X. This property is at the origin of the definition of ω as the *angular velocity* of the event horizon.

Higher dimensional generalisations of the Kerr metric have been constructed by Myers and Perry.[76]

In the examples discussed so far the black hole event horizon is a connected hypersurface in space-time. In fact,[13, 25] there are no *static vacuum* solutions with several black holes, consistently with the intuition that gravity is an attractive force. However, static multi black holes become possible in presence of electric fields. The list of known examples is exhausted by the *Majumdar-Papapetrou* black holes, in which the metric g and the electromagnetic potential A take the form

$$g = -u^{-2}dt^2 + u^2(dx^2 + dy^2 + dz^2),\tag{13}$$

$$A = u^{-1}dt,\tag{14}$$

with some nowhere vanishing function u. Einstein–Maxwell equations read then

$$\frac{\partial u}{\partial t} = 0,\qquad \frac{\partial^2 u}{\partial x^2} + \frac{\partial^2 u}{\partial y^2} + \frac{\partial^2 u}{\partial z^2} = 0.\tag{15}$$

Standard MP black holes are obtained if the coordinates x^μ of (13)–(14) cover the range $\mathbb{R} \times (\mathbb{R}^3 \setminus \{\vec{a}_i\})$ for a finite set of points $\vec{a}_i \in \mathbb{R}^3$, $i = 1, \ldots, I$, and if the function u has the form

$$u = 1 + \sum_{i=1}^{I} \frac{m_i}{|\vec{x} - \vec{a}_i|},\tag{16}$$

for some positive constants m_i. It has been shown by Hartle and Hawking[54] that every standard MP space–time can be analytically extended to an electro–vacuum space–time with a non–empty black hole region. Higher-dimensional generalisations of the MP black holes, with very similar properties, have been found by Myers.[75]

1.2. $\Lambda \neq 0$

So far we have assumed a vanishing cosmological constant Λ. However, there is interest in solutions with $\Lambda \neq 0$: Indeed, there is strong evidence that we live in a universe with $\Lambda > 0$. On the other hand, space-times with a negative cosmological constant appear naturally in many models of theoretical physics, e.g. in string theory.

In space-time dimension four, examples are given by the *generalised Kottler* and the *generalised Nariai* solutions

$$ds^2 = -\left(k - \frac{2m}{r} - \frac{\Lambda}{3}r^2\right)dt^2 + \frac{dr^2}{k - \frac{2m}{r} - \frac{\Lambda}{3}r^2} + r^2 d\Omega_k^2 \,, \quad k = 0, \pm 1 \,,$$

$$\tag{17}$$

$$ds^2 = -\left(\lambda - \Lambda r^2\right)dt^2 + \frac{dr^2}{\lambda - \Lambda r^2} + |\Lambda|^{-1}d\Omega_k^2 \,, \quad k = \pm 1 \,, \quad k\Lambda > 0 \,, \lambda \in \mathbb{R}$$

$$\tag{18}$$

where $d\Omega_k^2$ denotes a metric of constant Gauss curvature k on a two-dimensional compact manifold ^2M. These are static solutions of the vacuum Einstein equation with a cosmological constant Λ. The parameter $m \in \mathbb{R}$ is related to the *Gibbons-Hawking mass* of the foliation $t = $ const, $r = $ const.

As an example of the analysis in this context, consider the metrics (17) with $k = 0$ and $\Lambda = -3$:

$$ds^2 = -\left(r^2 - \frac{2m}{r}\right)dt^2 + \frac{dr^2}{r^2 - \frac{2m}{r}} + r^2(d\varphi^2 + d\psi^2) \,, \tag{19}$$

with φ and ψ parameterising S^1. If $m > 0$ there is a coordinate singularity at $r = (2m)^{1/3}$; an extension can be constructed as in (3) by replacing the coordinate t with

$$v = t + f(r) \,, \quad f' = \frac{1}{r^2 - \frac{2m}{r}} \,. \tag{20}$$

This leads to a smooth Lorentzian metric for all $r > 0$,

$$ds^2 = -\left(r^2 - \frac{2m}{r}\right)dv^2 + 2dvdr + r^2(d\varphi^2 + d\psi^2) \,. \tag{21}$$

We have now an exterior region $r > (2m)^{1/3}$, a black hole event horizon at $r = (2m)^{1/3}$, and a black hole region for $r < (2m)^{1/3}$.

Similarly when $\lambda\Lambda > 0$ the metrics (18) have an exterior region defined by the condition $r > \sqrt{\lambda/\Lambda}$. A procedure similar to the above leads to an extension across an event horizon $r = \sqrt{\lambda/\Lambda}$. Note that the asymptotic behavior of metrics (18) is rather different from that of metrics (17).

The Kottler examples can be generalised to higher dimensions as follows:[10] Let $M = \mathbb{R} \times (r_0, \infty) \times N^{n-1}$, with $N := N^{n-1}$ compact, and with metric of form:

$$g_m = -Vdt^2 + V^{-1}dr^2 + r^2 g_N \,, \tag{22}$$

where g_N is any Einstein metric, $Ric_{g_N} = \lambda g_N$, with g_N scaled so that $\lambda = \pm(n-2)$ or 0. Then for $V = V(r)$ given by

$$V = c + r^2 - (2m)/r^{n-2} , \tag{23}$$

with $c = \pm 1$ or 0 respectively, g_m is a static solution of the vacuum Einstein equations, with $Ric_{g_m} = -n g_m$. When appropriately extended, the resulting space-times possess an event horizon at the largest positive root r_0 of $V(r)$.

It turns out that the collection of static vacuum black holes with a negative cosmological constant is much richer than the one with $\Lambda = 0$. This is due to rather different asymptotic behavior of the solutions. An elegant way of capturing this asymptotic behavior, due to Penrose,[82] proceeds as follows (for notational simplicity we assume that $\Lambda < 0$ has been scaled as in (23)): Replacing in (22) the coordinate r by $x = 1/r$ one obtains $g_m = x^{-2} \tilde{g}_m$, where

$$\tilde{g}_m = -(1 + cx^2 - 2mx^n)dt^2 + \frac{dx^2}{1 + cx^2 - 2mx^n} + g_N . \tag{24}$$

We are interested in the metric $\tilde{g}_m$ for $r \geq r_0$ with some large r_0, this corresponds to x small, $0 < x \leq x_0 := 1/r_0$. The surprising fact is that

$\tilde{g}_m$ *extends by continuity to a smooth Lorentzian metric on the set* $x \in [0, x_0]$.

It is then natural to look for static vacuum metrics of the form $x^{-2}\tilde{g}$, with $\tilde{g}$ smoothly extending to the conformal boundary at infinity $\{x = 0\}$. Such metrics will be called *conformally compactifiable*. In Refs. 2 and 3 the following is shown: write $\tilde{g}|_{x=0}$ as $-\alpha^2 dt^2 + g_N$, where g_N is a Riemannian metric on N, with $\partial_t \alpha = \partial_t g_N = 0$. Then:

(1) Let $\mathring{g}_N$ be a Riemannian metric, with sectional curvatures equal to minus one, on the compact manifold N. Then for all t-independent (α, g_N) close enough to $(1, \mathring{g}_N)$ there exists an associated static, vacuum, conformally compactifiable black hole metric.

(2) In space-time dimension $n+1 = 4$, for all compact N the set of (α, g_N) corresponding to conformally compactifiable static vacuum black holes contains an infinite dimensional manifold.

All metrics presented so far in this section were static. A family of rotating stationary solutions, generalising the Myers-Perry solutions to $\Lambda \neq 0$, can be found in Ref. 53.

Rather surprisingly, when $\Lambda < 0$ there exist static vacuum black holes in space-time dimension three,[e] discovered by Bañados, Teitelboim and Zanelli.[5] The static, circularly symmetric, vacuum solutions take the form

$$ds^2 = -\left(\frac{r^2}{\ell^2} - m\right)dt^2 + \left(\frac{r^2}{\ell^2} - m\right)^{-1}dr^2 + r^2 d\phi^2 \,, \qquad (25)$$

where m is related to the total mass and $\ell^2 = -1/\Lambda$. For $m > 0$, this can be extended, as in (3) with $V^2 = r^2/\ell^2 - m$, to a black hole space-time with event horizon located at $r_H = \ell\sqrt{m}$. There also exist rotating counterparts of (25), discussed in the reference just given.

1.3. Black strings and branes

Consider any vacuum black hole solution $(\mathcal{M}, g)$, and let (N, h) be a Riemannian manifold with a Ricci flat metric, $\text{Ric}(h) = 0$. Then the space-time $(\mathcal{M} \times N, g \oplus h)$ is again a vacuum space-time, containing a black hole region in the sense used so far. (Similarly if $\text{Ric}(g) = \sigma g$ and $\text{Ric}(h) = \sigma h$ then $\text{Ric}(g \oplus h) = \sigma g \oplus h$.) Objects of this type are called *black strings* when $\dim N = 1$, and *black branes* in general. Due to lack of space they will not be discussed here, see Refs. 70, 80 and references therein.

2. Model Independent Concepts

We now describe a general framework for the notions used in the previous sections. The mathematical notion of black hole is meant to capture the idea of a region of space-time which *cannot be seen by "outside observers"*. Thus, at the outset, one assumes that there exists a family of physically preferred observers in the space-time under consideration. When considering isolated physical systems, it is natural to define the "exterior observers" as observers which are "very far" away from the system under consideration. The standard way of making this mathematically precise is by using conformal completions, already mentioned above: A pair $(\tilde{\mathcal{M}}, \tilde{g})$ is called a *conformal completion at infinity*, or simply *conformal completion*, of $(\mathcal{M}, g)$ if $\tilde{\mathcal{M}}$ is a manifold with boundary such that:

(1) $\mathcal{M}$ is the interior of $\tilde{\mathcal{M}}$,
(2) there exists a function Ω, with the property that the metric $\tilde{g}$, defined as $\Omega^2 g$ on $\mathcal{M}$, extends by continuity to the boundary of $\tilde{\mathcal{M}}$, with the extended metric remaining of Lorentzian signature,

---

[e]There are no such vacuum black holes with $\Lambda > 0$, or with $\Lambda = 0$ and degenerate horizons.[58]

(3) Ω is positive on $\mathscr{M}$, differentiable on $\widetilde{\mathscr{M}}$, vanishes on the boundary

$$\mathscr{I} := \widetilde{\mathscr{M}} \setminus \mathscr{M} \, ,$$

with $d\Omega$ *nowhere vanishing* on $\mathscr{I}$.

(In the example (24) we have $\Omega = x$, and $\mathscr{I} = \{x = 0\}$.) The boundary $\mathscr{I}$ of $\widetilde{\mathscr{M}}$ is called Scri, a phonic shortcut for "script I". The idea here is the following: forcing Ω to vanish on $\mathscr{I}$ ensures that $\mathscr{I}$ lies infinitely far away from any physical object — a mathematical way of capturing the notion "very far away". The condition that $d\Omega$ does not vanish is a convenient technical condition which ensures that $\mathscr{I}$ is a smooth three-dimensional hypersurface, instead of some, say, one- or two-dimensional object, or of a set with singularities here and there. Thus, $\mathscr{I}$ is an idealised description of a family of observers at infinity.[f]

To distinguish between various points of $\mathscr{I}$ one sets

$\mathscr{I}^{+} = \{$points in $\mathscr{I}$ which are to the future of the physical space-time$\}$,

$\mathscr{I}^{-} = \{$points in $\mathscr{I}$ which are to the past of the physical space-time$\}$.

(Recall that a point p is to the future, respectively to the past, of q if there exists a future directed, respectively past directed, causal curve from q to p. Causal curves are curves γ such that their tangent vector $\dot{\gamma}$ is causal everywhere, $g(\dot{\gamma}, \dot{\gamma}) \leq 0$.) One then defines the black hole region $\mathscr{B}$ as

$$\mathscr{B} := \{\text{the set of points in } \mathscr{M} \text{ from which}$$
$$\text{no future directed causal curve in } \widetilde{\mathscr{M}} \text{ meets } \mathscr{I}^{+}\} \, . \, (26)$$

By definition, points in the black hole region cannot thus send information to $\mathscr{I}^{+}$; equivalently, observers on $\mathscr{I}^{+}$ cannot see points in $\mathscr{B}$. The *white hole* region $\mathscr{W}$ is defined by changing the time orientation in (26).

In order to obtain a meaningful definition of black hole, one needs to assume further that $\mathscr{I}^{+}$ satisfies a few regularity conditions. For example, if we consider the standard conformal completion of Minkowski space-time, then of course $\mathscr{B}$ will be empty. However, one can remove points from that completion, obtaining sometimes a new completion with a non-empty black hole region. (Think of a family of observers who stop to exist at time $t = 0$, they will never be able to see any event with $t > 0$, leading to a black hole region with respect to this family.) We shall return to this question shortly.

---

[f]We note that the behavior of the metric in the asymptotic region for the black strings and branes of Section 1.3 is *not* compatible with this framework.

A key notion related to the concept of a black hole is that of *future* ($\mathscr{E}^+$) and *past* ($\mathscr{E}^-$) *event horizons*,

$$\mathscr{E}^+ := \partial\mathscr{B} , \qquad \mathscr{E}^- := \partial\mathscr{W} . \tag{27}$$

Under mild assumptions, event horizons in stationary space-times with matter satisfying the *null energy condition*,

$$T_{\mu\nu}\ell^\mu\ell^\nu \geq 0 \quad \text{for all null vectors } \ell^\mu, \tag{28}$$

are smooth null hypersurfaces, analytic if the metric is analytic.[28] This is, however, not the case in the non-stationary case: roughly speaking, event horizons are non-differentiable at end points of their generators. In Ref. 29 a horizon has been constructed which is non-differentiable on a dense set. The best one can say in general is that event horizons are Lipschitz,[83] semi-convex[28] topological hypersurfaces.

In order to develop a reasonable theory one also needs a regularity condition for the interior of space-time. This has to be a condition which does not exclude singularities (otherwise the Schwarzschild and Kerr black holes would be excluded), but which nevertheless guarantees a well-behaved exterior region. One such condition, assumed in all the results described below, is the existence in $\mathscr{M}$ of an asymptotically flat space-like hypersurface $\mathscr{S}$ with compact interior region. This means that $\mathscr{S}$ is the union of a finite number[g] of *asymptotically flat ends* $\mathscr{S}_{ext}$, each diffeomorphic to $\mathbb{R}^n \setminus B(0,R)$, and of a compact region $\mathscr{S}_{int}$. Further, either $\mathscr{S}$ has no boundary, or the boundary of $\mathscr{S}$ lies on $\mathscr{E}^+ \cup \mathscr{E}^-$. To make things precise, for any spacelike hypersurface let g_{ij} be the induced metric, and let K_{ij} denote its extrinsic curvature. A space–like hypersurface $\mathscr{S}_{ext}$ diffeomorphic to $\mathbb{R}^n$ minus a ball will be called an α-*asymptotically flat* end, for some $\alpha > 0$, if the fields (g_{ij}, K_{ij}) satisfy the fall–off conditions

$$|g_{ij}-\delta_{ij}|+\cdots+r^k|\partial_{\ell_1\dots\ell_k}g_{ij}|+r|K_{ij}|+\cdots+r^k|\partial_{\ell_1\dots\ell_{k-1}}K_{ij}| \leq Cr^{-\alpha} , \tag{29}$$

for some constants C, $k \geq 1$. The fall-off rate is typically determined either by requiring that the leading deviations from flatness are identical to those in the Tangherlini solution (1) with V given by (7), or that the fall-off rate be the same as in (7) (which leads to $\alpha = n - 2$), or by requiring a well-defined ADM mass (which leads to $\alpha > (n - 2)/2$).

---

[g]There is no loss of generality in assuming that there is only one such region, if $\mathscr{S}$ is allowed to have a trapped or marginally trapped boundary. However, it is often more convenient to work with hypersurfaces without boundary.

In dimension $3 + 1$ there exists a canonical way of constructing a conformal completion with good global properties for stationary space-times which are asymptotically flat in the sense of (29) for some $\alpha > 0$, and which are vacuum sufficiently far out in the asymptotic region, as follows: Equation (29) and the stationary Einstein equations can be used[9] to prove a complete asymptotic expansion of the metric in terms of powers of $1/r$.[h] The analysis in Refs. 37 and 40 shows then the existence of a smooth conformal completion at null infinity. This conformal completion is referred to as the *standard completion* and will be assumed from now on. It coincides with the completion constructed in the last section for the metrics (22).

As already pointed out, an analysis along the lines of Beig and Simon[9] has only been performed so far in dimension $3 + 1$, and it is not clear what happens in general, because the proofs use an identity which is wrong in other dimensions. On one hand there sometimes exist smooth conformal completions — we have just constructed some in the previous section. On the other hand, it is known that the hypothesis of smoothness of the conformal completion is overly restrictive in odd space-time dimensions in general[i], though it could conceivably be justifiable for stationary solutions. Whatever the case, we shall follow the $\mathscr{I}$ approach here, and we refer the reader to Ref. 19 for a discussion of further drawbacks of this approach, and for alternative proposals.

Returning to the event horizon $\mathscr{E} = \mathscr{E}^+ \cup \mathscr{E}^-$, it is not very difficult to show that every Killing vector field X is necessarily tangent to $\mathscr{E}$: indeed, since $\mathscr{M}$ is invariant under the flow of X, so is $\mathscr{I}^+$, and therefore also $I^-(\mathscr{I}^+)$, and therefore also its boundary $\mathscr{E}^+ = \partial I^-(\mathscr{I}^+)$. Similarly for $\mathscr{E}^-$. Hence X is tangent to $\mathscr{E}$. the Since both $\mathscr{E}^\pm$ are null hypersurfaces, it follows that X is either null or spacelike on $\mathscr{E}$. This leads to a preferred class of event horizons, called *Killing horizons*. By definition, a Killing horizon associated with a Killing vector K is a *null hypersurface* which coincides with a connected component of the set

$$\mathcal{H}(K) := \{ p \in \mathscr{M} \, : \, g(K,K)(p) = 0 \, , \, K(p) \neq 0 \} \, . \tag{30}$$

---

[h] In higher dimensions it is straightforward to prove an asymptotic expansion of stationary vacuum solutions in terms of $\ln^j r / r^i$.

[i] In even space-time dimension smoothness of $\mathscr{I}$ might fail because of logarithmic terms in the expansion.[31, 65] In odd space-time dimensions the situation is (seemingly) even worse, because of half-integer powers of $1/r$[57]

A simple example is provided by the "boost Killing vector field" $K = z\partial_t + t\partial_z$ in Minkowski space-time: $\mathcal{H}(K)$ has four connected components

$$\mathcal{H}_{\epsilon\delta} := \{t = \epsilon z\,, \delta t > 0\}\,, \quad \epsilon, \delta \in \{\pm 1\}\,.$$

The closure $\overline{\mathcal{H}}$ of $\mathcal{H}$ is the set $\{|t| = |z|\}$, which is not a manifold, because of the crossing of the null hyperplanes $\{t = \pm z\}$ at $t = z = 0$. Horizons of this type are referred to as *bifurcate Killing horizons*, with the set $\{K(p) = 0\}$ being called the *bifurcation surface* of $\mathcal{H}(K)$. The bifurcate horizon structure in the Kruskal-Szekeres-Schwarzschild space-time can be clearly seen in Figure 1.

The Vishveshwara-Carter lemma[16, 90] shows that if a Killing vector K in an $(n+1)$–dimensional space-time is hypersurface-orthogonal, $K^\flat \wedge dK^\flat = 0$, then the set $\mathcal{H}(K)$ defined in (30) is a union of smooth null hypersurfaces, with K being tangent to the null geodesics threading $\mathcal{H}$, and so is indeed a union of Killing horizons. It has been shown by Carter[16] that the same conclusion can be reached in asymptotically flat, vacuum, four-dimensional space-times if the hypothesis of hypersurface-orthogonality is replaced by that of existence of two linearly independent Killing vector fields. The proof proceeds via an analysis of the orbits of the isometry group in four-dimensional asymptotically flat manifolds, together with Papapetrou's orthogonal-transitivity theorem, and does not generalise to higher dimensions without further hypotheses.

In stationary-axisymmetric space-times a Killing vector K *tangent to the generators* of a Killing horizon $\mathcal{H}$ can be normalised so that $K = X + \omega Y$, where X is the Killing vector field which asymptotes to a time translation in the asymptotic region, and Y is the Killing vector field which generates rotations in the asymptotic region. The constant ω is called the *angular velocity of the Killing horizon* $\mathcal{H}$.

On a Killing horizon $\mathcal{H}(K)$ one necessarily has

$$\nabla^\mu(K^\nu K_\nu) = -2\kappa K^\mu. \tag{31}$$

Assuming that the horizon is bifurcate (Ref. 61, p. 59), or that the so-called *dominant energy condition* holds (this means that $T_{\mu\nu}X^\mu X^\nu \geq 0$ for all timelike vector fields X) (Ref. 56, Theorem 7.1), it can be shown that κ is constant (recall that Killing horizons are always connected in our terminology), it is called *the surface gravity of* $\mathcal{H}$. A Killing horizon is called *degenerate* when $\kappa = 0$, and non–degenerate otherwise; by an abuse of terminology one similarly talks of degenerate black holes, *etc.* In Kerr space-times we have $\kappa = 0$ if and only if $m = a$. All horizons in the multi-black hole Majumdar-Papapetrou solutions (13)-(16) are degenerate.

A fundamental theorem of Boyer shows that degenerate horizons are closed. This implies that a horizon $\mathcal{H}(K)$ such that K has zeros in $\overline{\mathcal{H}}$ is non-degenerate, and is of bifurcate type, as described above. Further, a *non-degenerate* Killing horizon with *complete* geodesic generators always contains zeros of K in its closure. However, it is not true that existence of a non-degenerate horizon implies that of zeros of K: take the Killing vector field $z\partial_t + t\partial_z$ in Minkowski space-time from which the 2-plane $\{z = t = 0\}$ has been removed. The universal cover of that last space-time provides a space-time in which one cannot restore the points which have been artificially removed, without violating the manifold property.

The *domain of outer communications* (d.o.c.) of a black hole space-time is defined as

$$\langle\langle \mathcal{M} \rangle\rangle := \mathcal{M} \setminus \{\mathcal{B} \cup \mathcal{W}\} \, . \tag{32}$$

Thus, $\langle\langle \mathcal{M} \rangle\rangle$ is the region lying outside of the white hole region and outside of the black hole region; it is the region which can both be seen by the outside observers and influenced by those.

The subset of $\langle\langle \mathcal{M} \rangle\rangle$ where X is spacelike is called the *ergoregion*. In the Schwarzschild space-time $\omega = 0$ and the ergoregion is empty, but neither of these is true in Kerr with $a \neq 0$.

A very convenient method for visualising the global structure of space-times is provided by the *Carter-Penrose diagrams*. An example of such a diagram is presented in Figure 1.

A corollary of the *topological censorship theorem* of Friedman, Schleich and Witt[43, 46, 47] is that d.o.c.'s of regular black hole space-times satisfying the dominant energy condition are simply connected.[45, 50] This implies that connected components of event horizons in stationary, asymptotically flat, four-dimensional space-times have $\mathbb{R} \times S^2$ topology.[12, 35] The restrictions in higher dimension are less stringent,[14, 48] in particular in space-time dimension five an $\mathbb{R} \times S^2 \times S^1$ topology is allowed. A vacuum solution with this horizon topology has been indeed found by Emparan and Reall.[42]

Space-times with good causality properties can be sliced by families of spacelike surfaces $\mathscr{S}_t$, this provides an associated slicing $\mathscr{E}_t = \mathscr{S}_t \cap \mathscr{E}$ of the event horizon. It can be shown that the area of the $\mathscr{E}_t$'s is well defined,[28] this is not a completely trivial statement in view of the poor differentiability properties of $\mathscr{E}$. A key theorem of Hawking[55] (compare Ref. 28) shows that, in suitably regular asymptotically flat space-times, the area of $\mathscr{E}_t$'s is a monotonous function of t. This property carries over to black-hole regions associated to *null-convex* families of observers, as in Ref. 19.

Vacuum or electrovacuum regions with a *timelike* Killing vector can be endowed with an analytic chart in which the metric is analytic. This result has often been misinterpreted as holding up-to-the horizon. However, rather mild global conditions forbid timelike Killing vectors on event horizons. The Curzon metric, studied by Scott and Szekeres[88] provides an example of failure of analyticity at degenerate horizons. One-sided analyticity at *static non-degenerate* vacuum horizons has been proved recently.[26] It is expected that the result remains true for stationary Killing horizons, but the proof does not generalise in any obvious way.

3. Classification of Asymptotically Flat Stationary Black Holes ("No hair theorems")

We confine attention to the "outside region" of black holes, the domain of outer communications (32). For reasons of space we only consider vacuum solutions; there is a similar theory for electro-vacuum black holes.[17, 18, 23, 24, 95] There also exists a somewhat less developed theory for black hole spacetimes in the presence of nonabelian gauge fields.[91]

Based on the facts below, it is expected that the d.o.c.'s of appropriately regular, stationary, asymptotically flat four-dimensional vacuum black holes are isometrically diffeomorphic to those of Kerr black holes.

(1) The *rigidity theorem* (Hawking[44, 55]): event horizons in regular, *non–degenerate*, stationary, *analytic*, four-dimensional vacuum black holes are either *Killing horizons* for X, or there exists a second Killing vector in $\langle\langle \mathscr{M} \rangle\rangle$. The proof does not seem to generalise to higher dimensions without further assumptions.

(2) The *Killing horizons theorem* (Sudarsky-Wald[89]): *non–degenerate* stationary vacuum black holes such that the *event horizon is the union of Killing horizons of* X are *static*. Both the proof in Ref. 89, and that of existence of maximal hypersurfaces needed there,[34] are valid in any space dimensions $n \geq 3$.

(3) The Schwarzschild black holes exhaust the family of *static* regular vacuum black holes (Israel,[60] Bunting – Masood-ul-Alam,[13] Chruściel[25]). The proof in Ref. 25 carries over immediately to all space dimensions $n \geq 3$ (compare Refs. 52, 87), with the proviso of validity of the rigidity part of the Riemannian positive energy theorem.[j]

---

[j]The proofs of this last theorem, known at the time of writing of this work, require the existence of a spin structure in space dimensions larger than eleven,[41] though the result is expected to hold without any restrictions.

(4) The Kerr black holes satisfying

$$m^2 > a^2 \qquad (33)$$

exhaust the family of *non–degenerate, stationary–axisymmetric,* vacuum, *connected, four-dimensional* black holes. Here m is the total ADM mass, while the product am is the total ADM angular momentum. The framework for the proof has been set-up by Carter, and the statement above is due to Robinson.[86] The Emparan-Reall metrics[42] show that there is no uniqueness in higher dimensions, even if three commuting Killing vectors are assumed; see, however, Ref. 74.

The above results are collectively known under the name of *no hair theorems*, and they have *not* provided the final answer to the problem so far even in four dimensions: First, there are no *a priori* reasons known for the analyticity hypothesis in the rigidity theorem. Next, degenerate horizons have been completely understood in the static case only.

In all results above it has been assumed that the metric approaches the Minkowski one in the asymptotic region. Anderson[1] has shown that, under natural regularity hypothesis, the only alternative concerning the asymptotic behavior for *static* $(3+1)$–dimensional vacuum black holes are "small ends", as defined in his work. Solutions with this last behavior have been constructed by Korotkin and Nicolai,[63] and it would be of interest to prove that there are no others. In higher dimension other asymptotic behaviors are possible, examples are given by the metrics (22) with $V = c - (2m)/r^{n-2}$, and g_N as described there.

Yet another key open question is that of existence of *non-connected* regular stationary-axisymmetric vacuum black holes. The following result is due to Weinstein:[93] Let $\partial \mathscr{S}_a$, $a = 1, \ldots, N$ be the connected components of $\partial \mathscr{S}$. Let $X^b = g_{\mu\nu} X^\mu dx^\nu$, where X^μ is the Killing vector field which asymptotically approaches the unit normal to $\mathscr{S}_{ext}$. Similarly set $Y^b = g_{\mu\nu} Y^\mu dx^\nu$, Y^μ being the Killing vector field associated with rotations. On each $\partial \mathscr{S}_a$ there exists a constant ω_a such that the vector $X + \omega_a Y$ is tangent to the generators of the Killing horizon intersecting $\partial \mathscr{S}_a$. The constant ω_a is called the angular velocity of the associated Killing horizon. Define

$$m_a = -\tfrac{1}{8\pi} \int_{\partial \mathscr{S}_a} *dX^b \,, \qquad (34)$$

$$L_a = -\tfrac{1}{4\pi} \int_{\partial \mathscr{S}_a} *dY^b \,. \qquad (35)$$

Such integrals are called *Komar integrals*. One usually thinks of L_a as the angular momentum of each connected component of the black hole. Set

$$\mu_a = m_a - 2\omega_a L_a . \tag{36}$$

Weinstein shows that one necessarily has $\mu_a > 0$. The problem at hand can be reduced to a *harmonic map* equation, also known as the *Ernst equation*, involving a singular map from $\mathbb{R}^3$ with Euclidean metric δ to the two-dimensional hyperbolic space. Let $r_a > 0$, $a = 1, \ldots, N-1$, be the distance in $\mathbb{R}^3$ along the axis between neighboring black holes as measured with respect to the (unphysical) metric δ. Weinstein proves that for *non-degenerate* regular black holes the inequality (33) holds, and that the metric on $\langle\langle \mathcal{M} \rangle\rangle$ is determined up to isometry by the $3N - 1$ parameters

$$(\mu_1, \ldots, \mu_N, L_1, \ldots, L_N, r_1, \ldots, r_{N-1}) \tag{37}$$

just described, with $r_a, \mu_a > 0$. These results by Weinstein contain the no-hair theorem of Carter and Robinson as a special case. Weinstein also shows that for every $N \geq 2$ and for every set of parameters (37) with $\mu_a, r_a > 0$, there exists a solution of the problem at hand. It is known that for some sets of parameters (37) the solutions will have "strut singularities" between some pairs of neighboring black holes,[69, 71, 77, 94] but the existence of the "struts" for all sets of parameters as above is not known, and is one of the main open problems in our understanding of stationary–axisymmetric electro–vacuum black holes. The existence and uniqueness results of Weinstein remain valid when strut singularities are allowed in the metric at the outset, though such solutions do not fall into the category of regular black holes discussed so far.

Some of the results above have been generalised to $\Lambda \neq 0$.[4, 11, 33, 49, 84]

4. Dynamical Black Holes: Robinson-Trautman Metrics

The only known family of vacuum, singularity-free (in the sense described in the previous section), dynamical black holes, with exhaustive understanding of the global structure to the future of a Cauchy surface, is provided by the *Robinson-Trautman* (RT) metrics.

By definition, the Robinson–Trautman space–times can be foliated by a null, hypersurface orthogonal, shear free, expanding geodesic congruence. It has been shown by Robinson and Trautman that in such a space–time there always exists a coordinate system in which the metric takes the form

$$ds^2 = -\Phi \, du^2 - 2du \, dr + r^2 e^{2\lambda} \mathring{g}_{ab} \, dx^a \, dx^b, \quad x^a \in {}^2M, \quad \lambda = \lambda(u, x^a), \tag{38}$$

$$\mathring{g}_{ab} = \mathring{g}_{ab}(x^a), \quad \Phi = \frac{R}{2} + \frac{r}{12m}\Delta_g R - \frac{2m}{r}, \quad R = R(g_{ab}) \equiv R(e^{2\lambda}\mathring{g}_{ab}),$$

m is a constant which is related to the total Bondi mass of the metric, R is the Ricci scalar of the metric $g_{ab} \equiv e^{2\lambda}\mathring{g}_{ab}$, and $(^2M, \mathring{g}_{ab})$ is a smooth Riemannian manifold which we shall assume to be a two-dimensional sphere (other topologies are considered in Ref. 21).

For metrics of the form (38), the Einstein vacuum equations reduce to a single parabolic evolution equation for the two-dimensional metric $g = g_{ab}dx^a dx^b$:

$$\partial_u g = \frac{\Delta R}{12m}g \,. \tag{39}$$

This is equivalent to a non-linear fourth order parabolic equation for the conformal factor λ. The Schwarzschild metric provides an example of a time-independent solution.

The Cauchy data for an RT metric consist of $\lambda_0(x^a) \equiv \lambda(u = u_0, x^a)$. Equivalently, one prescribes a metric $g_{\mu\nu}$ of the form (38) on the null hypersurface $\{u = u_0, x^a \in {}^2M, r \in (0, \infty)\}$. Note that this hypersurface extends up to a curvature singularity at $r = 0$, where the scalar $R_{\alpha\beta\gamma\delta}R^{\alpha\beta\gamma\delta}$ diverges as r^{-6}. This is a 'white hole singularity", familiar from all known black hole spaces-times.

It is proved in Ref. 20 that, for $m > 0$, every such initial λ_0 leads to a black hole space-time. More precisely, one has the following: For any $\lambda_0 \in C^\infty(S^2)$ there exists a Robinson–Trautman space-time $(\mathcal{M}, g)$ with a "half-complete" $\mathcal{I}^+$, the global structure of which is shown in Figure 5. Moreover, there exist an infinite number of non-isometric vacuum

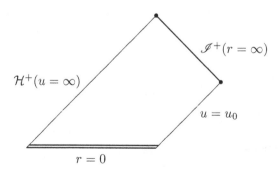

Fig. 5.   The global structure of RT space-times with $m > 0$ and spherical topology.

Robinson–Trautman C^5 extensions[k] of $(\mathcal{M}, g)$ through $\mathscr{H}^+$, which are obtained by gluing to $(\mathcal{M}, g)$ any other Robinson–Trautman spacetime with the same mass parameter m, as shown in Figure 6. Each such extension leads to a black-hole space-time, in which $\mathscr{H}^+$ becomes a black hole event horizon. (There also exist an infinite number of C^{117} vacuum RT extensions of $(\mathcal{M}, g)$ through $\mathscr{H}^+$ — one such extension can be obtained by gluing a copy of $(\mathcal{M}, g)$ to itself. Somewhat surprisingly, no extensions of C^{123} differentiability class exist in general.)

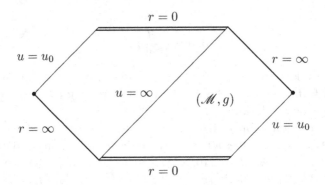

Fig. 6.   Vacuum RT extensions beyond $\mathcal{H}^+$

5. Initial Data Sets Containing Trapped, or Marginally Trapped, Surfaces

Let $\mathscr{T}$ be a *compact, $(n-1)$–dimensional, spacelike* submanifold in a $(n+1)$–dimensional space-time $(\mathcal{M}, g)$. We assume that there is a continuous choice ℓ of a field of future directed null normals to $\mathscr{T}$, which will be referred to as the *outer one*. Let e_i, $i = 1, \cdots, n-1$ be a local ON frame on $\mathscr{T}$, one sets

$$\theta_+ = \sum_{i=1}^{n-1} g(\nabla_{e_i} \ell, e_i) \, .$$

Then $\mathscr{T}$ will be called *future outer trapped* if $\theta_+ > 0$, and *marginally future outer trapped* if $\theta_+ = 0$. A marginally trapped surface lying within a spacelike hypersurface is often referred to as an *apparent horizon*.

---

[k]By this we mean that the metric can be C^5 extended beyond $\mathscr{H}^+$; the extension can actually be chosen to be of $C^{5,\alpha}$ differentiability class, for any $\alpha < 1$.

It is a folklore theorem in general relativity that, under appropriate global conditions, existence of a future outer trapped or marginally trapped surface implies that of a non-empty black hole region. So one strategy in constructing black hole space-times is to find initial data which will contain trapped, or marginally trapped, surfaces[6, 8, 38, 39, 68, 72]

It is useful to recall how apparent horizons are detected using initial data: let $(\mathscr{S}, g, K)$ be an initial data set, and let $S \subset \mathscr{S}$ be a compact embedded two-dimensional two-sided submanifold in $\mathscr{S}$. If n^i is the field of outer normals to S and H is the outer mean extrinsic curvature[l] of S within $\mathscr{S}$ then, in a convenient normalisation, the divergence θ_+ of future directed null geodesics normal to S is given by

$$\theta_+ = H + K_{ij}(g^{ij} - n^i n^j) \,. \tag{40}$$

In the time-symmetric case θ_+ reduces thus to H, and S is trapped if and only if $H < 0$, marginally trapped if and only if $H = 0$. Thus, in this case apparent horizons correspond to compact minimal surfaces within $\mathscr{S}$.

It should be emphasised that the existence of disconnected apparent horizons within an initial data set does not guarantee, as of the time of writing this work, a multi-black-hole spacetime, because our understanding of the long time behavior of solutions of Einstein equations is way too poor. Some very partial results concerning such questions can be found in Ref. 32.

5.1. *Brill-Lindquist initial data*

Probably the simplest examples are the time-symmetric initial data of Brill and Lindquist. Here the space-metric at time $t = 0$ takes the form

$$g = \psi^{4/(n-2)} \left((dx^1)^2 + \ldots + (dx^n)^2 \right) \,, \tag{41}$$

with

$$\psi = 1 + \sum_{i=1}^{I} \frac{m_i}{2|\vec{x} - \vec{x}_i|^{n-2}} \,.$$

The positions of the poles $\vec{x}_i \in \mathbb{R}^n$ and the values of the mass parameters $m_i \in \mathbb{R}$ are arbitrary. If all the m_i are positive and sufficiently small, then for each i there exists a small minimal surface with the topology of a sphere which encloses $\vec{x}_i$.[32] From Ref. 62, in dimension $3+1$ the associated maximal globally hyperbolic development possesses a $\mathscr{I}^+$ which is complete

---

[l]We use the definition that gives $H = 2/r$ for round spheres of radius r in three-dimensional Euclidean space.

to the past. However $\mathscr{I}^+$ cannot be smooth,[64] and it is not known how large it is to the future. One expects that the intersection of the event horizon with the initial data surface will have more than one connected component for sufficiently small values of $m_i/|\vec{x}_k - \vec{x}_j|$, but this is not known.

5.2. The "many Schwarzschild" initial data

There is a well-known special case of (41), which is the space-part of the Schwarzschild metric centred at $\vec{x}_0$ with mass m :

$$g = \left(1 + \frac{m}{2|\vec{x} - \vec{x}_0|^{n-2}}\right)^{4/(n-2)} \delta , \tag{42}$$

where δ is the Euclidean metric. Abusing terminology in a standard way, we call (42) simply the Schwarzschild metric. The sphere $|\vec{x}-\vec{x}_0| = m/2$ is minimal, and the region $|\vec{x} - \vec{x}_0| < m/2$ corresponds to the second asymptotic region. This feature of the geometry, as connecting two asymptotic regions, is sometimes referred to as the *Einstein-Rosen bridge*, see Figure 2.

Now fix the radii $0 \le 4R_1 < R_2 < \infty$. Denoting by $B(\vec{a}, R)$ the open coordinate ball centred at $\vec{a}$ with radius R, choose points

$$\vec{x}_i \in \Gamma_0(4R_1, R_2) := \begin{cases} B(0, R_2) \setminus \overline{B(0, 4R_1)} , & R_1 > 0 \\ B(0, R_2) , & R_1 = 0 , \end{cases}$$

and radii r_i, $i = 1, \ldots, 2N$, so that the closed balls $\overline{B(\vec{x}_i, 4r_i)}$ are all contained in $\Gamma_0(4R_1, R_2)$ and are pairwise disjoint. Set

$$\Omega := \Gamma_0(R_1, R_2) \setminus \left(\cup_i \overline{B(\vec{x}_i, r_i)} \right) . \tag{43}$$

We assume that the $\vec{x}_i$ and r_i are chosen so that Ω is invariant with respect to the reflection $\vec{x} \to -\vec{x}$. Now consider a collection of nonnegative mass parameters, arranged into a vector as

$$\vec{M} = (m, m_0, m_1, \ldots, m_{2N}),$$

where $0 < 2m_i < r_i$, $i \ge 1$, and in addition with $2m_0 < R_1$ if $R_1 > 0$ but $m_0 = 0$ if $R_1 = 0$. We assume that the mass parameters associated to the points $\vec{x}_i$ and $-\vec{x}_i$ are the same. The remaining entry m is explained below.

Given this data, it follows from the work in Refs.[27,36] that there exists a $\delta > 0$ such that if

$$\sum_{i=0}^{2N} |m_i| \le \delta , \tag{44}$$

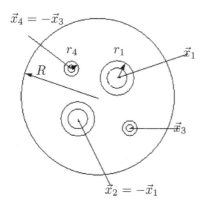

Fig. 7. "Many Schwarzschild" initial data with four black holes. The initial data are *exactly* Schwarzschild within the four innermost circles and outside the outermost one. The free parameters are R, $(\vec{x}_1, r_1, m_1)$, and $(\vec{x}_3, r_3, m_3)$, with sufficiently small m_a's. We impose $m_2 = m_1$, $r_2 = r_1$, $m_4 = m_3$ and $r_4 = r_3$.

then there exists a number

$$m = \sum_{i=0}^{2N} m_i + O(\delta^2)$$

and a C^∞ metric $\hat{g}_{\vec{M}}$ which is a solution of the time-symmetric vacuum constraint equation

$$R(\hat{g}_{\vec{M}}) = 0 \,,$$

such that:

(1) On the punctured balls $B(\vec{x}_i, 2r_i) \setminus \{\vec{x}_i\}$, $i \geq 1$, $\hat{g}_{\vec{M}}$ is the Schwarzschild metric, centred at $\vec{x}_i$, with mass m_i;
(2) On $\mathbb{R}^n \setminus \overline{B(0, 2R_2)}$, $\hat{g}_{\vec{M}}$ agrees with the Schwarzschild metric centred at 0, with mass m;
(3) If $R_1 > 0$, then $\hat{g}_{\vec{M}}$ agrees on $B(0, 2R_1) \setminus \{0\}$ with the Schwarzschild metric centred at 0, with mass m_0.

By point (1) above each of the spheres $|\vec{x} - \vec{x}_i| = m_i/2$ is an apparent horizon.

A key feature of those initial data is that we have complete control of the *space-time metric* within the domains of dependence of $B(\vec{x}_i, 2r_i) \setminus \{\vec{x}_i\}$ and of $\mathbb{R}^n \setminus \overline{B(0, 2R_2)}$, where the space-time metric is a Schwarzschild metric.

Because of the high symmetry, one expects that "all black holes will eventually merge", so that the event horizon will be a connected hypersurface in space-time.

5.3. *Black holes and gluing methods*

A recent alternate technique for gluing initial data sets is given in Refs. 59. In this approach, general initial data sets on compact manifolds or with asymptotically Euclidean or hyperboloidal ends are glued together to produce solutions of the constraint equations on the connected sum manifolds. Only very mild restrictions on the original initial data are needed. The neck regions produced by this construction are again of Schwarzschild type. The overall strategy of the construction is similar to that used in many previous gluing constructions in geometry. Namely, one takes a family of approximate solutions to the constraint equations and then attempts to perturb the members of this family to exact solutions. There is a parameter η which measures the size of the neck, or gluing region; the main difficulty is caused by the tension between the competing demands that the approximate solutions become more nearly exact as $\eta \to 0$ while the underlying geometry and analysis become more singular. In this approach, the conformal method of solving the constraints is used, and the solution involves a conformal factor which is exponentially close to one (as a function of η) away from the neck region. It has been shown[30] that the deformation can actually be localised near the neck in generic situations.

Consider, now, an asymptotically flat time-symmetric initial data set, to which several other time-symmetric initial data sets have been glued by this method. If the gluing regions are made small enough, the existence of a non-trivial minimal surface, hence of an apparent horizon, follows by standard results. This implies the existence of a black hole region in the maximal globally hyperbolic development of the data.

It is shown in Ref. 32 that the intersection of the event horizon with the initial data hypersurface will have more than one connected component for several families of glued initial data sets.

Acknowledgements

The author is grateful to Robert Beig for many useful discussions, to Kayll Lake for comments and electronic figures, and to Jean-Philippe Nicolas for figures. The presentation here borrows heavily on Ref. 7, and can be thought of as an extended version of that work.

Work partially supported by a Polish Research Committee grant 2 P03B 073 24.

References

1. M.T. Anderson, *On the structure of solutions to the static vacuum Einstein equations*, Annales H. Poincaré **1** (2000), 995–1042, gr-qc/0001018.
2. M.T. Anderson, P.T. Chruściel, and E. Delay, *Non-trivial, static, geodesically complete vacuum space-times with a negative cosmological constant*, JHEP **10** (2002), 063, 22 pp., gr-qc/0211006.
3. M.T. Anderson, P.T. Chruściel, and E. Delay, *Non-trivial, static, geodesically complete vacuum space-times with a negative cosmological constant II: $n \geq 5$*, Proceedings of the Strasbourg Meeting on AdS-CFT correspondence (Berlin, New York) (O. Biquard and V. Turaev, eds.), IRMA Lectures in Mathematics and Theoretical Physics, de Gruyter, in press, gr-qc/0401081.
4. L. Andersson and M. Dahl, *Scalar curvature rigidity for asymptotically locally hyperbolic manifolds*, Annals of Global Anal. and Geom. **16** (1998), 1–27, dg-ga/9707017.
5. M. Bañados, C. Teitelboim, and J. Zanelli, *Black hole in three-dimensional spacetime*, Phys. Rev. Lett. **69** (1992), 1849–1851, hep-th/9204099.
6. R. Beig, *Generalized Bowen-York initial data*, S. Cotsakis *et al.*, Eds., Mathematical and quantum aspects of relativity and cosmology. Proceedings of the 2nd Samos meeting. Berlin: Springer Lect. Notes Phys. **537**, 55–69, 2000.
7. R. Beig and P.T. Chruściel, *Stationary black holes*, Encyclopedia of Mathematical Physics (2005), in press, gr-qc/0502041.
8. R. Beig and N.Ó Murchadha, *Vacuum spacetimes with future trapped surfaces*, Class. Quantum Grav. **13** (1996), 739–751.
9. R. Beig and W. Simon, *On the multipole expansion for stationary space-times*, Proc. Roy. Soc. London A **376** (1981), 333–341.
10. D. Birmingham, *Topological black holes in anti-de Sitter space*, Class. Quantum Grav. **16** (1999), 1197–1205, hep-th/9808032.
11. W. Boucher, G.W. Gibbons, and G.T. Horowitz, *Uniqueness theorem for anti-de Sitter spacetime*, Phys. Rev. **D30** (1984), 2447–2451.
12. S.F. Browdy and G.J. Galloway, *Topological censorship and the topology of black holes.*, Jour. Math. Phys. **36** (1995), 4952–4961.
13. G. Bunting and A.K.M. Masood–ul–Alam, *Nonexistence of multiple black holes in asymptotically euclidean static vacuum space-time*, Gen. Rel. Grav. **19** (1987), 147–154.
14. M. Cai and G.J. Galloway, *On the topology and area of higher dimensional black holes*, Class. Quantum Grav. **18** (2001), 2707–2718, hep-th/0102149.
15. B. Carter, *Global structure of the Kerr family of gravitational fields*, Phys. Rev. **174** (1968), 1559–1571.
16. B. Carter, *Killing horizons and orthogonally transitive groups in space–time*, Jour. Math. Phys. **10** (1969), 70–81.
17. B. Carter, *Black hole equilibrium states*, Black Holes (C. de Witt and B.

de Witt, eds.), Gordon & Breach, New York, London, Paris, 1973, Proceedings of the Les Houches Summer School.

18. B. Carter, *The general theory of the mechanical, electromagnetic and thermodynamic properties of black holes*, General relativity (S.W. Hawking and W. Israel, eds.), Cambridge University Press, Cambridge, 1979, pp. 294–369.

19. P.T. Chruściel, *Black holes*, Proceedings of the Tübingen Workshop on the Conformal Structure of Space-times, H. Friedrich and J. Frauendiener, Eds., Springer Lecture Notes in Physics **604**, 61–102 (2002), gr-qc/0201053.

20. P.T. Chruściel, *Semi-global existence and convergence of solutions of the Robinson-Trautman (2-dimensional Calabi) equation*, Commun. Math. Phys. **137** (1991), 289–313.

21. P.T. Chruściel, *On the global structure of Robinson–Trautman space–times*, Proc. Roy. Soc. London **A 436** (1992), 299–316.

22. P.T. Chruściel, *On completeness of orbits of Killing vector fields*, Class. Quantum Grav. **10** (1993), 2091–2101, gr-qc/9304029.

23. P.T. Chruściel, *"No Hair" Theorems – folklore, conjectures, results*, Differential Geometry and Mathematical Physics (J. Beem and K.L. Duggal, eds.), Cont. Math., vol. 170, AMS, Providence, 1994, pp. 23–49, gr-qc/9402032.

24. P.T. Chruściel, *Uniqueness of black holes revisited*, Helv. Phys. Acta **69** (1996), 529–552, Proceedings of Journés Relativistes 1996, Ascona, May 1996, N. Straumann,Ph. Jetzer and G. Lavrelashvili (Eds.), gr-qc/9610010.

25. P.T. Chruściel, *The classification of static vacuum space–times containing an asymptotically flat spacelike hypersurface with compact interior*, Class. Quantum Grav. **16** (1999), 661–687, gr-qc/9809088.

26. P.T. Chruściel, *On analyticity of static vacuum metrics at non-degenerate horizons*, Acta Phys. Pol. **B36** (2005), 17–26, gr-qc/0402087.

27. P.T. Chruściel and E. Delay, *On mapping properties of the general relativistic constraints operator in weighted function spaces, with applications*, Mém. Soc. Math. de France. **94** (2003), 1–103, gr-qc/0301073v2.

28. P.T. Chruściel, E. Delay, G. Galloway, and R. Howard, *Regularity of horizons and the area theorem*, Annales Henri Poincaré **2** (2001), 109–178, gr-qc/0001003.

29. P.T. Chruściel and G.J. Galloway, *Horizons non–differentiable on dense sets*, Commun. Math. Phys. **193** (1998), 449–470, gr-qc/9611032.

30. P.T. Chruściel, J. Isenberg, and D. Pollack, *Initial data engineering*, Commun. Math. Phys. (2005), in press, gr-qc/0403066.

31. P.T. Chruściel, M.A.H. MacCallum, and D. Singleton, *Gravitational waves in general relativity. XIV: Bondi expansions and the "polyhomogeneity" of Scri*, Phil. Trans. Roy. Soc. London A **350** (1995), 113–141.

32. P.T. Chruściel and R. Mazzeo, *On "many-black-hole" vacuum spacetimes*, Class. Quantum Grav. **20** (2003), 729–754, gr-qc/0210103.

33. P.T. Chruściel and W. Simon, *Towards the classification of static vacuum spacetimes with negative cosmological constant*, Jour. Math. Phys. **42** (2001), 1779–1817, gr-qc/0004032.

34. P.T. Chruściel and R.M. Wald, *Maximal hypersurfaces in stationary asymp-*

totically flat space–times, Commun. Math. Phys. **163** (1994), 561–604, gr–qc/9304009.

35. P.T. Chruściel and R.M. Wald, *On the topology of stationary black holes*, Class. Quantum Grav. **11** (1994), L147–152.

36. J. Corvino, *Scalar curvature deformation and a gluing construction for the Einstein constraint equations*, Commun. Math. Phys. **214** (2000), 137–189.

37. S. Dain, *Initial data for stationary space-times near space-like infinity*, Class. Quantum Grav. **18** (2001), 4329–4338, gr-qc/0107018.

38. S. Dain, *Initial data for two Kerr-like black holes*, Phys. Rev. Lett. **87** (2001), 121102, gr-qc/0012023.

39. S. Dain, *Trapped surfaces as boundaries for the constraint equations*, Class. Quantum Grav. **21** (2004), 555–574, Corrigendum ib. p. 769, gr-qc/0308009.

40. T. Damour and B. Schmidt, *Reliability of perturbation theory in general relativity*, Jour. Math. Phys. **31** (1990), 2441–2453.

41. A. Degeratu and M. Stern, *The positive mass conjecture for non-spin manifolds*, math.DG/0412151.

42. R. Emparan and H.S. Reall, *A rotating black ring in five dimensions*, Phys. Rev. Lett. **88** (2002), 101101, hep-th/0110260.

43. J.L. Friedman, K. Schleich, and D.M. Witt, *Topological censorship*, Phys. Rev. Lett. **71** (1993), 1486–1489, erratum **75** (1995) 1872.

44. H. Friedrich, I. Rácz, and R.M. Wald, *On the rigidity theorem for spacetimes with a stationary event horizon or a compact Cauchy horizon*, Commun. Math. Phys. **204** (1999), 691–707, gr-qc/9811021.

45. G.J. Galloway, *On the topology of the domain of outer communication*, Class. Quantum Grav. **12** (1995), L99–L101.

46. G.J. Galloway, *A "finite infinity" version of the FSW topological censorship*, Class. Quantum Grav. **13** (1996), 1471–1478.

47. G.J. Galloway, K. Schleich, D.M. Witt, and E. Woolgar, *Topological censorship and higher genus black holes*, Phys. Rev. **D60** (1999), 104039, gr-qc/9902061.

48. G.J. Galloway, K. Schleich, D.M. Witt, and E. Woolgar, *The AdS/CFT correspondence conjecture and topological censorship*, Phys. Lett. **B505** (2001), 255–262, hep-th/9912119.

49. G.J. Galloway, S. Surya, and E. Woolgar, *On the geometry and mass of static, asymptotically AdS spacetimes, and the uniqueness of the AdS soliton*, Commun. Math. Phys. **241** (2003), 1–25, hep-th/0204081.

50. G.J. Galloway and E. Woolgar, *The cosmic censor forbids naked topology*, Class. Quantum Grav. **14** (1996), L1–L7, gr-qc/9609007.

51. D. Garfinkle and S.G. Harris, *Ricci fall-off in static and stationary, globally hyperbolic, non-singular spacetimes*, Class. Quantum Grav. **14** (1997), 139–151, gr-qc/9511050.

52. G.W. Gibbons, D. Ida, and T. Shiromizu, *Uniqueness and non-uniqueness of static black holes in higher dimensions*, Phys. Rev. Lett. **89** (2002), 041101.

53. G.W. Gibbons, H. Lu, Don N. Page, and C.N. Pope, *Rotating black holes in higher dimensions with a cosmological constant*, Phys. Rev. Lett. **93** (2004), 171102, hep-th/0409155.

54. J.B. Hartle and S.W. Hawking, *Solutions of the Einstein–Maxwell equations with many black holes*, Commun. Math. Phys. **26** (1972), 87–101.

55. S.W. Hawking and G.F.R. Ellis, *The large scale structure of space-time*, Cambridge University Press, Cambridge, 1973.

56. M. Heusler, *Black hole uniqueness theorems*, Cambridge University Press, Cambridge, 1996.

57. S. Hollands and R.M. Wald, *Conformal null infinity does not exist for radiating solutions in odd spacetime dimensions*, (2004), gr-qc/0407014.

58. D. Ida, *No black hole theorem in three-dimensional gravity*, Phys. Rev. Lett. **85** (2000), 3758–3760, gr-qc/0005129.

59. J. Isenberg, R. Mazzeo, and D. Pollack, *Gluing and wormholes for the Einstein constraint equations*, Commun. Math. Phys. **231** (2002), 529–568, gr-qc/0109045.

60. W. Israel, *Event horizons in static vacuum space-times*, Phys. Rev. **164** (1967), 1776–1779.

61. B.S. Kay and R.M. Wald, *Theorems on the uniqueness and thermal properties of stationary, nonsingular, quasi-free states on space-times with a bifurcate horizon*, Phys. Rep. **207** (1991), 49–136.

62. S. Klainerman and F. Nicolò, *On local and global aspects of the Cauchy problem in general relativity*, Class. Quantum Grav. **16** (1999), R73–R157.

63. D. Korotkin and H. Nicolai, *A periodic analog of the Schwarzschild solution*, (1994), gr-qc/9403029.

64. J.A. Valiente Kroon, *On the nonexistence of conformally flat slices in the Kerr and other stationary spacetimes*, (2003), gr-qc/0310048.

65. J.A. Valiente Kroon, *Time asymmetric spacetimes near null and spatial infinity. I. Expansions of developments of conformally flat data*, Class. Quantum Grav. **21** (2004), 5457–5492, gr-qc/0408062.

66. K. Lake, http://grtensor.org/blackhole.

67. K. Lake, *An explicit global covering of the Schwarzschild-Tangherlini black holes*, Jour. Cosm. Astr. Phys. **10** (2003), 007 (5 pp.).

68. S. Lęski, *Two black hole initial data*, (2005), gr-qc/0502085.

69. Y. Li and G. Tian, *Regularity of harmonic maps with prescribed singularities*, Commun. Math. Phys. **149** (1992), 1–30.

70. J.M. Maldacena, *Black holes in string theory*, Ph.D. thesis, Princeton, 1996, hep-th/9607235.

71. V.S. Manko and E. Ruiz, *Exact solution of the double-Kerr equilibrium problem*, Class. Quantum Grav. **18** (2001), L11–L15.

72. D. Maxwell, *Solutions of the Einstein constraint equations with apparent horizon boundary*, (2003), gr-qc/0307117.

73. C.W. Misner, K. Thorne, and J.A. Wheeler, *Gravitation*, Freeman, San Fransisco, 1973.

74. Y. Morisawa and D. Ida, *A boundary value problem for the five-dimensional stationary rotating black holes*, Phys. Rev. **D69** (2004), 124005, gr-qc/0401100.

75. R.C. Myers, *Higher dimensional black holes in compactified space-times*, Phys. Rev. **D35** (1987), 455–466.

76. R.C. Myers and M.J. Perry, *Black holes in higher dimensional space-times*, Ann. Phys. **172** (1986), 304–347.

77. G. Neugebauer and R. Meinel, *Progress in relativistic gravitational theory using the inverse scattering method*, Jour. Math. Phys. **44** (2003), 3407–3429, gr-qc/0304086.

78. J.-P. Nicolas, *Dirac fields on asymptotically flat space-times*, Dissertationes Math. (Rozprawy Mat.) **408** (2002), 1–85.

79. B. O'Neill, *The geometry of Kerr black holes*, A.K. Peters, Wellesley, Mass., 1995.

80. A.W. Peet, *TASI lectures on black holes in string theory*, Boulder 1999: Strings, branes and gravity (J. Harvey, S. Kachru, and E. Silverstein, eds.), World Scientific, Singapore, 2001, hep-th/0008241, pp. 353–433.

81. N. Pelavas, N. Neary, and K. Lake, *Properties of the instantaneous ergo surface of a Kerr black hole*, Class. Quantum Grav. **18** (2001), 1319–1332, gr-qc/0012052.

82. R. Penrose, *Zero rest-mass fields including gravitation*, Proc. Roy. Soc. London **A284** (1965), 159–203.

83. R. Penrose, *Techniques of differential topology in relativity*, SIAM, Philadelphia, 1972, (Regional Conf. Series in Appl. Math., vol. 7).

84. J. Qing, *On the uniqueness of AdS spacetime in higher dimensions*, Annales Henri Poincaré **5** (2004), 245–260, math.DG/0310281.

85. I. Rácz and R.M. Wald, *Global extensions of spacetimes describing asymptotic final states of black holes*, Class. Quantum Grav. **13** (1996), 539–552, gr-qc/9507055.

86. D.C. Robinson, *Uniqueness of the Kerr black hole*, Phys. Rev. Lett. **34** (1975), 905–906.

87. M. Rogatko, *Uniqueness theorem of static degenerate and non-degenerate charged black holes in higher dimensions*, Phys. Rev. **D67** (2003), 084025, hep-th/0302091.

88. S.M. Scott and P. Szekeres, *The Curzon singularity. II: Global picture*, Gen. Rel. Grav. **18** (1986), 571–583.

89. D. Sudarsky and R.M. Wald, *Mass formulas for stationary Einstein Yang-Mills black holes and a simple proof of two staticity theorems*, Phys. Rev. **D47** (1993), 5209–5213, gr-qc/9305023.

90. C.V. Vishveshwara, *Generalization of the "Schwarzschild Surface" to arbitrary static and stationary metrics*, Jour. Math. Phys. **9** (1968), 1319–1322.

91. M.S. Volkov and D.V. Gal'tsov, *Gravitating non-Abelian solitons and black holes with Yang–Mills fields*, Phys. Rep. **319** (1999), 1–83, hep-th/9810070.

92. M. Walker, *Bloc diagrams and the extension of timelike two–surfaces*, Jour. Math. Phys. **11** (1970), 2280–2286.

93. G. Weinstein, *The stationary axisymmetric two–body problem in general relativity*, Commun. Pure Appl. Math. **XLV** (1990), 1183–1203.

94. G. Weinstein, *On the force between rotating coaxial black holes*, Trans. of the Amer. Math. Soc. **343** (1994), 899–906.

95. G. Weinstein, *N-black hole stationary and axially symmetric solutions of the Einstein/Maxwell equations*, Commun. Part. Diff. Eqs. **21** (1996), 1389–1430.

CHAPTER 5

THE PHYSICAL BASIS OF BLACK HOLE ASTROPHYSICS

RICHARD H. PRICE

Department of Physics & Astronomy and CGWA
University of Texas at Brownsville, Brownsville, TX 78520, USA
rprice@phys.utb.edu

Once considered only mathematical odditites, black holes are now recognized as important astrophysical objects. This article focuses on the physics of black holes that relates to their astrophysical roles.

1. Introduction

We seem to hear about black holes in statements like those in a political season. There are claims that seem to be completely incompatible with other claims. A few of these seeming contradictions follow.

I. Frozen or continuous collapse?

We hear that as a particle approaches a black hole its motion continually slows, and in a finite time it does not reach the black hole. In this same sense, a spherical star that has lost its pressure support slows its contraction, and only asymptotically becomes a black hole.

On the other hand, we hear that there is nothing singular about the surface of a black hole. A particle falls through the surface in a finite time and no singular physics is involved in this passage. Similarly, a collapsing star becomes a black hole in a finite time, and no singular forces are involved in the black hole formation.

II. Observing the unobservable black hole

We frequently hear about observations of luminosity from black holes, such as x-ray emission from stellar mass black holes. We hear that black holes are, in a way that is incompletely understood, the energy source for active galactic nuclei.

Yet we hear that nothing can escape from a black hole.

III. Newtonian-like points, or spacetime regions

We hear about black holes in clusters and binaries moving as if they themselves were particles,

Yet we hear that black holes are not objects, that they are regions of spacetime.

IV. Black holes: simple or exceedingly unsimple

We hear about black hole oscillations, and hear that state-of-the art supercomputer codes are being used to study the dynamics of the black hole formed from binary inspiral of compact objects.

Yet we hear that black holes are exceedingly simple. They have only three properties: mass, angular momentum and (though astrophysically irrelevant) electrical charge.

Part of the resolution of the contradictions is the different contexts in which statements are made. But part is due to the fact that relativity and common usage of language are often in conflict. For example, the key to most of the paradoxes of special relativity is the fact that physical time is not absolute. In black hole spacetimes the meaning of time is even more slippery, and it is the nature of relativistic time that leads to several of the contradictions.

The purpose of this article is to explain enough of the underlying physics of black holes so the reader can understand and appreciate the role played by black holes in astrophysics, and will see the contradictions in the above statements to be illusory.

This article does not deal with black holes as the very interesting mathematical solutions that they are[a]; they are treated from that point of view in the accompanying article by P. Chruściel. The slippery meaning of time warns us, however, that it is impossible to have any real understanding of black hole interactions without dealing with some of the mathematical description of black holes, so we will deal with some.

Just as it does not focus on mathematical details, neither does this article give much detail of astrophysical observations and models. But some discussions of black hole astrophysical models, and the observations that motivate them, is not only interesting, it also helps to clarify the physics, and to remind the reader that these regions of exotic spacetime curvature have a reality.

---

[a]Nor does this article deal with such black hole quantum phenomena as Hawking radiation, which is of no consequence for astrophysical-sized holes.

2. Stationary Black Hole Spacetimes

For many purposes special relativity is best understood with the Minkowskian 4-dimensional metric[1,2]. Coordinates x, y, z, t are laid out in space and time in a way that can very precisely be dictated, and the geometry of spacetime is described by the "distance" formula, or "metric," a spacetime equivalent of the Pythagorean formula for differential displacements,

$$ds^2 = -c^2 dt^2 + dx^2 + dy^2 + dz^2 . \tag{1}$$

It is crucial to understand that a coordinate transformation can lead to a very different appearance for the metric. In the case of Minkowski spacetime (gravity-free spacetime) there are preferred types of coordinates, the Minkowski coordinates, in which the metric takes the simple form in Eq. (1). Tranformations can be carried out from one Minkowski system to another, but can also be made to a nonpreferred system in which the metric takes on a very different appearance. As an example of this we could take the relatively simple transformation

$$t = T \quad x = X(T/\mathcal{T})^{1/2} \quad y = Y \quad z = Z , \tag{2}$$

where $\mathcal{T}$ is a positive constant with the dimensionality of time (and is introduced to maintain dimensional consistency). In these new T, X, Y, Z coordinates the metric becomes

$$ds^2 = - \left[1 - \frac{X^2}{4c^2 T\mathcal{T}} \right] c^2 dT^2 + \frac{X}{\mathcal{T}} dT dX + 4 \frac{T}{\mathcal{T}} dX^2 + dY^2 + dZ^2 . \tag{3}$$

This formula has a completely different character from that in Eq. (1). In particular the formula suggests that the geometry it describes is dependent on time T. There is another, more subtle ugliness in Eq. (3): the $dT dX$ cross term. This term means that motion in the $+X$ direction is physically different from motion in the $-X$ direction.

We feel intuitively, of course, that the simple spacetime of special relativity does not change in time, that it is stationary (unchanging in time) and isotropic (so that $+x$ is the same as $-x$), but Eq. (3) shows that we must be careful about how we state this. The correct way is: There exists a coordinate system in which the metric is stationary and isotropic.

As long as we are being careful, it is good to be careful about the difference between a "time" coordinate and a "space" coordinate. If we consider a slice of spacetime with $dT = 0$, we get

$$ds^2 = 4 \frac{T}{\mathcal{T}} dX^2 + dY^2 + dZ^2 . \tag{4}$$

The condition for this constant T slice to be spatial, and hence for T to be a time coordinate, is that the value of ds^2 be positive for any displacement $\{dX, dY, dZ\}$. But in Eq. (4) we see that this is the case only for $T > 0$. Thus T is a time coordinate only for $T > 0$. (One might object that negative T is prohibited by the form of the transformation in Eq. (2). But here we should view Eq. (3) as a metric given to us for analysis. The connection to the Minkowski formula is the ultimate clarification of Eq. (3), but such connections are not always available and, if available, are usually not obvious.)

The example in Eq. (3) has prepared us to understand the Schwarzchild geometry which has the formula

$$ds^2 = -\left(1 - \frac{2GM}{rc^2}\right) c^2 dt^2 + \left(1 - \frac{2GM}{rc^2}\right)^{-1} dr^2 + r^2 d\theta^2 + r^2 \sin^2\theta \, d\phi^2 \,.$$

$$(5)$$

Here G is the universal gravitational constant and M is a parameter with the dimensions of mass. If the M parameter is set to zero, Eq. (5) takes the form of the Minkowski metric expressed with spherical polar spatial coordinates, a simple transformation from Eq. (1).

For $M > 0$, it turns out that Eq. (5) cannot be transformed to the Minkowski metric, and therefore does not represent gravity-free spacetime. It does, however, represent a very fundamental solution in Einstein's theory. That theory, general relativity, consists of partial differential equations that must be satisfied by the metric functions, the functions, such as $1 - 2GM/rc^2$ in Eq. (5), that appear in the metric. The partial differential equations, the field equations of general relativity, connect these functions to the nongravitational energy and momentum content of spacetime. The Schwarzschild metric is a solution of Einstein's equation discovered almost 90 years ago[3], for vacuum, i.e., for spacetime in which there is no matter, no energy and no fields, except for the gravitational field itself which is encoded in the geometry. In Einstein's theory, furthermore, the Schwarzschild metric turns out to be the unique spherically symmetric vacuum solution that is asymptotically flat. (It approaches the Minkowski metric as $r \to \infty$.) It represents, therefore, the spacetime outside a spherically symmetric gravitating body. In that sense it plays the same role as $\Phi = 1/r$ in electromagnetic theory. This solution appears to be stationary. That is, there exist coordinates (the t, r, θ, ϕ coordinates) in which the metric is independent of the time coordinate. This is intuitively satisfying; like electromagnetic waves, gravitational waves are transverse, and like electromagnetic waves, gravitational waves cannot be spherically symmetric. There would then seem to be nothing that can be dynamical in a spherically symmetric gravitational spacetime, and the mathematics of the theory confirms this.

The familiar large-r features of the Schwazschild metric disappear at small r. The scale for "small" is the "Schwarzchild radius" of the spacetime,

$$r_g \equiv 2GM/c^2. \tag{6}$$

At $r < r_g$, all connection with Newtonian physics breaks down. The constant t slices, in this region, are not spacelike; the coefficient of dr^2 is negative, so the slice is not a spatial slice, and t is not a time coordinate. In fact, it is rather clear in Eq. (5) that in this $r < r_g$ inner region t is a spatial coordinate and r is a time coordinate.

We now have an awkward situation. We have a coordinate system in which the metric is independent of one of the coordinates t, but that is not a time coordinate for $r < r_g$. In that region the spacetime has a symmetry (it is invariant with respect to translations in t) but it is not stationary. We could, of course, search for a better coordinate system, and coordinate systems can be found (such coordinate systems as the Kruskal, Novikov, and Eddington-Finkelstein systems[4]) whose constant time slices are everywhere spatial. Though these coordinates are well suited to expressing the geometry both for $r > r_g$ and for $r < r_g$, these coordinates are ill suited to express the stationary nature of the solution. In such coordinates either the metric has explicit time dependence (Novikov and Kruskal systems) or has a cross term (Eddington-Finkelstein system). We have a choice then: we can have a system that expresses the inherent simplicity of the spacetime, or we can have a system that does not introduce unnecessary coordinate artifacts.

Another rather important stationary vacuum spacetime is given by the two-parameter (M and a) Kerr metric

$$ds^2 = -\left(1 - \frac{2GMr}{\Sigma}\right)c^2 dt^2 - \frac{4GMar\sin^2\theta}{\Sigma}\,dt\,d\phi$$

$$+\frac{\Sigma}{\Delta}dr^2 + \Sigma d\theta^2 + \frac{\sin^2\theta}{\Sigma}\left[\left(r^2 + a^2\right)^2 - a^2\Delta\sin^2\theta\right]d\phi^2 \tag{7}$$

where

$$\Delta \equiv r^2 - r_g r + a^2 \qquad \Sigma \equiv r^2 + a^2\cos^2\theta. \tag{8}$$

Here again $r_g \equiv 2GM/c^2$, where M is the mass parameter for the spacetime. It is useful to notice that the Schwarzschild metric is the special $a = 0$ case of the Kerr metric. When we set $dt = 0$ we get a three-dimensional geometry that is spatial for $\Delta > 0$, but Δ is negative for $r_- < r < r_+$, where

$$r_\pm = \tfrac{1}{2}r_g \pm \sqrt{\left(\tfrac{1}{2}r_g\right)^2 - a^2} \ . \tag{9}$$

As was the case for the Schwarzschild geometry, here again the t coordinate is convenient, but is not "time" in all regions of spacetime.

The parameter M in the Schwarzschild metric is mass. One justification for this is that we can use the nonvacuum equations of general relativity to construct solutions for spacetimes around stationary spherical massive objects, such as perfect fluid stars. Outside the material of these objects the spacetime is empty and the metric must be the Schwarzschild metric. Joining that solution smoothly to the interior nonvacuum solution leads to an expression for M that is (aside from some relativistic ambiguities) the sum of all the mass and energy (including gravitational binding energy) inside the object.

The fact that the a parameter in Eq. (7) multiplies the $dtd\phi$ cross term correctly suggests that it is related to rotation. It is in fact a measure of the angular momentum of the gravitational source, but the connection to a source requires some care. There is no unique vacuum metric around a rotating object and, in fact, no solution is known for a realistic nonvacuum interior metric that joins smoothly to an exterior Kerr metric. We do know this, however: if we compute the total angular momentum of a rotating object, then the vacuum spacetime exterior to that object has a form that at large r agrees with the large-r form of the Kerr metric if the total angular momentum J and the Kerr parameter a are related by

$$a = J/Mc . \tag{10}$$

3. Particles and Fields Near Black Holes

3.1. *Particle worldlines*

Though the use of curved spacetime to represent gravity seems unnecessarily abstract, the connection is compelling. The bridge between the two ideas is the equivalence principle, the fact that all particles experience the same acceleration by gravity. This means that to specify the gravitational trajectory, the worldline, of a particle you need only to specify its initial location and velocity, not its mass, its strangeness, or any other particle property[b].

---

[b] It is being assumed here that "point" particles cannot have spin. Classically, particles with angular momentum must have nonzero size.

The physics of gravity is therefore not a study of particles, but of particle paths through spacetime, and paths are determined by the geometry they live in.

In gravity-free spacetime the worldline of a particle must be straight. In metric theories of gravitation, like Einstein's general relativity, the generalization is that the worldline must be as straight as possible for a free particle (a particle moving under gravity, but free of any "real" forces). The straighest possible path, called a geodesic, is mathematically the path that has a locally unchanging unit tangent. (The derivative of the unit tangent along the path is zero.)

For particle paths, of course, we are interested primarily in timelike curves for which infinitesimally separated points have $ds^2 < 0$. This condition of timelike separation tells us that a particle can follow the path without ever going faster than the speed of light. For these curves a parameter that is both physically and geometrically appealing is the proper time τ along the particle worldline, defined by $d\tau = \sqrt{-ds^2}$. The physical importance follows from the fact that τ is the time measured by clocks comoving with the particle. A massless particle, like a photon, must move at the speeed of light. Its worldline, called a null worldline, is characterized by $ds = 0$ for nearby points, so proper time is not meaningful. Instead, we can notice that the 4-momentum $\vec{p}$ of a particle of mass m is given, in terms of spacetime displacement $d\vec{x}$ by $\vec{p} = m\,d\vec{x}/d\tau$. We can choose to parameterize a particle worldline with a parameter λ such that the 4-momentum of the massless particle is $\vec{p} = d\vec{x}/d\lambda$. Such a parameter, called an affine parameter, can loosely be thought of as the $m \to 0$ limit of τ/m.

For the Schwarzschild spacetime of Eq. (5), any free orbit is equivalent to an equatorial orbit, so we can fix $\theta = \pi/2$ and can describe gravitational motion by specifying $t(\tau), r(\tau), \phi(\tau)$ along the worldline. It turns out that the geodesic equations are equivalent to the following constants of motion:

$$r^2 \frac{d\phi}{d\tau} = \mathcal{L} \qquad \left(1 - \frac{2GM}{rc^2}\right)\frac{dt}{d\tau} = \mathcal{E} \tag{11}$$

$$\left(1 - \frac{2GM}{rc^2}\right)\left(\frac{dt}{d\tau}\right)^2 - \left(1 - \frac{2GM}{rc^2}\right)^{-1}\left(\frac{dr}{cd\tau}\right)^2 - r^2\left(\frac{d\phi}{cd\tau}\right)^2 = 1. \tag{12}$$

The constants $\mathcal{E}$ and $\mathcal{L}$ here play the role of conserved particle energy and particle angular momentum. (For discussion of $\mathcal{L}, \mathcal{E}$, and why they are constant, see Sec. 25.5 of the text by Misner et al.[4].)

Though these equations do determine the path through spacetime, it is important to remember that τ is a quantity that is meaningful only on the worldline. It has no direct meaning to our time, that is to the time of the observers. The more familiar "common sense" description of physics is to specify $r(t)$ and $\phi(t)$ in terms of a universal time t that applies everywhere. In general there is nothing like universal time in a curved spacetime, but the spacetimes of Eqs. (5) and (7) are independent of t, and thus t does have a favored status.

3.2. *Radial orbits*

In this duality of time descriptions we begin to see a source of apparently contradictory statements about the spacetimes in Eqs. (5) and (7). For radial infall ($\mathcal{L} = 0$) of a particle in the Schwarzschild spacetime the equations above give us $dr/dt \propto 1 - r_g/r$, so the particle reaches $r = r_g$ only asymptotically at $t \to \infty$, as shown on the right in Fig. 1.

The equations of radial infall, on the other hand, tell us

$$\frac{dr}{d\tau} = -\sqrt{\mathcal{E}^2 - 1 + r_g/r} \; , \tag{13}$$

and $r = r_g$ is an ordinary point of the differential equation. A plot of the trajectory from this point of view is shown on the left in Fig. 1, with events, A, B, C indicated. This plot shows that the worldline does not end at r_g, but continues, as indicated by the later events events, D, E.

The "particle" that is radially falling in could be the surface of a star, if we consider the pressure support of the star to be negligible so that the surface is freely falling. The collapse of this star is frozen, as on the right in Fig. 1. As described with global time t the collapse never quite reaches $r = r_g$. On the other hand, to an ill-fated observer sitting on the star, the passage through $r = r_g$ is an event of no noticeable significance.

The problem clearly lies with the coordinate t, and the very form of Eq. (5) is evidence of a problem. This coordinate, so useful as a global coordinate for $r > r_g$, is a singular coordinate at $r = r_g$. The appearance of bad geometry, and hence of misleading physics, at $r = r_g$ is an artifact of the very convenient set of coordinates. In other, less convenient coordinates, there is no singularity at $r = r_g$.

Before we dismiss the t coordinate as a completely bad idea, we should repeat that τ is not enough. We need to connect physical events with the time at which they are observed and t *is* the appropriate coordinate for that. (For one thing it is the clock time of distant astronomers in weak gravitational fields and small velocities.) It is the coordinate with which we can address the question "what do the astronomers see" as the particle, or

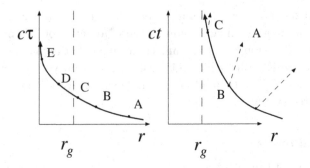

Fig. 1. Radial infall in the Schwarzschild spacetime; the proper time versus global coordinate time description.

stellar surface moves inward. In order for there to be something for them to see, we have added (dashed) lines representing outgoing photons carrying information to the curious observer.[c] As we approach $r = r_g$ along the worldline, the value of dr/dt approaches zero for our outward photons.

This trend becomes even more pronounced for $r < r_g$. Events subsequent to C, events such as D, simply cannot send information outward. This is best understood in the fact that t becomes spacelike and r becomes timelike in the $r < r_g$ region. In that region, the future turns out to be in the decreasing r direction. Since all particles or signals, whether freely falling or not, must move to the future, no information can move from $r < r_g$ to $r > r_g$. The distant oberservers, in fact *any* observers who remain in the outer $r > r_g$ region of spacetime, are doomed to ignorance about what goes on in the inner $r < r_g$ region. Any astronomer in the inner region, of course, is even more tragically doomed. She must ineluctably move to decreasing r until reaching $r = 0$ in a finite time, where she can confirm that there are *true* (geometric, not coordinate-induced) singularities in the spacetime there, singularities that exist in any coordinate system, and that correspond to infinite gravitational tidal forces on freely falling particles.

Because an absolutely fundamental principle (motion to the future) prevents events in the inner region from being seen by observers in the outer region, the dividing surface, $r = r_g$, is called an event horizon or simply horizon. What is crucially important about the horizon is that the spacetime

---

[c]It is important to understand that there is no meaning to a picture of these photon worldlines on the left in Fig. 1. The τ coordinate has no meaning off the worldline of the infalling particle.

geometry is smooth at the horizon, and physics is completely nonsingular, but the horizon represents a boundary in spacetime between (interior) events that cannot communicate to the asymptotically flat universe, and those (exterior) events that can. This role accounts for the singularities that appear in the metric when we try to give a description that is suitable only for the exterior events.

There is an important property of the event horizon that will help us understand them in more complicated contexts. The horizon could be traced out by a set of photon worldlines. If a photon is emitted, in the outward direction, exactly on the event horizon, it will stay on the event horizon. (A photon emitted in any other direction will go inward.) In a much simpler context, Minkowski spacetime, a simple expanding spherical shell of photons has the same property. Worldlines can cross into it, but not out of it. What is exotic about the Schwarschild horizon is that it is not expanding. As a result of strong gravity, the shell of outgoing photons always has the same area.

Since light (in the broadest sense) cannot escape from inside or on the horizon, the horizon and its interior can emit nothing, and they are called a black hole.

In Eq. (7) the Kerr metric is written in a system called Boyer-Lindquist coordinates. A comparison of the coefficients of the dr^2 term in Eq. (7), and in (5) suggests that $\Delta = 0$, may be a horizon. Though the argument is weak, the conclusion turns out to be correct. The two roots $r_\pm$ of Δ given in Eq. (9) are in fact horizons across which worldlines can cross in only one direction. It is the so-called outer horizon corresponding to the root $r_+ \equiv r_g/2 + \sqrt{rg^2/4 - a^2}$ that is astrophysically relevant, since it is a boundary to an asymptotically flat outer region, the universe of weak fields and astronomical observations. The inner horizon r_- is a one-way barrier to travel between inner regions of the Kerr spacetime. Such inner regions can have physical reality only if the Kerr hole was cast into the original spacetime of the universe. If the Kerr hole formed as the result of the collapse of a rotating astrophysical object, then Eq. (7) only describes the spacetime outside the material of that object. The interior regions of the Kerr spacetime are replaced by the spacetime of the astrophysical material.

In addition to the outer horizon, the Kerr spacetime has another very nonclassical feature, a surface called the ergosphere that is outside r_+ (except at the poles where it coincides with r_+). Outside the ergosphere a particle can have any angular motion, i.e., can have a positive, negative or zero value of $d\phi/dt$. A rotational motion is encoded in Eq. (7) that can be thought of as a dragging of space along with the rotation of the hole. This dragging becomes so strong inside the ergosphere that $d\phi/dt > 0$ for any worldline. (Equivalently: the future ineluctably has an increase in ϕ just as the future, inside the horizon, ineluctably has a decrease in r). The ergosphere, at

$$r = r_g/2 + \sqrt{r_g^2/4 - a^2 \cos^2 \theta}, \qquad (14)$$

has explicit angular dependence. It is not spherically symmetric, but neither is the horizon; the θ-independence of the horizon is an artifact of a particular convenient coordinate system.

3.3. Nonradial orbits

Much of the astrophysics of black holes involves the trajectories of particles and photons in the neighborhood of a hole. In principle, these trajectories are not properties of black holes since they apply to the empty spacetime outside any spherically symmetric object. In practice, the interesting features are relevant almost exclusively to black holes. Interesting non-Newtonian features of orbits are significant only for r comparable to r_g. The surfaces of most astrophysical bodies are at a radius much larger than their r_g, so the exterior spacetime starts at $r \gg r_g$. The exception to this is neutron stars which are almost as compact as black holes of equivalent mass.

Fortunately, most of the interesting features of particle trajectories can be found in the Schwarzschild spacetime, so everything we need is in principle contained in the three rather simple equations in Eqs. (11) and (12). For definiteness, we limit attention to massive particles and we combine the equations of Eq. (12) in the form

$$\mathcal{E}^2 = \left(\frac{dr}{cd\tau} \right)^2 + V(r) \qquad (15)$$

where

$$V(r) \equiv 1 - \frac{r_g}{r} + \frac{\mathcal{L}^2}{c^2 r^2} - \frac{r_g \mathcal{L}^2}{c^2 r^3}. \qquad (16)$$

This equation, along with Eq. (11), completely determines the orbital shapes and dynamics. It is interesting that, aside from the re-identification of constants, the Newtonian counterpart of this equation differs only in that it is missing the last term, the r^{-3} term. To see the effect of that term we plot out the "effective potential" $V(r)$ for several values of $\mathcal{L}$ in

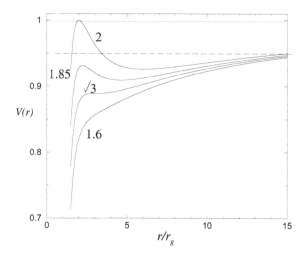

Fig. 2.   Effective potential for motion in the Schwarzschild spacetime. Curves are labeled with the values of $\mathcal{L}/r_g c$. The dashed line at $V = 0.95$ and the dotted line at $V = 0.998$ illustrate turning points.

Fig. 2. Since $V(r)$ plays the role of a potential in Eq. (15), we can infer the nature of particle orbits from Fig. 2. For $r/r_g \gg 1$ the orbit is determined by the usual effects of gravitational attraction $(-r_g/r)$, and centrifugal replulsion $(+\mathcal{L}^2/c^2 r^2)$. The extra non-Newtonian term at the end of Eq. (16) represents the inescapably strong pull of gravity on a particle close to the hole.

Consider, for example, a particle with $\mathcal{E}^2 = 0.95$. The dashed line in Fig. 2 shows that the particle will have turning points approximately at $r/r_g = 3.4$ and 15.0, which turn out to be not very different from the Newtonian turning points. This particle would then be in a very eccentric bound orbit which, when combined with the equation for $d\phi/d\tau$ in Eq. (11), can be shown not to be a closed orbit, but rather has the nature of a precessing ellipse. The turning point at $r/r_g = 1.56$ is a new, completely non-Newtonian feature. It is, of course, not a feature of the eccentric orbit between the turning points at $r/r_g = 3.4$ and 15.0. Rather it applies to a particle that is created very close to the hole, moving outward with $\mathcal{E} = 0.95$ and $\mathcal{L}/r_g c = 2$. This particle will reach a maximum of r equal to $1.56\,r_g$ where it turns, and begins its spiral into the hole.

The effect of the non-Newtonian peaks of the potential has an important consequence for the shape of orbits. For a particle with $\mathcal{L}/r_g c = 2$, the peak of the potential is at $V = 1$. If this particle happens to have $\mathcal{E}^2 = 0.998$

(shown as the dotted line in Fig. 2) it will spend a great deal of proper time τ near the turning point at $r/r_g = 2.14$, just to the right of the peak. Since Eq. (11) tells us that $d\phi/d\tau$ is finite at that turning point we conclude that the particle will go through a large range of ϕ near $r/r_g = 2.14$. In principle, a particle can come in from a large distance (a far-away turning point or infinity) and can orbit the hole at small radius an arbitrarily large number of times before departing for large distances (if $\mathcal{E}^2$ is below the potential peak) or into the hole (for $\mathcal{E}^2$ above the peak). This pattern is sometimes called the "zoom-whirl" character of orbits near the hole, and is of importance in understanding the waveform of gravitational radiation that would be emitted from a compact object spiralling near a supermassive black hole in a galactic nucleus.

There are several additional important features of relativistic orbits that can be inferred from Fig. 2. One is the contrast with the difficulty of a falling particle hitting a Newtonian point source. In Newtonian theory the centrifugal barrier dominates at small r except in the unique case of radial motion and $\mathcal{L} = 0$. To hit the center, a particle must be aimed right at the center. In the Schwarzschild geometry, the particle must only have sufficient energy to surmount the energy barrier. In the case $\mathcal{L}/r_g c < \sqrt{3}$, there is no barrier, and the particle must fall into the hole.

In Newtonian gravity centrifugal and gravitational forces can balance at any radius, so circular orbits exist at any radius. As Fig. 2 shows, this is not true for orbits in the Schwarzschild spacetime. If $\mathcal{L}/r_g c$ is less than $\sqrt{3}$, no circular orbit is possible; relativistic gravitational attraction overwhelms centrifugal force. The minimum orbit, commonly referred to as the "ISCO" (innermost stable circular orbit) is at $r = 3r_g$, the point of inflection of the $\mathcal{L}/r_g c = \sqrt{3}$ potential.

Orbits in the Kerr spacetime differ from Schwarzschild orbits in a few important ways. One difference is that for motion in the orbital plane the details of the orbits depend on whether the particle is circulating in the prograde direction (the same direction as the hole's rotation) or retrograde (opposite to hole's rotation). It is of particular interest that for prograde motion the radius of the ISCO is smaller than that for a Schwarzschild hole of the same mass. In the so-called extreme[d] Kerr case, with $a = M$ the prograde ISCO is at $r = r_g/2$. This means that progrde particle motion can explore the spacetime considerably closer to the horizon. This, and

---

[d]For $a > M$ the Kerr metric no longer describes a hole; see the accompanying article by P. Chruściel.

many other details of particle orbits around Kerr orbits are of interest in connection with sources of gravitational waves. Compact objects orbiting supermassive, rapidly rotating holes in the galactic nuclei can be powerful sources of gravitational waves and wonderful probes of spacetime geometry.

A feature of Kerr orbits that is absent in the Schwarzschild case is that $\theta(\tau)$ is important; not all orbits are equivalent to equatorial orbits. Interestingly there exists a third constant of motion that encodes the θ dynamics. Unlike $\mathcal{E}$ and $\mathcal{L}$, this third constant, called the Carter constant[5], is not based on a simple symmetry of the spacetime. Even with this additional constant, the Kerr orbits are complicated enough that attention has usually been focused on equatorial orbits. Recently, the advent of gravitational wave antennas has changed that, and there is great interest in Kerr orbits, including non-equatorial orbits.

The story of photon orbits is slightly simpler than that for massive particles. Only two modifications are needed in Eqs. (11) and (12). Proper time τ must be replaced by affine parameter λ and the right hand side of Eq. (12) must be set to zero, rather than to unity. The effect on the orbital equations are, again, that τ is replaced by λ in Eq. (15), and that $1 - r_g/r$, the first two terms on the right of Eq. (16) must be omitted. From the resulting simple diagram analogous to Fig. 2, we find that there is a circular orbit at only one radius, $r = 3r_g/2$ and that this orbit is unstable. Photons can have zoom-whirl orbits with an arbitary number of orbits near $r = 3r_g/2$.

3.4. Field dynamics

Electric and magnetic fields near black holes can play a part in the astrophysics of black holes. The way in which these fields behave is determined by the nature of the event horizon. For vacuum fields, fields whose sources are not near the hole, the role of the horizon is "simply" to be a boundary condition for the dynamics of those fields, and that boundary condition is that field changes can propagate only into the horizon, and not out of the horizon.

In this sense, the horizon is the same as spatial infinity. In fact, in the partial differential equations describing the field dynamics in the Schwarzschild background is remarkably simple

$$-\frac{\partial^2 \psi}{\partial t^2} + \frac{\partial^2 \psi}{\partial r^{*2}} + \left(1 - \frac{r_g}{r}\right) \frac{1}{r^2} \left[\frac{1}{\sin\theta}\frac{\partial}{\partial\theta}\left(\sin\theta\frac{\partial\psi}{\partial\theta}\right) + \frac{1}{\sin\theta^2}\frac{\partial^2\psi}{\partial\phi^2}\right] = 0.$$

(17)

Here ψ is r^2 times the radial component of either the electric or magnetic field. The coordinates t, r, θ, ϕ are those in Eq. (5), and r^* is

$$r^* \equiv r + r_g \ln (r_g/r - 1).\tag{18}$$

This "tortoise" coordinate, introduced by Wheeler[6], asymptotically approaches r for $r \gg r_g$, but moves the horizon to $r^* = -\infty$. Equation (17) shows that far from the hole the ψ field satisfies the familiar three dimensional wave equation, but near the horizon ψ acts as if it were a one dimensional field propagating to the infinitely distant horizon. This succintly represents the role of the horizon. It acts as if it is a radiative infinity, but because it has finite area and curvature it has effects not found at true spatial infinity.

There are equations very similar to Eq. (17) for massless fields of any spin both for the Schwarzschild and for the Kerr spacetimes. One interesting application of these equations is that we can search for single frequency solutions with the physically appropriate boundary conditions of ingoing propagation at the horizon and outgoing propagation at infinity. These solutions are called quasinormal modes[7] of the hole because they have some of the flavor of the normal modes of a mechanical system. Since these modes lose energy to radiation, they are strongly damped and their frequencies are complex. As discussed in Sec. 5, such modes are likely to be important, and possibly dominant, in the gravitational wave generation by black holes.

Much more difficult than the free field question is that of fields coupled to sources falling through the horizon. Questions about this process arise when highly conducting plasma falls through the horizon. Viewed in the frame of the falling plasma, there can be no singularly large forces at the horizon. (Locally the horizon is a perfectly ordinary place) so magnetic flux trapped in the plasma must be carried through the horizon. This suggests that magnetic field lines will link the interior of the horizon with distant astrophysical regions in which the magnetic field lines are anchored. Viewed in the stationary frame of external observers, however, the plasma never makes it through the horizon, and falling plasma and magnetic flux accumulate outside the horizon, with magnetic field pressure increasing, especially in the case of a rotating hole.

The astrophysical process has been understood only relatively recently[8]. What is relevant to the astrophysics, of course, is the interaction of the hole the fields and the distant plasma. We must, therefore, find a description that is useful to the external observers. From their point of view it is true that the magnetic field lines do in fact accumulate just outside the horizon, and the dynamics of the strong fields leads to complicated field line connections. But all this is happening very close to the horizon, a distance away that is small compared to length scales characterisitic of the hole and the large scale magnetic field. The complex magnetic field dynamics, in fact, has some of the character of a boundary layer phenomenon, and can be treated in the same spirit. The results of the complex boundary phenomena can be averaged over small time and space scales. The electromagnetic effects of the horizon can then be represented by a material membrane slightly outside the horizon and imbued with electromagnetic properties that represent the action of the horizon. In this membrane viewpoint[9] the horizon has a surface resistivity of 377 ohms per square, or ohm-cm per centimeter. (It is no coincidence that this is the radiation resistance of empty space.) The complex near-horizon magnetic reconnection is then replaced by a simple smooth picture of ohmic dissipation due to currents driven in the conducting membrane.

4. Observational Black Holes

4.1. *General considerations*

We have emphasized that nothing can "get out" of a black hole. But this is only true in the sense that no signal or information can propagate outward from the dark side of the event horizon to the distant universe. Something *can* get out: the gravitational pull of the hole can "get out" (or maybe simply "be out") since it is not a signal, it is an (almost) unchanging property of the hole. Similarly, the dragging of inertial frames is related to the (almost) unchanging angular momentum of a hole. (Why almost? Electrical charge is absolutely conserved in Maxwellian theory. The nonlinear nature of relativistic gravity, however, allows gravitational fields to carry off mass and angular momentum.)

These (almost) unchanging characteristics of the hole may not represent information flowing outward from the dark side, but they exert a profound influence very close to the hole, and it this influence that gives rise to observable phenomena, and hence to our ability to observe black holes. For the most part (but not entirely!) the mystical spacetime properties of a black hole are not what ultimately create distinct observable signatures of a black hole. For the most part what is observationally interesting about a black hole is that it's so small. It represents a great deal of mass compacted to a size comparable to r_g. Thus astrophysical gas, particles, light etc., can get within a distance on the order of r_g, and experience an extremely strong gravitational pull. This could not happen with an ordinary astrophysical gravitating object because the surface of the object itself would be much larger than r_g.

A point that is not to be missed in this is that for the most part the qualitative features of black hole astrophysics can be understood with Newtonian gravitational theory. Quantative features, and some qualitative features of black holes, however, cannot be treated with Newtonian theory. One example of what can be quantitatively important is the effect of spacetime on the radiation reaching a distant astronomer. In fact, three different effects enter the picture. The first is the that each photon propagating out of a deep gravitational well will lose energy in its fight against the gravitational pull. In a stationary spacetime, like that of Eq. (5) or (7), it turns out that the effect is given by the coefficient of the $c^2 dt^2$ term in the formula for the metric geometry. Thus photons emitted from some location at radius r_{source} will arrive at the distant astronomer with only a fraction $\sqrt{1 - r_g/r_{\text{source}}}$ of its original energy. A somewhat related effect is the relation of time at r_{source} and to that for the distant astonomer. If photons are emitted at r_{source} with a time separation Δt, they will be captured by the distant astronomer with a time separation $\Delta t / \sqrt{1 - r_g/r_{\text{source}}}$. (This can be seen to be related to the idea that the period of oscillation of the photon, with its reduced energy, must be decreased.) Yet another effect is the distortion of the pattern of emission, an effect that falls into the general category of photon orbits. All these radiative influences involve the factor $\sqrt{1 - r_g/r_{\text{source}}}$, and all tend to act to decrease the radiation received by the astronomer. As a practical matter then, the astronomer is prevented not only from seeing inside the black hole, but also is prevented from seeing radiation from the region close to the hole.

These effects on radiation are gravitational equivalents of effects that are familiar (doppler shift, time dilation, relativistic beaming) in special relativity. The special nature of black holes can enter in more idiosyncratic ways involving the unique nature of the event horizon. Such phenomena include zoom-whirl orbits, quasinormal oscillations and the membrane-like electromagnetic properties of the horizon.

If we were to make the broadest categorization of the ways in which black holes can be observed, it would be: their influence on light, their influence on astrophysical gas or plasmas, and their influence on nearby compact astrophysical objects. In the first category is gravitational lensing, discussed briefly above. In the second is the phenomenon of black hole accretion and the production of luminous signals. In the third category are such phenomena as the observation of hidden members of binary pairs and star motion near galactic centers.

We can deal quickly with the first category. A gravitating object (galaxy, black hole,...) that lies near the line of sight from a distant galaxy or quasar can bend the light rays from that source and distort the location and shape of that source. It might seem that black holes, with their ability to deflect light through an arbitrarily large angle are a likely source of this gravitational lensing. Black hole lensing, however, requires an improbable alignment of distant source and black hole lens, though evidence of such an alignment has been detected[10].

4.2. *Black hole mass ranges*

In discussing astrophysical black holes it is important to identify two very different possible astrophysical populations. One is stellar mass holes with masses on the order of the solar mass $M_\odot \approx 2 \times 10^{33}$ grams. Such holes can be formed as the endpoint of stellar evolution or as the result of accretion onto neutron stars. The second population is supermassive holes, on the order of $10^6 M_\odot$ to $10^9 M_\odot$; these are believed to be at the center of many, perhaps all, massive galaxies. There is also some evidence of an additional "intermediate" mass range[11] of 100s to 1000s of $M_\odot$.

The physics of the black holes themselves has an extrememly simple dependence on mass. For a Schwarzschild hole all scales are set by the mass, and there is only one length scale, r_g which itself is proportional to the mass. (In the case of a Kerr hole there is also a dimensionless parameter a/r_g that describes the importance of rotational effects.) All gravitational effects can be described in terms of the mass. Thus, for example, the near-horizon gravitational acceleration $GM/r_g^2 \sim M^{-1}$, and the near-horizon tidal force is $GM/r_g^3 \sim M^{-2}$. The characteristic density of a hole, its mass divided by $(4\pi/3)r_g^3$ scales as M^{-2}. Somewhat counter-intuitively, the smaller the black hole mass is, the more exotic is its near-horizon environment. For example, the characterisitic density turns out to be around nuclear densities for stellar mass black holes, reminding us that stellar mass black holes are not much more compact than neutron stars. For large supermassive black holes, by contrast, the density is small compared to terrestrial materials.

This observation would suggest that stellar black holes have observational signatures that are more easily distinguished from those of less exotic objects, while supermassive black holes cannot be easily identified. The opposite is true. Neutron stars are nearly as compact as stellar mass black holes, and in interpreting the nature of an observation neutron stars often can be ruled out only because astrophysical theory dictates a maximum mass of a neutron star of only a few $M_\odot$. To preclude neutron star explanations, stellar mass black hole candidates must usually be larger than this limit. By contrast, there is no other astrophysical object that we know of that can imitate a supermassive black hole, a mass of (say) $10^8 M_\odot$ confined in a region of size $\sim 10^{13}$ cm. A massive galaxy, with about ten times the mass of such a black hole, is larger in diameter by a factor around 10^{10}.

4.3. Accretion, dynamos and luminosity

The fact that black holes are tiny on astrophysical scales means that gas falling into a black hole will undergo enormous compressional heating, and the hot plasma will radiate. An order of magnitude estimate can be made based on the idea of conversion of gravitational energy to thermal energy, and suggests that temperatures could reach more than 10^{12} K as the plasma approaches the hole. Well before that temperature is reached, however, the plasma will lose thermal energy via radiation, constraining the increase in temperature and therefore suppressing compressional heating. Gravitational energy can therefore be thought of as being converted not to the random motions of thermal energy, but to radial kinetic energy, and wastefully carried into the hole. Calculations taking into account the interaction of compressional heating and radiative cooling show that plasma inflowing toward a black hole of a few solar masses will tend to reach temperatures on the order of 10^9 K, and the dominant radiation will therefore be in the form of X-rays. In this manner, in principle, we could detect black holes as point-like sources of X-rays coming, of course, not from inside the black hole, but from the region very near the hole. Radial infall is a good introductory lesson in how black holes can be luminous sources, but radial infall is extremely inefficient; only a small fraction of the available gravitational energy appears as outgoing luminosity. The hope for observability lies with a more efficient method, and a more plausible astrophysical scenario.

As we saw in the discussion of particle orbits, to fall into the hole a particle must have angular momentum per unit mass-energy on the order of $c r_g$. Typical astrophysical velocities are of order of $10^{-3}c$ and typical astrophysical sizes are at least of the order of the earth's orbital radius 10^{13} cm (compared to $r_g \sim 10^5$ cm for a solar mass black hole). Unless the gas starts out implausibly devoid of angular momentum, in order to approach the black hole it must find a way to shed a great deal of angular momentum. For the purposes of producing an observable luminous signal, this is very good.

Rather than fall radially, the gas will spiral into an orbiting structure called an accretion disk[12]. Figure 3. shows an artist's conception of a black hole accretion disk being formed from the mass being pulled off an ordinary star that is the binary companion of the black hole.

As a good approximation, the gravitational forces are so strong compared to fluid forces that each bit of fluid moves on a nearly circular orbit at a relativistically-corrected Keplerian angular velocity. The bit of fluid gradually spirals in to the inner edge of the accretion disk at a location determined (according to present thinking) by magnetic instabilities in the disk. The voyage of the gas from large radii to this inner edge is a gradual process that efficiently converts gravitational binding energy to outgoing radiation.

Fig. 3.   Artist's conception of an accretion disk. (Melissa Weiss CXC, NASA).

The mechanism for this conversion is based on the disk's differential rotation. The almost-circular orbit nature of the fluid motion means that angular rotation is faster at smaller radius. Viscosity in the fluid will slow the rotation rate of the inner gas, thereby transporting angular momentum outward, allowing the gas gradually to spiral inward. The process involves viscous heating of the gas which requires radiation for cooling. The details of the disk's structure, temperature profile, luminous signature, and so forth, all depend on the microphysics of the viscosity, and there is no secure model for this. The viscosity may be based on the highly conducting plasma being threaded by tangled magnetic fields. In any case, some features of the models are relatively insensitive to the details; one of these is the nature of the characteristic radiation. Again it follows from a simple order-of-magnitude calculation that the temperature must be on the order of 10^9 K, and the radiation will be predominately in the form of X-rays.

When we move from stellar mass holes to supermassive holes, we move from physical models that miss details, to physical models themselves that are still uncertain. But we also move to astrophysical black holes whose existence is essentially certain (or at least establishment dogma). Some galaxies have active galactic nuclei (AGNs), tiny central regions that emit luminosity comparable to galactic luminosity, but are incredibly small on an astrophysical scale[13]. From a number of arguments, it is known that the engine – the source of energy – in these AGNs consists of a mass so compact that it must involve a supermassive black hole. The mechanism for producing the radiation must certainly be different from that of a black hole accretion disk, since AGNs are characterized by radio jets, narrow regions of strong radio emission emerging from a central source. Theorists believe that the mechanism for these complicated sources probably involves a large scale magnetic field anchored in a rapidly rotating supermassive hole. It is believed that this large scale magnetic field is created by the accretion of plasma carrying magnetic flux. The magnetic field then links the horizon to distant regions of the surrounding plasma. With the horizon playing the role of a rapidly rotating conducting membrane, the hole and magnetic field act like a kind of dynamo. The EMFs on rotating field lines create strong currents that flow to the distant plasma linked by the field lines. The production of these currents, in effect, converts the rotational energy of the black hole into radio and other forms of radiation.

There had been a distinct difference between the charactersitics and models of stellar mass holes and of supermassive holes, but a few years ago this difference faded. A class of objects called "microquasars"[14] was discovered that seemed to be stellar mass X-ray black holes that also have the radio jets of AGNs. The tentative understanding of these microquasars is that they are accretion disk objects that can build up large scale magnetic fields and behave like AGNs scaled down by a linear factor of 10^8 or so.

Figure 3 is meant to illustrate a possible structure for microquasar XTE J1550-564.

4.4. *Exotic orbits, galactic centers, and gravitational waves*

A direct method has been used to "prove" the existenece of a black hole in our own galaxy: The motions of stars in the central core of the galaxy are observed to be so rapid that there is no reasonable alternative model for the central region. An exciting way of detecting the presence of black holes lies only a few years in the future with the detection of gravitational waves[15] from black hole systems. Gravitational waves are physically meaningul (as opposed to coordinate induced) perturbations of the spacetime metric. Their effect is a propagating transverse stretching and contraction of space.

For a discussion of gravitational waves and the prospects for their detection, see the accompanying article by P. Saulson. For our purposes here we will rely on the fact that to zeroth order the properties of gravitational waves may be inferred by analogy to electromagnetism. To generate strong electromagnetic waves one wants to have a violent time dependence of an electrical charge distribution. This can be done on a small scale (even atomic scale) thanks to charge separation. Nothing like charge separation works for gravity; there's only one sign of gravitating mass. To have a rapidly changing mass distribution one needs to move astrophysical objects, and they can be moved only with gravitational fields. The strongest accelerations can be reached with the strongest gravitational pulls (and the shortest Keplerian times). This points to black holes, as can be seen in an order of magnitude expression for the luminosity (energy per time) emitted by a gravitational wave source is

$$\text{G.W. luminosity} \sim \frac{G}{c^5} \frac{M^2 v^4}{T^2}. \tag{19}$$

Here M is the characteristic mass that is moving; v is its velocity, and T is the time scale (e.g. an orbital period) on which it is moving. For a given mass, evidently one wants the most rapid motion, and this can best be done with the highly relativistic motions near a black hole. One very interesting source would be a particle mass m_p in an orbit of radius a around a much more massive object of mass M. If the Keplerian expressions for v and T are used we conclude that the luminosity is of order $G^4 m_p^2 M^3/c^5 a^5$. Clearly we want a to be as small as possible, which means that we must have mass M maximally compressed, i.e., it should be a black hole, so that a can be of order of the gravitational radius of GM/c^2, and the luminosity can be of order $(c^5/G)(m_p/M)^2$.

These "particle" orbits around supermassive black holes will be a nearly certainly detectible source for the space-based gravitational wave detector LISA scheduled to be launched in around a decade. But searching for them and fully understanding them will require the solution to a detailed theoretical problem. The particle orbits of Sec. 3 ignore radiation reaction associated with the generation of the outgoing gravitational waves. This is actually a perturbative effect that is justifiably ignored in a discussion, as in Sec. 3, of a small number of orbits, but is important if the particle motion is going to be tracked through many orbits of the inspiral of the particle, itself driven by gravitational radition reaction. Computations of this radiation reaction are usually carried out using an adiabatic approximation; the inspiral is adjusted from one nearly circular orbit to the next by subtracting the orbital energy lost to radiation. This method is effective except for the nonequatorial obits in the Kerr geometry. In that case radiation of energy and angular momentum does not tell us how to evolve the angle θ at which the orbits tilt away from the equatorial plane. Theorists are just begining to be able to handle this question with reasonably efficient computations[16].

That bothersome factor of $(m_p/M)^2$ in the gravitational luminosity means a reduction of 10^{-12} or worse. To achieve a truly stupendous rate of gravitational wave generation the "particle" must be replaced by a second black hole with a mass roughly comparable to the first, so that the source has a luminosity on the order of $c^5/G \approx 10^{59}$ ergs/sec. Our best guess is that the actual rate is only a few percent of this, but that power is still much more than the luminosity of a galaxy (10^{43} ergs/sec) or the most powerful astrophysical events detected (gamma ray bursts with 10^{54} ergs/sec).

This spectacular rate of energy loss must come from somewhere, and the somewhere is the orbital energy of the two holes. The holes therefore must spiral toward each other and eventually merge. In dealing with this situation we must take a very different viewpoint on black holes.

5. Dynamical Black Holes

Up to this point we have treated black holes as "backgrounds" for physics. Such physics as accretion and particle motion happened in a black hole spacetime background. When we start to talk about two black holes whirling around each other, that viewpoint is not adequate. The black holes themselves become participants. Nothing makes this clearer than a consideration of the endpoint of the process of binary black hole inspiral, the whirling of the two black holes around each other. Intuition dictates (correctly) that the endpoint is a merger of the black holes. Here, black holes, both the two original holes and the final hole, are very much in the foreground of the physics, not the background.

Section 2 has introduced black holes through examples of the Schwarzschild and Kerr spacetimes, which are stationary, unchanging in time (at least outside the event horizon). These stationary black holes are not sufficient if the black holes are to be participants; we now must consider a more general event horizon, a surface, dynamical or not, through which particles and light from the exterior universe can pass only inward. Astrophysical horizons, in fact, never really stationary. Since they are formed from astrophysical events, the horizons were not always present; they must have a dynamical history.

For a more thorough and careful general discussion of event horizons, see Sec. 2 of the accompanying article by P. Chruściel. Specific model solutions for dynamical horizons are given in Sec. 4 of that article. Here we will give only enough of a description to give an idea of the role of dynamical horizons in astrophyics. The best way to do this is to go back to the picture of the stationary horizon in the Schwarzschild spacetime as traced out by a set of radially outgoing photons that make no progress moving to larger radii. A black hole horizon may be thought of as a set of "outgoing" photons that never reach asymptotically flat regions of spacetime, and that never collapse inward. In the case of the Schwarzschild horizon the photons, in a sense, stay at the same place. The photons defining a dynamical horizon do not.

The cartoon on the left in Fig. 4 gives a very simple example: the head-on collision of two neutron stars. Time is taken to increase upward, so that each horizontal slice represents all of space at a single time. At time A in the cartoon of the neutron star collision, the neutron stars are moving slowly toward each other, but no horizon exists. On that time slice there are no points in space from which escape is impossible. At time B, however, the point midway between the two neutron stars becomes one from which escape is impossible. The region of no escape grows, engulfing both neutron stars, and continues to grow until, at time slice C it is approximately in the stationary form that it will only approach asymptotically at infinite time. It is this asymptotic spacetime that is described by the Schwarzschild geometry.

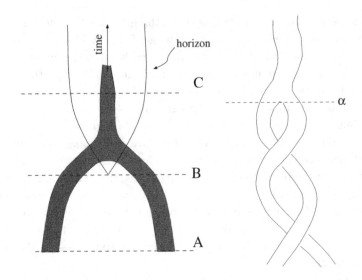

Fig. 4. Cartoons of horizon dynamics. On the left, two neutron stars undergo a head on collsion forming a black hole. On the right two "orbiting horizons" merge to form a final, larger horizon.

On the right in Fig. 4 is a cartoon, in the same spirit, of the binary motion and merger of two black holes. Here at early times there is already horizon structure; the disjoint horizons of each of the individual holes. A point on that slice is either in one black hole or in the other, or outside both. At time slice α, the two horizons, and the two no-escape regions, merge, forming a larger no-escape region, that is, forming a larger, spinning black hole.

We know that the mass and angular momentum of the final black hole will be somewhat less than the mass and energy of the original binary pair. The excess goes into outgoing gravitational waves. The greatest power generated in the inspiral-merger process comes near the very end, in the last orbit and the merger. A nonnegligible fraction of the mass energy of the system will be radiated in a very short time, but – for now – that is all we can say. The highly dynamical asymmetric character of the process means that only numerical computations can track the process, and the computational task has proved to be an enormous challenge. For the last decade this challenge has been the central focus of the developing field of "numerical relativity," but the ability to compute radiation from the last few orbits and merger of the binary black hole is still several years off. For a discussion of the importance and difficulty of these computation see the accompanying article on numerical relativity by P. Laguna.

This is not to say that we know nothing quantitative about dyanamical black holes. Since the early 70's we have learned a great deal about black hole dynamics by considering perturbations of the the stationary black hole spacetimes[17]. This approach leads to linearized equations for the perturbations that differ from Eq. (17) only in minor details. For these perturbations also, there are strongly damped quasinormal oscillations. As an example, for the dominant (least damped quadrupole) quasinormal oscillation of a non-rotating hole, the real and imaginary parts of the frequency are $0.74737 \, c/r_g$ and $0.17793 \, c/r_g$ respectively.

At first, "small perturbations" would seem to have nothing to do with the strong deviations in the late stages of black hole mergers pictured in Fig. 4. This is not the case for two reasons. First, as with many perturbation computations it turns out that the results are more robust than their derivation would suggest, and they apply even when the perturbations are moderately large. Second, and more important, in the late stages of merger, somewhere between times B and C in the head-on collision of neutron stars, and somewhat after α in the binary black hole inspiral, The violent large-deviation dynamics is all hidden inside the final horizon. What generates outgoing radiation is the oscillation of the spacetime outside that final horizon. Numerical evolution studies of the final merger show, in fact, that the very complicated process has a rather simple signature: the bulk of the radiation is in the form of the least damped quasinormal frequency.

In general, the two pieces of information contained in the real and imaginary part allow us to infer the two parameters, mass and angular momentum, of the hole. When (not if) gravitational waves are detected, this connection of black hole properties and observable characteristics of gravitational waves may be a key to understanding the sources of the waves.

6. Conclusion

We started by pointing to contradictory statements about black holes. We can now show that the contradicitions are illusory.

I. Frozen or continuous collapse?

This noncontradiction highlights the most fundamental potential confusion about holes: the varieties of "time" with which black hole processes can be described. With a time coordinate that makes manifest the stationary nature of stationary black hole spacetimes, a star will never collapse to form a hole. In terms of the proper time that describes the physics local to the collapsing surface, passage through the event horizon is continuous and smooth.

II. Observing the unobservable black hole

No information can propagate outward through the event horizon, but the compactness of a black hole means that the gravitational influence of the hole on its near environment gives rise to exotic phenomena (bending of light rays, generation of X-rays, gravitational waves...).

III. Newtonian-like points, or spacetime regions

It is a matter of distance from the hole. In a study of the properties of spacetime very near a black hole, the focus is on the defining properties, the nonsingular one-way nature of the event horizon. At distances several times r_g, however, the main influence of the hole is its $1/r^2$ pull. It therefore acts on its environment like a Newtonian gravitational source and simple arguments about relative frames and momentum conservation at large distances show that the hole must move like a Newtonian object.

IV. Black holes: simple or exceedingly unsimple

The stationary Schwarzschild and Kerr spacetimes are very clean and simple. Dynamical black holes are sufficiently complicated that an understanding of them can come only through numerical relativity, and that understanding is still some time in the future.

The history of the black hole has very much paralelled the place of Einstein's theory in science. The Schwarzschild geometry dates to the very start of general relativity. There was limited interest in general relativity, and limited understanding of black holes among most astronomers until the early 60s. New astronomical discoveries, starting with quasars, opened the possibility that Einstein's theory had an astrophysical role beyond cosmology. Now, almost a half century later, the transition is complete. It is now all but taken for granted that supermassive black holes lie at the centers of many, perhaps all, galaxies, and that certain point-like X-ray sources are evidence of stellar mass holes.

Acknowledgments

This work was supported by National Science Foundation grant PHY0244605, and NASA grant ATP03-0001-0027. I thank Frederick Jenet for useful comments on the manuscript.

References

1. J. B. Hartle, *Gravity: An Introduction to Einstein's General Relativity* (Addison Wesley, San Francisco, 2003).
2. S. Carroll, *Spacetime and Geometry: An Introduction to General Relativity* (Addison Wesley, San Francisco, 2004).
3. K. Schwarzschild, *Sitzber. Deut. Akad. Wiss. Berlin, Kl. Math.-Phys. Tech.* 189, (1916).
4. C. W. Misner, K. S. Thorne and J. A. Wheeler, *Gravitation* (Freeman, San Francisco, 1973), Sec. 31.4.
5. B. Carter, *Phys. Rev.* **174**, 1559 (19968).
6. J. A. Wheeler, *Phys. Rev.* **97**,511 (1955).
7. K. D. Kokkotas and B. G. Schmidt, Living Rev. Rel. **2**, 2(1999): cited on March 26, 2005.
8. R. D. Blandford and R. L. Znajek, *Mon. Not. Roy. Astron. Soc.* **179**, 433 (1977).
9. *Black Holes:The Membrane Paradigm*, edited by K. S. Thorne, R. H. Price, and D. A. Macdonald (Yale University Press, London, 1986).
10. D. P. Bennett et al., *American Astronomical Society, 195th AAS Meeting, #37.07; Bulletin of the American Astronomical Society*, **31**, 1422 (1999).
11. C. M. Miller, *Astrophys. J.*, **618**, 426 (2005).
12. M. A. Abramowicz ed. it Theory of Black Hole Accretion Discs (Cambridge University Press, Cambridge, 1998).
13. J. H. Krolik Active Galactic Nuclei (Princeton University Press, Princeton, 1998)
14. Ribó, M. and Paredes, J. M. and Martí, J. and Casares, J. and Bloom, J. S. and Falco, E. E. and Ros, E. and Massi, M., "Results of a search for new microquasars in the Galaxy," Revista Mexicana de Astronomia y Astrofisica Conference Series, *Revista Mexicana de Astronomia y Astrofísica Conference Series*, 23 (2004).
15. S. Rowan and J. Hough, "Gravitational Wave Detection by Interferometry (Ground and Space)," *Living Rev. Relativity 2*, (2000), 3. [Online article]: cited on 28 March, 2005, http://relativity.livingreviews.org/Articles/lrr-2000-3/index.html
16. E. Poisson, *Living Rev. Relativity*, **7**, 6 (2004).
17. M. Sasaki and H. Tagoshi, "Analytic Black Hole Perturbation Approach to Gravitational Radiation", *Living Reviews in Relativity*, **6**, 6 (2003).

CHAPTER 6

PROBING SPACE-TIME THROUGH NUMERICAL SIMULATIONS

PABLO LAGUNA

Institute for Gravitational Physics and Geometry
Center for Gravitational Wave Physics
Departments of Astronomy & Astrophysics and Physics
Penn State University
University Park, PA 16802, USA
pablo@astro.psu.edu

Einstein's equations describe gravity using an elegant but complicated set of equations. Finding astrophysically relevant solutions to these equations requires the most sophisticated numerical algorithms and powerful supercomputers available. The search for astrophysical solutions has made numerical relativity one of the most active areas of research in gravitational physics. Of particular interest in numerical relativity has been simulating the inspiral and coalescence of compact binaries involving black holes and neutrons stars. The outcome from these simulations will bring general relativity into harmony with the observations of gravitational radiation that are expected to take place in the immediate future. This article highlights current progress in numerical relativity. It also attempts to envision the future of this field and its integration with gravitational wave astronomy.

1. Multi-Faceted Numerical Relativity

A new era in astronomy will begin once gravitational wave interferometers such as LIGO, GEO, VIRGO, TAMA and, in the future, LISA detect *first light*. These detectors will give us a revolutionary view of the Universe, complementary to the electromagnetic perspective. In this new astronomy, the messengers are gravitational waves, ripples in the fabric of spacetime. These waves will have encoded detailed knowledge of the coherent, bulk motions of matter and the vibrations in the curvature of spacetime produced by a vast class of astrophysical sources; compact object binaries, supernovae, spinning neutron stars, gamma ray bursts and stochastic backgrounds are just a few examples of these sources. The detection of gravitational waves is a formidable undertaking, requiring innovative engineering, powerful data analysis tools and careful theoretical modelling. Among the sources of gravitational radiation, binary systems consisting of black holes and/or neutron stars are expected to be dominant. The ultimate goal of source modelling is to develop *generic* numerical codes capable of modelling the inspiral, merger, and ringdown of a compact object binary. Over the last couple of decades, advances in numerical algorithms and computer hardware have bring us closer to this goal.

General relativity and singularities come hand in hand. The mathematical study of singularities has been mostly done analytically. An analytic approach has obvious limitations; they are either restricted to over simplified situations or are only able to produce broad general conclusions. Numerical simulations are becoming an important tool in the exploration of the properties of singularities. In particular, the numerical work by Berger and collaborators[14] on investigations of naked singularities, chaos of the Mixmaster singularity and singularities in spatially inhomogeneous cosmologies has already demonstrated the great potential that a numerical approach has in producing detailed understanding of singularities in physically realistic situations.

The question of whether or not there is a minimum finite black hole mass in the gravitational collapse of smooth, asymptotically flat initial data lead to what is perhaps currently the most exciting and elegant result in numerical relativity. Choptuik used powerful adaptive mesh refinement methods to prove that the mass is infinitesimal.[23] In addition, Choptuik's work discovered completely unexpected effects. One of them is the scaling relation

$$M \approx C \, |p - p_*|^\gamma \tag{1}$$

with M the black hole mass and p a parameter characterizing the family of initial data. Another unexpected effect found by Choptuik is that the solution has a logarithmic scale-periodicity when $p \to p_*$. Finally, Choptuik found that the phenomena is *universal*, namely the "critical exponent" $\gamma \approx 0.37$ and the "critical echoing period" $\Delta \approx 3.44$ are the same for all one-parameter families of initial data. For a review of Choptuik's critical phenomena, its extensions and applications see Gundlach.[33]

Numerical relativity is now an area with many directions of research. The computational modelling of compact object binaries is, however, what is attracting most of the attention. This article focuses on this effort. It provides a basic overview and is intended for non-experts in numerical relativity. The following reviews are recommended for those interested in a more technical view:

Numerical Relativity and Compact Binaries by Thomas W. Baumgarte and Stuart L. Shapiro[12] This is comprehensive review with specific focus on traditional numerical relativity, namely those aspects of numerical relativity that have been designed for or applicable to the two body problem in numerical relativity. In my opinion this review is an excellent starting point for students and researchers interested in numerical relativity.

Initial Data for Numerical Relativity by Greg B. Cook.[25] This review gives an excellent introduction to the construction of Cauchy initial data for space-times containing binary systems of black holes and/or neutron stars. The emphasis is on those methods to obtain initial data representing binary systems in quasi-equilibrium orbits. The review also addresses the issue of the astrophysically content of initial data sets.

Hyperbolic Methods for Einstein Equations by Oscar A. Reula.[49] Reula's review is a most for those interested in the mathematical properties of evolution equations arising in 3+1 formulations of the Einstein equations. The review also examines the techniques to obtaining symmetric hyperbolic systems.

Characteristic Evolution and Matching by Jeff Winicour.[69] Characteristic formulations, although not widely used in numerical evolutions, have attractive properties for wave extraction. This review discusses the developments of characteristic codes and, in particular, the application of characteristic evolutions to Cauchy-characteristic matching.

Conformal Field Equations by Jörg Frauendiener.[28] This is a general review on the notion of conformal infinity. Of direct relevance to numerical relativity is the discussion on Friedrich's[29] conformal field equations and how this formulation of the field equations could provide a natural framework to study isolated systems and gravitational radiation.

2. Geometrodynamics and Numerical Evolutions

The most popular approach in numerical relativity is to view spacetime as the time history of the geometry of a space-like hypersurface or *Geometro-*

dynamics, as first referred by J.A. Wheeler.[44] Under the geometrodynamics point of view, the Einstein equations, $G_{\mu\nu} = 8\pi T_{\mu\nu}$ for the spacetime metric $g_{\mu\nu}$, can be reformulated or rewritten in a space+time or 3+1 structure by suitable projections on and orthogonal to the space-like hypersurfaces.

The Arnowitt-Deser-Misner[6] (ADM) equations are perhaps the simplest 3+1 formulation (see Appendix A). In the ADM formulation, the 10 Einstein equations for the spacetime metric $g_{\mu\nu}$ become 12 first order in time evolution equations for the metric h_{ij} and metric velocity K_{ij} or more formally the extrinsic curvature of the space-like hypersurfaces.[a] The remaining 4 equations are constraints (Hamiltonian and momentum) that the pair (h_{ij}, K_{ij}) must satisfy.

Obtaining long-term stable numerical evolutions of spacetimes containing black holes and neutron stars with codes based on the ADM equations has been extremely difficult. It is not completely clear the reasons behind the numerical problems with the ADM system. Most likely these problems have to do with the weakly hyperbolic properties of the system.[46] These impediments have motivated researchers to develop other 3+1 formulations of the Einstein equations.

The interest in numerical relativity on investigating 3+1 formulations has reinvigorated the field. Numerical relativity is no longer dominated by tour de force efforts. The field is receiving a balanced injection of formal mathematical ideas and creative engineering. This excitement has produced a zoo of formulations of the Einstein equations, formulations designed with the sole purpose of finding the silver bullet against numerical instabilities. The emphasis has been on manifestly hyperbolic formulations (see Reula's review[49]). These formulations are supported by mathematical theorems that provide valuable information, for instance, on the dependence of evolved data on initial conditions. Hyperbolicity alone is not enough. There are other crucial factors. For example, lower order non-liner terms could be responsible for triggering or facilitating the development of instabilities.[37] Only through numerical experimentation one is going to able to tell which formulation or formulations are more advantageous.

I will focus the discussion on three formulations. These are currently the most successful formulations used in 3D simulations involving non-trivial configurations of compact objects. The first of these formulations

---

[a]The following index notation has been adopted: 4D indices will be denoted with Greek letters and 3D indices with Latin letters. Latin indices are to be understood as pull-backs to the hypersurfaces in the foliation. Units are such that $G = c = 1$.

was originally introduced by Shibata and Nakamura,[54] and later reintroduced by Baumgarte and Shapiro.[10] This formulation is commonly known as the BSSN formulation. In general terms, the essence of the BSSN formulation is to work not with the pair (h_{ij}, K_{ij}) used by the ADM formulation but instead with $(\Phi, \hat{h}_{ij}, K, \hat{A}_{ij}, \widetilde{\Gamma}^i)$. The relation between ADM and BSSN variables are:

$$\Phi = \ln h^{1/12} \tag{2}$$

$$\hat{h}_{ij} = e^{-4\Phi} h_{ij} \tag{3}$$

$$K = K^i{}_i \tag{4}$$

$$\hat{A}_{ij} = e^{-4\Phi} A_{ij} \tag{5}$$

$$\widetilde{\Gamma}^i = -\partial_j \hat{g}^{ij}, \tag{6}$$

where $A_{ij} = K_{ij} - h_{ij} K/3$. Working explicitly with the scalars Φ and K is only one of the crucial steps in the BSSN equations. The other is the introduction of the connection variable $\widetilde{\Gamma}^i$. It is this variable that is mostly responsible for improving the hyperbolic properties of the BSSN system over those of the ADM system.[34]

The BSSN formulation has been quite successful. Codes based on these equations have produced simulations of binary neutron stars,[55,43,42] wobbling black holes,[60] boosted black holes,[59] distorted black holes,[3] black holes head on collisions,[5,27,58] black hole plunges[5] and binary black hole evolutions with angular momentum up to timescales of a single orbit.[21]

The second popular 3+1 formulation of the Einstein equation is the hyperbolic formulation developed by Kidder, Scheel and Teukolsky (KST).[39] Strictly speaking the KST system is a parameterized family of hyperbolic formulations. This formulation has had remarkable success in evolving single black holes. It is expected that in the near future this success will be translated to evolutions of binary black holes. The starting point in deriving the KST system is the introduction of a new auxiliary variable $d_{kij} \equiv \partial_k h_{ij}$, to eliminate second derivatives of the spatial metric. With this variable, the Einstein equations can be rewritten as:

$$\partial_t u = A^i \partial_i u + \rho \tag{7}$$

where u is a vector constructed from the evolution variables, the matrices $A^i(u)$ determine the hyperbolic properties of the system and $\rho(u)$ denotes the lower order terms, namely terms that do not contain derivatives of u. With a suitable choice of parameters, the KST system can be make symmetric hyperbolic.

Before continuing with the third formulation, it is important to point out that most of the 3+1 codes perform what are called free evolutions. That is, the Hamiltonian and momentum constraints enter only in the construction of initial data. Once the evolution starts, the constraints are only used for *quality* control purposes. Non-trivial free evolutions tend to be unstable. The community has adopted the name of *constraint violating modes* for the instabilities that plague these free evolutions. The only other type of instabilities that are not constraint violating would be *gauge mode* instabilities. Gauge mode instabilities are not generic. They require careful tuning since after all they represent a perfectly valid solution to the Einstein equations. The current view in the numerical relativity community is that if one is able to actively preserve the constraints during evolutions, constraint violating modes will be avoided. This observation has motivated the design of methodologies to project the evolved data back into the constrain surface. So far, these methods have not been completely successful.[36,35,20,66]

The third type of formulation is not new but has just recently received serious attention. The re-emergence of this formulation has been triggered by the great success by Pretorius[47] evolving binary black holes. This formulation is based on a generalization of the harmonic coordinates $\Box x^\mu \equiv g^{\alpha\beta}\nabla_\alpha\nabla_\beta x^\mu = 0$. When the harmonic coordinate condition is imposed, the principal part of the Einstein equations takes the form $g^{\alpha\beta}g_{\mu\nu,\alpha\beta}$. It was with this form of the Einstein equations that proofs of existence and uniqueness of solutions were obtained. In numerical relativity, harmonic decompositions of the Einstein equations have been largely ignored. Exceptions are the work by Garfinkle,[30] Szilagyi and Winicour[61] and the Z4-system.[17] The generalized harmonic coordinates are given by $\Box x^\mu = H^\mu$, with H_μ arbitrary source functions. These source functions H_μ are elevated to the rank of independent variables. That is, substitution of $\Box x^\mu = H^\mu$ into the Einstein equations $R_{\mu\nu} = 4\pi(2T_{\mu\nu} - g_{\mu\nu}T)$ yields

$$g^{\alpha\beta}g_{\mu\nu,\alpha\beta} + g^{\alpha\beta}{}_{,\mu}g_{\nu\alpha,\beta} + g^{\alpha\beta}{}_{,\nu}g_{\mu\alpha,\beta} + 2H_{(\mu,\nu)}$$

$$-2H_\alpha\Gamma^\alpha{}_{\mu\nu} + 2\Gamma^\alpha{}_{\beta\mu}\Gamma^\beta{}_{\alpha\mu} = -8\pi(2T_{\mu\nu} - g_{\mu\nu}T). \tag{8}$$

Promoting the source functions H_μ to independent variables implies that the quantities $C^\mu \equiv \Box x^\mu - H^\mu$ become constraints. At the analytic level, $C^\mu = 0$; during numerical evolutions, however, this is not the case because of truncation errors. There is indication that these constraints are much less difficult to preserve than the Hamiltonian and momentum constraints in 3+1 formulations.

With the generalized harmonic formulation one looses the explicit geometrodynamics point of view; that is, one no longer deals with geometrical
quantities directly associated with the foliation (i.e. spatial metric, extrinsic curvature, lapse, shift). Instead, one works directly with the spacetime
metric components. Is this something to worry about? The answer is likely
no. After all, most of the 3+1 hyperbolic formulations introduce auxiliary
variables devoid of any geometrical interpretation. Also, there are other
non-3+1 formulations such as the characteristic formulation that have also
enjoyed relative success. Unfortunately, because of the difficulties that the
characteristic formulation has on handling caustics, characteristic codes
these days are mostly used in connection with gravitational wave extraction,
namely Cauchy-Characteristic matching.

There are signs that the next *revolution* in codes will be the generalized
harmonic codes. Preliminary work by Pretorius[48] indicates that these codes
have great potential. He has able to carry out eccentric binary black hole
orbits without encountering instabilities or the need of introducing a large
number of fine tuned parameters. It is too early to say if the generalized
harmonic codes are indeed the *silver bullet* against instabilities. What is
clear is that they are quickly positioning as the leading contestant.

3. Black Hole Excision: Space-time Surgery

What makes black hole evolutions unique is the presence of black hole singularities. These are not simple singularities, as the gravitational interaction
$1/r^2$-singularities in N-body simulations, singularities that are amenable
to smoothing. In modelling black holes, the singularities are not known a
priori. They are intimately related to the solution one seeks. It would seem
that there is no way around it! Not quite, the *Cosmic Censorship conjectures* states that these singularities will be inside a horizon, hidden from
external observers. If we are able to track the existence of a horizon, one
can in principle exclude the singularity from the physically relevant part of
the computational domain.

There are two generic approaches to handle black hole singularities in
numerical evolutions. The oldest approach was to *avoid* the black hole singularities. This is possible because of the freedom to foliate the space-time
arbitrarily. An example of a slicing condition that avoids crushing into the
singularities is the maximal slicing, defined by the condition that the trace
of the extrinsic curvature vanishes.[57] Slicing conditions that avoid singularities suffer, however, from what is known as *grid stretching*; that is, the

growth of the proper distance (i.e. metric functions) between grid points in the neighborhood of the black hole. This growth is entirely a coordinate effect; nonetheless, it increases seriously the demand for computational resources needed to resolve the steep gradients that eventually develop in the metric functions.[15] The increase of proper distance is a direct consequence of using the lapse function to simultaneously slow-down the evolution near the black hole while keeping a *normal* pace far away from the holes. In some instances, it is possible to construct a shift vector that minimizes the grid stretching.[21]

An alternative approach to singularity avoidance is *black hole excision*. The method involves removing the singularity from the computational domain without modifying the causal structure of the space-time. The geometrical interpretation of this procedure is quite simple.[64] Once inside the event horizon, the light-cones are tilted inwards into the singularity, so in principle it is possible to remove a region, inside the event horizon, containing the singularity without the necessity of imposing any boundary conditions in the boundary of the removed region. The data at the boundary of this region are complete characterized from information inside of the computational domain. This approach is similar to dealing with outflow supersonic boundaries in computational fluid dynamics. In the case of black holes, one takes advantage that information within the event horizon cannot propagate *upstream* and leak out of the horizon. It is then crucial that the discretization of the PDEs respects the causal structure of the space-time in such a way that events inside the horizon are causally disconnected from its exterior. The first implementations of the principles behind the discretization involved in black hole excision were carried out by Seidel and Suen[52] ("causal differencing") and Alcubierre and Schutz[53] (*causal reconnection*). Black hole excision these days has become to some extent a routine exercise.[4,56]

A potentially serious problem in black hole excision are gauge modes. The only modes that are constrained to remain within the horizon are those carrying physical information. Gauge modes are free to emerge from the horizon. In some cases one takes advantage of this freedom (e.g. superluminal shift vectors) to construct a suitable foliation. It seems then crucial that a formulation of Einstein equations in terms of characteristic fields (i.e. hyperbolic form) is needed in order to facilitate the identification of physical and non-physical characteristics and their corresponding propagation speeds. This observation was what triggered in part the development of hyperbolic formulations of Einstein equations. It is not clear the importance of the explicit knowledge of the field characteristics. Black hole excision has been performed with BSSN codes without explicit reference to the characteristic fields.

The essential point in black hole excision is to locate the event horizon. This task requires, however, the complete future development of the spacetime. To circumvent this obvious difficulty, an apparent horizon (outer most trapped surface) is used instead. If an apparent horizon is found, an event horizon exists and surrounds the former.[67] For excision to work, one only needs to know that the removed region is entirely contained within the horizon, thus the precise location of the event horizon is not required, finding apparent horizons suffices. Historically, because excision requires tracking the apparent horizon, the approach was incorrectly called Apparent Horizon Boundary Condition.[52] The name is incorrect in the sense that black hole excision does not require a boundary condition. The same equations that are applied in the interior of the domain are used at the boundary.

4. Initial Data: The Astrophysical Connection

We are still at the stage in which the primary focus of simulations of neutron stars and/or black hole binaries is to achieve evolutions timescales for which more refined calculations would be able to deliver astrophysically interesting results. This is a natural situation, not exclusive to numerical relativity. Progress is made in incremental steps. At the early stages, simplifications are needed at the expense of sacrificing the accuracy of the models.

Even if codes are able to handle long enough evolutions, the astrophysical content of a numerical simulation is completely determined by boundary conditions and initial data. In general relativity, constructing initial data is a non-trivial exercise. Data must satisfy the Hamiltonian and momentum constraints:

$$R + K^2 - K_{ij}K^{ij} = 0 \tag{9}$$

$$\nabla_j(K^{ij} - h^{ij}K) = 0. \tag{10}$$

The challenge in constructing initial data is separating the four components in (h_{ij}, K_{ij}) that are fixed by (9) and (10) from those that are freely specifiable. The pioneer work to develop such framework is due to Lichnerowicz, York and collaborators.[40,70] York in particular was fundamental in formulating the Initial Data Problem with a structure amenable to numerical relativity (see Appendix B for details).

Early work on multiple black hole initial data was mostly concerned with finding *any* solution, without paying close attention to whether these solutions were astrophysically relevant or not. A series of simplifying assumptions were introduced to facilitate the task. Specifically, if one assumes initial data with maximal embedding and a conformally flat 3-geometry, York's version of the constraint equations take the simple form

$$8\hat{\Delta}\phi = -\hat{A}_{ij}\hat{A}^{ij}\phi^{-7} \tag{11}$$

$$\hat{\nabla}_j\hat{A}^{ij} = 0\,, \tag{12}$$

where ϕ is the conformal factor and $\hat{A}^{ij}$ the trace-free, conformal extrinsic curvature. Notice that in this form, one can solve first independently the momentum constraint (12) for $\hat{A}_{ij}$ and then use this solution to solve the non-linear constraint (11) for the conformal factor ϕ.

Bowen and York showed[18,71] that a solution to equation (12) for N black holes, each with linear momentum P_i^A and angular momentum S_i^A, is given by

$$\hat{A}_{ij} = \frac{3}{2}\sum_{A=1}^{N}\left\{\frac{1}{r_A^2}\left[P_i^A n_j^A + P_j^A n_i^A - (\eta_{ij} - n_i^A n_j^A)P_k^A n_A^k\right]\right.$$
$$\left. + \frac{1}{r_A^3}\left[\epsilon_{ilk}S_A^l n_A^k n_j^A + \epsilon_{jlk}S_A^l n_A^k n_i^A\right]\right\}\,, \tag{13}$$

where n_i^A is the unit normal at the throat of the A-th black hole, r_A the distance to the center of the A-th hole, and η_{ij} is the flat-metric.

The Bowen-York solution (13) has been extremely useful. For instance, Brandt and Brügmann used the Bowen-York solution to introduce the so called puncture approach.[19] The puncture method solves (9) based on a decomposition of the conformal factor ϕ of the form

$$\phi = u + \frac{1}{p} \tag{14}$$

$$\tag{15}$$

such that

$$\frac{1}{p} = \sum_{A=1}^{N}\frac{\mathcal{M}_A}{r_A}\,, \tag{16}$$

with $\mathcal{M}_A$ the "bare" masses of the black holes. The advantage of using the decomposition (14) is that Eq. (9) reduces to

$$\hat{\Delta}u = -\frac{1}{8}\hat{A}_{ij}\hat{A}^{ij}(1/p + u)^{-7}\,. \tag{17}$$

It is not difficult to show that the r.h.s. of this equation is regular everywhere, thus also the solution u. There is no need to excise the black holes.

The singularity has been explicitly handled! Because of its simplicity, the puncture approach has become extremely popular.

The main focus these days is in constructing astrophysically realistic initial data representing black hole and/or neutron star binaries. Without these data, the predictive power of gravitational wave source simulations, with relevance to data analysis efforts, is seriously compromised. Initial data sets consisting of binaries in quasi-circular orbits are of particular interest. They are needed as a starting point of any evolution simulation aimed at computing realistic gravitational waveforms. For binary black holes, the pioneering work to locate what could be identified as quasi-circular orbits was done by Cook.[24] The method, called the effective potential method, is based on the potential

$$E_b = E - 2M \,, \tag{18}$$

with E_b the binding energy of the system, E the total energy and M the irreducible mass of each individual black hole, typically given by the area of the apparent horizons. Given a sequence of initial data sets with constant black hole mass M and constant total angular momentum J, quasi-circular orbits are then found from the condition

$$\left. \frac{\partial E_b}{\partial l} \right|_{M,J} = 0 \,, \tag{19}$$

with l the proper separation of the horizons. The angular velocity Ω is found from

$$\Omega = \left. \frac{\partial E_b}{\partial J} \right|_{M,l} \,. \tag{20}$$

Given this approach to identifying quasi-circular orbits, the innermost stable circular orbit (ISCO) is found by looking at the end point of a sequence of quasi-circular orbits.

An alternative approach was introduced by Gourgoulhon, Grandclément and Bonazzola (GGB).[31,32] In this work, quasi-circular orbits are identified by applying the condition that the space-time possesses an approximate helical Killing vector. The construction of these data sets requires, however, a modification to the standard York conformal approach, a method allowing certain control on how the initial data evolve in time. In other words, one needs to be able to set up a "thin slab" of the space-time. Such a method is called the thin-sandwich approach (see Cook's review[24] for details). The basis of the thin sandwich method is to include as part of the initial data the lapse function and shift vector. By adding the lapse and the shift to the problem, one can make use of the freedom intrinsic to these objects to satisfy the approximately helical killing vector condition $\partial_t h_{ij} \approx 0$. The procedure amounts to constructing a lapse function and shift vector which effectively determine a frame co-rotating with the binary. In order for this co-rotating frame to match an approximate stationary spacetime, GGB suggest that quasi-circular orbits are those for which the total ADM and Komar masses agree. With this construct, GGB were able to identify an ISCO. It is interesting to point out that the approximate helical Killing vector sequences yield an ISCO identification in better agreement with post-Newtonian results [26] when compared with the effective potential sequences. The reasons for discrepancies are not well understood.

The future effort on constructing initial data sets of binary compact objects will likely be concentrated on relaxing the simplifying assumptions, in particular the conformal flatness condition. A formal definition of the degree of astrophysical realism in a data set is difficult a task, in particular given the absence of numerical codes capable of starting evolutions at the point when post-Newtonian approximations breakdown. Constructing initial data becomes an educated guess exercise. Recently there have been interesting attempts to produce more realistic initial data. For example, Tichy and collaborators[65] constructed initial data based on post-Newtonian expansions. More recently, Yunes and collaborators[72] introduced an approximate metric for a binary black hole spacetime that could be used as the freely specifiable data to construct initial data. This approximate metric consists of a post-Newtonian metric for a binary system asymptotically matched to a perturbed Schwarzschild metric for each hole.

It is not clear how sensitive the late evolution of a binary system would be to the initial data. It could very well be that nature is forgiven against *small* mistakes. That a certain amount of spurious radiation is tolerated, radiation that could be flushed away quick enough. Of course all of this is dependent on the questions we are trying to answer. As the quality of the observational data improves, the demand for higher accuracy in the simulations will also increase.

5. Gauge Conditions

Once initial data, evolution equations, black hole excision and robust numerical algorithms are in place, suitable coordinates must be chosen before any numerical evolution could take place. In the context of 3+1 formulations, a choice of coordinates amounts to a recipe for fixing the lapse function α and the shift vector β^i. The choice of α and β^i determines the foliation. Even though, the geometry of a spacetime does not depend on the foliation, in numerical relativity, it is of fundamental importance to prescribe lapse functions and shift vectors that generate foliations covering as much as possible the future development of the initial data. In particular, for the evolution of black holes, the construction of a foliation must be such that singularities and the artificial growth of metric functions due to gauge effects are avoided.

Gauge conditions can be classified into three general types.[38] If α and β^i are determined from four conditions $F_\mu(x^\nu, \alpha, \beta^i) = 0$, the gauge is called *fixed*. An example of this type of gauge is the so-called geodesic slicing ($\alpha = 1$, $\beta^i = 0$), where coordinate observers are free-falling. Because of this free-falling property, geodesic slicing should be avoided in the presence of singularities. Another example of fixed gauges is the synchronous gauge ($\alpha = \alpha(t)$, $\beta^i = 0$).

The second type of gauges are called *algebraic*. The four conditions in this case depend in general in the spatial metric h_{ij}, extrinsic curvature K_{ij} and their derivatives, namely $F_\mu(x^\nu, \alpha, \beta^i, h_{ij}, K_{ij}, \partial_\nu h_{ij} \ldots) = 0$. An example of this type of gauges is the popular slicing condition $\alpha = 1 + \ln h$ often called "1+log" slicing.

Finally, we have *differential* gauges, namely those gauge conditions involving PDEs for α and β^i. They are the most popular and effective gauge conditions. Within the differential gauges, one has those consisting of elliptic equations and those involving evolution equations, both hyperbolic or parabolic. Examples of differential elliptic gauges are respectively the maximal slicing and the minimal distortion shift:

$$\Delta\alpha = \alpha\,K_{ij}\,K^{ij} \tag{21}$$

$$(\Delta_L\beta)^i = 2\,A^{ij}\,\nabla_j\alpha + \frac{4}{3}\alpha\,\nabla^i K\,. \tag{22}$$

Although the computational cost of solving elliptic equations during an evolution step has been reduced in recent years, there is still hesitation using these differential elliptic gauges. An alternative has been to turn these elliptic equations into parabolic equations. For instance, the maximal slicing conditions (21) can be rewritten as:

$$\partial_\lambda\alpha = \Delta\alpha - \alpha\,K_{ij}\,K^{ij}\,, \tag{23}$$

with λ playing the role of a temporal parameter. As $\partial_\lambda\alpha \to 0$, this gauge conditions approached the maximal slicing conditions. Conditions (23) is also known as the "K-driver" condition.[9] A similar approach involving the minimal distortion shift (22) is called the "Γ-driver" condition.[2] Modifications to the Γ and K driver conditions have played a key role in simulations of both neutron star[55,43] and black hole[21] binaries in which the numerical evolutions is performed in a co-rotating frame with the binary.

Finally, there is a class of differential gauges that are becoming increasingly popular. These are generalizations of the *harmonic gauge* $\Box x^\mu = 0$ mentioned in Sec. 2 in connection with formulations of the Einstein equations. In the context of 3+1 formulations, the harmonic gauge takes the form

$$(\partial_t - \beta^j \partial_j)\,\alpha = -\alpha^2\, K \tag{24}$$

$$(\partial_t - \beta^j \partial_j)\,\beta^i = -\alpha^2\,(\nabla^i \ln \alpha + h^{jk}\,\Gamma^i_{jk})\,. \tag{25}$$

A commonly used modification to (24) is obtained by setting $\beta^i = 0$:

$$\partial_t \alpha = -\alpha^2\, f(\alpha)\, K\,. \tag{26}$$

Depending on the choice of $f(\alpha)$, this condition reduces to geodesic, harmonic or 1+log slicings.

In spite of the success of the gauge conditions here mentioned, development of new gauge conditions is very likely to continue as we consider space-times of increasing complexity. An interesting possibility could the use of hybrid gauge conditions, namely smooth superpositions of different gauge conditions.

6. Extracting Observables: Connecting with Data Analysis

Numerical relativity is in a transition stage, a transition from proof of concept simulations aimed at testing stability properties of formulations, excision methodology, gauge conditions, etc. to a phase in which one is able to extract or compute information about observables. More and more the attention has been shifted to improving the computation of gravitational waveforms as well as intrinsic properties of the compact objects such as mass and angular momentum.

Regarding wave extraction, until recently, the simplest and most popular method has been via the Zerilli-Moncrief formalism.[45,1] This method relies, however, on the assumption that, in the region where the wave extraction takes place, the space-time can be approximated as perturbations of a single, non-rotating black hole. Therefore, the extraction requires an estimate of the background black hole mass. An alternative is to use the Teukolsky formalism.[63] This approach has been used as basis for the *Lazarus* approach to wave extraction. In general terms the Lazarus approach consists of attaching at the end of a non-linear numerical simulation a single black hole close-limit approximation. This method made it possible to extend the short lived of numerical simulations. The need for using this approach is decreasing as the non-linear codes are capable to produce long term and stable simulations. Another wave extraction approach that is becoming increasingly popular is based on the Newman-Penrose formalism. Because it requires taking second derivatives, one must be careful when using this approach to avoid spurious numerical noise.[27,58] Just recently, Beetle and Burko[13] introduced a background-independent scalar curvature invariant that has the potential to become a very useful tool for wave extraction.

An approach with great potential is the Cauchy-characteristic extraction.[69] By matching an evolution based on outgoing characteristics to a Cauchy one, one can in principle compute gravitational waves at future null infinity. Unfortunately, this approach has been difficult to implement. The problems are not of conceptual nature but technical. It requires a highly non-trivial infrastructure in the interface between the Cauchy and characteristic domains. Currently, the most successful implementations have involved non-linear waves[16] and with a linearized Einstein harmonic system.[62]

Waveforms are certainly a very important byproduct in a simulation. However, equally relevant is to be able to extract physical information by other means. The *isolated* and *dynamical-horizon* framework developed by Ashtekar and collaborators[7] provides tools to extract information such as the mass and angular momentum of coalescing compact binaries. There is increase usage of this framework in numerical relativity. For instance, in the case of simulations of rotating neutron-star collapse to a Kerr black hole, the isolated and dynamical horizon framework has yield accurate and robust estimates of the mass and angular momentum of the resulting black hole.[8] Similarly, the framework was also applied with great success at the black hole produced in the merger from dynamical evolutions of quasi-circular binary black hole data.[5]

Finally, it is important to mention that the extraction and quality of physical information from numerical simulations depends strongly on proper treatment of outer boundary conditions. The Lazarus approach is an attempt to *push* away the location of the outer boundary and minimize the effects of reflections from the outer boundaries. The Cauchy-perturbative matching is another example of moving the location of the outer boundary as far as possible from the sources of radiations.[50] Independently of the type of boundary conditions used, it is important that those boundary conditions ensure constraint propagation.[51]

7. Imagining the Future

"It is tough to make predictions, especially about the future."[b] Nonetheless, given the current pace and direction of research in numerical relativity, extrapolating into the future and imagining what the field will look like in five or even ten years is not a futile exercise. For instance, we knew that when the day that adaptive or fixed mesh refinement technology arrived, it would open the door to consider simulations of orbiting compact objects at separations significantly beyond the innermost stable circular orbit. This has just recently become a reality.[48]

In my opinion, in the next few years, we will witness the development of codes capable of producing the last few orbits, merger and ringdown of compact object binaries. There will not be a unique formalism, numerical method or gauge condition. As results are produced, researchers will be able to borrow and adapt methodologies to their favorite approaches.

We will also be able to answer the question about what degree of astrophysically accuracy will be needed in initial data. Put in other words, can we construct initial data in which the "errors" due to our ignorance are quickly radiated away without a significant effect on the outcome of the simulation? This is related to the issue of estimating the level of errors, including numerical, that can be tolerated without seriously compromising the utility of numerical relativity results for data analysis.

Another area that is very likely to experience intense activity is the creation of open source problem solving environments. The Cactus[22] and Lorane[41] infrastructures are prime examples. Although there are a significant number of groups in numerical relativity currently using Cactus or at least its flesh, there has been some degree of resistance to fully embrace it. Reasons cited are the limitations in flexibility and degree of complexity. The next generation of problem solving environments will likely accommodate a broader range of users. It will be mature enough, so researcher will be able to do science with the infrastructure as it stands. At the same time, it will provide a suitable platform for researchers to carry out exploratory and innovative approaches to numerical relativity.

There is energy, and it is not *dark energy*, that is accelerating the "Universe of Numerical Relativity."

---

[b]Quote attributed to Niels Bohr and also to Yogi Berra

Acknowledgments

Work supported in part by NSF grant PHY-0244788 and the Center for Gravitational Wave Physics funded by the National Science Foundation under Cooperative Agreement PHY-0114375.

Appendix A. The ADM Formulation

This Appendix is intended as a brief introduction to the standard approach in numerical relativity of rewriting the Einstein equations in a space+time (3+1) form. Among all the 3+1 formulations developed, I will only discuss here the Arnowitt-Deser-Misner (ADM)[6] formulation. A word of caution is needed at this point. It is not recommendable to develop numerical codes based on the ADM formulation. These codes are notable because of their susceptibility to instabilities. However, from a pedagogical point of view, the ADM formulation, because of its simplicity, provides a good introduction to the subject of 3+1 formulations of the Einstein equations. For details and discussion of other formulations, I recommend the review by Baumgarte and Shapiro[12] and at the more mathematical level the review by Reula[49]. Readers interested on the big picture should skip this section.

A 3+1 decomposition of the Einstein equations begins with the construction of a foliation. The foliation consists of 3D level, space-like hypersurfaces of a scalar function. Given this foliation, the space-time metric can be rewritten as $g_{\mu\nu} = h_{\mu\nu} - n_\mu n_\nu$, where $h_{\mu\nu}$ is the intrinsic metric of the space-like hypersurfaces in the foliation and n^μ their time-like unit normal. The metric $h_{\mu\nu}$ only describes the internal geometry of the hypersurfaces. The embedding of these hypersurfaces in the 4D space-time is characterized by the extrinsic curvature $K_{\mu\nu}$, defined as $K_{\mu\nu} = -\nabla_{(\mu} n_{\nu)}$, where ∇_μ is the covariant derivative induced by the 3-metric $h_{\mu\nu}$. The intrinsic metric and the extrinsic curvature pair $(h_{\mu\nu}, K_{\mu\nu})$ represents the dynamical quantities in Einstein's theory.

If one introduces a 3+1 basis of vectors, one can equivalently write the 4-metric as the line element:

$$ds^2 = -\alpha^2 dt^2 + h_{ij}(dx^i + \beta^i dt)(dx^j + \beta^j dt). \tag{A.1}$$

In this line element, the scalar α, called the lapse function, represents the freedom of choosing arbitrarily time coordinates. The vector β^i, called the shift vector, contains the freedom of relabeling coordinate points in the space-like hypersurfaces. The lapse function and the shift vector are kinematical variables.

The evolution equations for the pair (h_{ij}, K_{ij}) are obtained from the definition of the extrinsic curvature and the (space, space) components of Einstein's equations. For vacuum spacetimes, these equations read

$$\partial_o h_{ij} = -2\alpha K_{ij} \,, \tag{A.2}$$

$$\partial_o K_{ij} = -\nabla_i \nabla_j \alpha + \alpha(R_{ij} + K K_{ij} - 2K_{ik}K^k{}_j), \tag{A.3}$$

where R_{ij} is the 3-Ricci tensor, R is its trace, K is the trace of K_{ij}, and $\partial_o \equiv \partial_t - \pounds_\beta$. Notice that because $h_{\mu\nu} n^\mu = 0$ and $K_{\mu\nu} n^\mu = 0$, we have restricted our attention to the pair (h_{ij}, K_{ij}).

The evolution equations (A.2) and (A.3) are commonly known as the ADM or $\dot{g}$ and $\dot{K}$ equations. However, it is important to point out that Eqs. (A.2) and (A.3) as they stand are not found in the original ADM work.[6] The first derivation of the $\dot{g}$ and $\dot{K}$ equations was done by York.[70]

The remaining (time, time) and (time, space) components of Einstein's equations yield the

Hamiltonian Constraint: $R + K^2 - K_{ij}K^{ij} = 0 \,,$ (A.4)

Momentum Constraint: $\nabla_j(K^{ij} - g^{ij}K) = 0 \,.$ (A.5)

Notice that the Hamiltonian and momentum constraint equations do not involve second time derivatives of the 3-metric nor the lapse function and shift vector. These equations represent conditions on the choices of initial data for the pair (h_{ij}, K_{ij}). At the continuum level, the equations (A.2) and (A.3) ensure that these constraints are preserved during the evolution.

The 3+1 approach naturally splits the task of solving Einstein's equations into:

- *The Initial Data Problem:* Construction of the initial values of (h_{ij}, K_{ij}) that satisfies the Hamiltonian (A.4) and momentum (A.5) constraints.
- *The Evolution Problem:* Evolution of the pair (h_{ij}, K_{ij}) using equations (A.2) and (A.3), respectively.

Appendix B. York's Conformal Approach

For pedagogical reasons, I will only review the York's conformal method,[70] also known as the conformal transverse traceless (CTT) decomposition. Generally speaking, the main challenge in constructing initial data is identifying which four *pieces* within the twelve components of the initial data, (h_{ij}, K_{ij}), are to be solved from the four constraint equations (A.4) and (A.5). The remaining eight pieces are freely specifiable. In some particular instances,[68,11] it is clear which metric or extrinsic curvature components are fixed by the constraints; however, for a general situation, such as finding initial data for collision of black holes, this choice is unclear.

The starting point of York's CTT method is a conformal transformation of the 3-metric of the initial hypersurface

$$h_{ij} = \phi^4 \hat{h}_{ij}, \tag{B.1}$$

where conformal quantities are denoted with hats. Next, the extrinsic curvature is decomposed into its trace and trace-free parts:

$$K_{ij} = A_{ij} + \frac{1}{3} h_{ij} K, \tag{B.2}$$

This decomposition is followed by a conformal transformation of the trace-free part of the extrinsic curvature:

$$A_{ij} = \phi^{-2} \hat{A}_{ij}. \tag{B.3}$$

The above conformal transformation is chosen, so the divergence of A^{ij}, which enters in the momentum constraint, has the following simple transformation property:

$$\nabla_j A^{ij} = \phi^{-10} \hat{\nabla}_j \hat{A}^{ij}, \tag{B.4}$$

where $\hat{\nabla}_i$ is the covariant derivative associated with the conformal 3-metric $\hat{h}_{ij}$. Finally, $\hat{A}_{ij}$ is split into its transverse and longitudinal parts,

$$\hat{A}^{ij} = \hat{A}^{ij}_\star + (\hat{l} W)^{ij}, \tag{B.5}$$

where by construction

$$\hat{\nabla}_j \hat{A}^{ij}_\star = 0, \tag{B.6}$$

and

$$(\hat{l} W)^{ij} \equiv \hat{\nabla}^i W^j + \hat{\nabla}^j W^i - \frac{2}{3} \hat{g}^{ij} \hat{\nabla}_k W^k. \tag{B.7}$$

Notice that the vector W^i is the generator of the longitudinal part of the extrinsic curvature $\hat{A}_{ij}$. With the transformations and decompositions above, the Hamiltonian and momentum constraints take the form

$$8 \hat{\Delta} \phi = \hat{R} \phi - \hat{A}_{ij} \hat{A}^{ij} \phi^{-7} + \frac{2}{3} K^2 \phi^5 \tag{B.8}$$

$$(\hat{\Delta}_l W)^i = \frac{2}{3} \phi^6 \hat{\nabla}^i K, \tag{B.9}$$

respectively, where $(\hat{\Delta}_l W)^i \equiv \hat{\nabla}_j (\hat{l} W)^{ij}$ and $\hat{\Delta} \equiv \hat{\nabla}^i \hat{\nabla}_i$. Constructing initial data sets consists then of first freely specifying the conformal metric

$\hat{h}_{ij}$, trace of the extrinsic curvature K and the transverse-traceless part of the extrinsic curvature $\hat{A}_{ij}^*$. Once these data have been specified, one solves Eqs. (B.8) and (B.9) for the conformal factor ϕ and the vector W^i. Notice that these equations are coupled via the terms involving $\hat{A}_{ij}$ in (B.8) and ϕ in (B.9).

References

1. A. Abrahams, D. Bernstein, D. Hobill, E. Seidel, and L. Smarr. *Phys. Rev. D*, 45:3544, 1992.
2. G. Alcubierre, M. Allen, B. Bruegmann, E. Seidel, and W.-M Suen. *Phys. Rev. D*, 62:124011, 2000.
3. M. Alcubierre, B. Bernd Bruegmann, D. Pollney, E. Seidel, and R. Takahashi. Black hole excision for dynamic black holes. *Phys. Rev. D*, 64:061501, 2001.
4. M. Alcubierre and B. Bruegmann. *Phys. Rev. D*, 63:104006, 2001.
5. M. Alcubierre, B. Bruegmann, P. Diener, D., F.S. Guzman, I. Hawke, S. Hawley, F. Herrmann, M. Michael Koppitz, D. Pollney, E. Seidel, and J. Thornburg. Dynamical evolution of quasi-circular binary black hole data. *gr-qc/0411149*, 2004.
6. R. Arnowitt, S. Deser, and C.W. Misner. The dynamics of general relativity. In L. Witten, editor, *Gravitation: Introduction to Current Research*, New York, 1962. Willey.
7. A. Ashtekar and B. Krishnan. Isolated and dynamical horizons and their applications. *Living Rev. Rel.*, 7:10, 2004.
8. L. Baiotti, I. Hawke, P.J. Montero, F. Loeffler, L. Rezzolla, N. Stergioulas, J.A. Font, and E. Seidel. Three-dimensional relativistic simulations of rotating neutron-star collapse to a kerr black hole. *Phys. Rev. D.*, 71:024035, 2005.
9. J. Balakrishna, G. Daues, E. Seidel, W.-M. Suen, M. Tobias, and E. Wang. *Class. Quant. Grav.*, 13:L135, 1996.
10. T. Bamgarte and Shapiro S.L. *Phys. Rev. D*, 59:024007, 1999.
11. J.M. Bardeen and T. Piran. General relativistic axisymmetric rotating systems: Coordinates and equations. *Phys. Reports*, 96:20, 1983.
12. T.W. Baumgarte and S.L. Shapiro. Numerical relativity and compact binaries. *Physics Reports*, 376:41, 2003.
13. C. Beetle and L.M. Burko. A radiation scalar for numerical relativity. *Phys. Rev. Lett.*, 89:271101, 2002.
14. K.B. Berger. Numerical approaches to spacetime singularities. *Living Rev. Rel.*, 5:1, 2002.
15. D.H. Bernstein, D.W. Hobill, and L.L. Smarr. Black hole spacetimes: Testing numerical relativity. In C.R. Evans, L.S. Finn, and D.W. Hobill, editors, *Frontiers in Numerical Relativity*, pages 57–73, Cambridge, 1989. Cambridge Univ. Press.
16. N. Bishop, R. Gomez, P. Holvorcem, R. Matznerr, P. Papadopoulos, and J. Winicour. *J. Comp. Phys.*, 136:140, 1997.

17. C. Bona, C. Ledvinka, C. Palenzuela, and M. Zacek. General-covariant evolution formalism for numerical relativity. *Phys. Rev D*, 68:041501, 2003.
18. J.M. Bowen and J.W. York. Time-asymetric initial data for black holes and black-hole collisions. *Phys. Rev. D*, 21:2047, 1980.
19. S. Brandt and B. Bruegmann. *Phys. Rev. Lett.*, 78:3606, 1997.
20. O. Brodbeck, S. Frittelli, P. Huebner, and O.A. Reula. Einstein's equations with asymptotically stable constraint propagation. *J. Math. Phys.*, 40:909, 1999.
21. B. Bruegmann, W. Tichy, and N. Jansen. Numerical simulation of orbiting black holes. *Phys. Rev. Lett.*, 92:211101, 2004.
22. Cactus. www.cactuscode.org.
23. M.W. Choptuik. Universality and scaling in gravitational collapse of a massless scalar field. *Phys. Rev. Lett.*, 70:9, 1993.
24. G. Cook. *Phys. Rev. D*, 50:5025, 1994.
25. G. Cook. Initial data for numerical relativity. *Living Rev. Rel.*, 3:5, 2000.
26. T. Damour, E. Gourgoulhon, and P. Grandclement. Circular orbits of corotating binary black holes: comparison between analytical and numerical results. *Phys. Rev. D*, 66:024007, 2002.
27. D.R. Fiske, J.G. Baker, J.R. van Meter, D.-I. Choi, and J.M. Centrella. Wave zone extraction of gravitational radiation in three-dimensional numerical relativity. *gr-qc/0503100*, 2005.
28. Jörg Frauendiener. Conformal infinity. *Living Rev. Rel.*, 7:1, 2004.
29. H. Friedrich. Hyperbolic reductions for einstein's field equations. *Class. Quantum Grav.*, 13:1451, 1996.
30. D. Garfinkle. Harmonic coordinate methods for simulating generic singularities. *Phys. Rev. D*, 65:044029, 2002.
31. E. Gourgoulhon, P. Grandclement, and Bonazzola S. *Phys. Rev. D*, 65:044020, 2002.
32. P. Grandclement, E. Gourgoulhon, and Bonazzola S. *Phys. Rev D*, 65:044021, 2002.
33. C. Gundlach. Critical phenomena in gravitational collapse. *Living Rev. Rel.*, 2:4, 1999.
34. C. Gundlach and J.M. Martin-Garcia. Symmetric hyperbolicity and consistent boundary conditions for second-order einstein equations. *Phys. Rev. D*, 70:044032, 2004.
35. C. Gundlach, J.M. Martin-Garcia, G. Calabrese, and I. Hinder. Constraint damping in the z4 formulation and harmonic gauge. *gr-qc/0504114*, 2005.
36. M. Holst, L. Lindblom, R. Owen, H.P. Pfeiffer, M.A. Scheel, and L.E. Kidder. Optimal constraint projection for hyperbolic evolution systems. *Phys. Rev. D*, 70:084017, 2004.
37. B. Kelly, P. Laguna, K. Lockitch, J. Pullin, E. Schnetter, D. Shoemaker, and M Tiglio. A cure for unstable numerical evolutions of single black holes: adjusting the standard adm equations. *Phys. Rev D*, 64:084013, 2001.
38. A.M. Khokhlov and I.D. Novikov. Gauge stability of 3+1 formulations of general relativity. *Class. Quant. Grav.*, 19:827, 2002.
39. L.E. Kidder, M.A. Scheel, and S.A. Teukolsky. *Phys. Rev D*, 64:064017, 2001.

40. A. Lichnerowicz. *J. Math. Pure Appl.*, 23:37, 1944.
41. Lorane. www.lorene.obspm.fr.
42. P. Marronetti, M.D. Duez, S.L. Shapiro, and T.W. Baumgarte. Dynamical determination of the innermost stable circular orbit of binary neutron stars. *Phys. Rev. Lett.*, 92:141101, 2004.
43. M. Miller, P. Gressman, and W.-M. Suen. Towards a realistic neutron star binary inspiral: Initial data and multiple orbit evolution in full general relativity. *Phys. Rev. D*, 69:064026, 2004.
44. C.W. Misner, K.S. Thorne, and J.A. Wheeler. *Gravitation.* Freeman, San Francisco, 1973.
45. V. Moncrief. *Ann. Phys.*, 88:323, 1974.
46. G. Nagy, O.E. Ortiz, and O.A. Reula. Strongly hyperbolic second order einstein's evolution equations. *Phys. Rev D*, 70:044012, 2004.
47. F. Pretorius. Numerical relativity using a generalized harmonic decomposition. *Class. Quant. Grav.*, 22:425, 2005.
48. F. Pretorius. Toward binary black hole simulations in numerical relativity. *talk at BIRS Numerical Relativity Workshop*, 2005.
49. O. Reula. Hyperbolic methods for einstein's equations. *Living Rev. Rel.*, 1:3, 1998.
50. L. Rezzolla, A.M. Abrahams, R.A. Matzner, M.E. Rupright, and S.L. Shapiro. Cauchy-perturbative matching and outer boundary conditions: computational studies. *Phys. Rev. D*, 59:064001, 1999.
51. O. Sarbach and M. Tiglio. Boundary conditions for einstein's field equations: Analytical and numerical analysis. *gr-qc/0412115*, 2004.
52. E. Seidel and W.-M. Suen. *Phys. Rev. Lett.*, 69:1845, 1992.
53. E. Seidel and W.-M. Suen. *J. Comp. Phys.*, 112:44, 1994.
54. M. Shibata and T. Nakamura. *Phys. Rev. D*, 52:5428, 1995.
55. M. Shibata, K. Taniguchi, and K. Uryu. Merger of binary neutron stars with realistic equations of state in full general relativity. *Phys. Rev. D*, 71:084021, 2005.
56. D. Shoemaker, K. Smith, U. Sperhake, P. Laguna, E. Schnetter, and Fiske D. Moving black holes via singularity excision. *Class. Quant. Grav.*, 20:3729, 2003.
57. L. Smarr and J.W. York. Kinematical conditions in the construction of spacetime. *Phys. Rev. D*, 17:2529, 1978.
58. U. Sperhake, B. Kelly, P. Laguna, K.L. Smith, and E. Schnetter. Black hole head-on collisions and gravitational waves with fixed mesh-refinement and dynamic singularity excision. *gr-qc/0503071*.
59. U. Sperhake and P. Laguna. Boosted black holes with dynamical excision. *in preparation*.
60. U. Sperhake, K.L. Smith, B. Kelly, P. Laguna, and D. Shoemaker. Impact of densitized lapse slicings on evolutions of a wobbling black hole. *Phys. Rev D*, 69:024012, 2004.
61. B. Szilagyi and Winicour J. Well-posed initial-boundary evolution in general relativity. *Phys. Re. D*, 68:041501, 2003.
62. B. Szilagyi, B. Schmidt, and J. Winicour. *Phys. Rev. D*, 65:064015, 2002.

63. S.A. Teukolsky. *Astrophys. J*, 185:635, 1973.

64. J. Thornburg. *Class. Quantum Grav.*, 4:1119, 1987.

65. W. Tichy, B. Bruegmann, M. Campanelli, and P. Diener. Binary black hole initial data for numerical relativity based on post-newtonian data. *Phys. Rev. D*, 61:104015, 2000.

66. M. Tiglio. Dynamical control of the constraints growth in free evolutions of the einstein's equations. *gr-qc/0304062*, 2003.

67. R.W. Wald. *General Relativity*. The University of Chicago Press, Chicago, 1984.

68. J. Wilson. A numerical method for relativistic hydrodynamics. In L. Smarr, editor, *Sources of Gravitational Radiation*, page 423, Cambridge, 1979. Cambridge Univ. Press.

69. J. Winicour. Characteristic evolution and matching. *Living Rev. Rel.*, 1:5, 1998.

70. J.M. York. Kinematic and dynamics in general relativity. In L. Smarr, editor, *Sources of Gravitational Radiation*, page 83, Cambridge, 1979. Cambridge Univ. Press.

71. J.M. York. Initial data for n black holes. *Physica*, 124A:629, 1984.

72. N. Yunes, W. Tichy, B.J. Owen, and B. Bruegmann. Binary black hole initial data from matched asymptotic expansions. *qr-qc/0503011*.

CHAPTER 7

UNDERSTANDING OUR UNIVERSE: CURRENT STATUS AND OPEN ISSUES

T. PADMANABHAN

IUCAA P.O. Box 4, Pune University Campus,
Ganeshkhind, Pune - 411 007, India
nabhan@iucaa.ernet.in

Last couple of decades have been the golden age for cosmology. High quality data confirmed the broad paradigm of standard cosmology but have thrusted upon us a preposterous composition for the universe which defies any simple explanation, thereby posing probably the greatest challenge theoretical physics has ever faced. Several aspects of these developments are critically reviewed, concentrating on conceptual issues and open questions.

1. Prologue: Universe as a Physical System

Attempts to understand the behaviour of our universe by applying the laws of physics lead to difficulties which have no parallel in the application of laws of physics to systems of more moderate scale — like atoms, solids or even galaxies. We have only one universe available for study, which itself is evolving in time; hence, different epochs in the past history of the universe are unique and have occurred only once. Standard rules of science, like repeatability, statistical stability and predictability cannot be applied to the study of the entire universe in a naive manner.

The obvious procedure will be to start with the current state of the universe and use the laws of physics to study its past and future. Progress in this attempt is limited because our understanding of physical processes at energy scales above 100 GeV or so lacks direct experimental support. What is more, cosmological observations suggest that nearly 95 per cent of the matter in the universe is of types which have not been seen in the laboratory; there is also indirect, but definitive, evidence to suggest that nearly 70 per cent of the matter present in the universe exerts negative pressure.

These difficulties — which are unique when we attempt to apply the laws of physics to an evolving universe — require the cosmologists to proceed in a multi faceted manner. The standard paradigm is based on the idea that the universe was reasonably homogeneous, isotropic and fairly featureless — except for small fluctuations in the energy density — at sufficiently early times. It is then possible to integrate the equations describing the universe forward in time. The results will depend on only a small number (about half a dozen) of parameters describing the composition of the universe, its current expansion rate and the initial spectrum of density perturbations. Varying these parameters allows us to construct a library of evolutionary models for the universe which could then be compared with observations in order to restrict the parameter space. We shall now describe some of the details in this approach.

2. The Cosmological Paradigm

Observations show that the universe is fairly homogeneous and isotropic at scales larger than about $150h^{-1}$ Mpc, where 1 Mpc $\approx 3 \times 10^{24}$ cm is a convenient unit for extragalactic astronomy and $h \approx 0.7$ characterizes[1] the current rate of expansion of the universe in dimensionless form. (The mean distance between galaxies is about 1 Mpc while the size of the visible universe is about $3000h^{-1}$ Mpc.) The conventional — and highly successful — approach to cosmology separates the study of large scale ($l \gtrsim 150h^{-1}$ Mpc) dynamics of the universe from the issue of structure formation at smaller scales. The former is modeled by a homogeneous and isotropic distribution of energy density; the latter issue is addressed in terms of gravitational instability which will amplify the small perturbations in the energy density, leading to the formation of structures like galaxies.

In such an approach, the expansion of the background universe is described by a single function of time $a(t)$ which is governed by the equations (with $c = 1$):

$$\frac{\dot{a}^2 + k}{a^2} = \frac{8\pi G\rho}{3}; \qquad d(\rho a^3) = -pda^3 \tag{1}$$

The first one relates expansion rate to the energy density ρ and $k = 0, \pm 1$ is a parameter which characterizes the spatial curvature of the universe. The second equation, when coupled with the equation of state $p = p(\rho)$ which relates the pressure p to the energy density, determines the evolution of energy density $\rho = \rho(a)$ in terms of the expansion factor of the universe.

In particular if $p = w\rho$ with (at least, approximately) constant w then, $\rho \propto a^{-3(1+w)}$ and (if we further assume $k = 0$, which is strongly favoured by observations) the first equation in Eq.(1) gives $a \propto t^{2/[3(1+w)]}$. We will also often use the redshift $z(t)$, defined as $(1 + z) = a_0/a(t)$ where the subscript zero denotes quantities evaluated at the present moment.

It is convenient to measure the energy densities of different components in terms of a *critical energy density* (ρ_c) required to make $k = 0$ at the present epoch. (Of course, since k is a constant, it will remain zero at all epochs if it is zero at any given moment of time.) From Eq.(1), it is clear that $\rho_c = 3H_0^2/8\pi G$ where $H_0 \equiv (\dot{a}/a)_0$ — called the Hubble constant — is the rate of expansion of the universe at present. The variables $\Omega_i \equiv \rho_i/\rho_c$ will give the fractional contribution of different components of the universe (i denoting baryons, dark matter, radiation, etc.) to the critical density. Observations then lead to the following results:

(1) Our universe has $0.98 \lesssim \Omega_{tot} \lesssim 1.08$. The value of Ω_{tot} can be determined from the angular anisotropy spectrum of the cosmic microwave background radiation (CMBR; see Section 5) and these observations (combined with the reasonable assumption that $h > 0.5$) show[2,3] that we live in a universe with critical density, so that $k = 0$.

(2) Observations of primordial deuterium produced in big bang nucleosynthesis (which took place when the universe was about few minutes in age) as well as the CMBR observations show[4] that the *total* amount of baryons in the universe contributes about $\Omega_B = (0.024 \pm 0.0012)h^{-2}$. Given the independent observations[1] which fix $h = 0.72 \pm 0.07$, we conclude that $\Omega_B \cong 0.04 - 0.06$. These observations take into account all baryons which exist in the universe today irrespective of whether they are luminous or not. *Combined with previous item we conclude that most of the universe is non-baryonic.*

(3) Host of observations related to large scale structure and dynamics (rotation curves of galaxies, estimate of cluster masses, gravitational lensing, galaxy surveys ..) all suggest[5] that the universe is populated by a non-luminous component of matter (dark matter; DM hereafter) made of weakly interacting massive particles which *does* cluster at galactic scales. This component contributes about $\Omega_{DM} \cong 0.20 - 0.35$ and has the simple equation of state $p_{DM} \approx 0$. (In the relativistic theory, the pressure $p \propto mv^2$ is negligible compared to energy density $\rho \propto mc^2$ for non relativistic particles.). The second equation in Eq.(1), then gives $\rho_{DM} \propto a^{-3}$ as the universe expands which arises from the evolution of number density of particles: $\rho = nmc^2 \propto n \propto a^{-3}$.

(4) Combining the last observation with the first we conclude that there must be (at least) one more component to the energy density of the universe contributing about 70% of critical density. Early analysis of several observations[6] indicated that this component is unclustered and has negative pressure. This is confirmed dramatically by the supernova observations (see Ref. 7; for a critical look at the data, see Ref. 8). The observations suggest that the missing component has $w = p/\rho \lesssim -0.78$ and contributes $\Omega_{DE} \cong 0.60 - 0.75$. The simplest choice for such *dark energy* with negative pressure is the cosmological constant which is a term that can be added to Einstein's equations. This term acts like a fluid with an equation of state $p_{DE} = -\rho_{DE}$; the second equation in Eq.(1), then gives $\rho_{DE} = $ constant as universe expands.

(5) The universe also contains radiation contributing an energy density $\Omega_R h^2 = 2.56 \times 10^{-5}$ today most of which is due to photons in the CMBR. The equation of state is $p_R = (1/3)\rho_R$; the second equation in Eq.(1), then gives $\rho_R \propto a^{-4}$. Combining it with the result $\rho_R \propto T^4$ for thermal radiation, it follows that $T \propto a^{-1}$. Radiation is dynamically irrelevant today but since $(\rho_R/\rho_{DM}) \propto a^{-1}$ it would have been the dominant component when the universe was smaller by a factor larger than $\Omega_{DM}/\Omega_R \simeq 4 \times 10^4 \Omega_{DM} h^2$.

(6) Together we conclude that our universe has (approximately) $\Omega_{DE} \simeq 0.7, \Omega_{DM} \simeq 0.26, \Omega_B \simeq 0.04, \Omega_R \simeq 5 \times 10^{-5}$. All known observations are consistent with such an — admittedly weird — composition for the universe.

Using $\rho_{NR} \propto a^{-3}, \rho_R \propto a^{-4}$ and $\rho_{DE}=$constant we can write Eq.(1) in a convenient dimensionless form as

$$\frac{1}{2}\left(\frac{dq}{d\tau}\right)^2 + V(q) = E \tag{2}$$

where $\tau = H_0 t, a = a_0 q(\tau), \Omega_{\mathrm{NR}} = \Omega_B + \Omega_{\mathrm{DM}}$ and

$$V(q) = -\frac{1}{2}\left[\frac{\Omega_R}{q^2} + \frac{\Omega_{\mathrm{NR}}}{q} + \Omega_{DE} q^2\right]; \quad E = \frac{1}{2}\left(1 - \Omega_{\mathrm{tot}}\right). \tag{3}$$

This equation has the structure of the first integral for motion of a particle with energy E in a potential $V(q)$. For models with $\Omega = \Omega_{\mathrm{NR}} + \Omega_{DE} = 1$, we can take $E = 0$ so that $(dq/d\tau) = \sqrt{V(q)}$. Based on the observed composition of the universe, we can identify three distinct phases in the evolution of the universe when the temperature is less than about 100 GeV. At high redshifts (small q) the universe is radiation dominated and $\dot{q}$ is independent of the other cosmological parameters. Then Eq.(2) can be easily integrated to give $a(t) \propto t^{1/2}$ and the temperature of the universe decreases as $T \propto t^{-1/2}$. As the universe expands, a time will come when ($t = t_{\rm eq}$, $a = a_{\rm eq}$ and $z = z_{\rm eq}$, say) the matter energy density will be comparable to radiation energy density. For the parameters described above, $(1 + z_{eq}) = \Omega_{NR}/\Omega_R \simeq 4 \times 10^4 \Omega_{DM} h^2$. At lower redshifts, matter will dominate over radiation and we will have $a \propto t^{2/3}$ until fairly late when the dark energy density will dominate over non relativistic matter. This occurs at a redshift of $z_{\rm DE}$ where $(1 + z_{\rm DE}) = (\Omega_{\rm DE}/\Omega_{\rm NR})^{1/3}$. For $\Omega_{\rm DE} \approx 0.7, \Omega_{\rm NR} \approx 0.3$, this occurs at $z_{\rm DE} \approx 0.33$. In this phase, the velocity $\dot{q}$ changes from being a decreasing function to an increasing function leading to an accelerating universe (see Fig.2). In addition to these, we believe that the universe probably went through a rapidly expanding, inflationary, phase very early when $T \approx 10^{14}$ GeV; we will say more about this in Section 4. (For a textbook description of these and related issues, see e.g. Ref. 9.)

3. Growth of Structures in the Universe

Having discussed the dynamics of the smooth universe, let us turn our attention to the formation of structures. In the conventional paradigm for the formation of structures in the universe, some mechanism is invoked to generate small perturbations in the energy density in the very early phase of the universe. These perturbations then grow due to gravitational instability and eventually form the structures which we see today. Such a scenario is constrained most severely by CMBR observations at $z \approx 10^3$. Since the perturbations in CMBR are observed to be small ($10^{-5} - 10^{-4}$ depending on the angular scale), it follows that the energy density perturbations were small compared to unity at the redshift of $z \approx 10^3$.

The central quantity one uses to describe the growth of structures is the *density contrast* defined as $\delta(t, \mathbf{x}) = [\rho(t, \mathbf{x}) - \rho_{\rm bg}(t)]/\rho_{\rm bg}(t)$ which characterizes the fractional change in the energy density compared to the background. Since one is often interested in the statistical description of structures in the universe, it is conventional to assume that δ (and other related quantities) are elements of a statistical ensemble. Many popular models of structure formation suggest that the initial density perturbations in the early universe can be represented as a Gaussian random variable with zero mean and a given initial power spectrum. The latter quantity is defined through the relation $P(t, k) = < |\delta_k(t)|^2 >$ where $\delta_{\mathbf{k}}$ is the Fourier transform of $\delta(t, \mathbf{x})$ and $< ... >$ indicates averaging over the ensemble. The two-point correlation function $\xi(t, x)$ of the density distribution is defined as the Fourier transform of $P(t, \mathbf{k})$ over $\mathbf{k}$.

When the $\delta \ll 1$, its evolution can be studied by linear perturbation theory and each of the spatial Fourier modes $\delta_{\mathbf{k}}(t)$ will grow independently. Then the power spectra $P(k,t) = <|\delta_{\mathbf{k}}(t)|^2>$ at two different times in the linear regime are related by $P(k,t_f) = \mathcal{F}^2(k,t_f,t_i,\mathrm{bg})P(k,t_i)$ where $\mathcal{F}$ (called transfer function) depends only on the parameters of the background universe (denoted generically as "bg") but *not* on the initial power spectrum. The form of $\mathcal{F}$ is essentially decided by two factors: (i) The relative magnitudes of the proper wavelength of perturbation $\lambda_{\mathrm{prop}}(t) \propto a(t)$ and the Hubble radius $d_H(t) \equiv H^{-1}(t) = (\dot{a}/a)^{-1}$ and (ii) whether the universe is radiation dominated or matter dominated. At sufficiently early epochs, the universe will be radiation dominated and the proper wavelength $\lambda_{\mathrm{prop}}(t) \propto a \propto t^{1/2}$ will be larger than $d_H(t) \propto t$. The density contrast of such modes, which are bigger than the Hubble radius, will grow[9] as a^2 until $\lambda_{\mathrm{prop}} = d_H(t)$. (See the footnote on page 184.) When this occurs, the perturbation at a given wavelength is said to enter the Hubble radius. If $\lambda_{\mathrm{prop}} < d_H$ and the universe is radiation dominated, the matter perturbation does not grow significantly and increases at best only logarithmically.[9,10] Later on, when the universe becomes matter dominated for $t > t_{\mathrm{eq}}$, the perturbations again begin to grow. (Some of these details depend on the gauge chosen for describing the physics but, of course, the final observable results are gauge independent; we shall not worry about this feature in this article.)

It follows from this description that modes with wavelengths greater than $d_{\mathrm{eq}} \equiv d_H(t_{\mathrm{eq}})$ — which enter the Hubble radius only in the matter dominated epoch — continue to grow at all times; modes with wavelengths smaller than d_{eq} suffer lack of growth (in comparison with longer wavelength modes) during the period $t_{\mathrm{enter}} < t < t_{\mathrm{eq}}$. This fact distorts the shape of the primordial spectrum by suppressing the growth of small wavelength modes (with $k > k_{\mathrm{eq}} = 2\pi/d_{\mathrm{eq}}$ that enter the Hubble radius in the radiation dominated phase) in comparison with longer ones, with the transition occurring at the wave number k_{eq} corresponding to the length scale $d_{\mathrm{eq}} = d_H(z_{\mathrm{eq}}) = (2\pi/k_{\mathrm{eq}}) \approx 13(\Omega_{\mathrm{DM}}h^2)^{-1}\mathrm{Mpc}$. Very roughly, the shape of $\mathcal{F}^2(k)$ can be characterized by the behaviour $\mathcal{F}^2(k) \propto k^{-4}$ for $k > k_{\mathrm{eq}}$ and $\mathcal{F}^2 \approx 1$ for $k < k_{\mathrm{eq}}$. The spectrum at wavelengths $\lambda \gg d_{\mathrm{eq}}$ is undistorted by the evolution since $\mathcal{F}^2$ is essentially unity at these scales.

We will see in the next section that inflationary models generate an initial power spectrum of the form $P(k) \propto k$. The evolution described above will distort it to the form $P(k) \propto k^{-3}$ for $k > k_{\mathrm{eq}}$ and leave it undistorted with $P \propto k$ for $k < k_{\mathrm{eq}}$. The power per logarithmic band in the wavenumber, $\Delta^2 \propto k^3 P(k)$, is approximately constant for $k > k_{eq}$ (actually increasing as $\ln k$ because of the logarithmic growth in the radiation dominated phase) and decreases as $\Delta^2 \propto k^4 \propto \lambda^{-4}$ at large wavelengths. It follows that Δ^2 is a monotonically decreasing function of the wavelength with more power at small length scales.

When $\delta_k \approx 1$, linear perturbation theory breaks down at the spatial scale corresponding to $\lambda = 2\pi/k$. Since there is more power at small scales, smaller scales go non-linear first and structure forms hierarchically. (Observations suggest that, in today's universe scales smaller than about $8h^{-1}$ Mpc are non-linear; see Fig.1) As the universe expands, the over-dense region will expand more slowly compared to the background, will reach a maximum radius, contract and virialize to form a bound nonlinear halo of dark matter. The baryons in the halo will cool and undergo collapse in a fairly complex manner because of gas dynamical processes. It seems unlikely that the baryonic collapse and galaxy formation can be understood by analytic approximations; one needs to do high resolution computer simulations to make any progress.[11]

The non linear evolution of the *dark matter halos* is somewhat different and worth mentioning because it contains the fascinating physics of statistical mechanics of self gravitating systems.[12] The standard instability of gravitating systems in a *static* background is moderated by the presence of a background expansion and it is possible to understand various features of nonlinear evolution of dark matter halos using different analytic approximations.[13] Among these, the existence of certain nonlinear scaling relations — which allows one to compute nonlinear power spectrum from linear power spectrum by a nonlocal scaling relation — seems to be most intriguing[14]. If $\bar{\xi}(x, t)$ is the mean correlation function of dark matter particles and $\bar{\xi}_L(x, t)$ is the same quantity computed in the linear approximation, then, it turns out that $\bar{\xi}(x, t)$ can be expressed as a *universal* function of $\bar{\xi}_L(x, t)$ in the form $\bar{\xi}(x, t) = U[\bar{\xi}_L(l, t)]$ where $x = l[1 + U[\bar{\xi}_L(l, t)]]^{-1/3}$. Incredibly enough, the form of U can be determined by theory[15] and thus allows one to understand several aspects of nonlinear clustering analytically. This topic has interesting connections with renormalisation group theory, fluid turbulence etc. and deserves the attention of wider community of physicists.

4. Inflation and Generation of Initial Perturbations

We saw that the two length scales which determine the evolution of perturbations are the Hubble radius $d_H(t) \equiv (\dot{a}/a)^{-1}$ and $\lambda(t) \equiv \lambda_0 a(t)$. Using their definitions and Eq.(1), it is easy to show that if $\rho > 0, p > 0$, then $\lambda(t) > d_H(t)$ for sufficiently small t.

This result leads to a major difficulty in conventional cosmology. Normal physical processes can act coherently only over length scales smaller than the Hubble radius. Thus any physical process leading to density perturbations at some early epoch, $t = t_i$, could only have operated at scales smaller than $d_H(t_i)$. But most of the relevant astrophysical scales (corresponding to clusters, groups, galaxies, etc.) were much bigger than $d_H(t)$ at sufficiently early epochs. Therefore, it is difficult to understand how any physical process operating in the early universe could have led to the seed perturbations in the early universe.

One way of tacking this difficulty is to arrange matters such that we have $\lambda(t) < d_H(t)$ at sufficiently small t. Since we cannot do this in any model which has both $\rho > 0, p > 0$ we need to invoke some exotic physics to get around this difficulty. The standard procedure is to make $a(t)$ increase rapidly with t (for example, exponentially or as $a \propto t^n$ with $n \gg 1$, which requires $p < 0$) for a brief period of time. Such a rapid growth is called "inflation" and in conventional models of inflation,[16] the energy density during the inflationary phase is provided by a scalar field with a potential $V(\phi)$. If the potential energy dominates over the kinetic energy, such a scalar field can act like an ideal fluid with the equation of state $p = -\rho$ and lead to $a(t) \propto e^{Ht}$ during inflation. Fig. (4) shows the behaviour of the Hubble radius and the wavelength $\lambda(t)$ of a generic perturbation (line AB) for a universe which underwent exponential inflation. In such a universe, it is possible for *quantum* fluctuations of the scalar field at A (when the perturbation scale leaves the Hubble radius) to manifest as *classical* perturbations at B (when the perturbation enters the Hubble radius). We will now briefly discuss these processes.

Consider a scalar field $\phi(t, \mathbf{x})$ which is nearly homogeneous in the sense that we can write $\phi(t, \mathbf{x}) = \phi(t) + \delta\phi(t, \mathbf{x})$ with $\delta\phi \ll \phi$. Let us first ignore the fluctuations and consider how one can use the mean value to drive a rapid expansion of the universe. The Einstein's equation (for $k = 0$) with the field $\phi(t)$ as the source can be written in the form

$$\frac{\dot{a}^2}{a^2} = H^2(t) = \frac{1}{3M_{\mathrm{Pl}}^2}\left[\frac{1}{2}\dot{\phi}^2 + V(\phi)\right] \tag{4}$$

where $V(\phi)$ is the potential for the scalar field and $M_{\text{Pl}} \equiv (8\pi G)^{-1/2} \approx 2.4 \times 10^{18}$ GeV in units with $\hbar = c = 1$. Further, the equation of motion for the scalar field in an expanding universe reduces to

$$\ddot{\phi} + 3H\dot{\phi} = -\frac{dV}{d\phi} \tag{5}$$

The solutions of Eqs. (4), (5) giving $a(t)$ and $\phi(t)$ will depend critically on the form of $V(\phi)$ as well as the initial conditions. Among these solutions, there exists a subset in which $a(t)$ is a rapidly growing function of t, either exponentially or as a power law $a(t) \propto t^n$ with an arbitrarily large value of n. It is fairly easy to verify that the solutions to Eqs. (4), (5) can be expressed in the form

$$V(t) = 3H^2 M_{Pl}^2 \left[1 + \frac{\dot{H}}{3H^2}\right]; \quad \phi(t) = \int dt [-2\dot{H}M_{Pl}^2]^{1/2} \tag{6}$$

Equation (6) completely solves the (reverse) problem of finding a potential $V(\phi)$ which will lead to a given $a(t)$. For example, power law expansion of the universe $[a \propto t^n]$ can be generated by using a potential $V \propto \exp[-\sqrt{(2/n)}(\phi/M_{Pl})]$.

A more generic way of achieving this is through potentials which allow what is known as *slow roll-over*. Such potentials have a gently decreasing form for $V(\phi)$ for a range of values for ϕ allowing $\phi(t)$ to evolve very slowly. Assuming a sufficiently slow evolution of $\phi(t)$ we can ignore: (i) the $\ddot{\phi}$ term in equation Eq. (5) and (ii) the kinetic energy term $\dot{\phi}^2$ in comparison with the potential energy $V(\phi)$ in Eq. (4). In this limit, Eq. (4),Eq. (5) become

$$H^2 \simeq \frac{V(\phi)}{3M_{\text{Pl}}^2}; \quad 3H\dot{\phi} \simeq -V'(\phi) \tag{7}$$

The validity of slow roll over approximation thus requires the following two parameters to be sufficiently small:

$$\epsilon(\phi) = \frac{M_{\text{Pl}}^2}{2} \left(\frac{V'}{V}\right)^2; \quad \eta(\phi) = M_{\text{Pl}}^2 \frac{V''}{V} \tag{8}$$

The end point for inflation can be taken to be the epoch at which ϵ becomes comparable to unity. If the slow roll-over approximation is valid until a time $t = t_{\text{end}}$, the amount of inflation can be characterized by the ratio $a(t_{\text{end}})/a(t)$. If $N(t) \equiv \ln[a(t_{\text{end}})/a(t)]$, then Eq. (7) gives

$$N \equiv \ln \frac{a(t_{\text{end}})}{a(t)} = \int_t^{t_{\text{end}}} H \, dt \simeq \frac{1}{M_{\text{Pl}}^2} \int_{\phi_{\text{end}}}^{\phi} \frac{V}{V'} \, d\phi \tag{9}$$

This provides a general procedure for quantifying the rapid growth of $a(t)$ arising from a given potential.

Let us next consider the spectrum of density perturbations which are generated from the quantum fluctuations of the scalar field.[17] This requires the study of quantum field theory in a time dependent background which is non-trivial. There are several conceptual issues (closely related to the issue of general covariance of quantum field theory and the particle concept[18]) in obtaining a c-number density perturbation from inherently quantum fluctuations. We shall not discuss these issues and will adopt a heuristic approach, as follows:

In the deSitter spacetime with $a(t) \propto \exp(Ht)$, there is a horizon in the spacetime and associated temperature $T = (H/2\pi)$ — just as in the case of black holes.[19] Hence the scalar field will have an intrinsic rms fluctuation $\delta\phi \approx T = (H/2\pi)$ in the deSitter spacetime at the scale of the Hubble radius. This will cause a time shift $\delta t \approx \delta\phi/\dot\phi$ in the evolution of the field between patches of the universe of size about H^{-1}. This, in turn, will lead to an rms fluctuation $\Delta = (k^3 P)^{1/2}$ of amplitude $\delta a/a = (\dot a/a)\delta t \approx H^2/(2\pi\dot\phi)$ at the Hubble scale. Since the wavelength of the perturbation is equal to Hubble radius at A (see Fig. (4)), we conclude that the rms amplitude of the perturbation when it leaves the Hubble radius is: $\Delta_A \approx H^2/(2\pi\dot\phi)$. Between A and B (in Fig. (4)) the wavelength of the perturbation is bigger than the Hubble radius and one can show[a] that $\Delta(at\ A) \approx \Delta(at\ B)$ giving $\Delta(at\ B) \approx H^2/(2\pi\dot\phi)$. Since this is independent of k, it follows that all perturbations enter the Hubble radius with constant power per decade. That is $\Delta^2(k, a) \propto k^3 P(k, a)$ is independent of k when evaluated at $a = a_{enter}(k)$ for the relevant mode.

From this, it follows that $P(k, a) \propto k$ at constant a. To see this, note that if $P \propto k^n$, then the power per logarithmic band of wave numbers is $\Delta^2 \propto k^3 P(k) \propto k^{(n+3)}$. Further, when the wavelength of the mode is larger than the Hubble radius, during the radiation dominated phase, the perturbation grows (see footnote) as $\delta \propto a^2$ making $\Delta^2 \propto a^4 k^{(n+3)}$. The epoch a_{enter} at which a mode enters the Hubble radius is determined by the

---

[a]For a single component universe with $p = w\rho \propto a^{-3(1+w)}$, we have from Eq. (1), the result $\ddot a = -(4\pi G/3)(1 + 3w)\rho a$. Perturbing this relation to $a \to a + \delta a$ and using $a = (t/t_0)^{2/(3+3w)}$ we find that δa satisfies the equation $t^2\ddot{\delta a} = m\delta a$ with $m = (2/9)(1+3w)(2 + 3w)(1 + w)^{-2}$. This has power law solutions $\delta a \propto t^p$ with $p(p - 1) = m$. The growing mode corresponds to the density contrast $\delta \propto (\delta a/a)$ which is easily shown to vary as $\delta \propto (\rho a^2)^{-1}$. In the inflationary, phase, ρ=const., $\delta \propto a^{-2}$; in the radiation dominated phase, $\rho \propto a^{-4}, \delta \propto a^2$. The result follows from these scalings.

relation $2\pi a_{enter}/k = d_H$. Using $d_H \propto t \propto a^2$ in the radiation dominated phase, we get $a_{enter} \propto k^{-1}$ so that

$$\Delta^2(k, a_{enter}) \propto a_{enter}^4 k^{(n+3)} \propto k^{(n-1)} \tag{10}$$

So if the power $\Delta^2 \propto k^3 P$ per octave in k is independent of scale k, at the time of entering the Hubble radius, then $n = 1$. In fact, a *prediction* that the initial fluctuation spectrum will have a power spectrum $P = Ak^n$ with $n = 1$ was made by Harrison and Zeldovich,[20] years before inflationary paradigm, based on general arguments of scale invariance. Inflation is one possible mechanism for generating such scale invariant perturbations.

As an example, consider the case of $V(\phi) = \lambda\phi^4$, for which Eq.(9) gives $N = H^2/2\lambda\phi^2$ and the amplitude of the perturbations at Hubble scale is:

$$\Delta \simeq \frac{H^2}{\dot{\phi}} \simeq \frac{3H^3}{V'} \simeq \lambda^{1/2} N^{3/2} \tag{11}$$

If we want inflation to last for reasonable amount of time ($N \gtrsim 60$, say) and $\Delta \approx 10^{-5}$ (as determined from CMBR temperature anisotropies; see Section 5), then we require $\lambda \lesssim 10^{-15}$. This has been a serious problem in virtually any reasonable model of inflation: *The parameters in the potential need to be extremely fine tuned to match observations.*

It is not possible to obtain n strictly equal to unity in realistic models, since scale invariance is always broken at some level. In a wide class of inflationary models this deviation is given by $(1 - n) \approx 6\epsilon - 2\eta$ (where ϵ and η are defined by Eq. (8)); this deviation $(1 - n)$ is obviously small when the slow roll over approximation ($\epsilon \ll 1, \eta \ll 1$) holds.

The same mechanism that produces density perturbations (which are scalar) will also produce gravitational wave (tensor) perturbations of some magnitude δ_{grav}. Since both the scalar and tensor perturbations arise from the same mechanism, one can relate the amplitudes of these two and show that $(\delta_{grav}/\delta)^2 \approx 12.4\epsilon$; clearly, the tensor perturbations are small compared to the scalar perturbations. Further, for generic inflationary potentials, $|\eta| \approx |\epsilon|$ so that $(1 - n) \approx 4\epsilon$ giving $(\delta_{grav}/\delta)^2 \approx \mathcal{O}(3)(1 - n)$. This is a relation between three quantities all of which are (in principle) directly observable and hence it can provide a test of the underlying model if and when we detect the stochastic gravitational wave background.

Finally, we mention the possibility that inflationary regime might act as a magnifying glass and bring the transplanckian regime of physics within the scope of direct observations.[21] To see how this could be possible, note that a scale λ_0 today would have been $\lambda_f \equiv \lambda_0(a_f/a_0) = \lambda_0(T_0/T_f) = 3\lambda_0 \times 10^{-27}$

at the end of inflation and $\lambda_i = \lambda_f \exp(-N) \simeq \lambda_f e^{-70}$ at the beginning of inflation, for typical numbers used in the inflationary scenario. This gives $\lambda_i \approx 3L_P(\lambda_0/1 \text{ Mpc})$ showing that most of the astrophysically relevant scales were smaller than Planck length, $L_P \equiv (G\hbar/c^3)^{1/2} \simeq 10^{-33}$ cm, during the inflation! Phenomenological models which make specific predictions regarding transplanckian physics (like dispersion relations,[22] for example) can then be tested using the signature they leave on the pattern of density perturbations which are generated.

5. Temperature Anisotropies of the CMBR

When the universe cools through $T \approx 1$ eV, the electrons combine with nuclei forming neutral atoms. This 're'combination takes place at a redshift of about $z \approx 10^3$ over a redshift interval $\Delta z = 80$. Once neutral atoms form, the photons decouple from matter and propagate freely from $z = 10^3$ to $z = 0$. This CMB radiation, therefore, contains fossilized signature of the conditions of the universe at $z = 10^3$ and has been an invaluable source of information.

If physical process has led to inhomogeneities in the $z = 10^3$ spatial surface, then these inhomogeneities will appear as temperature anisotropies $(\Delta T/T) \equiv S(\theta, \phi)$ of the CMBR in the sky today where (θ, ϕ) denotes two angles in the sky. It is convenient to expand this quantity in spherical harmonics as $S(\theta, \phi) = \sum a_{lm} Y_{lm}(\theta, \phi)$. If $\mathbf{n}$ and $\mathbf{m}$ are two directions in the sky with an angle α between them, the two-point correlation function of the temperature fluctuations in the sky can be expressed in the form

$$\mathcal{C}(\alpha) \equiv \langle S(\mathbf{n})S(\mathbf{m}) \rangle = \sum_l \frac{(2l+1)}{4\pi} C_l P_l(\cos\alpha); \quad C_l = \langle |a_{lm}|^2 \rangle \quad (12)$$

Roughly speaking, $l \propto \theta^{-1}$ and we can think of the (θ, l) pair as analogue of $(\mathbf{x}, \mathbf{k})$ variables in 3-D.

The primary anisotropies of the CMBR can be thought of as arising from three different sources (even though such a separation is gauge dependent). (i) The first is the gravitational potential fluctuations at the last scattering surface (LSS) which will contribute an anisotropy $(\Delta T/T)^2_\phi \propto k^3 P_\phi(k)$ where $P_\phi(k) \propto P(k)/k^4$ is the power spectrum of gravitational potential ϕ. (The gravitational potential satisfies $\nabla^2 \phi \propto \delta$ which becomes $k^2 \phi_k \propto \delta_k$ in Fourier space; so $P_\phi \equiv \langle |\phi_k|^2 \rangle \propto k^{-4} \langle |\delta_k|^2 \rangle \propto P(k)/k^4$.) This anisotropy arises because photons climbing out of deeper gravitational wells lose more energy on the average. (ii) The second source is the Doppler shift of the frequency of the photons when they are last scattered by moving electrons on the LSS. This is proportional to $(\Delta T/T)_D^2 \propto k^3 P_v$ where $P_v(k) \propto P/k^2$ is the power spectrum of the velocity field. (The velocity field is given by $\mathbf{v} \simeq \mathbf{g}t \propto t\nabla\phi$ so that, in Fourier space, $v_k \sim k\phi_k$ and $P_v = |v_k|^2 \propto k^2 P_\phi \propto k^{-2}P$.) (iii) Finally, we also need to take into account the intrinsic fluctuations of the radiation field on the LSS. In the case of adiabatic fluctuations, these will be proportional to the density fluctuations of matter on the LSS and hence will vary as $(\Delta T/T)_{\text{int}}^2 \propto k^3 P(k)$. Of these, the velocity field and the density field (leading to the Doppler anisotropy and intrinsic anisotropy described in (ii) and (iii) above) will oscillate at scales smaller than the Hubble radius at the time of decoupling since pressure support due to baryons will be effective at small scales. At large scales, for a scale invariant spectrum with $P(k) \propto k$, we get:

$$\left(\frac{\Delta T}{T}\right)_\phi^2 \propto \text{const}; \quad \left(\frac{\Delta T}{T}\right)_D^2 \propto k^2 \propto \theta^{-2}; \quad \left(\frac{\Delta T}{T}\right)_{\text{int}}^2 \propto k^4 \propto \theta^{-4} \quad (13)$$

where $\theta \propto \lambda \propto k^{-1}$ is the angular scale over which the anisotropy is measured. The fluctuations due to gravitational potential dominate at large scales while the sum of intrinsic and Doppler anisotropies will dominate at small scales. Since the latter two are oscillatory, we will expect an oscillatory behaviour in the temperature anisotropies at small angular scales. The typical value for the peaks of the oscillation are at about 0.3 to 0.5 degrees depending on the details of the model.

The above analysis is valid if recombination was instantaneous; but in reality the thickness of the recombination epoch is about $\Delta z \simeq 80$. Further, the coupling between the photons and baryons is not completely 'tight'. It can be shown that[9] these features will heavily damp the anisotropies at angular scales smaller than about 0.1 degree.

The fact that different processes contribute to the structure of angular anisotropies makes CMBR a valuable tool for extracting cosmological information. To begin with, the anisotropy at very large scales directly probes modes which are bigger than the Hubble radius at the time of decoupling and allows us to directly determine the primordial spectrum. The CMBR observations are *consistent* with the inflationary model for the generation of perturbations leading to $P = Ak^n$ and gives $A \simeq (28.3h^{-1}Mpc)^4$ and $n = 0.97 \pm 0.023$. (The first results[23] were from COBE and later results, especially from WMAP, have reconfirmed[2] them with far greater accuracy).

As we move to smaller scales we are probing the behaviour of baryonic gas coupled to the photons. The pressure support of the gas leads

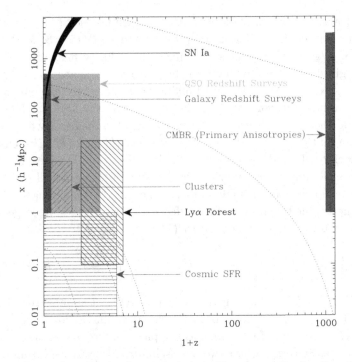

Fig. 1.   Different observations of cosmological significance and the length scales and redshift ranges probed by them. The broken (thin) lines in the figure are contours of $\Delta(k = 2\pi/x, a) = (5, 1.69, 1, 10^{-2}, 10^{-5})$ from bottom to top. Figure courtesy: T. Roy Choudhury.

to modulated acoustic oscillations with a characteristic wavelength at the $z = 10^3$ surface. Regions of high and low baryonic density contrast will lead to anisotropies in the temperature with the same characteristic wavelength (which acts as a standard ruler) leading to a series of peaks in the temperature anisotropy that have been detected. The angles subtended by these acoustic peaks will depend on the geometry of the universe and provides a reliable procedure for estimating the cosmological parameters. Detailed computations[9] show that: (i) The multipole index l corresponding to the first acoustic peak has a strong, easily observable, dependence on Ω_{tot} and scales as $l_p \approx 220\Omega_{\text{tot}}^{-1/2}$ if there is *no* dark energy and $\Omega_{tot} = \Omega_{NR}$. (ii) But if both non-relativistic matter and dark energy is present, with $\Omega_{\text{NR}} + \Omega_{DE} = 1$ and $0.1 \lesssim \Omega_{\text{NR}} \lesssim 1$, then the peak has only a very weak dependence on Ω_{NR} and $l_p \approx 220\Omega_{NR}^{0.1}$. Thus the observed location of the peak (which is around $l \sim 220$) can be used to infer that $\Omega_{\text{tot}} \simeq 1$. More precisely, the current observations show that $0.98 \lesssim \Omega_{tot} \lesssim 1.08$; combining with $h > 0.5$, this result implies the existence of dark energy.

The heights of acoustic peaks also contain important information. In particular, the height of the first acoustic peak relative to the second one depends sensitively on Ω_B and the current results are consistent with that obtained from big bang nucleosynthesis.[4]

Fig.1 summarises the different observations of cosmological significance and the range of length scales and redshift ranges probed by them. The broken (thin) lines in the figure are contours of $\Delta(k = 2\pi/x, a) = (5, 1.69, 1, 10^{-2}, 10^{-5})$ from bottom to top. Clearly, the regions where $\Delta(k, a) > 1$ corresponds to those in which nonlinear effects of structure formation is important. Most of the astrophysical observations — large scale surveys of galaxies, clusters and quasars, observations of intergalactic medium (Ly-α forest), star formation rate (SFR), supernova (SN Ia) data etc. — are confined to $0 < z \lesssim 7$ while CMBR allows probing the universe around $z = 10^3$. Combining these allows one to use the long "lever arm" of 3 decades in redshift and thus constrain the parameters describing the universe effectively.

6. The Dark Energy

It is rather frustrating that we have no direct laboratory evidence for nearly 96% of matter in the universe. (Actually, since we do not quite understand the process of baryogenesis, we do not understand Ω_B either; all we can *theoretically* understand now is a universe filled entirely with radiation!). Assuming that particle physics models will eventually (i) explain Ω_B and Ω_{DM} (probably arising from the lightest supersymmetric partner) as well as (ii) provide a viable model for inflation predicting correct value for A, one is left with the problem of understanding Ω_{DE}. While the issues (i) and (ii) are by no means trivial or satisfactorily addressed, the issue of dark energy is lot more perplexing, thereby justifying the attention it has received recently.

The key observational feature of dark energy is that — treated as a fluid with a stress tensor $T_b^a = \text{dia}(\rho, -p, -p, -p)$ — it has an equation of state $p = w\rho$ with $w \lesssim -0.8$ at the present epoch. The spatial part **g** of the geodesic acceleration (which measures the relative acceleration of two geodesics in the spacetime) satisfies an *exact* equation in general relativity given by $\nabla \cdot \mathbf{g} = -4\pi G(\rho + 3p)$. As long as $(\rho + 3p) > 0$, gravity remains attractive while $(\rho + 3p) < 0$ can lead to repulsive gravitational

effects. In other words, dark energy with sufficiently negative pressure will
accelerate the expansion of the universe, once it starts dominating over the
normal matter. This is precisely what is established from the study of high
redshift supernova, which can be used to determine the expansion rate of
the universe in the past.[7] Figure 2 presents the supernova data as a phase
portrait[8] of the universe. It is clear that the universe was decelerating at
high redshifts and started accelerating when it was about two-third of the
present size.

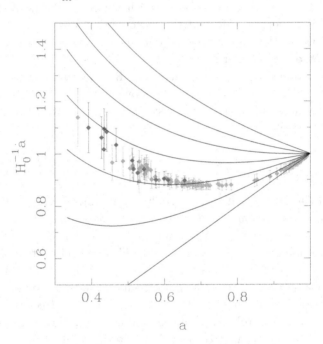

Fig. 2. The "velocity" $\dot{a}$ of the universe is plotted against the "position" a in the form
of a phase portrait. The different curves are for models parameterized by the value of
$\Omega_{DM}(= \Omega_m)$ keeping $\Omega_{tot} = 1$. The top-most curve has $\Omega_m = 1$ and the bottom-
most curve has $\Omega_m = 0$ and $\Omega_{DE} = 1$. The in-between curves show universes which
were decelerating in the past and began to accelerate when the dark energy started
dominating. The supernova data clearly favours such a model.

The simplest model for a fluid with negative pressure is the cosmological
constant[24] with $w = -1, \rho = -p =$ constant. If the dark energy is indeed
a cosmological constant, then it introduces a fundamental length scale in

the theory $L_\Lambda \equiv H_\Lambda^{-1}$, related to the constant dark energy density ρ_{DE} by $H_\Lambda^2 \equiv (8\pi G \rho_{DE}/3)$. In classical general relativity, based on the constants G, c and L_Λ, it is not possible to construct any dimensionless combination from these constants. But when one introduces the Planck constant, $\hbar$, it is possible to form the dimensionless combination $H_\Lambda^2(G\hbar/c^3) \equiv (L_P^2/L_\Lambda^2)$. Observations demand $(L_P^2/L_\Lambda^2) \lesssim 10^{-120}$ requiring enormous fine tuning. What is more, the energy density of normal matter and radiation would have been higher in the past while the energy density contributed by the cosmological constant does not change. Hence we need to adjust the energy densities of normal matter and cosmological constant in the early epoch very carefully so that $\rho_\Lambda \gtrsim \rho_{NR}$ around the current epoch. Because of these conceptual problems associated with the cosmological constant, people have explored a large variety of alternative possibilities. Though none of them does any better than the cosmological constant, we will briefly describe them in view of the popularity these models enjoy.

The most popular alternative to cosmological constant uses a scalar field ϕ with a suitably chosen potential $V(\phi)$ so as to make the vacuum energy vary with time. The hope is that, one can find a model in which the current value can be explained naturally without any fine tuning. We will discuss two possibilities based on the lagrangians:

$$L_{\text{quin}} = \frac{1}{2}\partial_a\phi\partial^a\phi - V(\phi); \quad L_{\text{tach}} = -V(\phi)[1 - \partial_a\phi\partial^a\phi]^{1/2} \qquad (14)$$

Both these lagrangians involve one arbitrary function $V(\phi)$. The first one, L_{quin}, which is a natural generalization of the lagrangian for a non-relativistic particle, $L = (1/2)\dot{q}^2 - V(q)$, is usually called quintessence.[25] When it acts as a source in Friedman universe, it is characterized by a time dependent $w(t) = (1 - (2V/\dot{\phi}^2))(1 + (2V/\dot{\phi}^2))^{-1}$.

The structure of the second lagrangian in Eq. (14) can be understood by an analogy with a relativistic particle with position $q(t)$ and mass m which is described by the lagrangian $L = -m\sqrt{1 - \dot{q}^2}$. We can now construct a field theory by upgrading $q(t)$ to a field ϕ and treating the mass parameter m as a function of ϕ [say, $V(\phi)$] thereby obtaining the second lagrangian in Eq. (14). This provides a rich gamut of possibilities in the context of cosmology.[26,27] This form of scalar field arises in string theories[28] and is called a tachyonic scalar field. (The structure of this lagrangian is similar to those analyzed previously in a class of models[29] called *K-essence*.) The stress tensor for the tachyonic scalar field can be written as the sum of a pressure less dust component and a cosmological constant. This suggests a possibility[26] of providing a unified description of both dark matter and dark energy using the same scalar field. (It is possible to construct more complicated scalar field lagrangians with even $w < -1$ describing what is called *phantom* matter; there are also alternatives to scalar field models, based on brane world scenarios. We shall not discuss either of these.)

Since the quintessence or the tachyonic field has an undetermined function $V(\phi)$, it is possible to choose this function in order to produce a given $H(a)$. To see this explicitly, let us assume that the universe has two forms of energy density with $\rho(a) = \rho_{\text{known}}(a) + \rho_\phi(a)$ where $\rho_{\text{known}}(a)$ arises from any known forms of source (matter, radiation, ...) and $\rho_\phi(a)$ is due to a scalar field. Let us first consider quintessence. Here, the potential is given implicitly by the form[30,26]

$$V(a) = \frac{1}{16\pi G} H(1 - Q) \left[6H + 2aH' - \frac{aHQ'}{1 - Q} \right] \tag{15}$$

$$\phi(a) = \left[\frac{1}{8\pi G} \right]^{1/2} \int \frac{da}{a} \left[aQ' - (1 - Q) \frac{d \ln H^2}{d \ln a} \right]^{1/2} \tag{16}$$

where $Q(a) \equiv [8\pi G \rho_{\text{known}}(a)/3H^2(a)]$ and prime denotes differentiation with respect to a. Given any $H(a), Q(a)$, these equations determine $V(a)$ and $\phi(a)$ and thus the potential $V(\phi)$. *Every quintessence model studied in the literature can be obtained from these equations.*

Similar results exists for the tachyonic scalar field as well.[26] For example, given any $H(a)$, one can construct a tachyonic potential $V(\phi)$ which is consistent with it. The equations determining $V(\phi)$ are now given by:

$$\phi(a) = \int \frac{da}{aH} \left(\frac{aQ'}{3(1 - Q)} - \frac{2}{3} \frac{aH'}{H} \right)^{1/2} \tag{17}$$

$$V = \frac{3H^2}{8\pi G} (1 - Q) \left(1 + \frac{2}{3} \frac{aH'}{H} - \frac{aQ'}{3(1 - Q)} \right)^{1/2} \tag{18}$$

Again, Eqs. (17) and (18) completely solve the problem. Given any $H(a)$, these equations determine $V(a)$ and $\phi(a)$ and thus the potential $V(\phi)$. A wide variety of phenomenological models with time dependent cosmological constant have been considered in the literature all of which can be mapped to a scalar field model with a suitable $V(\phi)$.

While the scalar field models enjoy considerable popularity (one reason being they are easy to construct!) they have not helped us to understand the nature of the dark energy at a deeper level because of several shortcomings: (1) They completely lack predictive power. As explicitly demonstrated

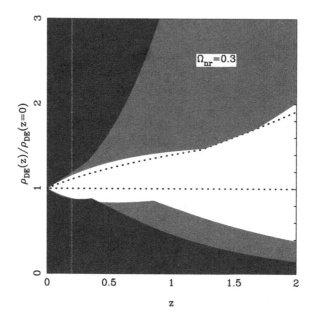

Fig. 3. Constraints on the possible variation of the dark energy density with redshift. The darker shaded region is excluded by SN observations while the lighter shaded region is excluded by WMAP observations. It is obvious that WMAP puts stronger constraints on the possible variations of dark energy density. The cosmological constant corresponds to the horizontal line at unity. The region between the dotted lines has $w > -1$ at all epochs.

above, virtually every form of $a(t)$ can be modeled by a suitable "designer" $V(\phi)$. (2) These models are degenerate in another sense. Even when $w(a)$ is known/specified, it is not possible to proceed further and determine the nature of the scalar field lagrangian. The explicit examples given above show that there are *at least* two different forms of scalar field lagrangians (corresponding to the quintessence or the tachyonic field) which could lead to the same $w(a)$. (See Ref. 8 for an explicit example of such a construction.) (3) All the scalar field potentials require fine tuning of the parameters in order to be viable. This is obvious in the quintessence models in which adding a constant to the potential is the same as invoking a cosmological constant. So to make the quintessence models work, *we first need to assume the cosmological constant is zero!* (4) By and large, the potentials used in the literature have no natural field theoretical justification. All of them are non-renormalisable in the conventional sense and have to be interpreted as a low energy effective potential in an ad-hoc manner.

One key difference between cosmological constant and scalar field models is that the latter lead to a $w(a)$ which varies with time. If observations have demanded this, or even if observations have ruled out $w = -1$ at the present epoch, then one would have been forced to take alternative models seriously. However, all available observations are consistent with cosmological constant ($w = -1$) and — in fact — the possible variation of w is strongly constrained[31] as shown in Figure 3.

Given this situation, we shall take a closer look at the cosmological constant as the source of dark energy in the universe.

7. ...For the Snark was a Boojum, You See

If we assume that the dark energy in the universe is due to a cosmological constant then we are introducing a *second* length scale, $L_\Lambda = H_\Lambda^{-1}$, into the theory (in addition to the Planck length L_P) such that $(L_P/L_\Lambda) \approx 10^{-60}$. Such a universe will be asymptotically deSitter with $a(t) \propto \exp(t/L_\Lambda)$ at late times. Figure 4 summarizes several peculiar features of such a universe.[32,33]

Using the the Hubble radius $d_H \equiv (\dot{a}/a)^{-1}$, we can distinguish between three different phases of such a universe. The first phase is when the universe went through a inflationary expansion with $d_H = $ constant; the second phase is the radiation/matter dominated phase in which most of the standard cosmology operates and $d_H \propto t$ increases monotonically; the third phase is that of re-inflation (or accelerated expansion) governed by the cosmological constant in which d_H is again a constant. The first and last phases are (approximately) time translation invariant; that is, $t \to t+$ constant is an (approximate) invariance for the universe in these two phases. The universe satisfies the perfect cosmological principle and is in steady state during these phases! In fact, one can easily imagine a scenario in which the two deSitter phases (first and last) are of very long duration. If $\Omega_\Lambda \approx 0.7, \Omega_{DM} \approx 0.3$ the final deSitter phase *does* last forever; as regards the inflationary phase, one can view it as lasting for an arbitrarily long (though finite) duration.

Given the two length scales L_P and L_Λ, one can construct two energy densities $\rho_P = 1/L_P^4$ and $\rho_\Lambda = 1/L_\Lambda^4$ in natural units ($c = \hbar = 1$). The first is, of course, the Planck energy density while the second one also has a natural interpretation. The universe which is asymptotically deSitter has a horizon and associated thermodynamics[19] with a temperature $T = H_\Lambda/2\pi$ and the corresponding thermal energy density $\rho_{thermal} \propto T^4 \propto$

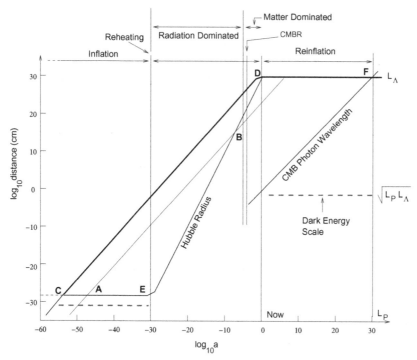

Fig. 4.   The geometrical structure of a universe with two length scales L_P and L_Λ corresponding to the Planck length and the cosmological constant. See text for detailed description of the figure.

$1/L_\Lambda^4 = \rho_\Lambda$. Thus L_P determines the *highest* possible energy density in the universe while L_Λ determines the *lowest* possible energy density in this universe. As the energy density of normal matter drops below this value, the thermal ambience of the deSitter phase will remain constant and provide the irreducible 'vacuum noise'. Note that the dark energy density is the the geometric mean $\rho_{DE} = \sqrt{\rho_\Lambda \rho_P}$ between the two energy densities. If we define a dark energy length scale L_{DE} such that $\rho_{DE} = 1/L_{DE}^4$ then $L_{DE} = \sqrt{L_P L_\Lambda}$ is the geometric mean of the two length scales in the universe. The figure 4 also shows the L_{DE} by broken horizontal lines.

While the two deSitter phases can last forever in principle, there is a natural cut off length scale in both of them which makes the region of physical relevance to be finite.[32] In the the case of re-inflation in the late universe, this happens (at point F) when the temperature of the CMBR radiation drops below the deSitter temperature. The universe will be essentially dominated by the vacuum thermal noise of the deSitter phase for

$a > a_F$. One can easily determine the dynamic range of DF to be

$$\frac{a_F}{a_D} \approx 2\pi T_0 L_\Lambda \left(\frac{\Omega_\Lambda}{\Omega_{DM}}\right)^{1/3} \approx 3 \times 10^{30} \qquad (19)$$

A natural bound on the duration of inflation arises for a different reason. Consider a perturbation at some given wavelength scale which is stretched with the expansion of the universe as $\lambda \propto a(t)$. (See the line marked AB in Fig.4.) If there was no re-inflation, *all* the perturbations will 're-enter' the Hubble radius at some time (the point B in Fig.4). But if the universe undergoes re-inflation, then the Hubble radius 'flattens out' at late times and some of the perturbations will *never* reenter the Hubble radius ! This criterion selects the portion of the inflationary phase (marked by CE) which can be easily calculated to be:

$$\frac{a_E}{a_C} = \left(\frac{T_0 L_\Lambda}{T_{\text{reheat}} H_{in}^{-1}}\right) \left(\frac{\Omega_\Lambda}{\Omega_{DM}}\right)^{1/3} = \frac{(a_F/a_D)}{2\pi T_{\text{reheat}} H_{in}^{-1}} \cong 10^{25} \qquad (20)$$

where we have assumed a GUTs scale inflation with $E_{\text{GUT}} = 10^{14}$ GeV $\cong T_{\text{reheat}}$ and $\rho_{in} = E_{\text{GUT}}^4$ giving $2\pi H_{in}^{-1} T_{\text{reheat}} = (3\pi/2)^{1/2}(E_P/E_{\text{GUT}}) \approx 10^5$. For a Planck scale inflation with $2\pi H_{in}^{-1} T_{\text{reheat}} = \mathcal{O}(1)$, the phases CE and DF are approximately equal. The region in the quadrilateral CEDF is the most relevant part of standard cosmology, though the evolution of the universe can extend to arbitrarily large stretches in both directions in time. This figure is definitely telling us something regarding the time translation invariance of the universe ('the perfect cosmological principle') and — more importantly — *about the breaking of this symmetry*, and it deserves more attention than it has received.

   Let us now turn to several other features related to the cosmological constant. A *non-representative* sample of attempts to understand/explain the cosmological constant include those based on QFT in curved space time,[34] those based on renormalisation group arguments,[35] quantum cosmological considerations,[36] various cancellation mechanisms[37] and many others. A study of these (failures!) reveals the following:

   (a) If observed dark energy is due to cosmological constant, we need to explain the small value of the dimensionless number $\Lambda(G\hbar/c^3) \approx 10^{-120}$. The presence of $G\hbar$ clearly indicates that we are dealing with quantum mechanical problem, coupled to gravity. Any purely classical solution (like a classically decaying cosmological constant) will require (hidden or explicit) fine-tuning. At the same time, this is clearly an infra-red issue, in the sense that the phenomenon occurs at extremely low energies!.

(b) In addition to the zero-point energy of vacuum fluctuations (which *must* gravitate[38]) the phase transitions in the early universe (at least the well established electro-weak transition) change the ground state energy by a large factor. *It is necessary to arrange matters so that gravity does not respond to such changes.* Any approach to cosmological constant which does not take this factor into account is fundamentally flawed.

(c) An immediate consequence is that, the gravitational degrees of freedom which couple to cosmological constant must have a special status and behave in a manner different from other degrees of freedom. (The non linear coupling of matter with gravity has several subtleties; see eg. Ref. 39.) If, for example, we have a theory in which the source of gravity is $(\rho+p)$ rather than $(\rho + 3p)$, then cosmological constant will not couple to gravity at all. Unfortunately it is not possible to develop a covariant theory of gravity using $(\rho+p)$ as the source. But we can achieve the same objective in different manner. Any metric g_{ab} can be expressed in the form $g_{ab} = f^2(x)q_{ab}$ such that $\det q = 1$ so that $\det g = f^4$. From the action functional for gravity

$$A = \frac{1}{2\kappa} \int \sqrt{-g}\, d^4x (R - 2\Lambda) = \frac{1}{2\kappa} \int \sqrt{-g}\, d^4x R - \frac{\Lambda}{\kappa} \int d^4x f^4(x) \quad (21)$$

it is obvious that the cosmological constant couples *only* to the conformal factor f. So if we consider a theory of gravity in which $f^4 = \sqrt{-g}$ is kept constant and only q_{ab} is varied, then such a model will be oblivious of direct coupling to cosmological constant and will not respond to changes in bulk vacuum energy. If the action (without the Λ term) is varied, keeping $\det g = -1$, say, then one is lead to a *unimodular theory of gravity* with the equations of motion $R_{ab} - (1/4)g_{ab}R = \kappa(T_{ab} - (1/4)g_{ab}T)$ with zero trace on both sides. Using the Bianchi identity, it is now easy to show that this is equivalent to a theory with an *arbitrary* cosmological constant. That is, cosmological constant arises as an (undetermined) integration constant in this model.[40] Unfortunately, we still need an extra physical principle to fix its value.

(d) The conventional discussion of the relation between cosmological constant and the zero point energy is too simplistic since the zero point energy has no observable consequence. The observed non trivial features of the vacuum state arise from the *fluctuations* (or modifications) of this vacuum energy. (This was, in fact, known fairly early in the history of cosmological constant problem; see, e.g., Ref.41). If the vacuum probed by the gravity can readjust to take away the bulk energy density $\rho_P \simeq L_P^{-4}$, quantum *fluctuations* can generate the observed value ρ_{DE}. One of the simplest models[42] which achieves this uses the fact that, in the semiclassical limit, the wave function describing the universe of proper four-volume $\mathcal{V}$ will vary as $\Psi \propto \exp(-iA_0) \propto \exp[-i(\Lambda_{\text{eff}}\mathcal{V}/L_P^2)]$. If we treat $(\Lambda/L_P^2, \mathcal{V})$ as conjugate variables then uncertainty principle suggests $\Delta\Lambda \approx L_P^2/\Delta\mathcal{V}$. If the four volume is built out of Planck scale substructures, giving $\mathcal{V} = NL_P^4$, then the Poisson fluctuations will lead to $\Delta\mathcal{V} \approx \sqrt{\mathcal{V}}L_P^2$ giving $\Delta\Lambda = L_P^2/\Delta\mathcal{V} \approx 1/\sqrt{\mathcal{V}} \approx H_0^2$. (This idea can be made more quantitative[42,43].)

In fact, it is *inevitable* that in a universe with two length scale L_Λ, L_P, the vacuum fluctuations will contribute an energy density of the correct order of magnitude $\rho_{DE} = \sqrt{\rho_\Lambda\rho_P}$. The hierarchy of energy scales in such a universe has[32,44] the pattern

$$\rho_{\text{vac}} = \frac{1}{L_P^4} + \frac{1}{L_P^4}\left(\frac{L_P}{L_\Lambda}\right)^2 + \frac{1}{L_P^4}\left(\frac{L_P}{L_\Lambda}\right)^4 + \cdots \tag{22}$$

The first term is the bulk energy density which needs to be renormalized away (by a process which we do not understand at present); the third term is just the thermal energy density of the deSitter vacuum state; what is interesting is that quantum fluctuations in the matter fields *inevitably generate* the second term. A rigorous calculation[44] of the dispersion in the energy shows that the *fluctuations* in the energy density $\Delta\rho$, inside a region bounded by a cosmological horizon, is given by

$$\Delta\rho_{\text{vac}} \propto L_P^{-2}L_\Lambda^{-2} \propto \frac{H_\Lambda^2}{G} \tag{23}$$

The numerical coefficient will depend on the precise nature of infrared cutoff radius (like whether it is L_Λ or $L_\Lambda/2\pi$ etc.). But *one cannot get away from* a fluctuation of magnitude $\Delta\rho_{vac} \simeq H_\Lambda^2/G$ that will exist in the energy density inside a sphere of radius H_Λ^{-1} if Planck length is the UV cut off. Since observations suggest that there is indeed a ρ_{vac} of *similar* magnitude in the universe, it seems natural to identify the two, after subtracting out the mean value for reasons which we do not understand. This approach explains why there is a *surviving* cosmological constant which satisfies $\rho_{DE} = \sqrt{\rho_\Lambda\rho_P}$ but not why the leading term in Eq. (22) should be removed.

8. Deeper Issues in Cosmology

It is clear from the above discussion that 'parametrised cosmology', which attempts to describe the evolution of the universe in terms of a small number of parameters, has made considerable progress in recent years. Having done this, it is tempting to ask more ambitious questions, some of which we will briefly discuss in this section.

There are two obvious questions a cosmologist faces every time (s)he gives a popular talk, for which (s)he has no answer! The first one is: *Why do the parameters of the universe have the values they have?* Today, we have no clue why the real universe follows one template out of a class of models all of which are permitted by the known laws of physics (just as we have no idea why there are three families of leptons with specified mass ratios etc.) Of the different cosmological parameters, $\Omega_{DM}, \Omega_B, \Omega_R$ as well as the parameters of the initial power spectrum A, n should arise from viable particle physics models which actually says something about phenomenology. (Unfortunately, these research areas are not currently very fashionable.) On the other hand, it is not clear how we can understand Ω_{DE} without a reasonably detailed model for quantum gravity. *In fact, the acid test for any viable quantum gravity model is whether it has something nontrivial to say about* Ω_{DE}; all the current candidates have nothing to offer on this issue and thus fail the test.

The second question is: *How (and why!) was the universe created and what happened before the big bang ?* The cosmologist giving the public lecture usually mumbles something about requiring a quantum gravity model to circumvent the classical singularity — but we really have no idea!. String theory offers no insight; the implications of loop quantum gravity for quantum cosmology have attracted fair mount of attention recently[45] but it is fair to say we still do not know how (and why) the universe came into being.

What is not often realised is that *certain aspects of this problem transcends the question of technical tractability of quantum gravity* and can be presented in more general terms. Suppose, for example, one has produced some kind of theory for quantum gravity. Such a theory is likely to come with a single length scale L_P. Even when one has the back drop of such a theory, it is not clear how one hopes to address questions like: (a) Why is our universe much bigger than L_P which is the only scale in the problem i.e., why is the mean curvature of the universe much smaller than L_P^{-2}? (b) How does the universe, treated as a dynamical system, evolve *spontaneously* from a quantum regime to classical regime? (c) How does one obtain the notion of a cosmological arrow of time, starting from timeless or at least time symmetric description?

One viable idea regarding these issues seems to be based on vacuum instability which describes the universe as an unstable system with an unbounded Hamiltonian. Then it is possible for the expectation value of spatial curvature to vary monotonically as, say, $< R > \propto L_P^{-2}(t/t_P)^{-\alpha}$ with some index α, as the universe expands in an unstable mode. Since the conformal factor of the metric has the 'wrong' sign for the kinetic energy term, this mode will become semiclassical first. Even then it is not clear how the arrow of time related to the expanding phase of the universe arises; one needs to invoke decoherence like arguments to explain the classical limit[46] and the situation is not very satisfactory. (For an alternative idea, regarding the origin of the universe as a quantum fluctuation, see ref. 47).

While the understanding of such 'deeper' issues might require the *details* of the viable model for quantum gravity, one should not ignore the alternative possibility that we are just not clever enough. It may turn out that certain obvious (low energy) features of the universe, that we take for granted, contain clues to the full theory of quantum gravity (just as the equality of inertial and gravitational masses, known for centuries, was turned on its head by Einstein's insight) if only we manage to find the right questions to ask. To illustrate this point, consider an atomic physicist who solves the Schrodinger equation for the electrons in the helium atom. (S)he will discover that, of all the permissible solutions, only half (which are antisymmetric under the exchange of electrons) are realized in nature though the Hamiltonian of the helium atom offers no insight for this feature. This is a low energy phenomenon the explanation of which lies deep inside relativistic field theory.

In this spirit, there are at least two peculiar features of our universe which are noteworthy:

(i) The first one, which is already mentioned, is the fact that our universe seemed to have evolved *spontaneously* from a quantum regime to classical regime bringing with it the notion of a cosmological arrow of time.[46] This is not a generic feature of dynamical systems (and is connected with the fact that the Hamiltonian for the system is unbounded from below).

(ii) The second issue corresponds to the low energy vacuum state of matter fields in the universe and the notion of the particle — which is an excitation around this vacuum state. Observations show that, in the classical limit, we do have the notion of an inertial frame, vacuum state and a notion of the particle such that the particle at rest in this frame will see, say, the CMBR as isotropic. This is nontrivial, since the notion of classical particle arises after several limits are taken: Given the formal quantum state $\Psi[g, matter]$ of gravity and matter, one would first proceed to a limit of quantum fields in curved background, in which the gravity is treated as a c-number[48]. Further taking the $c \to \infty$ limit (to obtain quantum mechanics) and $\hbar \to 0$ limit (to obtain the classical limit) one will reach the notion of a particle in classical theory. Miraculously enough, the full quantum state of the universe seems to finally select (in the low energy classical limit) a local inertial frame, such that the particle at rest in this frame will see the universe as isotropic — rather than the universe as accelerating or rotating, say. This is a nontrivial constraint on $\Psi[g, matter]$, especially since the vacuum state and particle concept are ill defined in the curved background[18]. One can show that this feature imposes special conditions on the wave function of the universe[49] in simple minisuperspace models but its wider implications in quantum gravity are unexplored.

References

1. W. Freedman et al., *Ap.J.* **553**, 47 (2001); J.R. Mould et al., *Ap.J.* **529**, 786 (2000).

2. P. de Bernardis et al., *Nature* **404**, 955 (2000); A. Balbi et al., *Ap.J.* **545**, L1 (2000); S. Hanany et al., *Ap.J.* **545**, L5 (2000); T.J. Pearson et al., *Ap.J.* **591**, 556-574 (2003); C.L. Bennett et al, *Ap. J. Suppl.* **148**, 1 (2003); D. N. Spergel et al., *Ap.J.Suppl.* **148**, 175 (2003); B. S. Mason et al., *Ap.J.* **591**, 540-555 (2003). For a recent summary, see e.g., L. A. Page,astro-ph/0402547.

3. For a review of the theory, see e.g., K.Subramanian, astro-ph/0411049.

4. For a review of BBN, see S.Sarkar, *Rept.Prog.Phys.* **59**, 1493-1610 (1996); G.Steigman, astro-ph/0501591. The consistency between CMBR observations and BBN is gratifying since the initial MAXIMA-BOOMERANG data gave too high a value as stressed by T. Padmanabhan and Shiv Sethi, *Ap. J.* **555**, 125 (2001), [astro-ph/0010309].

5. For a critical discussion of the current evidence, see P.J.E. Peebles, astro-ph/0410284.

6. G. Efstathiou et al., *Nature* **348**, 705 (1990); J. P. Ostriker, P. J. Steinhardt, *Nature* **377**, 600 (1995); J. S. Bagla et al., *Comments on Astrophysics* **18**, 275 (1996),[astro-ph/9511102].

7. S.J. Perlmutter et al., *Astrophys. J.* **517**, 565 (1999); A.G. Reiss et al., *Astron. J.* **116**, 1009 (1998); J. L. Tonry et al., *ApJ* **594**, 1 (2003); B. J. Barris,*Astrophys.J.* **602**, 571-594 (2004); A. G.Reiss et al., *Astrophys.J.* **607**, 665-687(2004).

8. T. Roy Choudhury,T. Padmanabhan, *Astron.Astrophys.* **429**, 807 (2005), [astro-ph/0311622]; T. Padmanabhan,T. Roy Choudhury, *Mon. Not. Roy. Astron. Soc.* **344**, 823 (2003) [astro-ph/0212573].

9. See e.g., T. Padmanabhan, *Structure Formation in the Universe*, (Cambridge University Press, Cambridge, 1993); T. Padmanabhan, *Theoretical Astrophysics, Volume III: Galaxies and Cosmology*, (Cambridge University Press, Cambridge, 2002); J.A. Peacock, *Cosmological Physics*, (Cambridge University Press, Cambridge, 1999).

10. Mészaros P. *Astron. Ap.* **38**, 5 (1975).

11. Pablo Laguna, in this volume; For a pedagogical description, see J.S. Bagla,

astro-ph/0411043; J.S. Bagla, T. Padmanabhan, *Pramana* **49**, 161-192 (1997), [astro-ph/0411730].

12. For a review, see T.Padmanabhan, *Phys. Rept.* **188** , 285 (1990); T. Padmanabhan, (2002)[astro-ph/0206131].

13. Ya.B. Zeldovich, A*stron.Astrophys.* **5**, 84-89,(1970); Gurbatov, S. N. et al,*MNRAS* **236**, 385 (1989); T.G. Brainerd et al., *Astrophys.J.* **418**, 570 (1993); J.S. Bagla, T.Padmanabhan, *MNRAS* **266**, 227 (1994), [gr-qc/9304021]; *MNRAS,* **286** , 1023 (1997), [astro-ph/9605202]; T.Padmanabhan, S.Engineer, *Ap. J.* **493**, 509 (1998), [astro-ph/9704224]; S. Engineer et.al., *MNRAS* **314** , 279-289 (2000), [astro-ph/9812452]; for a recent review, see T.Tatekawa, [astro-ph/0412025].

14. A. J. S. Hamilton et al., *Ap. J.* **374**, L1 (1991),; T.Padmanabhan et al., *Ap. J.* **466**, 604 (1996), [astro-ph/9506051]; D. Munshi et al., *MNRAS,* **290**, 193 (1997), [astro-ph/9606170]; J. S. Bagla, et.al., *Ap.J.* **495**, 25 (1998), [astro-ph/9707330]; N.Kanekar et al., *MNRAS* , **324**, 988 (2001), [astro-ph/0101562].

15. R. Nityananda, T. Padmanabhan, *MNRAS* **271**, 976 (1994), [gr-qc/9304022]; T. Padmanabhan, *MNRAS* **278**, L29 (1996), [astro-ph/9508124].

16. D. Kazanas, *Ap. J. Letts.* **241**, 59 (1980); A. A. Starobinsky, *JETP Lett.* **30**, 682 (1979); *Phys. Lett.* **B 91**, 99 (1980); A. H. Guth, *Phys. Rev. D* **23**, 347 (1981); A. D. Linde, *Phys. Lett.* **B 108**, 389 (1982); A. Albrecht, P. J. Steinhardt, *Phys. Rev. Lett.* **48**,1220 (1982); for a review, see e.g.,J. V. Narlikar and T. Padmanabhan, *Ann. Rev. Astron. Astrophys.* **29**, 325 (1991).

17. S. W. Hawking, *Phys. Lett.* **B 115**, 295 (1982); A. A. Starobinsky, *Phys. Lett.* **B 117**, 175 (1982); A. H. Guth, S.-Y. Pi, *Phys. Rev. Lett.* **49**, 1110 (1982); J. M. Bardeen et al., *Phys. Rev.* **D 28**, 679 (1983); L. F. Abbott,M. B. Wise, *Nucl. Phys.* **B 244**, 541 (1984).

18. S. A. Fulling, *Phys. Rev.* **D7** 2850–2862 (1973); W.G. Unruh, *Phys. Rev.* **D14**, 870 (1976); T. Padmanabhan, T.P. Singh, *Class. Quan. Grav.* 4, 1397 (1987); L. Sriramkumar, T. Padmanabhan , *Int. Jour. Mod. Phys.* **D 11**,1 (2002) [gr-qc/9903054].

19. G. W. Gibbons and S.W. Hawking, *Phys. Rev.* **D 15**, 2738 (1977); T.Padmanabhan, *Mod.Phys.Letts.* **A 17**, 923 (2002), [gr-qc/0202078]; *Class. Quant. Grav.*, **19**, 5387 (2002),[gr-qc/0204019]; for a recent review see T. Padmanabhan, *Phys. Reports* **406**, 49 (2005) [gr-qc/0311036].

20. E.R. Harrison, *Phys. Rev.* **D1**, 2726 (1970); Zeldovich Ya B. *MNRAS* **160**, 1p (1972).

21. One of the earliest attempts to include transplanckian effects is in T. Padmanabhan, *Phys. Rev. Letts.* **60**, 2229 (1988); T. Padmanabhan et al., *Phys. Rev.* **D 39** , 2100 (1989). A sample of more recent papers are J.Martin, R.H. Brandenberger, *Phys.Rev.* **D63**, 123501 (2001); A. Kempf, *Phys.Rev.* **D63** , 083514 (2001); U.H. Danielsson, *Phys.Rev.* **D66**, 023511 (2002); A. Ashoorioon et al., *Phys.Rev.* **D71** 023503 (2005); R. Easther, astro-ph/0412613; for a more extensive set of references, see L. Sriramkumar et al.,[gr-qc/0408034].

22. S. Corley, T. Jacobson, *Phys.Rev.* **D54**, 1568 (1996); T. Padmanabhan, *Phys.Rev.Lett.* **81**, 4297-4300 (1998), [hep-th/9801015]; *Phys.Rev.* **D59** 124012 (1999), [hep-th/9801138]; A.A. Starobinsky, *JETP Lett.* **73** 371-374 (2001); J.C. Niemeyer, R. Parentani, *Phys.Rev.* **D64** 101301 (2001); J. Kowalski-Glikman, *Phys.Lett.* **B499** 1 (2001); G.Amelino-Camelia, *Int.J.Mod.Phys.* **D11** 35 (2002).

23. G.F. Smoot. et al., *Ap.J.* **396**, L1 (1992),; T. Padmanabhan, D. Narasimha, *MNRAS* **259**, 41P (1992); G. Efstathiou et al., (1992), MNRAS, **258**, 1.

24. P. J. E. Peebles and B. Ratra, *Rev. Mod. Phys.* **75**, 559 (2003); S. M. Carroll, *Living Rev. Rel.* **4**, 1 (2001); T. Padmanabhan, *Phys. Rept.* **380**, 235 (2003) [hep-th/0212290]; V. Sahni and A. A. Starobinsky, *Int. J. Mod. Phys.* **D 9**, 373 (2000); J. R. Ellis, *Phil. Trans. Roy. Soc. Lond.* **A 361**, 2607 (2003).

25. One of the earliest papers was B. Ratra, P.J.E. Peebles, *Phys.Rev.* **D37**, 3406 (1988). An extensive set of references are given in T.Padmanabhan, [astro-ph/0411044] and in the reviews cited in the previous reference.

26. T. Padmanabhan, Phys. Rev. **D 66**, 021301 (2002) [hep-th/0204150]; T. Padmanabhan and T. R. Choudhury, *Phys. Rev.* **D 66**, 081301 (2002) [hep-th/0205055]; J. S. Bagla, H. K. Jassal and T. Padmanabhan, *Phys. Rev.* **D 67**, 063504 (2003) [astro-ph/0212198].

27. For a sample of early work, see G. W. Gibbons, *Phys. Lett.* **B 537**, 1 (2002); G. Shiu and I. Wasserman, *Phys. Lett.* **B 541**, 6 (2002); D. Choudhury et al., *Phys. Lett.* **B 544**, 231 (2002); A. V. Frolov et al.,*Phys. Lett.* **B 545**, 8 (2002); M. Sami, *Mod.Phys.Lett.* **A 18**, 691 (2003). More extensive set of references are given in T.Padmanabhan, [astro-ph/0411044].

28. A. Sen, JHEP **0204** 048 (2002), [hep-th/0203211].

29. An early paper is C. Armendariz-Picon et al., *Phys. Rev.* **D 63**, 103510 (2001). More extensive set of references are given in T.Padmanabhan, [astro-ph/0411044]

30. G.F.R. Ellis and M.S.Madsen, *Class.Quan.Grav.* **8**, 667 (1991).

31. H.K. Jassal et al., *MNRAS* **356**, L11-L16 (2005), [astro-ph/0404378]

32. T.Padmanabhan Lecture given at the *Plumian 300 - The Quest for a Concordance Cosmology and Beyond* meeting at Institute of Astronomy, Cambridge, July 2004; [astro-ph/0411044].

33. J.D. Bjorken, (2004) astro-ph/0404233.

34. E. Elizalde and S.D. Odintsov, *Phys.Lett.* **B321**, 199 (1994); **B333**, 331 (1994); N.C. Tsamis,R.P. Woodard, *Phys.Lett.* **B301**, 351-357 (1993); E. Mottola, *Phys.Rev.* **D31**, 754 (1985).

35. I.L. Shapiro, *Phys.Lett.* **B329**, 181 (1994); I.L. Shapiro and J. Sola, *Phys.Lett.* **B475**, 236 (2000); *Phys.Lett.* **B475** 236-246 (2000); hep-ph/0305279; astro-ph/0401015; I.L. Shapiro et al.,hep-ph/0410095; Cristina Espana-Bonet, et.al., *Phys.Lett.* **B574** 149-155 (2003); *JCAP* **0402**, 006 (2004); F.Bauer,gr-qc/0501078.

36. T. Mongan, *Gen. Rel. Grav.* **33** 1415 (2001), [gr-qc/0103021]; *Gen.Rel.Grav.* **35** 685-688 (2003); E. Baum, *Phys. Letts.* **B 133**, 185 (1983); T. Padmanabhan, *Phys. Letts.* **A104** , 19 (1984); S.W. Hawking, *Phys. Letts.* **B 134**, 403 (1984); Coleman, S., *Nucl. Phys.* **B 310**, p. 643 (1988).

37. A.D. Dolgov, in *The very early universe: Proceeding of the 1982 Nuffield Workshop at Cambridge*, ed. G.W. Gibbons, S.W. Hawking and S.T.C. Sikkos (Cambridge University Press, 1982), p. 449; S.M. Barr, *Phys. Rev.* D 36, 1691 (1987); Ford, L.H., *Phys. Rev.* D 35, 2339 (1987); Hebecker A. and C. Wetterich, *Phy. Rev. Lett.* 85, 3339 (2000); hep-ph/0105315; T.P. Singh, T. Padmanabhan, *Int. Jour. Mod. Phys.* A 3, 1593 (1988); M. Sami, T. Padmanabhan, *Phys. Rev.* D 67, 083509 (2003), [hep-th/0212317].

38. R. R. Caldwell, astro-ph/0209312; T.Padmanabhan, *Int.Jour. Mod.Phys.* A 4, 4735 (1989), Sec.6.1.

39. T. Padmanabhan (2004) [gr-qc/0409089].

40. A. Einstein, Siz. Preuss. Acad. Scis. (1919), translated as "Do Gravitational Fields Play an essential Role in the Structure of Elementary Particles of Matter," in The *Principle of Relativity*, by edited by A. Einstein et al. (Dover, New York, 1952); J. J. van der Bij et al., *Physica* A116, 307 (1982); F. Wilczek, *Phys. Rep.* 104, 111 (1984); A. Zee, in *High Energy Physics*, proceedings of the 20th Annual Orbis Scientiae, Coral Gables, (1983), edited by B. Kursunoglu, S. C. Mintz, and A. Perlmutter (Plenum, New York, 1985); W. Buchmuller and N. Dragon, *Phys.Lett.* B207, 292, (1988); W.G. Unruh, *Phys.Rev.* D 40 1048 (1989).

41. Y.B. Zel'dovich, *JETP letters* 6, 316 (1967); *Soviet Physics Uspekhi* 11, 381 (1968).

42. T. Padmanabhan, *Class.Quan.Grav.* 19, L167 (2002), [gr-qc/0204020]; for an earlier attempt, see D. Sorkin, *Int.J.Theor.Phys.* 36, 2759-2781 (1997); for related ideas, see Volovik, G. E., gr-qc/0405012; J. V. Lindesay et al., astro-ph/0412477; Yun Soo Myung, hep-th/0412224; E.Elizalde et al., hep-th/0502082.

43. G.E. Volovik, *Phys.Rept.* 351, 195-348 (2001); T. Padmanabhan, *Int.Jour.Mod.Phys.* D 13, 2293-2298 (2004), [gr-qc/0408051]; [gr-qc/0412068].

44. T. Padmanabhan, [hep-th/0406060]; Hsu and Zee, [hep-th/0406142].

45. Martin Bojowald, this volume.

46. T. Padmanabhan, *Phys.Rev.* D39, 2924 (1989); J.J. Hallwell, *Phys.Rev.* D39, 2912 (1989).

47. S.Dutta, T.Vachaspati, *Phys.Rev.* D71 (2005) 083507.

48. V.G. Lapchinsky, V.A. Rubakov, *Acta Phys.Polon.* B10, 1041 (1979); J. B. Hartle, *Phys. Rev.* D 37, 2818 (1988);D 38, 2985 (1988); T.Padmanabhan, *Class. Quan. Grav.* 6, 533 (1989); T.P. Singh, T. Padmanabhan, *Annals Phys.* 196, 296(1989).

49. T. Padmanabhan and T. Roy Choudhury, *Mod. Phys. Lett.* A 15, 1813-1821 (2000), [gr-qc/0006018].

CHAPTER 8

WAS EINSTEIN RIGHT? TESTING RELATIVITY AT THE CENTENARY

CLIFFORD M. WILL

Department of Physics and McDonnell Center for the Space Sciences,
Washington University, St. Louis, MO 63130, USA
cmw@wuphys.wustl.edu

We review the experimental evidence for Einstein's special and general relativity. A variety of high precision null experiments verify the weak equivalence principle and local Lorentz invariance, while gravitational redshift and other clock experiments support local position invariance. Together these results confirm the Einstein Equivalence Principle which underlies the concept that gravitation is synonymous with spacetime geometry, and must be described by a metric theory. Solar system experiments that test the weak-field, post-Newtonian limit of metric theories strongly favor general relativity. The Binary Pulsar provides tests of gravitational-wave damping and of strong-field general relativity. Recently discovered binary pulsar systems may provide additional tests. Future and ongoing experiments, such as the Gravity Probe B Gyroscope Experiment, satellite tests of the Equivalence principle, and tests of gravity at short distance to look for extra spatial dimensions could constrain extensions of general relativity. Laser interferometric gravitational-wave observatories on Earth and in space may provide new tests of gravitational theory via detailed measurements of the properties of gravitational waves.

1. Introduction

When I was a first-term graduate student some 36 years ago, it was said that the field of general relativity is "a theorist's paradise and an experimentalist's purgatory". To be sure, there were some experiments: Irwin Shapiro, then at MIT, had just measured the relativistic retardation of radar waves passing the Sun (an effect that now bears his name), Robert Dicke of Princeton was claiming that the Sun was flattened in an amount that would mess up general relativity's success with Mercury's perihelion advance, and Joseph Weber of the University of Maryland was just about to announce (40 years prematurely, as we now know) the detection of gravitational waves. Nevertheless the field was dominated by theory and by theorists. The field *circa* 1970 seemed to reflect Einstein's own attitudes: although he was not ignorant of experiment, and indeed had a keen insight into the workings of the physical world, he felt that the bottom line was the *theory*. As he once famously said, if experiment were to contradict the theory, he would have "felt sorry for the dear Lord".

Since that time the field has been completely transformed, and today at the centenary of Einstein's *annus mirabilis*, experiment is a central, and in some ways dominant component of gravitational physics. I know no better way to illustrate this than to cite the first regular article of the 15 June 2004 issue of Physical Review D: the author list of this "general relativity" paper fills an entire page, and the institution list fills most of another. This was one of the papers reporting results from the first science run of the LIGO laser interferometer gravitational-wave observatories, but it brings to mind papers in high-energy physics, not general relativity! The breadth of current experiments, ranging from tests of classic general relativistic effects such as the light bending and the Shapiro delay, to searches for short-range violations of the inverse-square law, to the operation of a space experiment to measure the relativistic precession of gyroscopes, to the construction and operation of gravitational-wave detectors, attest to the ongoing vigor of experimental gravitation.

Because of its elegance and simplicity, and because of its empirical success, general relativity has become the foundation for our understanding of the gravitational interaction. Yet modern developments in particle theory suggest that it is probably not the entire story, and that modification of the basic theory may be required at some level. String theory generally predicts a proliferation of scalar fields that could result in alterations of general relativity reminiscent of the Brans-Dicke theory of the 1960s. In the presence of extra dimensions, the gravity that we feel on our four-dimensional "brane" of a higher dimensional world could be somewhat different from a pure four-dimensional general relativity. Some of these ideas have motivated the possibility that fundamental constants may actually be dynamical variables, and hence may vary in time or in space. However, any theoretical speculation along these lines must abide by the best current empirical bounds. Decades of high-precision tests of general relativity have produced some very tight constraints. In this article I will review the experimental situation, and assess how well, after 100 years, Einstein got it right.

We begin in Sec. 2 with the "Einstein equivalence principle", which underlies the idea that gravity and curved spacetime are synonymous, and describe its empirical support. Section 3 describes solar system tests of gravity in terms of experimental bounds on a set of "parametrized post-Newtonian" (PPN) parameters. In Section 4 we discuss tests of general relativity using binary pulsar systems. Section 5 describes tests of gravitational theory that could be carried out using future observations of gravitational radiation. Concluding remarks are made in Section 6. For further discussion of topics in this chapter, and for references to the literature, the reader is referred to *Theory and Experiment in Gravitational Physics*[1] and to the "living" review articles[2,3,4].

2. The Einstein Equivalence Principle

The Einstein equivalence principle (EEP) is a powerful and far-reaching principle, which states that

- test bodies fall with the same acceleration independently of their internal structure or composition (Weak Equivalence Principle, or WEP),
- the outcome of any local non-gravitational experiment is independent of the velocity of the freely-falling reference frame in which it is performed (Local Lorentz Invariance, or LLI), and
- the outcome of any local non-gravitational experiment is independent of where and when in the universe it is performed (Local Position Invariance, or LPI).

The Einstein equivalence principle is the heart of gravitational theory, for it is possible to argue convincingly that if EEP is valid, then gravitation must be described by "metric theories of gravity", which state that (i) spacetime is endowed with a symmetric metric, (ii) the trajectories of freely falling bodies are geodesics of that metric, and (iii) in local freely falling reference frames, the non-gravitational laws of physics are those written in the language of special relativity.

General relativity is a metric theory of gravity, but so are many others, including the Brans-Dicke theory. In this sense, superstring theory is not metric, because of residual coupling of external, gravitation-like fields, to matter. Such external fields could be characterized as fields that do not vanish in the vacuum state (in contrast, say, to electromagnetic fields). Theories in which varying non-gravitational constants are associ-

ated with dynamical fields that couple to matter directly are also not metric
theories.

2.1. Tests of the weak equivalence principle

To test the weak equivalence principle, one compares the acceleration of
two laboratory-sized bodies of different composition in an external gravitational field. A measurement or limit on the fractional difference in acceler-
ation between two bodies yields a quantity $\eta \equiv 2|a_1 - a_2|/|a_1 + a_2|$, called
the "Eötvös ratio", named in honor of Baron von Eötvös, the Hungarian
physicist whose experiments carried out with torsion balances at the end
of the 19th century were the first high-precision tests of WEP[5]. Later classic experiments by Dicke and Braginsky[6,7] improved the bounds by several
orders of magnitude. Additional experiments were carried out during the
1980s as part of a search for a putative "fifth force", that was motivated in
part by a reanalysis of Eötvös' original data (the range of bounds achieved
during that period is shown schematically in the region labeled "fifth force"
in Figure 1).

In a torsion balance, two bodies of different composition are suspended
at the ends of a rod that is supported by a fine wire or fibre. One then looks
for a difference in the horizontal accelerations of the two bodies as revealed
by a slight rotation of the rod. The source of the horizontal gravitational
force could be the Sun, a large mass in or near the laboratory, or, as Eötvös
recognized, the Earth itself.

The best limit on η currently comes from the "Eöt-Wash" experiments
carried out at the University of Washington, which used a sophisticated
torsion balance tray to compare the accelerations of bodies of different composition toward the Earth, the Sun and the galaxy[8]. Another strong bound
comes from Lunar laser ranging (LLR), which checks the equality of free
fall of the Earth and Moon toward the Sun[9]. The results from laboratory
and LLR experiments are:

$$\eta_{\text{Eöt-Wash}} < 4 \times 10^{-13}, \quad \eta_{\text{LLR}} < 5 \times 10^{-13}. \tag{1}$$

In fact, by using laboratory materials whose composition mimics that of
the Earth and Moon, the Eöt-Wash experiments[8] permit one to infer an
unambiguous bound from Lunar laser ranging on the universality of acceleration of gravitational binding energy at the level of 1.3×10^{-3} (test of
the Nordtvedt effect – see Sec. 3.2 and Table 1.)

In the future, the Apache Point Observatory for Lunar Laser-ranging
Operation (APOLLO) project, a joint effort by researchers from the

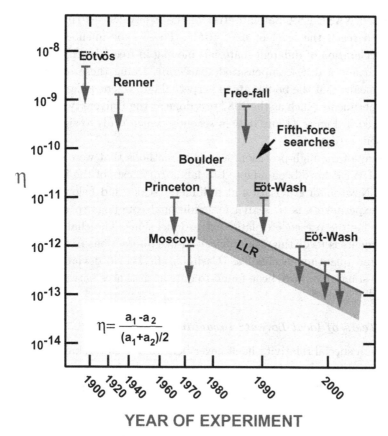

Fig. 1. Selected tests of the Weak Equivalence Principle, showing bounds on the fractional difference in acceleration of different materials or bodies. "Free-fall" and Eöt-Wash experiments, along with numerous others between 1986 and 1990, were originally performed to search for a fifth force. The dark line and the shading below it show evolving bounds on WEP for the Earth and the Moon from Lunar laser ranging (LLR).

Universities of Washington, Seattle, and California, San Diego, plans to use enhanced laser and telescope technology, together with a good, high-altitude site in New Mexico, to improve the Lunar laser-ranging bound by as much as an order of magnitude[10].

High-precision WEP experiments, can test superstring inspired models of scalar-tensor gravity, or theories with varying fundamental constants in which weak violations of WEP can occur via non-metric couplings. The project MICROSCOPE, designed to test WEP to a part in 10^{15} has been approved by the French space agency CNES for a possible 2008 launch. A proposed NASA-ESA Satellite Test of the Equivalence Principle (STEP) seeks to reach the level of $\eta < 10^{-18}$. These experiments will compare the acceleration of different materials moving in free-fall orbits around the Earth inside a drag-compensated spacecraft. Doing these experiments in space means that the bodies are in perpetual fall, whereas Earth-based free-fall experiments (such as the 1987 test done at the University of Colorado[11] indicated in Figure 1), are over in seconds, which leads to significant measurement errors.

Many of the high-precision, low-noise methods that were developed for tests of WEP have been adapted to laboratory tests of the inverse square law of Newtonian gravitation at millimeter scales and below. The goal of these experiments is to search for additional couplings to massive particles or for the presence of large extra dimensions. The challenge of these experiments is to distinguish gravitation-like interactions from electromagnetic and quantum mechanical (Casimir) effects. No deviations from the inverse square law have been found to date at distances between $10\,\mu$m and $10\,$mm[12,13,14,15,16].

2.2. *Tests of local Lorentz invariance*

Although special relativity itself never benefited from the kind of "crucial" experiments, such as the perihelion advance of Mercury and the deflection of light, that contributed so much to the initial acceptance of general relativity and to the fame of Einstein, the steady accumulation of experimental support, together with the successful merger of special relativity with quantum mechanics, led to its being accepted by mainstream physicists by the late 1920s, ultimately to become part of the standard toolkit of every working physicist. This accumulation included

- the classic Michelson-Morley experiment and its descendents[17,18,19,20],
- the Ives-Stillwell, Rossi-Hall and other tests of time-dilation[21,22,23],
- tests of the independence of the speed of light of the velocity of the source, using both binary X-ray stellar sources and high-energy pions[24,25],
- tests of the isotropy of the speed of light[26,27,28]

In addition to these direct experiments, there was the Dirac equation of quantum mechanics and its prediction of anti-particles and spin; later would come the stunningly successful relativistic theory of quantum electrodynamics.

On this 100th anniversary of the introduction of special relativity, one might ask "what is there to test?". Special relativity has been so thoroughly integrated into the fabric of modern physics that its validity is rarely challenged, except by cranks and crackpots. It is ironic then, that during the past several years, a vigorous theoretical and experimental effort has been launched, on an international scale, to find violations of special relativity. The motivation for this effort is not a desire to repudiate Einstein, but to look for evidence of new physics "beyond" Einstein, such as apparent violations of Lorentz invariance that might result from certain models of quantum gravity. Quantum gravity asserts that there is a fundamental length scale given by the Planck length, $L_p = (\hbar G/c^3)^{1/2} = 1.6 \times 10^{-33}$ cm, but since length is not an invariant quantity (Lorentz-FitzGerald contraction), then there could be a violation of Lorentz invariance at some level in quantum gravity. In brane world scenarios, while physics may be locally Lorentz invariant in the higher dimensional world, the confinement of the interactions of normal physics to our four-dimensional "brane" could induce apparent Lorentz violating effects. And in models such as string theory, the presence of additional scalar, vector and tensor long-range fields that couple to matter of the standard model could induce effective violations of Lorentz symmetry. These and other ideas have motivated a serious reconsideration of how to test Lorentz invariance with better precision and in new ways.

A simple way of interpreting some of these experiments is to suppose that a non-metric coupling to the electromagnetic interactions results in a change in the speed of electromagnetic radiation c relative to the limiting speed of material test particles c_0, in other words, $c \neq c_0$. In units where $c_0 = 1$, this would result in an action for charged particles and electromagnetic fields given, in a preferred reference frame (presumably that of the cosmic background radiation), by

$$I = -\sum_a m_{0a} \int (1 - v_a^2)^{1/2} dt + \sum_a e_a \int (-\Phi + \mathbf{A} \cdot \mathbf{v_a}) dt$$

$$+ \frac{1}{8\pi} \int (E^2 - c^2 B^2) d^3 x dt \,, \tag{2}$$

where $\Phi = -A_0$, $\mathbf{E} = -\nabla\Phi - \dot{\mathbf{A}}$, and $\mathbf{B} = \nabla \times \mathbf{A}$. This is sometimes called the "c^2" framework[29,30]; it is a special case of the "$TH\epsilon\mu$" framework of Lightman and Lee[31] for analysing non-metric theories of gravity, and of the "standard model extension" (SME) of Kostalecky and coworkers[32,33,34]. Such a Lorentz-non-invariant electromagnetic interaction would cause shifts

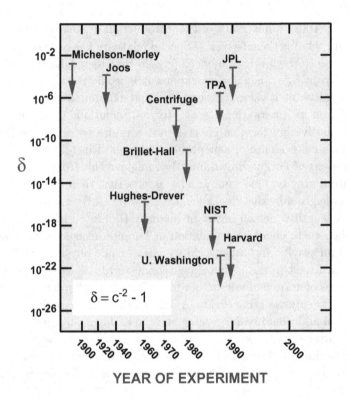

Fig. 2. Selected tests of local Lorentz invariance, showing bounds on the parameter $\delta = c^{-2} - 1$, where c is the speed of propagation of electromagnetic waves in a preferred reference frame, in units in which the limiting speed of test particles is unity.

in the energy levels of atoms and nuclei that depend on the orientation of the quantization axis of the state relative to our velocity in the rest-frame of the universe, and on the quantum numbers of the state, resulting in orientation dependence of the fundamental frequencies of such atomic clocks. The magnitude of these "clock anisotropies" would be proportional to $\delta \equiv |c^{-2} - 1|$.

The earliest clock anisotropy experiments were those of Hughes and Drever, although their original motivation was somewhat different[35,36]. Dramatic improvements were made in the 1980s using laser-cooled trapped atoms and ions[37,38,39]. This technique made it possible to reduce the broading of resonance lines caused by collisions, leading to improved bounds on δ shown in Figure 2 (experiments labelled NIST, U. Washington and Harvard, respectively).

The SME and other frameworks[40] have been used to analyse many new experimental tests of local Lorentz invariance, including comparisons of resonant cavities with atomic clocks, and tests of dispersion and birefringence in the propagation of high energy photons from astrophysical sources. Other testable effects of Lorentz invariance violation include threshold effects in particle reactions, gravitational Cerenkov radiation, and neutrino oscillations. Mattingly[4] gives a thorough and up-to-date review of both the theoretical frameworks and the experimental results, and describes possibilities for future tests of local Lorentz invariance.

2.3. *Tests of local position invariance*

Local position invariance, requires, among other things, that the internal binding energies of atoms be independent of location in space and time, when measured against some standard atom. This means that a comparison of the rates of two different kinds of clocks should be independent of location or epoch, and that the frequency shift of a signal sent between two identical clocks at different locations is simply a consequence of the apparent Doppler shift between a pair of inertial frames momentarily comoving with the clocks at the moments of emission and reception respectively. The relevant parameter in the frequency shift expression $\Delta f/f = (1+\alpha)\Delta U/c^2$, is $\alpha \equiv \partial \ln E_B/\partial(U/c^2)$, where E_B is the atomic or nuclear binding energy, and U is the external gravitational potential. If LPI is valid, the binding energy should be independent of the external potential, and hence $\alpha = 0$. The best bounds come from a 1976 rocket redshift experiment using Hydrogen masers, and a 1993 clock intercomparison experiment (a "null" redshift experiment)[41,42,43]. The results are:

$$\alpha_{\text{Maser}} < 2 \times 10^{-4}, \quad \alpha_{\text{Null}} < 10^{-4}. \tag{3}$$

Recent "clock comparison" tests of LPI were designed to look for possible variations of the fine structure constant on a cosmological timescale. An experiment done at the National Institute of Standards and Technology (NIST) in Boulder compared laser-cooled mercury ions with neutral cesium atoms over a two-year period, while an experiment done at the Observatory of Paris compared laser-cooled cesium and rubidium atomic fountains over five years; the results showed that the fine structure constant α is constant in time to a part in 10^{15} per year[44,45]. Plans are being developed to perform such clock comparisons in space, possibly on the International Space Station.

A better bound on $d\alpha/dt$ comes from analysis of fission yields of the Oklo natural reactor, which occurred in Africa 2 billion years ago, namely $(\dot{\alpha}/\alpha)_{\text{Oklo}} < 6 \times 10^{-17}$ per year[46]. These and other bounds on variations of constants, including reports (later disputed) of positive evidence for variations from quasar spectra, are discussed by Martins and others in Ref.[47].

3. Solar-System Tests

3.1. *The parametrized post-Newtonian framework*

It was once customary to discuss experimental tests of general relativity in terms of the "three classical tests", the gravitational redshift, which is really a test of the EEP, not of general relativity itself (see Sec. 2.3); the perihelion advance of Mercury, the first success of the theory; and the deflection of light, whose measurement in 1919 made Einstein a celebrity. However, the proliferation of additional tests as well as of well-motivated alternative metric theories of gravity, made it desirable to develop a more general theoretical framework for analysing both experiments and theories.

This "parametrized post-Newtonian (PPN) framework" dates back to Eddington in 1922, but was fully developed by Nordtvedt and Will in the period 1968 - 72. When we confine attention to metric theories of gravity, and further focus on the slow-motion, weak-field limit appropriate to the solar system and similar systems, it turns out that, in a broad class of metric theories, only the numerical values of a set of parameters vary from theory to theory. The framework contains ten PPN parameters: γ, related to the amount of spatial curvature generated by mass; β, related to the degree of non-linearity in the gravitational field; ξ, α_1, α_2, and α_3, which determine whether the theory violates local position invariance or local Lorentz invariance in *gravitational* experiments (violations of the Strong Equivalence Principle); and ζ_1, ζ_2, ζ_3 and ζ_4, which describe whether the theory has appropriate momentum conservation laws. For a complete exposition of the PPN framework see Ref. [1].

A number of well-known relativistic effects can be expressed in terms of these PPN parameters:

Deflection of light:

$$\Delta\theta = \left(\frac{1+\gamma}{2}\right)\frac{4GM}{dc^2}$$

$$= \left(\frac{1+\gamma}{2}\right) \times 1.7505\frac{R_\odot}{d} \text{ arcsec},\tag{4}$$

where d is the distance of closest approach of a ray of light to a body of mass M, and where the second line is the deflection by the Sun, with radius $R_\odot$.

Shapiro time delay:

$$\Delta t = \left(\frac{1+\gamma}{2}\right) \frac{4GM}{c^3} \ln\left[\frac{(r_1 + \mathbf{x}_1 \cdot \mathbf{n})(r_2 - \mathbf{x}_2 \cdot \mathbf{n})}{d^2}\right], \tag{5}$$

where Δt is the excess travel time of a round-trip electromagnetic tracking signal, $\mathbf{x}_1$ and $\mathbf{x}_2$ are the locations relative to the body of mass M of the emitter and receiver of the round-trip radar tracking signal (r_1 and r_2 are the respective distances) and $\mathbf{n}$ is the direction of the outgoing tracking signal.

Perihelion advance:

$$\begin{aligned}
\frac{d\omega}{dt} &= \left(\frac{2+2\gamma-\beta}{3}\right) \frac{GM}{Pa(1-e^2)c^2} \\
&= \left(\frac{2+2\gamma-\beta}{3}\right) \times 42.98\,\text{arcsec}/100\,\text{yr}, \tag{6}
\end{aligned}$$

where P, a, and e are the period, semi-major axis and eccentricity of the planet's orbit; the second line is the value for Mercury.

Nordtvedt effect:

$$\frac{m_G - m_I}{m_I} = \left(4\beta - \gamma - 3 - \frac{10}{3}\xi - \alpha_1 - \frac{2}{3}\alpha_2 - \frac{2}{3}\zeta_1 - \frac{1}{3}\zeta_2\right)\frac{|E_g|}{m_I c^2}, \tag{7}$$

where m_G and m_I are the gravitational and inertial masses of a body such as the Earth or Moon, and E_g is its gravitational binding energy. A non-zero Nordtvedt effect would cause the Earth and Moon to fall with a different acceleration toward the Sun.

Precession of a gyroscope:

$$\begin{aligned}
\Omega_{FD} &= -\frac{1}{2}\left(1 + \gamma + \frac{\alpha_1}{4}\right)\frac{G}{r^3 c^2}(\mathbf{J} - 3\mathbf{n}\,\mathbf{n}\cdot\mathbf{J}), \\
&= \frac{1}{2}\left(1 + \gamma + \frac{\alpha_1}{4}\right) \times 0.041\,\text{arcsec yr}^{-1}, \\
\Omega_{Geo} &= -\frac{1}{2}(1 + 2\gamma)\mathbf{v} \times \frac{Gm\mathbf{n}}{r^2 c^2}. \\
&= \frac{1}{3}(1 + 2\gamma) \times 6.6\,\text{arcsec yr}^{-1}, \tag{8}
\end{aligned}$$

where Ω_{FD} and Ω_{Geo} are the precession angular velocities caused by the dragging of inertial frames (Lense-Thirring effect) and by the geodetic effect, a combination of Thomas precession and precession induced by spatial curvature; J is the angular momentum of the Earth, and $\mathbf{v}$, $\mathbf{n}$ and r are the velocity, direction, and distance of the gyroscope. The second line in each case is the corresponding value for a gyroscope in polar Earth orbit at about 650 km altitude (Gravity Probe B, Sec. 3.3).

In general relativity, $\gamma = 1$, $\beta = 1$, and the remaining parameters all vanish.

3.2. *Bounds on the PPN parameters*

Four decades of experiments, ranging from the standard light-deflection and perihelion-shift tests, to Lunar laser ranging, planetary and satellite tracking tests of the Shapiro time delay, and geophysical and astronomical observations, have placed bounds on the PPN parameters that are consistent with general relativity. The current bounds are summarized in Table 1.

To illustrate the dramatic progress of experimental gravity since the dawn of Einstein's theory, Figure 3 shows a history of results for $(1 + \gamma)/2$, from the 1919 solar eclipse measurements of Eddington and his colleagues (which made Einstein a public celebrity), to modern-day measurements using very-long-baseline radio interferometry (VLBI), advanced radar tracking of spacecraft, and orbiting astrometric satellites such as Hipparcos. The most recent results include a measurement of the Shapiro delay using the *Cassini* spacecraft[48], and a measurement of the bending of light via analysis

Table 1.   Current limits on the PPN parameters.

Parameter	Effect	Limit	Remarks
$\gamma - 1$	(i) time delay	2.3×10^{-5}	Cassini tracking
	(ii) light deflection	3×10^{-4}	VLBI
$\beta - 1$	(i) perihelion shift	3×10^{-3}	$J_2 = 10^{-7}$ from
			helioseismology
	(ii) Nordtvedt effect	5×10^{-4}	$\eta = 4\beta - \gamma - 3$ assumed
ξ	Earth tides	10^{-3}	gravimeter data
α_1	orbital polarization	10^{-4}	Lunar laser ranging
			PSR J2317+1439
α_2	solar spin	4×10^{-7}	alignment of Sun
	precession		and ecliptic
α_3	pulsar acceleration	2×10^{-20}	pulsar $\dot{P}$ statistics
η^1	Nordtvedt effect	10^{-3}	Lunar laser ranging
ζ_1	–	2×10^{-2}	combined PPN bounds
ζ_2	binary motion	4×10^{-5}	$\ddot{P}_p$ for PSR 1913+16
ζ_3	Newton's 3rd law	10^{-8}	Lunar acceleration
ζ_4	–	–	not independent

^{1}Here $\eta = 4\beta - \gamma - 3 - 10\xi/3 - \alpha_1 - 2\alpha_2/3 - 2\zeta_1/3 - \zeta_2/3$

of VLBI data on 541 quasars and compact radio galaxies distributed over the entire sky[49].

The perihelion advance of Mercury, the first of Einstein's successes, is now known to agree with observation to a few parts in 10^3. Although there was controversy during the 1960s about this test because of Dicke's claims of an excess solar oblateness, which would result in an unacceptably large Newtonian contribution to the perihelion advance, it is now known from helioseismology that the oblateness is of the order of a few parts in 10^7, as expected from standard solar models, and too small to affect Mercury's orbit, within the experimental error.

Scalar-tensor theories of gravity are characterized by a coupling function $\omega(\phi)$ whose size is inversely related to the "strength" of the scalar field relative to the metric. In the solar system, the parameter $|\gamma-1|$, for example is equal to $1/(2 + \omega(\phi_0))$, where ϕ_0 is the value of the scalar field today outside the solar system. Solar-system experiments (primarily the Cassini results[48]) constrain $\omega(\phi_0) > 40000$.

Proposals are being developed for advanced space missions which will have tests of PPN parameters as key components, including GAIA, a high-precision astrometric telescope (successor to Hipparcos), which could measure light-deflection and γ to the 10^{-6} level[50], and the Laser Astrometric Test of Relativity (LATOR), a mission involving laser ranging to a pair of satellites on the far side of the Sun, which could measure γ to a part in 10^8, and could possibly detect second-order effects in light propagation[51].

3.3. *Gravity Probe-B*

The NASA Relativity Mission called Gravity Probe-B recently completed its mission to measure the Lense-Thirring and geodetic precessions of gyroscopes in Earth orbit[52]. Launched on April 20, 2004 for a 16-month mission, it consisted of four spherical fused quartz rotors coated with a thin layer of superconducting niobium, spinning at 70 - 100 Hz, in a spacecraft containing a telescope continuously pointed toward a distant guide star (IM Pegasi). Superconducting current loops encircling each rotor measured the change in direction of the rotors by detecting the change in magnetic flux through the loop generated by the London magnetic moment of the spinning superconducting film. The spacecraft orbited the Earth in a polar orbit at 650 km altitude. The proper motion of the guide star relative to distant quasars was measured before, during and after the mission using VLBI. The primary science goal of GPB was a one-percent measurement of the 41

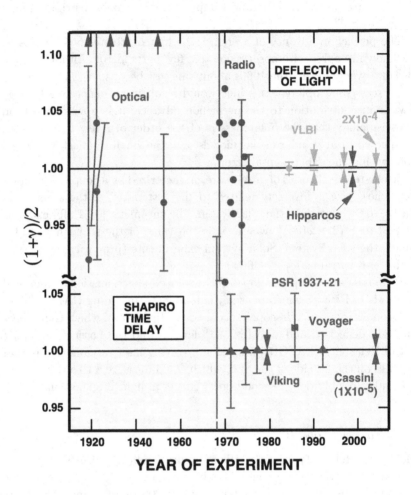

Fig. 3.   Measurements of the coefficient $(1 + \gamma)/2$ from observations of the deflection of light and of the Shapiro delay in propagation of radio signals near the Sun. The general relativity prediction is unity. "Optical" denotes measurements of stellar deflection made during solar eclipes, "Radio" denotes interferometric measurements of radio-wave deflection, and "VLBI" denotes Very Long Baseline Radio Interferometry. "Hipparcos" denotes the European optical astrometry satellite. Arrows denote values well off the chart from one of the 1919 eclipse expeditions and from others through 1947. Shapiro delay measurements using the Cassini spacecraft on its way to Saturn yielded tests at the 0.001 percent level, and light deflection measurements using VLBI have reached 0.02 percent.

milliarcsecond per year frame dragging or Lense-Thirring effect caused by the rotation of the Earth; its secondary goal was to measure to six parts in 10^5 the larger 6.6 arcsecond per year geodetic precession caused by space curvature [Eq. (8)].

A complementary test of the Lense-Thirring precession, albeit with about 10 times lower accuracy than the GPB goal, was reported by Ciufolini and Pavlis[53]. This experiment measured the precession of the orbital planes of two Earth-orbiting laser-ranged satellites called LAGEOS, using up-to-date models of the gravitational field of the Earth in an attempt to subtract the dominant Newtonian precession with sufficient accuracy to yield a measurement of the relativistic effect.

4. The Binary Pulsar

The binary pulsar PSR 1913+16, discovered in 1974 by Joseph Taylor and Russell Hulse, provided important new tests of general relativity, specifically of gravitational radiation and of strong-field gravity. Through precise timing of the pulsar "clock", the important orbital parameters of the system could be measured with exquisite precision. These included non-relativistic "Keplerian" parameters, such as the eccentricity e, and the orbital period (at a chosen epoch) P_b, as well as a set of relativistic "post-Keplerian" parameters. The first PK parameter, $\langle \dot{\omega} \rangle$, is the mean rate of advance of periastron, the analogue of Mercury's perihelion shift. The second, denoted γ' is the effect of special relativistic time-dilation and the gravitational redshift on the observed phase or arrival time of pulses, resulting from the pulsar's orbital motion and the gravitational potential of its companion. The third, $\dot{P}_b$, is the rate of decrease of the orbital period; this is taken to be the result of gravitational radiation damping (apart from a small correction due to galactic differential rotation). Two other parameters, s and r, are related to the Shapiro time delay of the pulsar signal if the orbital inclination is such that the signal passes in the vicinity of the companion; s is a direct measure of the orbital inclination $\sin i$. According to GR, the first three post-Keplerian effects depend only on e and P_b, which are known, and on the two stellar masses which are unknown. By combining the observations of PSR 1913+16 with the GR predictions, one obtains both a measurement of the two masses, and a test of GR, since the system is overdetermined. The results are[54]

$$m_1 = 1.4414 \pm 0.0002 M_\odot \ , \quad m_2 = 1.3867 \pm 0.0002 M_\odot \ ,$$
$$\dot{P}_b^{GR}/\dot{P}_b^{OBS} = 1.0013 \pm 0.0021 \ . \tag{9}$$

Table 2. Parameters of the binary pulsar PSR 1913+16.

Parameter	Symbol	Value[1] in PSR1913+16	Value[1] in J0737-3039
Keplerian Parameters			
Eccentricity	e	0.6171338(4)	0.087779(5)
Orbital Period	P_b (day)	0.322997448930(4)	0.1022525563(1)
Post-Keplerian Parameters			
Periastron advance	$\langle \dot{\omega} \rangle$ ($^\circ \mathrm{yr}^{-1}$)	4.226595(5)	16.90(1)
Redshift/time dilation	γ' (ms)	4.2919(8)	0.38(5)
Orbital period derivative	$\dot{P}_b$ (10^{-12})	$-2.4184(9)$	
Shapiro delay ($\sin i$)	s		0.9995($-32, +4$)

[1] Numbers in parentheses denote errors in last digit.

The results also test the strong-field aspects of GR in the following way: the neutron stars that comprise the system have very strong internal gravity, contributing as much as several tenths of the rest mass of the bodies (compared to the orbital energy, which is only 10^{-6} of the mass of the system). Yet in general relativity, the internal structure is "effaced" as a consequence of the Strong Equivalence Principle (SEP), a stronger version of EEP that includes *gravitationally* bound bodies and local *gravitational* experiments. As a result, the orbital motion and gravitational radiation emission depend *only* on the masses m_1 and m_2, and not on their internal structure. By contrast, in alternative metric theories, SEP is not valid in general, and internal-structure effects can lead to significantly different behavior, such as the emission of dipole gravitational radiation. Unfortunately, in the case of scalar-tensor theories of gravity, because the neutron stars are so similar in PSR 1913+16 (and in other double-neutron star binary pulsar systems), dipole radiation is suppressed by symmetry; the best bound on the coupling parameter $\omega(\phi_0)$ from PSR 1913+16 is in the hundreds.

However, the recent discovery of the relativistic neutron star/white dwarf binary pulsar J1141-6545, with a 0.19 day orbital period, may ultimately lead to a very strong bound on dipole radiation, and thence on scalar-tensor gravity[55,56]. The remarkable "double pulsar" J0737-3039 is a binary system with two detected pulsars, in a 0.10 day orbit seen almost edge on, with eccentricity $e = 0.09$, and a periastron advance of $17°$ per year. A variety of novel tests of relativity, neutron star structure, and pulsar magnetospheric physics will be possible in this system[57,58]. For a review of binary pulsar tests, see[3].

5. Gravitational-Wave Tests of Gravitation Theory

The detection of gravitational radiation by either laser interferometers or resonant cryogenic bars will, it is widely stated, usher in a new era of gravitational-wave astronomy[59,60]. Furthermore, it will yield new and interesting tests of general relativity (GR) in its radiative regime[61].

5.1. *Polarization of gravitational waves*

A laser-interferometric or resonant bar gravitational-wave detector measures the local components of a symmetric 3×3 tensor which is composed of the "electric" components of the Riemann tensor, R_{0i0j}. These six independent components can be expressed in terms of polarizations (modes with specific transformation properties under null rotations). Three are transverse to the direction of propagation, with two representing quadrupolar deformations and one representing an axisymmetric "breathing" deformation. Three modes are longitudinal, with one an axially symmetric stretching mode in the propagation direction, and one quadrupolar mode in each of the two orthogonal planes containing the propagation direction. General relativity predicts only the first two transverse quadrupolar modes, independently of the source, while scalar-tensor gravitational waves can in addition contain the transverse breathing mode. More general metric theories predict up to the full complement of six modes. A suitable array of gravitational antennas could delineate or limit the number of modes present in a given wave. If distinct evidence were found of any mode other than the two transverse quadrupolar modes of GR, the result would be disastrous for GR. On the other hand, the absence of a breathing mode would not necessarily rule out scalar-tensor gravity, because the strength of that mode depends on the nature of the source.

5.2. *Speed of gravitational waves*

According to GR, in the limit in which the wavelength of gravitational waves is small compared to the radius of curvature of the background spacetime, the waves propagate along null geodesics of the background spacetime, *i.e.* they have the same speed, c, as light. In other theories, the speed could differ from c because of coupling of gravitation to "background" gravitational fields. For example, in some theories with a flat background metric η, gravitational waves follow null geodesics of η, while light follows null geodesics of $\mathbf{g}^1$. In brane-world scenarios, the apparent speed of gravitational waves could differ from that of light if the former can propagate off the brane into the higher dimensional "bulk". Another way in which the speed of gravitational waves could differ from c is if gravitation were propagated by a massive field (a massive graviton), in which case v_g would be given by, in a local inertial frame,

$$\frac{v_g}{c} = \left(1 - \frac{m_g^2 c^4}{E^2}\right)^{1/2} \approx 1 - \frac{1}{2}\frac{c^2}{f^2\lambda_g^2},$$ (10)

where m_g, E and f are the graviton rest mass, energy and frequency, respectively, and $\lambda_g = h/m_g c$ is the graviton Compton wavelength ($\lambda_g \gg c/f$ assumed). An example of a theory with this property is the two-tensor massive graviton theory of Visser[62].

The most obvious way to test for a massive graviton is to compare the arrival times of a gravitational wave and an electromagnetic wave from the same event, *e.g.* a supernova. For a source at a distance D, the resulting bound on the difference $|1 - v_g/c|$ or on λ_g is

$$\left|1 - \frac{v_g}{c}\right| < 5 \times 10^{-17}\left(\frac{200\,\mathrm{Mpc}}{D}\right)\left(\frac{\Delta t}{1\,\mathrm{s}}\right),$$ (11)

$$\lambda_g > 3 \times 10^{12}\,\mathrm{km}\left(\frac{D}{200\,\mathrm{Mpc}}\frac{100\,\mathrm{Hz}}{f}\right)^{1/2}\left(\frac{1}{f\Delta t}\right)^{1/2},$$ (12)

where $\Delta t \equiv \Delta t_a - (1 + Z)\Delta t_e$ is the "time difference", where Δt_a and Δt_e are the differences in arrival time and emission time, respectively, of the two signals, and Z is the redshift of the source. In many cases, Δt_e is unknown, so that the best one can do is employ an upper bound on Δt_e based on observation or modelling.

However, there is a situation in which a bound on the graviton mass can be set using gravitational radiation alone[63]. That is the case of the inspiralling compact binary, the final stage of evolution of systems like the binary pulsar, in which the loss of energy to gravitational waves has brought the binary to an inexorable spiral toward a final merger. Because the frequency of the gravitational radiation sweeps from low frequency at the initial moment of observation to higher frequency at the final moment, the speed of the gravitational waves emitted will vary, from lower speeds initially to higher speeds (closer to c) at the end. This will cause a distortion of the observed phasing of the waves and result in a shorter than expected overall time Δt_a of passage of a given number of cycles. Furthermore, through the technique of matched filtering, the parameters of the compact binary can be measured accurately[64], and thereby the effective emission time Δt_e can be determined accurately.

Table 3. Potentially achievable bounds on λ_g from gravitational-wave observations of inspiralling compact binaries.

$m_1(M_\odot)$	$m_2(M_\odot)$	Distance (Mpc)	Bound on λ_g (km)
Ground-based (LIGO/VIRGO)			
1.4	1.4	300	4.6×10^{12}
10	10	1500	6.0×10^{12}
Space-based (LISA)			
10^7	10^7	3000	6.9×10^{16}
10^5	10^5	3000	2.3×10^{16}

A full noise analysis using proposed noise curves for the advanced LIGO ground-based detectors, and for the proposed space-based LISA antenna yields potentially achievable bounds that are summarized in Table 3. These potential bounds can be compared with the solid bound $\lambda_g > 2.8 \times 10^{12}$ km, derived from solar system dynamics, which limit the presence of a Yukawa modification of Newtonian gravity of the form $V(r) = (GM/r)\exp(-r/\lambda_g)$[65], and with the model-dependent bound $\lambda_g > 6 \times 10^{19}$ km from consideration of galactic and cluster dynamics[62].

5.3. *Tests of scalar-tensor gravity*

Scalar-tensor theories generically predict dipole gravitational radiation, in addition to the standard quadrupole radiation, which results in modifications in gravitational-radiation back-reaction, and hence in the evolution of the phasing of gravitational waves from inspiralling sources. The effects are strongest for systems involving a neutron star and a black hole. Double neutron star systems are less promising because the small range of masses near 1.4 $M_\odot$ with which they seem to occur results in suppression of dipole radiation by symmetry. Double black-hole systems turn out to be observationally identical in the two theories, because black holes by themselves cannot support scalar "hair" of the kind present in these theories. Dipole radiation will be present in black-hole neutron-star systems, however, and could be detected or bounded via matched filtering[66].

Interesting bounds could be obtained using observations of low-frequency gravitational waves by a space-based LISA-type detector. For example, observations of a $1.4M_\odot$ NS inspiralling to a $10^3 M_\odot$ BH with a signal-to-noise ratio of 10 could yield a bound on ω between 2.1×10^4

and 2.1×10^5, depending on whether spins play a significant role in the inspiral[67,68,69].

6. Conclusions

Einstein's relativistic triumph of 1905 and its follow-up in 1915 altered the course of science. They were triumphs of the imagination and of theory; experiment played a secondary role. In the past four decades, we have witnessed a second triumph for Einstein, in the systematic, high-precision experimental verification of his theories. Relativity has passed every test with flying colors. But the work is not done. Tests of strong-field gravity in the vicinity of black holes and neutron stars need to be carried out. Gammay-ray, X-ray and gravitational-wave astronomy will play a critical role in probing this largely unexplored aspect of general relativity.

General relativity is now the "standard model" of gravity. But as in particle physics, there may be a world beyond the standard model. Quantum gravity, strings and branes may lead to testable effects beyond standard general relativity. Experimentalists will continue a vigorous search for such effects using laboratory experiments, particle accelerators, space instrumentation and cosmological observations. At the centenary of relativity it could well be said that experimentalists have joined the theorists in relativistic paradise.

This work was supported in part by the US National Science Foundation, Grant No. PHY 03-53180.

References

1. C. M. Will, *Theory and Experiment in Gravitational Physics*, (Cambridge University Press, Cambridge, 1993).
2. C. M. Will, Living Rev. Relativ. **4**, 4 (2001) [Online article]: cited on 1 April 2005, http://www.livingreviews.org/lrr-2001-4.
3. I. H. Stairs Living Rev. Relativ. **6** 5 (2003) [On-line article] Cited on 1 April 2005, http://www.livingreviews.org/lrr-2003-5
4. D. Mattingly, Living Rev. Relativ. (submitted) (gr-qc/0502097).
5. R. V. Eötvös, V. Pekár and E. Fekete, Ann. d. Physik **68**, 11 (1922).
6. P. G. Roll, R. Krotkov and R. H. Dicke, Ann. Phys. (N.Y.) **26**, 442 (1964).
7. V. B. Braginsky and V. I. Panov, Sov. Phys. JETP **34**, 463 (1972).
8. S. Baessler, B. R. Heckel, E. G. Adelberger, J. H. Gundlach, U. Schmidt and H. E. Swanson, Phys. Rev. Lett. **83**, 3585 (1999).
9. J. G. Williams, X. X. Newhall and J. O. Dickey, Phys. Rev. D **53**, 6730 (1996).

10. J. G. Williams, S. Turyshev and T. W. Murphy, Int. J. Mod. Phys. D **13**, 567 (2004).
11. T. M. Niebauer, M. P. McHugh, and J. E. Faller, Phys. Rev. Lett. **59**, 609 (1987).
12. J. C. Long, H. W. Chan and J. C. Price, Nucl. Phys. B **539**, 23 (1999).
13. C. D. Hoyle, U. Schmidt, B. R. Heckel, E. G. Adelberger, J. H. Gundlach, D. J. Kapner and H. E. Swanson, Phys. Rev. Lett. **86**, 1418 (2001).
14. C. D. Hoyle, D. J. Kapner, B. R. Heckel, E. G. Adelberger, J. H. Gundlach, U. Schmidt, and H. E. Swanson, Phys. Rev. D **70**, 042004 (2004).
15. J. Chiaverini, S. J. Smullin, A. A. Geraci, D. M. Weld, A. Kapitulnik, Phys. Rev. Lett. **90**, 151101 (2003).
16. J. C. Long, H. W. Chan, A. B. Churnside, E. A. Gulbis, M. C. M. Varney and J. C. Price, Nature **421**, 922 (2003).
17. A. A. Michelson and E. W. Morley, Am. J. Sci. **134**, 333 (1887).
18. R. S. Shankland, S. W. McCuskey, F. C. Leone and G. Kuerti, Rev. Mod. Phys. **27**, 167 (1955).
19. T. S. Jaseja, A. Javan, J. Murray and C. H. Townes, Phys. Rev. **133**, A1221 (1964).
20. A. Brillet and J. L. Hall, Phys. Rev. Lett. **42**, 549 (1979).
21. H. E. Ives and G. R. Stilwell, J. Opt. Soc. Am. **28**, 215 (1938).
22. B. Rossi and D. B. Hall, Phys. Rev. **59**, 223 (1941).
23. F. J. M. Farley, J. Bailey, R. C. A. Brown, M. Giesch, H. Jöstlein, S. van der Meer, E. Picasso and M. Tannenbaum, Nuovo Cimento **45**, 281 (1966).
24. K. Brecher, Phys. Rev. Lett. **39**, 1051 (1977).
25. T. Alväger, F. J. M. Farley, J. Kjellman and I. Wallin, Phys. Lett. **12**, 260 (1977).
26. D. C. Champeney, G. R. Isaak and A. M. Khan, Phys. Lett. **7**, 241 (1963).
27. E. Riis, L.-U. A. Anderson, N. Bjerre, O. Poulson, S. A. Lee and J. L. Hall, Phys. Rev. Lett. **60**, 81 (1988); *ibid* **62**, 842 (1989).
28. T. P. Krisher, L. Maleki, G. F. Lutes, L. E. Primas, R. T. Logan, J. D. Anderson and C. M. Will, Phys. Rev. D **42**, 731 (1990).
29. M. P. Haugan and C. M. Will, Phys. Today **40**, 69 (May) (1987).
30. M. D. Gabriel and M. P. Haugan, Phys. Rev. D **41**, 2943 (1990).
31. A. P. Lightman and D. L. Lee, Phys. Rev. D **8**, 364 (1973).
32. D. Colladay and V. A. Kostalecky, Phys. Rev. D **55**, 6760 (1997).
33. D. Colladay and V. A. Kostalecky, Phys. Rev. D **58**, 116002 (1998).
34. V. A. Kostalecky and M. Mewes, Phys. Rev. D **66**, 056005 (2002).
35. V. W. Hughes, H. G. Robinson and V. Beltran-Lopez, Phys. Rev. Lett. **4**, 342 (1960).
36. R. W. P. Drever, Phil. Mag. **6**, 683 (1961).
37. J. D. Prestage, J. J. Bollinger, W. M. Itano and D. J. Wineland, Phys. Rev. Lett. **54**, 2387 (1985).
38. S. K. Lamoreaux, J. P. Jacobs, B. R. Heckel, F. J. Raab and E. N. Fortson, Phys. Rev. Lett. **57**, 3125 (1986).
39. T. E. Chupp, R. J. Hoare, R. A. Loveman, E. R. Oteiza, J. M. Richardson, M. E. Wagshul and A. K. Thompson, Phys. Rev. Lett. **63**, 1541 (1989).

40. T. Jacobson, S. Liberati, and D. Mattingly, Phys. Rev. D **67**, 124011 (2003).
41. R. F. C. Vessot, M. W. Levine, E. M. Mattison, E. L. Blomberg, T. E. Hoffman, G. U. Nystrom, B. F. Farrell, R. Decher, P. B. Eby, C. R. Baugher, J. W. Watts, D. L. Teuber and F. O. Wills, Phys. Rev. Lett. **45**, 2081 (1980).
42. A. Godone, C. Novero, P. Tavella, and K. Rahimullah, Phys. Rev. Lett. **71**, 2364 (1993).
43. J. D. Prestage, R. L. Tjoelker, and L. Maleki, Phys. Rev. Lett. **74**, 3511 (1995).
44. H. Marion, F. Pereira Dos Santos, M. Abgrall, S. Zhang, Y. Sortais, S. Bize, I. Maksimovic, D. Calonico, J. Grünert, C. Mandache, P. Lemonde, G. Santarelli, Ph. Laurent, A. Clairon and C. Salomon, Phys. Rev. Lett. **90**, 150801 (2003).
45. S. Bize, S. A. Diddams, U. Tanaka, C. E. Tanner, W. H. Oskay, R. E. Drullinger, T. E. Parker, T. P. Heavner, S. R. Jefferts, L. Hollberg, W. M. Itano, and J. C. Bergquist, Phys. Rev. Lett. **90**, 150802 (2003).
46. T. Damour and F. Dyson, Nucl. Phys. B **480**, 37 (1996).
47. C. J. A. P. Martins, editor, *The Cosmology of Extra Dimensions and Varying Fundamental Constants* (Kluwer Academic Publishers, The Netherlands, 2003); also published in Astrophys.Space Sci. **283**, 439 (2003).
48. B. Bertotti, L. Iess and P. Tortora, *Nature* **425**, 374 (2003).
49. S. S. Shapiro, J. L. Davis, D. E. Lebach, and J. S. Gregory, Phys. Rev. Lett. **92**, 121101 (2004).
50. GAIA: for information about the project, see http://astro.estec.esa.-nl/GAIA/.
51. S. G. Turyshev, M. Shao, and K. Nordtvedt, Jr., Class. Quantum Gravit. **21**, 2773 (2004).
52. For information about Gravity Probe B see http://einstein.stanford.edu.
53. I. Ciufolini and E. C. Pavlis, Nature **431**, 958 (2004).
54. J. M. Weisberg, J. H. Taylor, in *Binary Radio Pulsars*, edited by F. A. Rasio and I. H. Stairs (Astronomical Society of the Pacific Conference Series, Vol. 328), p. 25.
55. M. Bailes, S. M. Ord, H. S. Knight and A. W. Hotan, Astrophys. J. Lett. **595**, L52 (2003).
56. G. Esposito-Farèse, in *Proceedings of the 10th Marcel Grossmann Meeting*, in press (gr-qc/0402007).
57. A. G. Lyne, M. Burgay, M. Kramer, A. Possenti, R. N. Manchester, F. Camilo, M. A. McLaughlin, D. R. Lorimer, N. D'Amico, B. C. Joshi, J. Reynolds, P. C. C. Freire, Science **303**, 1153 (2004).
58. M. Kramer, D. R. Lorimer, A. G. Lyne, M. A. McLaughlin, M. Burgay, N. D'Amico, A. Possenti, F. Camilo, P. C. C. Freire, B. C. Joshi, R. N. Manchester, J. Reynolds, J. Sarkissian, I. H. Stairs, and R. D. Feldman, *Proceedings of The 22nd Texas Symposium on Relativistic Astrophysics*, submitted (astro-ph/0503386).
59. K. S. Thorne, In *300 Years of Gravitation*, edited by S.W. Hawking and W. Israel (Cambridge University Press, Cambridge, 1987), p. 330.
60. B. C. Barish and R. Weiss, Phys. Today **52**, 44 (October) (1999).

61. C. M. Will, Phys. Today **52**, 38 (October) (1999).
62. M. Visser, Gen. Relativ. and Gravit. **30**, 1717 (1998).
63. C. M. Will, Phys. Rev. D **57**, 2061 (1998).
64. C. Cutler, T. A. Apostolatos, L. Bildsten, L. S. Finn, É. E. Flanagan, D. Kennefick, D. M. Marković, A. Ori, E. Poisson, G. J. Sussman, and K. S. Thorne, Phys. Rev. Lett. **70**, 2984 (1993).
65. C. Talmadge, J.-P. Berthias, R. W. Hellings, and E. M. Standish, Phys. Rev. Lett. **61**, 1159 (1988).
66. C. M. Will, Phys. Rev. D **50**, 6058 (1994).
67. P. D. Scharre and C. M. Will, Phys. Rev. D **65**, 042002 (2002).
68. C. M. Will and N. Yunes, Class. Quantum Gravit. **21**, 4367 (2004).
69. E. Berti, A. Buonanno and C. M. Will, Phys. Rev. D **71**, 084025 (2005).

CHAPTER 9

RECEIVING GRAVITATIONAL WAVES

PETER R. SAULSON

Department of Physics, Syracuse University
201 Physics Building, Syracuse, NY 13244-1130, USA
saulson@physics.syr.edu

The quest for successful reception of gravitational waves is a long story, with the ending still to be written. In this article, I trace the story from thought experiment, through increasingly sophisticated design and prototyping stages, to the point where observations of unprecedented sensitivity are now getting under way. In the process, one can see the sort of effort that is required to transform beautiful and simple relativistic ideas into sensitive experiments that actually work.

1. Introduction

For this special issue on the 100th anniversary of the theory of relativity, I would like to sketch an inspiring and informative story of one path of its development, in a direction that still places relativity at the frontier of scientific exploration in 2005: the reception of gravitational waves. Rooted firmly in the prehistory of relativity (the Michelson-Morley experiment), it aims at validating key provisions of general relativity, and turning thence to exploring the universe in a brand new way. In telling this story, I hope to also illustrate a process that is explored too rarely, how a physical idea makes the transition from pure thought, through the stage of a thought experiment, and then through the various stages of accommodation to the realities of measurement in a world that has a lot going on beside the experiment one hopes to carry out.

In telling this story, it will not be possible to follow all of the historical byways, nor to pursue all of the physics lessons that one could draw from it. I will be somewhat ruthless about focusing on a single story line. For a more balanced treatment of the history (and much insight as well), the reader is invited to read both Marcia Bartusiak's *Einstein's Unfinished Symphony*[1] (aimed at general readers but telling the story fully and richly) and *Gravity's Shadow* by Harry Collins[2] (focusing on issues in the sociology of science, but using that focus to explore both the physics and the history of the subject.) For more depth on basic physics and technical issues, the author's *Fundamentals of Interferometric Gravitational Wave Detectors*[3] can be consulted.

The story line in this article will start with Einstein's original thoughts, and then turn to various thought experiments that explored the physical reality of gravitational waves. It is in that guise alone that I will treat the work of Joseph Weber; it is impossible in this brief article to do full justice to the wide range of his contributions to the field. Instead, I will concentrate narrowly on the development of interferometric detectors that are just now, in 2005, about to revolutionize the field. Even with that limitation, I will have to keep the focus narrowed mostly to the work of just one line of pioneers, those who most directly contributed to the development of just one of today's forefront detectors: the Laser Interferometer Gravitational Wave Observatory, or LIGO.[4]

2. Origin of the Idea of Gravitational Waves and Gravitational Wave Detectors

Einstein did not conceive of gravitational waves in 1905. But his work on relativity in that year did lay the foundation for gravitational waves to be described in 1916.[5] The most direct connection is this: one of the foundations of relativity is the idea that no signal can be transmitted faster than the speed of light. While Einstein did not immediately explore the connection with gravity, his thoughts soon turned to the problem of the un-relativistic behavior embodied in Newton's Law of Gravitation. The action at a distance that so troubled Newton himself is at the core of the conflict with relativity. If Newton's Law were literally true, a mass's gravitational field throughout all space would point directly at the mass, no matter what the state of motion of the mass. One could use this property to establish instantaneous gravitational communication across the universe; moving a mass rapidly back and forth would cause the gravitational field at large distances to point instantaneously to the different positions of the mass. A sufficiently sensitive gravimeter could read out a message.

General relativity solves this problem. By 1916, Einstein was able to show (although with a mistake that wasn't corrected until 1918) that his field equations, when specialized to the weak field case, could be cast in the form of a wave equation, with wave speed equal to c.

It wasn't at all obvious at first what the waves were. It took a long time for difficulties with gauge dependence of the wave solutions to be resolved. Along the way, Einstein and many of his contemporaries went through waves of doubt about the physical reality of gravitational waves. The oral history of physics contains stories of Einstein preparing a paper proving that gravitational waves were gauge objects only, only to pull the paper at the last moment.[6] What is clearly documented, by an odd kind of negative evidence, is Einstein's complete lack of conviction that gravitational waves would ever be detected. Einstein's technical and popular writings refer often to various experimental test of relativity; most famous are the three classic tests (precession of orbits, bending of light paths, and gravitational redshift). But I have been unable to find any mention anywhere in his writings of possible experiments to detect gravitational waves.

The layers of Einstein's doubts were only slowly peeled back, one by one. The proof that gravitational waves had an independent physical reality (i.e. that they were not just gauge artifacts) was made by showing that energy could be extracted from them. The thought experiment in question is usually attributed to Bondi.[7] The argument started by recognizing that a gravitational wave could cause relative motion between two spatially separated masses. (More on this below.) The physical reality of this motion, and thus of the gravitational wave that drove it, was made manifest by considering that the relative motion of the masses could produce sliding friction if there was a "rigid friction disk carried by one of them", against which the second mass was allowed to slide. Since energy would be turned locally into heat by the friction, it was clear that energy was delivered to the masses from the gravitational wave. Thus, gravitational waves carry energy, and must be physically real themselves.

Another proof of the reality of gravitational waves was provided by Joseph Weber, as part of his visionary program to make gravitational wave detection a reality.[8] (The authors of the proof described above were all theorists.) Weber's great insight was to write an explicit expression for the differential (or tidal) force due to a gravitational wave; then, it could be displayed as a driving force by being put on the right hand side of the equation of motion of, say, a simple harmonic oscillator. Writing a general relativistic gravitational effect as a force goes rather against the philosophy of relativity, but was a necessary step to allow simple manipulation of the gravitational wave's effects on an equal footing with elastic and frictional forces.

Weber's paper demonstration that gravitational waves were real was only the beginning of his contributions to the field. He went on to build and operate a number of gravitational wave detectors, resonant bars of aluminum which were implementations of his simple models. That is to say, he "reduced to practice" the idea of gravitational wave detection. The importance of this step cannot be emphasized too much. Notwithstanding the controversy that greeted Weber's claims to have actually seen gravitational waves, and the eventual rejection of those claims,[9] the whole field of gravitational wave detection owes its existence to Weber's work.

The line of thinking that led to LIGO and its cousins was prompted in part by Weber's work (and by his claimed detections), but physically it draws on simpler, more relativistic thinking. Rainer Weiss was teaching a relativity course at MIT in the late 1960's, and grappled with how best to find the measurable quantities among the many coordinate-dependent numbers in a calculation. A paper by Pirani[10] pointed Weiss to one key idea: the distance between freely-falling masses can be determined by measuring the round-trip travel time of light between them. The special role that freely-falling masses play in responding only to gravitational effects, and thus revealing the character of the space-time in which they are embedded, makes this an especially appealing kind of thought experiment. Weiss, like Weber a few years before him, was ready to make the next leap, that of finding a way to reduce his thought experiment to practice. In addition to the simplicity of the physics (no non-gravitational forces in the thought-experiment version), Weiss's alternative vision of how to detect gravitational waves was to offer important practical advantages as well.

(A note about the risk of historical over-simplification: Weber and a student, Robert Forward, also invented the same kind of gravitational wave detector as Weiss did, at about the same time. For a number of years, parallel development of the technology was carried out by Weiss at MIT and Forward at Hughes Research Lab.[11] There had also been an earlier mention of the same idea by Gertsenshtein and Pusovoit, although they weren't able to follow it up with laboratory development.[12])

3. Free-Mass Gravitational Wave Detectors

Let's follow the line of thinking that led to Weiss's (and Weber and Forward's) invention of interferometric gravitational wave detection. Imagine a plane in space in which a square grid has been marked out by a set of infinitesimal test masses. This is a prescription for embodying a section of a transverse traceless coordinate system, marking out coordinates by masses that are freely-falling (i.e. that feel no non-gravitational forces).

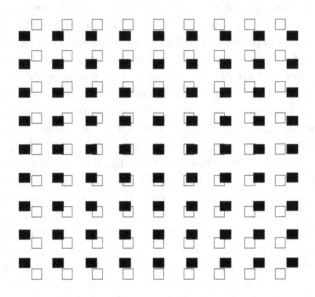

Fig. 1. An array of free test masses. The open squares show the positions of the masses before the arrival of the gravitational wave. The filled squares show the positions of the masses during the passage of a gravitational wave of the plus polarization.

Now imagine that a gravitational wave is incident on the set of masses, along a direction normal to the plane. Take this direction to be the z axis, and the masses to be arranged along the x and y axes. Then, if the wave has the polarization called h_+, it will cause equal and opposite shifts in the formerly equal x and y separations between neighboring masses in the grid. That is, for one polarity of the wave, the separations of the masses along the x direction will decrease, while simultaneously the separations along the y direction will increase. When the wave oscillates to opposite polarity, the opposite effect occurs.

If, instead, a wave of polarization $h_\times$ is incident on the set of test masses, then there will be (to first order in the wave amplitude) no changes in the distances between any mass and its nearest neighbors along the x and y directions. However, $h_\times$ is responsible for a similar pattern of distance changes between a mass and its neighbors along the diagonals of the grid.

There are several other aspects of the gravitational wave's deformation of the test system that are worth pondering. Firstly, the effect on any pair of neighbors in a given direction is identical to that on any other pair. The same *fractional* change occurs between other pairs oriented along the same direction, no matter how large their separation. This means that a larger *absolute* change in separation occurs, the larger is the original separation between two test masses. This property, that we can call "tidal" because of its similarity to the effect of ordinary gravitational tides, is exploited in in the design of interferometric detectors of gravitational waves.

Another aspect of this pattern that is worthy of note is that the distortion is uniform throughout the coordinate grid. This means that any one of the test masses can be considered to be at rest, with the others moving in relation to it. In other words, a gravitational wave does not cause any absolute acceleration, only relative accelerations between masses. This, too, is fully consistent with other aspects of gravitation as described by the general theory of relativity: a single freely-falling mass can not tell whether it is subject to a gravitational force. Only a measurement of relative displacements between freely-falling test masses (the so-called "geodesic deviation") can reveal the presence of a gravitational field.

4. A *Gedanken* Experiment to Detect a Gravitational Wave

To demonstrate the physical reality of gravitational waves, concentrate on three of the test masses, one chosen arbitrarily from the plane, along with its nearest neighbors in the $+x$ and $+y$ directions. Imagine that we have equipped the mass at the vertex of this "L" with a lamp that can be made to emit very brief pulses of light. Imagine also that the two masses at the ends of the "L" are fitted with mirrors aimed so that they will return the flashes of light back toward the vertex mass.

First, we will sketch how the apparatus can be properly set up, in the absence of a gravitational wave. Let the lamp emit a train of pulses, and observe when the reflected flashes of light are returned to the vertex mass by the mirrors on the two end masses. Adjust the distances from the vertex mass to the two end masses until the two reflected flashes arrive simultaneously.

Once the apparatus is nulled, let the lamp keep flashing, and wait for a burst of gravitational waves to arrive. When a wave of $\hat{h}_+$ polarization passes through the apparatus along the z axis, it will disturb the balance between the lengths of the two arms of the "L". Imagine that the gravita-

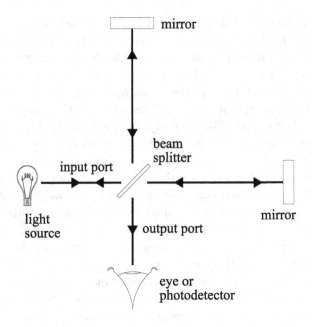

Fig. 2.   A schematic diagram of an apparatus that can detect gravitational waves. It
has the form of a Michelson interferometer.

tional wave has a waveform given by

$$h^{\mu\nu} = h(t)\widehat{h}_+.$$

To see how this space-time perturbation changes the arrival times of the
two returned flashes, let us carefully calculate the time it takes for light to
travel along each of the two arms.

First, consider light in the arm along the x axis. The interval between
two neighboring space-time events linked by the light beam is given by

$$\begin{aligned}
ds^2 = 0 &= g_{\mu\nu}dx^\mu dx^\nu \\
&= \left(\eta_{\mu\nu} + h_{\mu\nu}\right) dx^\mu dx^\nu \\
&= -c^2 dt^2 + \left(1 + h_{11}(t)\right) dx^2.
\end{aligned} \tag{1}$$

This says that the effect of the gravitational wave is to modulate the square
of the distance between two neighboring points of fixed coordinate separation dx (as marked, in this gauge, by freely-falling test particles) by a
fractional amount h_{11}.

We can evaluate the light travel time from the beam splitter to the end of the x arm by integrating the square root of Eq. 1

$$\int_0^{\tau_{out}} dt = \frac{1}{c} \int_0^L \sqrt{1 + h_{11}} dx \approx \frac{1}{c} \int_0^L \left(1 + \frac{1}{2} h_{11}(t)\right) dx, \qquad (2)$$

where, because we will only encounter situations in which $h \ll 1$, we've used the binomial expansion of the square root, and dropped the utterly negligible terms with more than one power of h. We can write a similar equation for the return trip

$$\int_{\tau_{out}}^{\tau_{rt}} dt = -\frac{1}{c} \int_L^0 \left(1 + \frac{1}{2} h_{11}(t)\right) dx. \qquad (3)$$

The total round trip time is thus

$$\tau_{rt} = \frac{2L}{c} + \frac{1}{2c} \int_0^L h_{11}(t) dx - \frac{1}{2c} \int_L^0 h_{11}(t) dx. \qquad (4)$$

The integrals are to be evaluated by expressing the arguments as a function just of the position of a particular wavefront (the one that left the beam-splitter at $t = 0$) as it propagates through the apparatus. That is, we should make the substitution $t = x/c$ for the outbound leg, and $t = (2L - x)/c$ for the return leg. Corrections to these relations due to the effect of the gravitational wave itself are negligible.

A similar expression can be written for the light that travels through the y arm. The only differences are that it will depend on h_{22} instead of h_{11} and will involve a different substitution for t.

If $2\pi f_{gw} \tau_{rt} \ll 1$, then we can treat the metric perturbation as approximately constant during the time any given flash is present in the apparatus. There will be equal and opposite perturbations to the light travel time in the two arms. The total travel time difference will therefore be

$$\Delta\tau(t) = h(t) \frac{2L}{c} = h(t) \tau_{rt0}, \qquad (5)$$

where we have defined $\tau_{rt0} \equiv 2L/c$.

If we imagine replacing the flashing lamp with a laser that emits a coherent beam of light, we can express the travel time difference as a phase shift by comparing the travel time difference to the (reduced) period of oscillation of the light, or

$$\Delta\phi(t) = h(t) \tau_{rt0} \frac{2\pi c}{\lambda}. \qquad (6)$$

Another way to say this is that the phase shift between the light that traveled in the two arms is equal to a fraction h of the total phase a light beam accumulates as it traverses the apparatus. This immediately says that the longer the optical path in the apparatus, the larger will be the phase shift due to the gravitational wave (until L reaches one quarter of the gravitational wavelength.)

Thus, this *gedanken* experiment has demonstrated that gravitational waves do indeed have physical reality, since they can (at least in principle) be measured. The measuring device that we invented turns out to be a Michelson interferometer. Furthermore, it suggests a straightforward interpretation of the dimensionless metric perturbation h. The gravitational wave amplitude gives the fractional change in the difference in light travel times along two perpendicular paths whose endpoints are marked by freely-falling test masses.

5. First Steps from Thought Experiment to Real Experiment

The essence of the argument given above is found in Weiss's first publication on the subject (actually only an internal MIT research progress report.)[13] But Weiss went much further than a thought experiment. The core of his 1972 paper was a serious account of how to construct an experiment, made with existing technology out of real materials, that could implement a good approximation to the thought experiment described above. He went on to list the noise sources that would limit the sensitivity of a gravitational wave detector built according to his design, a list that is remarkably complete even when read in 2005. The punch line of the whole argument is that it was easy to picture a gravitational wave detector that could perform substantially better than those with which Weber was claiming to see gravitational waves. Thus, either his discoveries could be followed up in greater detail (if they proved genuine), or much deeper searches could be carried out, if Weber's events weren't genuine (as eventually proved to be the case.)

It is important to have the basic scale of the measurement in mind, in order to appreciate the challenge that experimenters were facing. The first generation of Weber bars had a sensitivity to gravitational waves with strain amplitudes of order 10^{-16}. By the time one recognizes, with the benefit of hindsight, that Weber's results were erroneous and that one would have to look for much weaker signals before having a good chance of finding any, the sensitivity goal needs to be more like 10^{-21}, or perhaps even smaller. (The community consensus that this was roughly the required sensitivity was evident by the time of the Battelle/Seattle conference of

1978.[14]) This is a tremendous challenge to measurement; the extreme measures needed to meet the challenge are reflected in what Weiss proposed to build.

Let's look at the basic features of Weiss's proposal. Falling masses aren't easy to use in a terrestrial laboratory, so he substituted masses held so that their motion in the horizontal direction was that of a harmonic oscillator with low resonant frequency. (Weiss called for "horizontal seismometer suspensions", realized in current detectors as pendulums with very fine metal or silica fibers.) At frequencies above the resonance, the masses will respond as if they were free, while nevertheless staying in place on long time scales. For sensing the differential changes in test mass separation, Weiss called for a Michelson interferometer, as suggested in the thought experiment above. The nearly free test masses play the optical role of the mirrors of the interferometer, so that the phase difference between the arms fulfills Weiss's implementation of Pirani's prescription. The light source illuminating the interferometer was to be a powerful stable laser, which would enable very fine length comparisons to be made.

The dimensions of the interferometer are a crucial design parameter. The signal (i.e. the phase shift of the light in an arm) will grow as the separation of the masses grows, until the round trip light travel time reaches one half of the period of the gravitational wave. The simplest thing (and in many ways the best) would be to make a simple Michelson interferometer with one round trip for the light between beam splitter and end mirror. For a gravitational wave with a period of 1 ms, the arm length would be 75 km. At 10 ms, the length swells to 750 km. A nice idea, but hard to build on the Earth, and to afford to build. (For one thing, the space between the mirrors should be evacuated, so that there aren't spurious apparent length changes coming from variations in the index of refraction of air. The curvature of the Earth is also complicating factor for construction of straight paths of this length.)

The more practical thing to do is to place the test masses closer together than the ideal, but then to make up an optical path as long as the ideal by reflecting the light between them multiple times. Weiss proposed a practical optical folding scheme, based on Herriott delay lines. It worked, although most present detectors use a cleverer (although trickier) scheme proposed by Ron Drever. (See below.)

The length is a crucial design parameter because it is related to the signal to noise ratio. The signal is largest when the length is the longest (up to a quarter of a gravitational wavelength.) How does the noise go?

That depends on which noise source we are talking about. Measurement of phase shifts themselves depends on fluctuations in light power. If governed only by the statistical fluctuations in photon arrival rate ("shot noise") then one does best using the largest amount of light power. The signal to noise ratio is maximized by maximizing the optical path length and the light power, but doesn't depend on whether or not the optical path is folded to fit into a smaller physical length.

A different tradeoff occurs for most of the other noise sources that affect an interferometer. They are mainly effects that cause the test masses to move. For this kind of noise (generically called displacement noise), the more times the light bounces off a mirror, the greater the degradation of the signal to noise ratio. For this reason, the best interferometric gravitational wave detector is the longest one it is possible to build. This is the reason that each LIGO site is 4 km long.

Weiss produced a rather long list of displacement noise terms. The presence of each in significant amounts calls for the longest arms that are possible, but beyond that each is mitigated by a separate strategy. The length of the list gives one hint of why the successful implementation of gravitational wave detection is a challenge.

Here is Weiss's list:

- mechanical thermal noise, or Brownian motion, of the test masses and each of their internal modes of vibration
- radiation pressure noise from the light in the interferometer
- seismic noise
- "radiometer effect" noise, from residual gas in the vacuum being heated where light is absorbed on mirror surfaces
- cosmic ray impacts
- gravitational-gradient noise, or noise arising from the Newtonian gravitational force of moving massive bodies, from earth motion due to seismic noise, or from density fluctuations in the air around the interferometer
- forces from time-varying electric fields (such as the "patch effect") interacting with the test masses
- forces from time-varying magnetic fields interacting with any magnets on the test masses (for control) or with magnetic impurities.

This list defined much of the design work that followed Weiss's 1972 paper. His list was remarkably prescient, so much so that the few items not on it represent interesting stories in their own right. Below, we will compare the important noise sources at various future stages of design, and in the nearly completed interferometers of 2005, to Weiss's list.

It is important to mention one other feature of Weiss's proposed design. Weiss understood that, in choosing an interferometer as the measuring device, he had paid for high sensitivity with small dynamic range. That is, an interferometer gives an output proportional to the input motions only over a small range of motion; the output is actually a periodic function of the input. Furthermore, minimization of the shot noise and the other less fundamental (but still important) optical noise sources requires that the interferometer stay very close to a chosen operating point. Finally, even to keep the approximately-free masses properly aligned with one another so that the light forms an interferometer, it is necessary to be able to control the angles of the mirrors. As a result, it is necessary to equip the interferometer with an elaborate control system. Much of the careful engineering of a successful interferometer goes into a control system that can meet all of the requirements listed here, without in itself injecting noise that would dominate the experiment's noise budget. (This makes the experiment design bear a strong resemblance to the "null instrument" character of, for example, the Eötvös experiment designed by Weiss's postdoctoral mentor Robert Dicke.[15])

The arc of the story line is starting to become apparent here. At the level of basic physics, freely-falling masses are the natural way to make manifest the fundamental structure of space-time, and in particular to reveal the passage of a gravitational wave. At the level of a thought experiment, too, it is natural to measure the separation of freely-falling masses by bouncing light off of their mirrored surfaces. Once one tries to see how to make a precise measurement, though, suddenly the simplicity vanishes. Freely-falling masses are replaced by masses held in low-frequency harmonic oscillators. Those in turn need to rest on elaborate mechanical filters to reject seismic noise. And all of the optical components must have their positions and angles continuously controlled at low frequencies in order for the interferometer to function at all.

It is almost a wonder that systems like this can approximate in any way the ideal of a set of freely-falling masses. And yet they do, at least at high frequencies, where the suspension's forces are negligible and the effect of control forces on the masses is small or accounted for. But the contrast between the pure and simple world of theory and the complex world of a high-sensitivity experiment is striking.

6. Further Advances

Weiss's plan laid out a basic scheme for sensitive detectors of gravitational waves, but by itself the plan would not have sufficed to make what has come to be known as LIGO. Key additional features were contributed by a host of other physicists and engineers.

One very important contributor was Ron Drever. He had been drawn into gravitational wave detection by Weber's work. When his (and others') repetitions of Weber's experiment failed to yield any evidence for the signals that Weber claimed to see, Drever turned his attention to ways in which much more sensitive gravitational wave detectors could be built.

Drever had learned of the idea to make interferometric detectors, and began exploring this technology himself. But he soon became convinced that the beam-folding technology proposed by Weiss, the Herriott delay line, had numerous practical difficulties. One was that the mirrors had to be quite large in order to keep the various reflections spatially separated, of order 1 m in diameter for a version with kilometer-scale arm lengths. Another problem (first recognized by the Max Planck group working in Garching, Germany[16]) was that light could scatter from the intended path into other spots in the pattern, due to imperfections in the mirror surfaces. Time variations in this scattering, caused by fluctuations in the laser wavelength or by low-frequency motion of the mirrors, became a serious source of noise (not on Weiss's list.)

Drever proposed a solution that has been adopted by most of the large interferometers operating today. Instead of spatially-separated reflections from the mirrors, superimpose them all in one place. This sounds like non-sense until one realizes that a pair of mirrors can be operated as a Fabry-Perot cavity, in which the light reflected from the cavity (near resonance) changes phase as the length of the cavity changes, as though it had made a large number of discrete round trips.

Drever, in a set of lecture notes where he first explained this idea, is explicit about both the advantages and the drawbacks of this beam-folding scheme.[17]

The diameter of the cavity mirrors can be considerably smaller than that of delay-line mirrors.... This reduces the diameter of the vac-uum pipe required, and also may make it easier to keep mechanical resonances in the mirrors and their mountings high compared with the frequency of the gravity waves, thus minimizing thermal noise. The Fabry-Perot system has however, some obvious disadvantages too — particularly the requirement for the very precise control of the wavelength of the laser and of the lengths of the cavities. Indeed with long cavities of the high finesse desirable here exceptional short-term wavelength stability is required from the laser.

The heart of the difficulty is that, unlike a delay line, a Fabry-Perot cavity stores light because it is in itself an interferometer — the trapping of the light for many round trips comes about only by careful adjustment of the phases of the superposed beams. This can only occur when the wavelength of the light and the length of the cavity are in resonance, that is matched so that an integer number of waves fits into the cavity. Very near the resonance condition, the phase of the output light varies with mirror separation in the same way as the light that has traveled through a delay line. To achieve this condition, the light and the arm have to be locked together by a servo system. Drever's lecture goes on to describe the style of servo required, one that he and his group developed in conjunction with John Hall's group at the Joint Institute for Laboratory Astrophysics in Boulder, Colorado.[18] This servo design has its roots in an analogous microwave device developed by Robert Pound.[19]

While the essence of the difficulty was thus solved, in practice the use of Fabry-Perot cavities has additional complications. One is due to the fact that when a cavity is not very close to resonance, the phase of the output light has almost no dependence at all on the separation of the mirrors, thus making it very hard to generate the sort of signal necessary to acquire the lock on resonance in the first place. An additional level of complication comes when the arm cavities are assembled into a complete Michelson interferometer, since the interferometers within the interferometer need to be separately controlled without degrading the function of the main instrument. Solving these sorts of problems robustly proved to be challenging work. That it now has been done successfully in LIGO is a great engineering triumph.

Drever goes on to show another important improvement in sensitivity that can be achieved with Fabry-Perot technology. It starts from recognizing an opportunity in operating the interferometer so that, in the absence of a gravitational wave, no light exits the interferometer toward the output photodetector; instead, all of the light returns toward the laser. Weiss had proposed this operating point as a way of minimizing excess noise in the interferometer readout. Drever's insight was that, given the ultra-low levels of absorption then becoming available in mirror coatings, the power in that "waste" light was nearly as great as that of the fresh light arriving from the laser. Drever proposed an additional mirror to redirect the output light into the interferometer, in superposition with the new light. In effect, the whole interferometer becomes a single Fabry-Perot resonant cavity. This system, called power recycling, has an advantage in available effective light power that is dramatic; as implemented in LIGO, it is as if one has a laser 30 times more powerful than the one actually used. Thus, there is a large reduction in the shot noise (which shrinks as the light power grows) that is the fundamental limit to interferometer readout precision. This advantage is paid for by an additional layer of complexity of interferometer control, beyond that of just using Fabry-Perot cavities for the arms of the interferometer. Nevertheless, the challenge has been met, and the benefits achieved.

7. First Steps Toward Kilometer-Scale Interferometers

Laboratory-scale implementations of these ideas were pursued in several places: at MIT, Glasgow, Caltech, and Garching. The best of these efforts achieved a great deal of success in operating interferometers which performed about as well as their noise budgets would allow.[20] But none of them was able to catch up with the continuing progress in development of Weber-style detectors, now cooled with cryogens. Nor were these interferometers intended to do so. Instead, they were considered prototypes of devices that could truly implement Weiss's vision, interferometers with kilometer-scale arms. Indeed, the groups mentioned above came together as two of today's leading projects: the Glasgow and Garching groups formed the core of the GEO collaboration, while the MIT and Caltech groups eventually coalesced as the LIGO project. The Italian and French groups that formed the Virgo project, and the Japanese groups that built TAMA, also formed at about the same time. There is also an active effort in Australia.

Now, the arc of our story moves farther still from the beautiful idealization of pure theory or the thought experiment. Large interferometers are inherently big-science projects. Construction of a high precision scientific instrument kilometers on a side calls for careful engineering as more limitations of the practical world join the picture. (A new concept, cost, enters the picture for the first time, but it is beyond the scope of this article to discuss it.)

Each of the large projects mentioned above went through similar planning processes, at roughly the same time. For pedagogical purposes, I choose to focus on a single line of development, that of the U.S. LIGO Project.

In almost the same sense as the early table-top interferometers were prototypes of larger instruments, so too did the proposals for kilometer-scale interferometers have a prototype. This was the report called "A Study of a Long Baseline Gravitational Wave Antenna System", submitted to the U.S. National Science Foundation in October 1983.[21] (It has since its presentation been called the "Blue Book" because of the color of the cheap paper cover in which it was bound.) It was prepared as the product of a planning exercise funded by the National Science Foundation starting in 1981. The most novel feature is that, in addition to sections describing the physics of gravitational wave detection, it also contains extensive sections written by industrial consultants from Stone & Webster Engineering Corporation and from Arthur D. Little, Inc. These latter contributors were essential, because this document contains, for the first time anywhere, an extensive discussion of the engineering details specific to the problems of the construction and siting of a large interferometer. The report was presented by the MIT and Caltech groups at a meeting of the NSF's Advisory Council for Physics late in 1983. While not a formal proposal, it served as a sort of "white paper", suggesting the directions that subsequent proposals might (and in large measure did) take.

The industrial study was undertaken with the aim of identifying what design trade-offs would allow for a large system to be built at minimum cost, and to establish a rough estimate of that cost (along with cost scaling laws) so that the NSF could consider whether it might be feasible to proceed with a full-scale project. Before such an engineering exercise could be meaningful, though, it was necessary to define what was meant by "full-scale". The Blue Book approaches this question by first modeling the total noise budget as a function of frequency, then evaluating the model as a function of arm lengths ranging from 50 meters (not much longer than the Caltech prototype) to 50 km. The design space embodied in this model was then explored in a process guided by three principles:

- "The antenna should not be so small that the fundamental limits of performance can not be attained with realistic estimates of technical capability." This was taken to mean that the length ought to be long enough that one could achieve shot noise limited performance for laser power of 100 W, without being limited instead by displacement noise sources, over a band of interesting frequencies. The length resulting from this criterion strongly depended on whether one took that band to begin around 1 kHz (in which case

$L = 500$ m was adequate), 100 Hz (where $L = 5$ km was only approaching the required length), or lower still (in which case even $L = 50$ km would not suffice.) Evidently, this strictly physics-based criterion was too elastic to be definitive.

- "The scale of the system should be large enough so that further improvement of the performance by a significant factor requires cost increments by a substantial factor." In other words, the system should be long enough so that the cost is not dominated by the length-independent costs of the remote installation.

- "Within reason no choice in external parameters of the present antenna design should preclude future internal design changes which, with advances in technology, will substantially improve performance." This was a justification for investing in a large-diameter beam tube, and for making sure that the vacuum system could achieve pressures as low as 10^{-8} torr.

In an iterative process, rough application of these principles was used to set the scope of options explored by the industrial consultants. Then at the end of the process, the principles were used again to select a preferred design. Arm lengths as long as 10 km were explored, and tube diameters as large as 48 inches. An extensive site survey was also carried out by the consultants. It was aimed at establishing that sites existed that were suitable for a trenched installation (which put stringent requirements on flatness of the ground) of a 5 km interferometer. The survey covered Federal land across the United States, and a study of maps of all land in the Northeastern United States, along with North Carolina, Colorado, and Nebraska. Thirteen "suitable" sites were identified. Evaluation criteria also included land use (specifically that the site not be crossed by roads, railroads, or oil and gas pipelines), earthquake risk, drainage, and accessibility.

The site survey also attempted to identify possibilities of locating an interferometer in a subsurface mine, which would give a more stable thermal environment and perhaps also reduced seismic noise (if it were located deep enough, and if it were inactive.) No mines were found in the United States with two straight orthogonal tunnels even 2 km in length.

The conclusion of the exercise was a "proposed design" with the following features:

- Two interferometer installations separated by "continental" distances.
- Interferometer arm length of $L = 5$ km.

- Beam tubes of 48 inch diameter made of aluminum (chosen for an expected cost savings over stainless steel) pumped by a combination of Roots-blowers for roughing and ion-pumps for achieving and maintaining the high vacuum. A delay line interferometer would require a diameter of almost the proposed size. The large diameter would also allow the installation of multiple Fabry-Perot interferometers side by side.
- The proposed installation method was to enclose the tube in a 7' by 12' cover constructed of a "multi-plate pipe-arch", in turn installed 4 feet below grade in a trench that was subsequently back-filled with soil.

Note that there were no specific recommendations for the design of the interferometers themselves, beyond the "straw man" used for estimating the noise budget.

The Blue Book was presented by the MIT and Caltech groups to the National Science Foundation's Advisory Committee for Physics, and got a respectful reception. As a result, the MIT and Caltech research groups were encouraged to combine their forces to develop a complete specific design. Subsequently, both groups received funding with the eventual goal of a joint proposal for construction of a large interferometer system.

The years between the Blue Book's submission in 1983 and the approval of the construction of LIGO in 1991 were eventful ones. The Caltech and MIT groups worked together, adopting common management in 1987. Progress on laboratory prototypes was heartening. A 40-meter interferometer at Caltech achieved shot-noise limited sensitivity above 1 kHz, at a level of $h(f) \approx 2 \times 10^{-19}/\sqrt{\text{Hz}}$. This level of noise demonstrated that, at least at those high frequencies, there were no substantial displacement noise sources unaccounted for in the noise budget.

The 1989 proposal to build LIGO specified these key features:

- LIGO would consist of two widely separated sites under common management. This would allow searches for transient events to make use of coincidence techniques.
- The instruments would be Michelson interferometers with the Fabry-Perot system for beam folding in the arms and for power recycling.
- The two LIGO facilities would have arm lengths of 4 kilometers.
- One of the LIGO sites would contain two interferometers, one of full length and one only 2 kilometers long. This would provide several

benefits. One was that detection could be based on triple coincidence, not double, with obvious benefits in reduced false-alarm rates. The other advantage was that genuine gravitational wave signals should show the proper scaling with interferometer arm length, a signature not likely to be mimicked by most noise sources.

- Design parameters were chosen which would make LIGO by far the most sensitive gravitational wave detector ever operated, sufficient to get into the range where detection of astronomical signals becomes plausible (although not guaranteed.)

8. LIGO Moves Forward

Even with the approval of construction, much engineering and design work remained to be done. This is the kind of work that, although filled with creative solutions to physics challenges, is usually not well documented, at least in the regular refereed literature. To gain insight, one needs to turn to internal technical documents.

The *LIGO Science Requirements Document (SRD)*[22] is a useful summary of the official goals of the team building LIGO. Among other things, it tabulates the nominal values of the key parameters of the interferometer and its mirrors (arm length, laser wavelength, laser power, mirror reflectivities, mirror dimensions, quality factors of mirror and suspension modes, and performance expected from the seismic isolation system.) Then, by modeling the key noise mechanisms, the LIGO team predicted what limiting sensitivity (or "noise floor") could be expected from such an interferometer. Three power laws bound the performance. At the lowest frequencies (up to about 40 Hz) seismic noise sets the limit; the decrease with frequency is so steep that it came to be called the "seismic wall". At intermediate frequencies (roughly from 40 Hz to 150 Hz), the off-resonance thermal noise associated with the pendulum mode of the test masses set the limit on performance. Finally, above 150 Hz, the shot noise in the readout of the arm length difference in the interferometer set the limit to performance. A graph of this predicted noise floor came to be plotted on every graph showing performance of the interferometers, throughout the commissioning process.

A team of over one hundred engineers and physicists worked at Caltech and MIT (and eventually at the two sites at Hanford, Washington and Livingston, Louisiana) to design, build, install, and commission the three LIGO interferometers. The many-faceted challenges that they faced defy a simple linear exposition. But perhaps a few principal themes of their work can illustrate how this phase fits into the arc of the story that we are telling. After the theory, the thought experiments, the conceptual designs of the interferometers and the LIGO observatory system overall, this was the period in which the final choices of materials, dimensions, construction methods, and assembly plans had to be made, after which parts were built or bought, and the whole interferometer assembled. The stakes were high: real money was being spent, and choices needed to be right (or to be fixed if they weren't.)

One interesting document from this phase of work is the *Detector Subsystem Requirements*, written in 1996.[23] In the early part of this work, the division of the interferometer into subsystems is outlined. (See the explanatory diagram, Figure 3.) Each major subsystem was assigned an official TLA (three letter acronym.) The subsystems were:

- PSL, the pre-stabilized laser,
- IOO, input optics components,
- COC, core optics components (the main interferometer mirrors),
- COS, core optics support components,
- LSC, length sensing and control system,
- ASC, angular sensing and control system,
- SUS, mirror suspension system, and
- SEI, seismic isolation system.

Also in LIGO were the physical environment monitoring system (PEM, not shown in the diagram) and the control and data system (CDS, shown but not with its proper TLA.)

This diagram was constructed to show the interfaces (mechanical, optical, and electrical) of the various subsystems as they related to one another to form a whole LIGO interferometer. The fact that such a diagram was needed testifies to both the complexity of the whole system, and to the consequent need to parcel out design work to different teams, who nevertheless needed to ensure that all of the parts would work together.

The main function of the *Detector Subsystems Requirements* document was to deal with the various sources of "technical noise", noise processes that were considered amenable to careful engineering. Beyond a simple listing of known noise processes, the document assigned budgeted amounts of noise that were allowed from each process. The goal was to ensure that none of the technical noise sources would make a significant contribution to the final noise spectrum of the interferometers. Only seismic, pendulum ther-

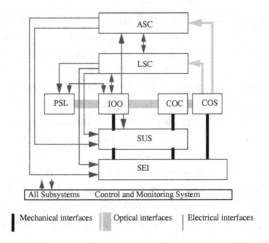

Fig. 3.   The subsystems of LIGO, and the interfaces between them.

mal, and shot noises were to be allowed to define the ultimate performance. "The subsystem designs are constrained to limit the effect of each technical noise source to a level lower than the allowed overall noise by a linear factor of at least 10." From this noise budgeting exercise, the requirements on the main parameters of the interferometer were seen to "flow down" into requirements on the various subsystems.

Another figure from the *Detector Subsystem Requirements* document shows the noise spectra from more noise sources than the three used to construct the "noise floor" of the Science Requirements Document. (See Figure 4.) In addition to seismic noise, pendulum thermal noise, and shot noise, the figure gives the expected contributions from

- thermal noise in the suspension top plate, pendulum wire "violin" modes, the vertical "bounce" mode of the pendulum suspension, the pitch and yaw motions of the mirrors, and in the internal vibrations of the mirrors,
- radiation pressure noise,
- Newtonian gravitational gradient noise from the environment (calculated for the Hanford seismic noise spectrum), and
- phase noise due to residual hydrogen gas in the vacuum system.

As of the date of this document, seismic noise transmitted through the vibration isolation stacks was expected to exceed the noise budget up to about 60 Hz; now it is known that the isolation systems perform better than that. Also, narrow peaks in the thermal noise spectra associated with wire "violin" resonances and the internal resonance of the mirrors show up above the shot noise limit at high frequencies, but this was entirely expected. The rest of the noise budget shows a complete success in restricting technical noise to low enough levels, at least at this stage of the design process.

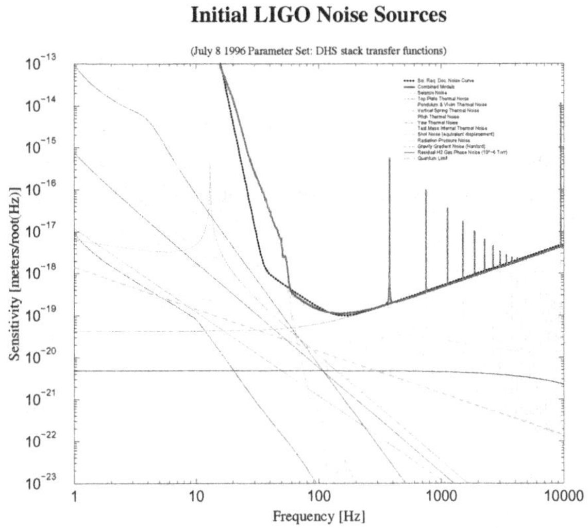

Initial LIGO Noise Sources

Fig. 4.   Noise budget for LIGO, from the Detector Subsystems Requirements document.

9. Construction, Installation, and Commissioning

Construction of LIGO got under way in 1994. Here are a few key dates:[24]

- 1994: Construction begins at Hanford, WA site.
- 1995: Construction begins at Livingston, LA site.
- 1998: Completion of buildings and of concrete foundations for 4 km arms.
- 1999: Completion of vacuum systems.
- 2000: Installation of interferometer parts close enough to complete that integrated testing could begin.

These milestones can serve as a reminder that LIGO was not only a physics experiment, but a large civil construction project, and a vacuum system construction project of notable magnitude as well. Realization of a gravitational wave detector involved a lot of cranes, bulldozers, cement trucks, and welding rigs.

The first operation of a LIGO interferometer occurred in October 2000. At that moment, it did not matter that the interferometer noise was orders of magnitude above the SRD noise floor. It was a triumph just to see that the tricky control problem of nested Fabry-Perot cavities had been successfully solved.

Soon, though, the next hard task began, "commissioning", that is making the interferometers work closer and closer to their design performance. The style of work was a mix of staged implementation of design features, debugging of problem circuits or out-of-spec optics, improvisation of fixes to unforeseen issues, and general problem-solving. Interferometer alignment proved to be more difficult than expected. Seismic noise due to logging near the Livingston site was strong enough to prevent interferometer operation during workdays, necessitating the addition of an advanced servo-based vibration isolation system. Absorption of light by the mirrors was less than expected in some of the mirrors and more than anticipated in others, so the figures of some mirrors needed to be adjusted by illuminating them with high power CO_2 lasers. The general spirit was to measure, model, improvise and then engineer solutions to both these large problems and others too numerous to mention here. The success that has been achieved testifies to both a technically strong team of physicists and engineers, and to good management that husbanded resources so that there was something left with which to solve the problems. Steady support of the funding agency, the National Science Foundation, was also crucial.

During the commissioning phase, there have been several pauses to operate the three LIGO interferometers as actual gravitational wave detectors, thus producing data that could be searched for signals. The first Science Run (S1) began on 23 August and ended on 9 September 2002. Since then, there have been three more Science Runs. Most have been carried out in coincidence with operation of the GEO600 interferometer in Germany; some have also been coordinated with the TAMA 300 m interferometer in Japan and with the ALLEGRO bar at Louisiana State University.

An account of the state of the LIGO and GEO instruments during the S1 run can be found in [24]. The article gives a detailed account of the state of the various subsystems of the interferometers. The noise was roughly two orders of magnitude above the SRD noise floor at high frequencies, and a larger factor above at frequencies below a few hundred Hz. (See

Figure 5.) All of the major noise sources were understood: at high frequency the limit was shot noise and electronic noise, high because at that time only a small fraction of the laser power could be used. Near 100 Hz was a band dominated by noise from the circuits that sent currents to the magnetic actuators on the mirrors; that circuit has since been re-engineered. At low frequency, excess noise came from optical lever servos used to control the interferometer alignment. This noise, too, has since been lowered dramatically.

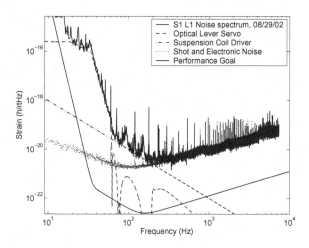

Fig. 5.   Strain noise spectrum of the Livingston 4 km interferometer during the S1 run. Also shown are the leading noise terms, and the SRD noise floor.

One other way in which the LIGO interferometers fell short of expectations during S1 was in their "duty cycle", that is the fraction of time that they stayed at their operating points for successfully collecting data. The 2 km and 4 km interferometers at Hanford achieved duty cycles of 73% and 57%, respectively, while the Livingston 4 km interferometer achieved a duty cycle of only 41%, because of the noise from logging. (GEO600 had dramatically better success, a duty cycle of 98%.) The lesson is that it is not just the sensitivity that requires hard work in order to achieve; the reliability needed for a functioning observatory also requires substantial effort. Since S1, the duty cycle of the LIGO interferometers has been improved substantially.

The data from the S1 run was analyzed for signals of four different classes: sinusoidal signals (such as would be expected from not-quite-round pulsars),[25] quasi-sinusoids sweeping up in frequency and amplitude (which are expected from inspiraling neutron star binary systems),[26] brief transients of a variety of waveforms (perhaps generated by supernovae),[27] and steady background noise (such as might be generated in the early universe, or by the superposition of many discrete sources.)[28]

The analysis was carried out by a group not yet mentioned in this article, the LIGO Scientific Collaboration, or LSC. It was formed in 1997 to carry out the scientific program of LIGO. It now has a membership of over 500 scientists and engineers; a strong core of its membership is employed at Caltech, MIT, or one of the two LIGO sites, but it also includes several hundred scientists from over thirty U.S. educational/research institutions and about 100 members from the nine institutions in Europe that make up the GEO collaboration. The LSC also carries out research and development on advanced interferometers. (See below.)

By the time of the fourth Science Run (S4, 22 February to 23 March 2005), the interferometers were running within about a factor of two of the noise floor of the *Science Requirements Document*. (See Figure 6.)[29] Further commissioning work subsequent to the S4 run has brought the noise level even closer to the SRD noise floor, including to within a few tens of percent in the shot noise limited region.[30]

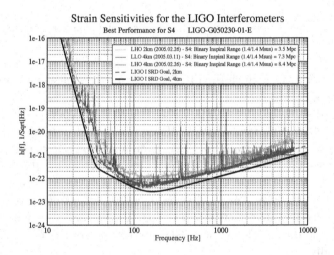

Fig. 6.   The noise spectra of the three LIGO interferometers during the S4 run. Note how close the performance has now come to the SRD noise floor.

Equally interesting from the point of view of this article's story arc is the noise budget that underlies and explains this latest measured noise. While it includes the "big three" (shot noise, pendulum thermal noise, and seismic noise), it needs to account for a long list of other technical noise contributions, including many not in the *Detector Subsystems Requirements* document. The technical noise terms most significant at this moment are noise coming from the sensors used to control mirror alignment. Next in importance is noise that comes from the control of the position of the interferometer's beam-splitter and the power recycling mirror. Various other sources of noise are less significant, but still make a measurable contribution at present. Once these "extra" noise terms are included, the measured spectrum is very well explained.

There are several lessons to draw from this progress. One is the triumph of engineering behind the accomplishment of performance so close to the SRD noise floor, as well as a triumph of physical understanding in the ability to model the noise budget so completely. Another, though, is the difficulty of realizing what seemed like such simple thought experiment in Pirani's paper. We are almost there, but looking back one can see how much hard work it took to get where we are.

10. Are We There Yet?

The title of this article is "Receiving Gravitational Waves". As I write, the LIGO Scientific Collaboration is finishing the papers describing the analysis of data from the S2 and S3 runs, and is already well into the analysis of data from our most recent run, S4. The LSC is also making plans for a long science run (of order one year in duration) that will constitute the fulfillment of LIGO's role as an observatory searching for gravitational waves.

But when will we (and our colleagues working on other detectors) actually receive gravitational waves? That is hard to predict. Although our published papers to date are only upper limits, it would not be out of the question for there to be a detection waiting for us in the data already recorded. Our long science run has a reasonable chance at a successful "reception", although according to most astronomical models we would have to be a bit lucky for that to occur.

Over the next several years, a world-wide network of interferometers (and several Weber-style bars of almost comparable sensitivity) will search for gravitational waves. GEO600 continues to approach its design sensitivity, and will continue to coordinate its running with LIGO. TAMA also has plans to continue improving and to continue coordinated running. Soon, too, Virgo's 3 km interferometer will come on line. It is expected have a sensitivity comparable to LIGO's. This set of detectors will constitute a powerful global network in the search for gravitational waves.

In case we aren't lucky enough to find signals at this level of sensitivity, there is a plan that ought to ensure success. The initial conception of LIGO (and the vision approved by the NSF) called for evolution in the installed interferometers. Almost since the start of the construction of LIGO, research has been moving forward on ways to build a better interferometer. One of the earliest and main roles of the LIGO Scientific Collaboration has been to coordinate this research and to pull it together into a coherent design. The outcome of that work was the proposal to build Advanced LIGO.[31] Expected to improve sensitivity by at least a factor of ten at high frequencies and to extend the observing band down as low as 10 Hz, Advanced LIGO will use the same buildings and vacuum system as initial LIGO, but will replace all of the interferometer components. At that new standard of performance, standard astronomical wisdom predicts that a multitude of signals will be within our grasp. (There are also plans for advanced interferometers under serious consideration in Japan and in Europe.)

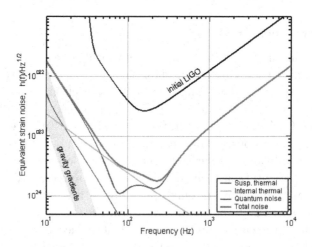

Fig. 7. A model of the noise spectrum of the "reference design" for Advanced LIGO, compared with the SRD noise floor of initial LIGO. There are several features worthy of note. Seismic noise will not be important above 10 Hz. At most frequencies shown, the spectrum is dominated by fundamental interferometer sensing noise (shot noise at high frequencies, radiation pressure noise at low frequencies. Thermal noise from the mirror's internal modes of vibration is important in the vicinity of 100 Hz.

In November 2004, the National Science Board approved Advanced LIGO. It is hoped that funding will be approved by the U.S. Congress within a few years, so that Advanced LIGO can begin searching the skies early in the next decade.

Acknowledgments

The author's research is supported by National Science Foundation Grant PHY-0140335. The Laser Interferometer Gravitational Wave Observatory is supported by the National Science Foundation under Cooperative Agreement PHY-0107417.

References

1. M. Bartusiak, *Einstein's Unfinished Symphony: Listening to the Sounds of Space-Time* (Joseph Henry Press, Washington, D.C., 2000).
2. H. Collins, *Gravity's Shadow: The Search for Gravitational Waves* (University of Chicago Press, Chicago, 2004).
3. P. R. Saulson, *Fundamentals of Interferometric Gravitational Wave Detectors* (World Scientific, Singapore, 1994).
4. B. Barish and R. Weiss, *Physics Today* **52** no. 10, 44 (1999).
5. A. Einstein, *Sitzungsberichte, Preussische Akademie der Wissenschaften* 688 (1916); A. Einstein, *Sitzungsberichte, Preussische Akademie der Wissenschaften* 154 (1918).
6. Peter Bergmann, private communication 1993.
7. H. Bondi, *Nature* **179**, 1072 (1957). In this article, Bondi credits I. Robinson with independent discovery of this proof, and also cites discussions at the Chapel Hill meeting on gravitation in 1957.
8. J. Weber, *General Relativity and Gravitational Waves* (Interscience, New York, 1961).
9. P. Kafka and L. Schnupp, *Astron. Astrophys.* **70**, 97 (1978).
10. F. A. E. Pirani, *Acta Phys. Polon.* **15**, 389 (1956).
11. G. E. Moss, L. R. Miller, and R. L. Forward, *Appl. Opt.* **10**, 2495 (1971); R. L. Forward, *Phys. Rev. D* **17**, 379 (1978).
12. M. E. Gertsenshtein and V. I. Pustovoit, *Sov. Phys. JETP* **16**, 433 (1962).
13. R. Weiss, *Quarterly Progress Report, MIT Research Lab of Electronics* **105**, 54 (1972).
14. L. L. Smarr, ed., *Sources of Gravitational Radiation* (Cambridge University Press, Cambridge, 1979).
15. P. G. Roll, R. Krotkov, and R. H. Dicke, *Ann. Phys. (N. Y.)* **26**, 442 (1964).
16. R. Schilling, L. Schnupp, W. Winkler, H. Billing, K. Maischberger, and A. Rüdiger, *J. Phys. E: Sci. Instrum.* **14**, 65 (1981).
17. R. W. P. Drever, in *Gravitational Radiation*, eds. N. Deruelle and T. Piran (North Holland, Amsterdam, 1983), p. 321.

18. R. W. Drever, J. L. Hall, F. V. Kowalski, J. Hough, G. M. Ford, A. J. Munley, and H. Ward, *Appl. Phys. B* **31**, 97 (1983).

19. R. V. Pound, *Rev. Sci. Instrum.* **17**, 490 (1946).

20. See, for example, D. Shoemaker, R. Schilling, L. Schnupp, W. Winkler, K. Maischberger, and A. Rüdiger, *Phys. Rev. D* **38**, 423 (1988).

21. P. S. Linsay, P. R. Saulson, R. Weiss, and S. Whitcomb, *A Study of a Long Baseline Gravitational Wave Antenna System* (MIT, Cambridge, Mass., 1983), unpublished.

22. A. Lazzarini and R. Weiss, *LIGO Science Requirements Document (SRD)*, LIGO-E950018-02-E (1995), unpublished.

23. D. Shoemaker, *Detector Subsystem Requirements*, LIGO-E960112-05-D (1996), unpublished.

24. B. Abbott *et al.* (The LIGO Scientific Collaboration), *Nucl. Instr. Meth. Phys. Res. A* **517**, 154 (2004).

25. B. Abbott *et al.* (The LIGO Scientific Collaboration), *Phys. Rev. D* **69** 082004 (2004).

26. B. Abbott *et al.* (The LIGO Scientific Collaboration), *Phys. Rev. D* **69** 122001 (2004).

27. B. Abbott *et al.* (The LIGO Scientific Collaboration), *Phys. Rev. D* **69** 102001 (2004).

28. B. Abbott *et al.* (The LIGO Scientific Collaboration), *Phys. Rev. D* **69** 122004 (2004).

29. http://www.ligo.caltech.edu/docs/G/G050230-01/G050230-01.pdf

30. LIGO Livingston Observatory electronic log, 19-20 April 2005.

31. The Advanced LIGO proposal can be found at http://www.ligo.caltech.edu/advLIGO/

CHAPTER 10

RELATIVITY IN THE GLOBAL POSITIONING SYSTEM

NEIL ASHBY

Department of Physics, University of Colorado
Boulder, CO 80309-0390, USA
Neil.Ashby@mobek.colorado.edu

The Global Positioning System (GPS) uses accurate, stable atomic clocks in satellites and on the ground to provide world-wide position and time determination. These clocks have gravitational and motional frequency shifts which are so large that, without carefully accounting for numerous relativistic effects, the system would not work. This article discusses the conceptual basis, founded on special and general relativity, for navigation using GPS. Relativistic principles and effects of practical importance include the constancy of the speed of light, the equivalence principle, the Sagnac effect, time dilation, gravitational frequency shifts, and relativity of synchronization. Experimental tests of relativity obtained with a GPS receiver aboard the TOPEX/POSEIDON satellite will be discussed. Recently frequency jumps arising from satellite orbit adjustments have been identified as relativistic effects.

1. Introduction

The "Space Segment" of the GPS consists of 24 satellites carrying atomic clocks. (Spare satellites and spare clocks in satellites exist.) There are four satellites in each of six orbital planes inclined at 55° with respect to earth's equatorial plane, distributed so that from any point on the earth, four or more satellites are almost always above the local horizon. The clocks provide accurate references for timing signals that are transmitted from each satellite. The signals can be thought of as sequences of events in space-time, characterized by positions and times of transmission. Associated with these events are messages specifying the transmission events' space-time coordinates; below I will discuss the system of reference in which these coordinates are given. Additional information contained in the messages includes an almanac for the entire satellite constellation, information about satellite vehicle health, and information from which Universal Coordinated Time as maintained by the U. S. Naval Observatory–UTC(USNO)–can be determined.

The GPS "Control Segment" includes a number of ground-based monitoring stations, which continually gather information from the satellites. These data are sent to a Master Control Station in Colorado Springs, CO, which analyzes the constellation and predicts the satellite ephemerides and clock behavior for the next few hours. This information is then uploaded into the satellites for retransmission to users. The "User Segment" consists of all users who, by receiving signals transmitted from the satellites, are able to determine their position, velocity, and the time on their local clocks.

The GPS is a navigation and timing system that is operated by the United States Department of Defense (DOD), and therefore is partly classified. Several organizations monitor GPS signals independently and provide services from which satellite ephemerides and clock behavior can be obtained. Position accuracies in the neighborhood of 5-10 cm are not unusual. Carrier phase measurements of the transmitted signals are commonly done to better than a millimeter.

GPS signals are received on earth at two carrier frequencies, L1 (154 × 10.23 MHz) and L2 (120 × 10.23 MHz). The L1 carrier is modulated by two types of pseudorandom noise codes, one at 1.023 MHz that repeats every millisecond–called the Coarse/Acquisition or C/A code–and an encrypted one at 10.23 MHz called the P-code. P-code receivers have access to both L1 and L2 frequencies and can correct for ionospheric delays, whereas civilian users only have access to the C/A code. An additional frequency for civilian users is being planned. There are thus two levels of positioning service available in real time, the Precise Positioning Service utilizing P-code, and the Standard Positioning Service using only C/A code. The DOD has the capability of dithering the transmitted signal frequencies and other signal characteristics, so that C/A code users would be limited in positioning accuracy to about ±100 meters. This is termed Selective Availability, or SA. SA was turned off by order of President Clinton in May, 2000.

Technology developments in the late 1960s and early 1970s that directly benefited the GPS included the Transit system, developed by the Johns Hopkins Applied Physics Laboratory, the Naval Research Laboratory's Timation satellites, and the U. S. Air Force Project 621B.[20] The Transit system relied on the Doppler shift of continuously transmitted tones from satellites in polar orbits to determine user position; development of

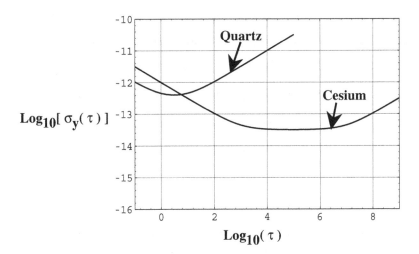

Fig. 1.   Allan deviations of Cesium clocks and quartz oscillators, plotted as a function of averaging time τ.

prediction algorithms for satellite ephemerides was an important contribution of the Transit program to the GPS. Space-qualified atomic clocks with exceptional stability were deployed by the Timation program; the Timation satellites provided precise time and time transfer between various points on earth's surface, and broadcast synchronized signals at various frequencies for phase ambiguity resolution and navigation. The 621B program demonstrated that broadcasting ranging signals based on pseudo-random noise had many advantages, including improved detection sensitivity of very weak signals, and resistance to jamming. These programs were merged in 1973 with the creation of the GPS Joint Program Office, with subsequent management and participation by all branches of the Military Services in GPS development.

Early in the development of the GPS many individuals recognized that relativistic effects, although tiny, were going to be important in the GPS, but it was unclear what should be done to account for them. Numerous conceptual and numerical errors, and claims that relativity was not incorporated properly into the GPS were made, resulting in a somewhat contentious and controversial situation. Conferences were held in 1979 and 1985 to review such issues. Although an Air Force Study in 1986 found nothing wrong, such claims continued resulting in an additional meeting in 1995. A flawed technical report[1] was issued as late as 1996 by an apparently authoritative source–the Aerospace Corporation, where the civilian counterpart of the JPO is located. One source of confusion was the Sagnac effect that makes it impossible to self-consistently synchronize clocks on the surface of the rotating earth by performing operations with electromagnetic signals or slowly moving portable clocks, like those that could successfully be used to synchronize clocks in an inertial frame. Thus a problem that had to be solved was effective synchronization of clocks fixed or slowly moving on earth's surface. Another source of confusion was whether the sun's gravitational potential would have a significant effect on clocks in GPS satellites.

The GPS is made possible by extremely accurate, stable atomic clocks. Figure 1 gives a plot of the Allan deviation $\sigma_y(\tau)$ for a high-performance Cesium clock, as a function of sample time τ. If an ensemble of clocks is initially synchronized, then when compared to each other after a time τ, the Allan deviation provides a measure of the rms fractional frequency deviation among the clocks due to intrinsic noise processes in the clocks. Frequency offsets and frequency drifts are additional systematic effects which must be accounted for separately. Also on Figure 1 is an Allan deviation plot for a Quartz oscillator such as is typically found in a GPS receiver. Quartz oscillators usually have better short-term stability performance characteristics than Cesium clocks, but after 100 seconds or so, Cesium has far better performance. In actual clocks there is a wide range of variation around the nominal values plotted in Figure 1. The most stable GPS clocks use Rubidium atoms and reach maximum stability levels of a few parts in 10^{15} after about ten days. What this means is that after initializing such a clock, and leaving it alone for ten days, it should be correct to within about 5 parts in 10^{15}, or 0.4 nanoseconds. Relativistic effects are huge compared to this.

The purpose of this article is to explain how relativistic effects are accounted for in the GPS. Although clock velocities are small and gravitational fields are weak near the earth, they give rise to significant relativistic effects. These effects include first- and second-order Doppler frequency shifts of clocks due to their relative motion, gravitational frequency shifts, and the Sagnac effect due to earth's rotation. If such effects are not accounted for properly, unacceptably large errors in GPS navigation and time transfer will result. In the GPS one can find many examples of the application of fundamental relativity principles. Also, experimental tests of relativity can be performed with GPS, although generally speaking these are not at a level of precision any better than previously existing tests.

The principles of position determination and time transfer in the GPS can be very simply stated. Let there be four synchronized atomic clocks that transmit sharply defined pulses from the positions $\mathbf{r}_j$ at times t_j, with $j = 1, 2, 3, 4$ an index labeling the different transmission events. Suppose that these four signals are received at position $\mathbf{r}$, at one and the same instant t. Then, from the principle of the constancy of the speed of light,

$$c^2(t - t_j)^2 = |\mathbf{r} - \mathbf{r}_j|^2, \quad j = 1, 2, 3, 4. \tag{1}$$

where the defined value of c is exactly 299792458 m/s. These four equations can be solved for the unknown space-time coordinates of the reception event, $\{\mathbf{r}, t\}$. Hence, the principle of the constancy of c finds application as the fundamental concept on which the GPS is based. Timing errors of one ns will lead to positioning errors of the order of 30 cm. Also, obviously, it is necessary to specify carefully the reference frame in which the transmitter clocks are synchronized, so that Eqs. (1) are valid.

The timing pulses in question can be thought of as places in the transmitted wave trains where there is a particular phase reversal of the circularly polarized electromagnetic signals. At such places the electromagnetic field tensor passes through zero and therefore provides relatively moving observers with sequences of events that they can agree on, at least in principle.

Eqs. (1) have an important reciprocity feature: if an event occurs at $\{\mathbf{r}, t\}$ near earth's surface that sends out electromagnetic pulses, and these are detected by four or more satellites at positions and times $\{\mathbf{r}_j, t_j\}$, then the position and time of the event can be determined; this is the (classified) nuclear event detection half of the GPS.

2. Reference Frames and the Sagnac Effect

Almost all users of GPS are at fixed locations on the rotating earth, or else are moving very slowly over earth's surface. This led to an early design decision to broadcast the satellite ephemerides in a model earth-centered, earth-fixed, reference frame (ECEF frame), in which the model earth rotates about a fixed axis with a defined rotation rate, $\omega_E = 7.2921151467 \times 10^{-5}$ rad s^{-1}. This reference frame is designated by the symbol WGS-84(G873).[2] For discussions of relativity, the particular choice of ECEF frame is immaterial. Also, the fact the earth truly rotates about a slightly different axis with a variable rotation rate has little consequence for relativity and I shall not go into this here. I shall simply regard the ECEF frame of GPS as closely related to, or determined by, the International Terrestrial Reference Frame established by the Bureau International des Poids et Mesures (BIPM) in Paris.

It should be emphasized that the transmitted navigation messages provide the user only with a function from which the satellite position can be calculated *in the ECEF* as a function of the transmission time. Usually, the satellite transmission times t_j are unequal, so the coordinate system in which the satellite positions are specified changes orientation from one transmission event to another. Therefore, to implement Eqs. (1), the receiver must generally perform a different rotation for each measurement made, into some common inertial frame, so that Eqs. (1) apply. After solving the propagation delay equations, a final rotation must usually be performed into the ECEF to determine the receiver's position. This can become exceedingly complicated and confusing. A technical note[3] discusses these issues in considerable detail.

Although the ECEF frame is of primary interest for navigation, many physical processes (such as electromagnetic wave propagation) are simpler to describe in an inertial reference frame. Certainly, inertial reference frames are needed to express Eqs. (1). In the ECEF frame used in the GPS, the unit of time is the SI second as realized by the clock ensemble of the U. S. Naval Observatory and the unit of length is the SI meter. This is important in the GPS because it means that local observations using GPS are insensitive to effects on the scales of length and time measurements due to other solar system bodies that are time-dependent.

Let us therefore consider the simplest instance of a transformation from an inertial frame, in which the space-time is Minkowskian, to a rotating frame of reference. Thus, ignoring gravitational potentials for the moment, the metric in an inertial frame in cylindrical coordinates is

$$-ds^2 = -(c\,dt)^2 + dr^2 + r^2 d\phi^2 + dz^2, \tag{2}$$

and the transformation to a coordinate system $\{t', r', \phi', z'\}$ rotating at the uniform angular rate ω_E is

$$t = t', \quad r = r', \quad \phi = \phi' + \omega_E t', \quad z = z'. \tag{3}$$

This results in the following well-known metric (Langevin metric) in the rotating frame:

$$-ds^2 = -\left(1 - \frac{\omega_E^2 r'^2}{c^2}\right)(cdt')^2 + 2\omega_E r'^2 d\phi' dt' + (d\sigma')^2, \tag{4}$$

where the abbreviated expression $(d\sigma')^2 = (dr')^2 + (r'd\phi')^2 + (dz')^2$ for the square of the coordinate distance has been introduced.

The time transformation $t = t'$ in Eqs. (3) is deceivingly simple. It means that in the rotating frame the time variable t' is really determined in the underlying inertial frame. It is an example of coordinate time. A similar concept is used in the GPS.

Now consider a process in which observers in the rotating frame attempt to use Einstein synchronization (that is, the principle of the constancy of the speed of light) to establish a network of synchronized clocks. Light travels along a null world line so we may set $ds^2 = 0$ in Eq. (4). Also, it is sufficient for this discussion to keep only terms of first order in the small parameter $\omega_E r'/c$. Then solving for (cdt'),

$$cdt' = d\sigma' + \frac{\omega_E r'^2 d\phi'}{c}. \tag{5}$$

The quantity $r'^2 d\phi'/2$ is just the infinitesimal area dA'_z in the rotating coordinate system swept out by a vector from the rotation axis to the light pulse, and projected onto a plane parallel to the equatorial plane. Thus, the total time required for light to traverse some path is

$$\int_{\text{path}} dt' = \int_{\text{path}} \frac{d\sigma'}{c} + \frac{2\omega_E}{c^2} \int_{\text{path}} dA'_z. \quad [\text{light}] \tag{6}$$

Observers fixed on the earth, who were unaware of earth rotation, would use just $\int d\sigma'/c$ to synchronize their clock network. Observers at rest in the underlying inertial frame would say that this leads to significant path-dependent inconsistencies, which are proportional to the projected area encompassed by the path. Consider, for example, a synchronization process that follows earth's equator once around eastwards. For earth, $2\omega_E/c^2 = 1.6227 \times 10^{-21}$ s/m^2 and the equatorial radius is $a_1 = 6,378,137$ m, so the area is $\pi a_1^2 = 1.27802 \times 10^{14}$ m^2 . Then, the last term in Eq. (6) is of magnitude $2\pi\omega_E a_1^2/c^2 = 207.4$ ns.

From the underlying inertial frame, this can be regarded as the additional travel time required by light to catch up to the moving reference point. Simple-minded use of Einstein synchronization in the rotating frame gives only $\int d\sigma'/c$, and thus leads to a significant error. Traversing the equator once eastward, the last clock in the synchronization path would lag the first clock by 207.4 ns. Traversing the equator once westward, the last clock in the synchronization path would lead the first clock by 207.4 ns.

In an inertial frame a portable clock can be used to disseminate time. The clock must be moved so slowly that changes in the moving clock's rate due to time dilation, relative to a reference clock at rest on earth's surface, are extremely small. On the other hand, observers in a rotating frame who attempt this find that the proper time elapsed on the portable clock is affected by earth's rotation. Factoring Eq. (4), the proper time increment $d\tau$ on the moving clock is given by

$$(d\tau)^2 = (ds/c)^2 = dt'^2 \left[1 - \left(\frac{\omega_E r'}{c} \right)^2 - \frac{2\omega_E r'^2 d\phi'}{c^2 dt'} - \left(\frac{d\sigma'}{cdt'} \right)^2 \right]. \quad (7)$$

For a slowly moving clock $(d\sigma'/cdt')^2 \ll 1$, so the last term in brackets in Eq. (7) can be neglected. Also, keeping only first order terms in the small quantity $\omega_E r'/c$, and solving for dt', leads to

$$\int_{\text{path}} dt' = \int_{\text{path}} d\tau + \frac{2\omega_e}{c^2} \int_{\text{path}} dA'_z. \quad [\text{portable clock}] \quad (8)$$

This should be compared with Eq. (6). Path-dependent discrepancies in the rotating frame are thus inescapable whether one uses light or portable clocks to disseminate time, while synchronization in the underlying inertial frame using either process is self-consistent.

Eqs. (6) and (8) can be reinterpreted as a means of realizing coordinate time $t' = t$ in the rotating frame, if after performing a synchronization process appropriate corrections of the form $+2\omega_E \int_{\text{path}} dA'_z/c^2$ are applied. This was recognized in the early 1980s by the Consultative Committee for the Definition of the Second and the International Radio Consultative Committee who formally adopted procedures incorporating such corrections for the comparison of time standards located far apart on earth's surface. For the GPS it means that synchronization of the entire system of ground-based and orbiting atomic clocks is performed in the local inertial frame, or ECI coordinate system.[4]

GPS can be used to compare times on two earth-fixed clocks when a single satellite is in view from both locations. This is the "common-view" method of comparison of Primary standards, whose locations on earth's surface are usually known very accurately in advance from ground-based surveys. Signals from a single GPS satellite in common view of receivers at the two locations provide enough information to determine the time difference between the two local clocks. The Sagnac effect is very important in making such comparisons, as it can amount to hundreds of nanoseconds. In 1984 GPS satellites 3, 4, 6, and 8 were used in simultaneous common view between three pairs of earth timing centers, to accomplish closure in performing an around-the-world Sagnac experiment. The centers were the National Bureau of Standards (NBS) in Boulder, CO, Physikalisch-Technische Bundesanstalt (PTB) in Braunschweig, West Germany, and

Tokyo Astronomical Observatory (TAO). The size of the Sagnac correction varied from 240 to 350 ns. Enough data were collected to perform 90 independent circumnavigations. The actual mean value of the residual obtained after adding the three pairs of time differences was 5 ns, which was less than 2 percent of the magnitude of the calculated total Sagnac effect.[5]

3. GPS Coordinate Time and TAI

In the GPS, the time variable $t' = t$ becomes a coordinate time in the rotating frame of the earth, which is realized by applying appropriate corrections while performing synchronization processes. Synchronization is thus performed in the underlying inertial frame in which self-consistency can be achieved.

With this understanding, I next describe the gravitational fields near the earth due to the earth's mass itself. Assume that earth's mass distribution is static; that there exists a locally inertial, non-rotating, freely falling coordinate system with origin at the earth's center of mass, and write an approximate solution of Einstein's field equations in isotropic coordinates:

$$-ds^2 = -(1 + \frac{2V}{c^2})(cdt)^2 + (1 - \frac{2V}{c^2})(dr^2 + r^2 d\theta^2 + r^2 \sin^2\theta d\phi^2). \quad (9)$$

where $\{r, \theta, \phi\}$ are spherical polar coordinates and where V is the Newtonian gravitational potential of the earth, given approximately by:

$$V = -\frac{GM_E}{r}\left[1 - J_2\left(\frac{a_1}{r}\right)^2 P_2(\cos\theta)\right]. \quad (10)$$

In Eq. (10), $GM_E = 3.986004418 \times 10^{14}$ m^{3}s^{-2} is the product of earth's mass times the Newtonian gravitational constant, $J_2 = 1.0826300 \times 10^{-3}$ is earth's quadrupole moment coefficient, and $a_1 = 6.3781370 \times 10^6$ is earth's equatorial radius.[a] The angle θ is the polar angle measured downward from the axis of rotational symmetry; P_2 is the Legendre polynomial of degree 2. In using Eq. (9), it is an adequate approximation to retain only terms of first order in the small quantity V/c^2. Higher multipole moment contributions to Eq. (10) have a very small effect for relativity in GPS. One additional expression for the invariant interval is needed: the transformation of Eq. (9) to a rotating, ECEF coordinate system by means of transformations equivalent to Eqs. (3). The transformations for spherical polar coordinates are like Eqs. (3), except that $\theta = \theta'$ replaces $z = z'$. Upon performing the

---

[a]WGS-84(G873) values of these constants are used in this article.

transformations, and retaining only terms of order $1/c^2$, the scalar interval becomes:

$$-ds^2 = -[1 + \tfrac{2V}{c^2} - \left(\tfrac{\omega_E r' \sin\theta'}{c}\right)^2](c\,dt')^2 + 2\omega_E r'^2 \sin^2\theta'\,d\phi'dt'$$

$$+ \left(1 - \tfrac{2V}{c^2}\right)(dr'^2 + r'^2 d\theta'^2 + r'^2 \sin^2\theta'\,d\phi'^2). \qquad (11)$$

To the order of the calculation, this result is a simple superposition of the metric, Eq. (9), with the corrections due to rotation expressed in Eq. (4).

The Earth's geoid. In Eqs. (9) and (11), the rate of coordinate time is determined by atomic clocks at rest at infinity. The rate of GPS coordinate time, however, is closely related to International Atomic Time (TAI), which is a time scale computed by the BIPM on the basis of inputs from hundreds of primary time standards, hydrogen masers, and other clocks from all over the world. In producing this time scale, corrections are applied to reduce the elapsed proper times on the contributing clocks to earth's geoid, a surface of constant effective gravitational equipotential at mean sea level in the ECEF.

Universal Coordinated Time (UTC) is another time scale, which differs from TAI by a whole number of leap seconds. These leap seconds are inserted every so often into UTC so that UTC continues to correspond to time determined by earth's rotation. Time standards organizations that contribute to TAI and UTC generally maintain their own time scales. For example, the time scale of the U. S. Naval Observatory, based on an ensemble of Hydrogen masers and Cs clocks, is denoted UTC(USNO). GPS time is steered so that, apart from the leap second differences, it stays within 100 ns UTC(USNO). Usually, this steering is so successful that the difference between GPS time and UTC(USNO) is less than about 40 ns. GPS equipment cannot tolerate leap seconds, as such sudden jumps in time would cause receivers to lose their lock on transmitted signals, and other undesirable transients would occur.

To account for the fact that reference clocks for the GPS are not at infinity, I shall consider the rates of atomic reference clocks at rest on the earth's geoid. These clocks move because of the earth's spin; also, they are at varying distances from the earth's center of mass since the earth is slightly oblate. In order to proceed one needs a model expression for the shape of this surface, and a value for the effective gravitational potential on this surface in the rotating frame.

For this calculation, I use Eq. (11) in the ECEF. For a clock at rest in the rotating frame, Eq. (11) reduces to:

$$-ds^2 = -(1 + \frac{2V}{c^2} - \frac{\omega_E^2 r'^2 \sin^2 \theta'}{c^2})(c\,dt')^2 = -(1 + \frac{2\Phi}{c^2})(c\,dt')^2, \quad (12)$$

with the potential V given by Eq. (10). Here Φ is the effective gravitational potential in the rotating frame. For a clock at rest on earth's geoid, the constancy of Φ determines the radius r' of the model geoid as a function of polar angle θ'. The numerical value of Φ_0 can be determined at the equator where $\theta' = \pi/2$ and $r' = a_1$. This gives

$$\begin{aligned}
\frac{\Phi_0}{c^2} &= -\frac{GM_E}{a_1 c^2} - \frac{GM_E J_2}{2a_1 c^2} - \frac{\omega_E^2 a_1^2}{2c^2} \\
&= -6.95348 \times 10^{-10} - 3.764 \times 10^{-13} - 1.203 \times 10^{-12} \\
&= -6.96927 \times 10^{-10}.
\end{aligned} \quad (13)$$

There are thus three distinct contributions to this effective potential: a simple $1/r$ contribution due to the earth's mass; a more complicated contribution from the quadrupole potential, and a centripetal term due to the earth's rotation. These contributions have been divided by c^2 in the above equation since the time increment on an atomic clock at rest on the geoid can be easily expressed thereby. In recent resolutions of the International Astronomical Union[6], a "Terrestrial Time" scale (TT) has been defined by adopting the value $\Phi_0/c^2 = 6.969290134 \times 10^{-10}$. Eq. (13) agrees with this definition to within the accuracy needed for the GPS.

From Eq. (11), for clocks on the geoid,

$$d\tau = ds/c = dt' \left(1 + \frac{\Phi_0}{c^2}\right). \quad (14)$$

Clocks at rest on the rotating geoid run slow compared to clocks at rest at infinity by about seven parts in 10^{10}. These effects sum to more than 10,000 times larger than the fractional frequency stability of a high-performance cesium clock. The shape of the geoid in this model can be obtained by setting $\Phi = \Phi_0$ and solving Eq. (12) for r' in terms of θ'. The first few terms in a power series in the variable $x' = \sin \theta'$ can be expressed as:

$$r' = 6356742.025 + 21353.642\,x'^2 + 39.832\,x'^4 + 0.798\,x'^6 + 0.003\,x'^8 \text{ m}. \quad (15)$$

This treatment of the gravitational field of the oblate earth is limited by the simple model of the gravitational field. Actually, what I have done is estimate the shape of the so-called "reference ellipsoid," from which the actual geoid is conventionally measured.

Better models can be found in the literature of geophysics.[7,8,9] The next term in the multipole expansion of the earth's gravity field is about a thousand times smaller than the contribution from J_2; although the actual shape of the geoid can differ from Eq. (15) by as much as 100 meters, the effects of such terms on timing in the GPS are small. Incorporating up to 20 higher zonal harmonics in the calculation affects the value of Φ_0 only in the sixth significant figure.

Observers at rest on the geoid define the unit of time in terms of the proper rate of atomic clocks. In Eq. (14), Φ_0 is a constant. On the left side of Eq. (14), $d\tau$ is the increment of proper time elapsed on a standard clock at rest, in terms of the elapsed coordinate time dt. Thus, the very useful result has emerged, that ideal clocks at rest on the geoid of the rotating earth all beat at the same rate. This is reasonable since the earth's surface is a gravitational equipotential surface in the rotating frame. (It is true for the actual geoid whereas I have constructed a model.) Considering clocks at two different latitudes, the one further north will be closer to the earth's center because of the flattening–it will therefore be more red shifted. However, it is also closer to the axis of rotation, and going more slowly, so it suffers less second-order Doppler shift. The earth's oblateness gives rise to an important quadrupole correction. This combination of effects cancels exactly on the reference surface.

Since all clocks at rest on the geoid beat at the same rate, it is advantageous to exploit this fact to redefine the rate of coordinate time. In Eq. (9) the rate of coordinate time is defined by standard clocks at rest at infinity. I want instead to define the rate of coordinate time by standard clocks at rest on the surface of the earth. Therefore, I shall define a new coordinate time t'' by means of a constant rate change:

$$t'' = (1 + \Phi_0/c^2)t' = (1 + \Phi_0/c^2)t. \tag{16}$$

The correction is about seven parts in 10^{10} (see Eq. (13)).

When this time scale change is made, the metric of Eq. (11) in the earth-fixed rotating frame becomes:

$$-ds^2 = -\left(1 + \frac{2(\Phi - \Phi_0)}{c^2}\right)(cdt'')^2 + 2\omega_E r'^2 \sin^2 \theta' d\phi' dt''$$
$$+ \left(1 - \frac{2V}{c^2}\right)(dr'^2 + r'^2 d\theta'^2 + r'^2 \sin^2 \theta' d\phi'^2), \tag{17}$$

where only terms of order c^{-2} have been retained. Whether I use dt' or dt'' in the Sagnac cross term makes no difference since the Sagnac term is very small anyway. The same time scale change in the non-rotating ECI metric,

Eq. (9), gives:

$$-ds^2 = -\left(1 + \frac{2(V - \Phi_0)}{c^2}\right)(cdt'')^2 + (1 - \frac{2V}{c^2})(dr^2 + r^2 d\theta^2 + r^2 \sin^2 \theta d\phi^2).$$

(18)

Eqs. (17) and (18) imply that the proper time elapsed on clocks at rest on the geoid (where $\Phi = \Phi_0$) is identical with the coordinate time t''. This is the correct way to express the fact that ideal clocks at rest on the geoid provide all of our standard reference clocks.

4. The Realization of Coordinate Time

I can now address the real problem of clock synchronization within the GPS. In the remainder of this article I shall drop the primes on t'' and just use the symbol t, with the understanding that the unit of time is referenced to UTC(USNO) on the rotating geoid, but with synchronization established in an underlying, locally inertial, reference frame. The metric Eq. (18) will henceforth be written:

$$-ds^2 = -\left(1 + \frac{2(V - \Phi_0)}{c^2}\right)(cdt)^2 + (1 - \frac{2V}{c^2})(dr^2 + r^2 d\theta^2 + r^2 \sin^2 \theta d\phi^2).$$

(19)

It is obvious that Eq. (19) contains within it the well-known effects of time dilation (the apparent slowing of moving clocks) and frequency shifts due to gravitation. Consequently path-dependent effects on orbiting GPS clocks must be accounted for.

On the other hand, according to General Relativity, the coordinate time variable t of Eq. (19) is valid in a coordinate patch large enough to cover the earth and the GPS satellite constellation. Eq. (19) is an approximate solution of the field equations near the earth, which include the gravitational fields due to earth's mass distribution. In this local coordinate patch, the coordinate time is single-valued. (It is not unique because there is still gauge freedom, but Eq. (19) represents a fairly simple and reasonable choice of gauge.) Therefore, it is natural to propose that the coordinate time variable t of Eqs. (19) and (17) be used as a basis for synchronization in the neighborhood of the earth.

To see how this works for a slowly moving atomic clock, solve Eq. (19) for dt as follows. First factor out $(cdt)^2$ from all terms on the right:

$$-ds^2 = -\left[1 + \frac{2(V - \Phi_0)}{c^2} - \left(1 - \frac{2V}{c^2}\right)\frac{dr^2 + r^2 d\theta^2 + r^2 \sin^2 \theta d\phi^2}{(cdt)^2}\right](cdt)^2.$$

(20)

Simplify by writing the velocity in the ECI coordinate system as

$$v^2 = \frac{dr^2 + r^2 d\theta^2 + r^2 \sin^2 \theta d\phi^2}{dt^2}. \tag{21}$$

Only terms of order c^{-2} need be kept, so the potential term modifying the velocity term can be dropped. Then, upon taking a square root, the proper time increment on the moving clock is approximately

$$d\tau = ds/c = \left[1 + \frac{(V - \Phi_0)}{c^2} - \frac{v^2}{2c^2}\right] dt. \tag{22}$$

Finally, solving for the increment of coordinate time and integrating along the path of the atomic clock,

$$\int_{\text{path}} dt = \int_{\text{path}} d\tau \left[1 - \frac{(V - \Phi_0)}{c^2} + \frac{v^2}{2c^2}\right]. \tag{23}$$

The proper time on the clock is thus corrected to give coordinate time.

Suppose for a moment there were no gravitational fields. Then picture an underlying non-rotating reference frame, a local inertial frame, unattached to the spin of the earth, but with its origin at the center of the earth. In this non-rotating frame, introduce a fictitious set of standard clocks, available anywhere, all synchronized by the Einstein synchronization procedure, and running at agreed upon rates such that synchronization is maintained. These clocks read the coordinate time t. Next, introduce the rotating earth with a set of standard clocks distributed around upon it, possibly roving around. Apply to each of the standard clocks a set of corrections based on the known positions and motions of the clocks, given by Eq. (23). This generates a "coordinate clock time" in the earth-fixed, rotating system. This time is such that at each instant the coordinate clock agrees with a fictitious atomic clock at rest in the local inertial frame, whose position coincides with the earth-based standard clock at that instant. Thus, coordinate time is equivalent to time that would be measured by standard clocks at rest in the local inertial frame.[10]

When the gravitational field due to the earth is considered, the picture is only a little more complicated. There still exists a coordinate time that can be found by computing a correction for gravitational red shift, given by the first correction term in Eq. (23).

5. Relativistic Effects on Satellite Clocks

For atomic clocks in satellites, it is most convenient to consider the motions as they would be observed in the local ECI frame. Then the Sagnac effect becomes irrelevant. (The Sagnac effect on moving ground-based receivers must still be considered.) Gravitational frequency shifts and second-order Doppler shifts must be taken into account together. The term Φ_0 in Eq. (23) includes the scale correction needed in order to use clocks at rest on the earth's surface as references. The quadrupole contributes to Φ_0 in the term $-GM_E J_2/2a_1$ in Eq. (23); there it contributes a fractional rate correction of -3.76×10^{-13}. This effect is large. Also, V is the earth's gravitational potential at the satellite. Fortunately, earth's quadrupole potential falls off very rapidly with distance and up until very recently its effect on satellite vehicle (SV) clock frequency has been neglected. This will be discussed in a later section; for the present I only note that the effect of earth's quadrupole potential on SV clocks is only about one part in 10^{14}.

Satellite orbits. Let us assume that the satellites move along Keplerian orbits. This is a good approximation for GPS satellites, but poor if the satellites are at low altitude. This assumption yields relations with which to simplify Eq. (23). Since the quadrupole (and higher multipole) parts of the earth's potential are neglected, in Eq. (23) the potential is $V = -GM_E/r$. Then the expressions can be evaluated using what is known about the Newtonian orbital mechanics of the satellites. Denote the satellite's orbit semimajor axis by a and eccentricity by e. Then the solution of the orbital equations is as follows:[11] the distance r from the center of the earth to the satellite in ECI coordinates is

$$r = a(1 - e^2)/(1 + e \cos f) = a(1 - e \cos E). \tag{24}$$

The semimajor axis of GPS satellites is $a = 27,561.75$ m, chosen so that a given satellite will appear in exactly the same place against the celestial sphere twice per day. The angle f, called the true anomaly, is measured from perigee along the orbit to the satellite's position. The true anomaly can be calculated in terms of the eccentric anomaly E, according to the relationships:

$$\cos f = \frac{\cos E - e}{1 - e \cos E}, \qquad \sin f = \sqrt{1 - e^2} \frac{\sin E}{1 - e \cos E}. \tag{25}$$

To find the eccentric anomaly E, one must solve the transcendental equation

$$E - e \sin E = \sqrt{\frac{GM_E}{a^3}}(t - t_p), \tag{26}$$

where t_p is the coordinate time of perigee passage.

In Newtonian mechanics, the gravitational field is conservative and total energy is conserved. Using the above equations for the Keplerian orbit, one can show that the total energy per unit mass of the satellite is:

$$\frac{1}{2}v^2 - \frac{GM_E}{r} = -\frac{GM_E}{2a}.$$ (27)

If I use Eq. (27) for v^2 in Eq. (23), then I get the following expression for the elapsed coordinate time on the satellite clock:

$$\Delta t = \int_{\text{path}} d\tau \left[1 + \frac{3GM_E}{2ac^2} + \frac{\Phi_0}{c^2} - \frac{2GM_E}{c^2}\left(\frac{1}{a} - \frac{1}{r}\right)\right].$$ (28)

The first two constant rate correction terms in Eq. (28) have the values:

$$\frac{3GM_E}{2ac^2} + \frac{\Phi_0}{c^2} = +2.5046 \times 10^{-10} - 6.9693 \times 10^{-10} = -4.4647 \times 10^{-10}.$$ (29)

The negative sign in this result means that the standard clock in orbit is beating too fast, primarily because its frequency is gravitationally blueshifted. In order for the satellite clock to appear to an observer on the geoid to beat at the chosen frequency of 10.23 MHz, the satellite clocks are adjusted lower in frequency so that the proper frequency is:

$$\left[1 - 4.4647 \times 10^{-10}\right] \times 10.23 \text{ MHz} = 10.229\ 999\ 995\ 43 \text{ MHz}.$$ (30)

This adjustment is either accomplished on the ground before the clock is placed in orbit or applied in the navigation message after the clocks are placed in orbit and measured.

Figure 2 shows the net fractional frequency offset of an atomic clock in a circular orbit, which is essentially the left side of Eq. (29) plotted as a function of orbit radius a, with a change of sign. Five sources of relativistic effects contribute in Figure 2. Several interesting orbit radii are marked. For a low earth orbiter such as the Space Shuttle, the velocity is so great that slowing due to time dilation is the dominant effect, while for a GPS satellite clock, the gravitational blue shift is greater. The effects cancel at $a \approx 9545$ km. The Global Navigation Satellite System GALILEO, which is currently being designed under the auspices of the European Space Agency, will have orbital radii of approximately 30,000 km.

There is an interesting story about this frequency offset. At the time of launch of the NTS-2 satellite (23 June 1977), which contained the first cesium atomic clock to be placed in orbit, it was recognized that orbiting clocks would require relativistic corrections, but there was uncertainty as to its magnitude as well as its sign. There were some who doubted that relativistic effects were truths that would need to be incorporated![12] A frequency synthesizer was built into the satellite clock system so that after launch, if in fact the rate of the clock in its final orbit was that predicted

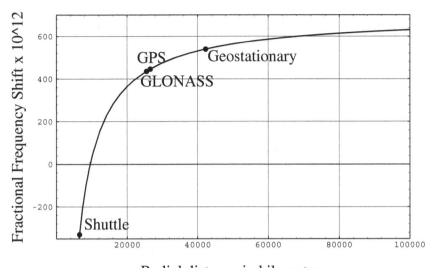

Radial distance in kilometers

Fig. 2.   Net fractional frequency shift of clock in a circular orbit.

by general relativity, then the synthesizer could be turned on, bringing the clock to the coordinate rate necessary for operation. After the cesium clock was turned on in NTS-2, it was operated for about 20 days to measure its clock rate before turning on the synthesizer.[13] The frequency measured during that interval was +442.5 parts in 10^{12} compared to clocks on the ground, while general relativity predicted +446.5 parts in 10^{12}. The difference was well within the accuracy capabilities of the orbiting clock. This then gave about a 1% verification of the combined second-order Doppler and gravitational frequency shift effects for a clock at 4.2 earth radii.

Additional small frequency offsets can arise from clock drift, environmental changes, and other unavoidable effects such as the inability to launch the satellite into an orbit with precisely the desired semimajor axis. The navigation message provides satellite clock frequency corrections for users so that in effect, the clock frequencies remain as close as possible to the frequency of the U. S. Naval Observatory's reference clock ensemble. Because of such procedures, it would now be difficult to test the frequency offset predicted by relativity.

When GPS satellites were first deployed, the specified factory frequency offset was slightly in error because the important contribution from earth's centripetal potential (see Eq. (13)) had been inadvertently omitted at one stage of the evaluation. Although GPS managers were made aware of this error in the early 1980s, eight years passed before system specifications were changed to reflect the correct calculation.[14] As understanding of the numerous sources of error in the GPS slowly improved, it eventually made sense to incorporate the correct relativistic calculation.

The eccentricity correction. The last term in Eq. (28) may be integrated exactly by using the following expression for the rate of change of eccentric anomaly with time, which follows by differentiating Eq. (26):

$$\frac{dE}{dt} = \frac{\sqrt{GM_E/a^3}}{1 - e\cos E}.$$ (31)

Also, since a relativistic correction is being computed, $ds/c \simeq dt$, so:

$$
\begin{aligned}
\int \left[\frac{2GM_E}{c^2} \left(\frac{1}{r} - \frac{1}{a} \right) \right] \frac{ds}{c} &\simeq \frac{2GM_E}{c^2} \int \left(\frac{1}{r} - \frac{1}{a} \right) dt \\
&= \frac{2GM_E}{ac^2} \int dt \left(\frac{e\cos E}{1 - e\cos E} \right) \\
&= \frac{2\sqrt{GM_E a}}{c^2} e\,(\sin E - \sin E_0) \\
&= +\frac{2\sqrt{GM_E a}}{c^2} e\sin E + \text{constant}.
\end{aligned}
$$ (32)

The constant of integration in Eq. (32) can be dropped since this term is lumped with other clock offset effects in the Master Control Station's estimate of the clock's behavior. The net correction for clock offset due to relativistic effects that vary in time is:

$$\Delta t_r = +4.4428 \times 10^{-10} \frac{\text{sec}}{\sqrt{\text{meter}}} e\sqrt{a}\sin E.$$ (33)

This correction must be made by the receiver; it is a correction to the coordinate time as transmitted by the satellite. For a satellite of eccentricity $e = 0.01$, the maximum size of this term is about 23 ns. The correction is needed because of a combination of effects on the satellite clock due to gravitational frequency shift and second-order Doppler shift, which vary due to orbit eccentricity.

It is not at all necessary, in a navigation satellite system, that the eccentricity correction be applied by the receiver. It appears that the clocks in the GLONASS satellite system do have this correction applied before broadcast. In fact historically, this was dictated in the GPS by the small amount of computing power available in the early GPS satellite vehicles.

It may now be too late to reverse this decision because of the investment that many dozens of receiver manufacturers have in their products. However, it does mean that receivers are supposed to incorporate the relativity correction; therefore, if appropriate data can be obtained in raw form from a receiver one can measure this effect. Such measurements are discussed next.

6. TOPEX/POSEIDON Relativity Experiment

At present, the frequencies of atomic clocks in replacement satellites are carefully measured after launch and then adjusted to the frequency that is required for operation. The largest remaining effect is the eccentricity effect, Eq. (33). It is intended that GPS receivers correct for this effect, so a receiver that can output data on transmission and reception events can be used to test whether the relativistic prediction agrees with experiment. The TOPEX satellite carries a six-channel receiver with a very good quartz oscillator to provide the time reference, and is in an orbit of radius 7,714 km and period 6745 seconds. The receiver motion is highly dynamic, as it passes under the GPS constellation eleven times per day. A stringent test of the relativistic prediction can therefore be performed.

The local quartz clock, which is a free-running oscillator subject to various noise and drift processes, can be in error by a large amount. So the first task is to determine the local clock time in terms of GPS time. For this purpose the six available channels in the receiver provide considerable redundancy. The trajectories of the TOPEX and GPS satellites were determined independently of the on-board clocks, by means of Doppler tracking from $\approx 10^2$ stations maintained by the Jet Propulsion Laboratory (JPL). Generally, at each time point during the experiment, observations were obtained from six (sometimes five) satellites. There is sufficient redundancy in the measurements to obtain good estimates of the TOPEX clock time and the rms error in this time due to measurement noise.

The rms deviation from the mean of the TOPEX clock time measurements is plotted in Figure 3 as a function of time. The average rms error is 29 cm, corresponding to about one ns of propagation delay. Much of this variation can be attributed to multipath effects–multiple reflections of the signals from objects in the neighborhood of the receiver's antenna.

With the local TOPEX clock time determined in terms of GPS time, the eccentricity effect from some GPS satellite clock of interest can be determined by using five of the receiver channels to determine the TOPEX

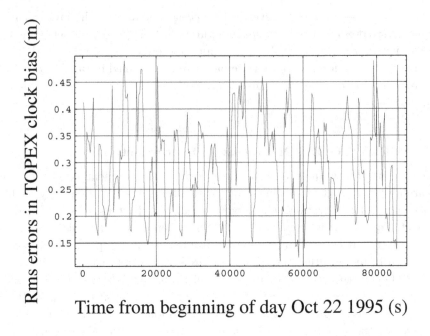

Time from beginning of day Oct 22 1995 (s)

Fig. 3. Rms deviation from mean of TOPEX clock bias determinations.

position and the sixth channel to measure the eccentricity effect on the sixth satellite clock. Strictly speaking, in finding the eccentricity effect this way for a particular satellite, one should not include data from that satellite in the determination of the clock bias. One can show, however, that the penalty for this is simply to increase the rms error by a factor of 6/5, to 35 cm. Figure 4 shows the measured eccentricity effect for SV #13, which has the largest eccentricity of the satellites that were tracked, $e = .01486$. The solid curve in Figure 4 is the theoretically predicted effect, from Eq. (33). While the agreement is fairly good, one can see some evidence of systematic bias during particular passes, where the rms error (plotted as vertical lines on the measured dots) is significantly smaller than the discrepancies between theory and experiment. For this particular satellite, the rms deviation between theory and experiment is 22 cm, which is about 2.2 % of the maximum magnitude of the effect, 10.2 meters.

Similar plots were obtained for 25 GPS satellites that were tracked during this experiment. For the entire constellation, the agreement between theory and experiment is within about 2.5%.

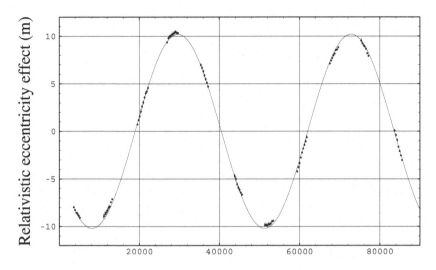

Time from beginning of day Oct 22 1995 (s)

Fig. 4. Comparison of predicted and measured eccentricity effect for SV #13.

7. Doppler Effect

Since orbiting clocks have had their rate adjusted so that they beat co-ordinate time, and since responsibility for correcting for the periodic relativistic effect due to eccentricity has been delegated to receivers, one must take extreme care in discussing the Doppler effect for signals transmitted from satellites. Even though second-order Doppler effects have been accounted for, for earth-fixed users there will still be a first-order (longitudinal) Doppler shift, which has to be dealt with by receivers. As is well known, in a static gravitational field coordinate frequency is conserved during propagation of an electromagnetic signal along a null geodesic. If one takes into account only the monopole and quadrupole contributions to earth's gravitational field, then the field is static and one can exploit this fact to discuss the Doppler effect.

Consider the transmission of signals from rate-adjusted transmitters orbiting on GPS satellites. Let the gravitational potential and velocity of the satellite be $V(\mathbf{r}_j) \equiv V_j$, and $\mathbf{v}_j$, respectively. Let the frequency of the satellite transmission, before the rate adjustment is done, be $f_0 = 10.23$ MHz. After taking into account the rate adjustment discussed previously, it is straightforward to show that for a receiver of velocity $\mathbf{v}_R$ and gravitational potential V_R (in ECI coordinates), the received frequency is

$$f_R = f_0 \left[1 + \frac{-V_R + \mathbf{v}_R^2/2 + \Phi_0 + 2GM_E/a + 2V_j}{c^2} \right] \frac{(1 - \mathbf{N} \cdot \mathbf{v}_R/c)}{(1 - \mathbf{N} \cdot \mathbf{v}_j/c)} , \quad (34)$$

where $\mathbf{N}$ is a unit vector in the propagation direction in the local inertial frame. For a receiver fixed on the earth's rotating geoid, this reduces to

$$f_R = f_0 \left[1 + \frac{2GM_E}{c^2} \left(\frac{1}{a} - \frac{1}{r} \right) \right] \frac{(1 - \mathbf{N} \cdot \mathbf{v}_R/c)}{(1 - \mathbf{N} \cdot \mathbf{v}_j/c)} . \quad (35)$$

The correction term in square brackets gives rise to the eccentricity effect. The longitudinal Doppler shift factors are not affected by these adjustments; they will be of order 10^{-5} while the eccentricity effect is of order $e \times 10^{-10}$.

8. Crosslink Ranging

In the "Autonav" mode of GPS operation, receivers on the satellites listen to signals from the other satellites and determine their own position and coordinate time by direct exchange of signals. The standard atomic clock in the transmitting satellite suffers a rate adjustment, and then needs an eccentricity correction to get the coordinate time. Then a signal is sent to another satellite which requires calculating a coordinate time of propagation possibly incorporating a relativistic time delay. There is then a further transformation of rate and another "$e \sin E$" correction to get the atomic time on the receiving satellite's clock. So that the rate adjustment does not introduce confusion into this analysis, I shall assume the rate adjustments are already accounted for and use the subscript 'S' to denote coordinate time measurements using rate-adjusted satellite clocks.

Then, let a signal be transmitted from satellite #i, at position $\mathbf{r}_i$ and having velocity $\mathbf{v}_i$ in ECI coordinates, at satellite clock time $T_S^{(i)}$, to satellite #j, at position $\mathbf{r}_j$ and having velocity $\mathbf{v}_j$. The coordinate time at which this occurs, apart from a constant offset, from Eq. (32) will be

$$T^{(i)} = T_S^{(i)} + \frac{2\sqrt{GMa_i}}{c^2} e_i \sin E_i. \quad (36)$$

The coordinate time elapsed during propagation of the signal to the receiver in satellite #j is in first approximation l/c, where l is the distance between transmitter at the instant of transmission, and receiver at the instant of reception: $\Delta T = T^{(j)} - T^{(i)} = l/c$. The Shapiro time delay corrections to this will be discussed in the next section. Finally, the coordinate time of arrival of the signal is related to the time on the receiving satellite's adjusted clock by the inverse of Eq. (36):

$$T_S^{(j)} = T^{(j)} - \frac{2\sqrt{GMa_j}}{c^2} e_j \sin E_j. \tag{37}$$

Collecting these results,

$$T_S^{(j)} = T_S^{(i)} + \frac{l}{c} - \frac{2\sqrt{GMa_j}}{c^2} e_j \sin E_j + \frac{2\sqrt{GMa_i}}{c^2} e_i \sin E_i. \tag{38}$$

In Eq. (38) the distance l is the actual propagation distance, in ECI coordinates, of the signal. This result contains all the relativistic corrections that need to be considered for direct time transfer by transmission of a time-tagged pulse from one satellite to another.

9. Frequency Shifts Induced by Orbit Changes

Improvements in GPS motivate attention to other small relativistic effects that have previously been too small to be explicitly considered. For SV clocks, these include frequency changes due to orbit adjustments, and effects due to earth's oblateness. For example, between July 25 and October 10, 2000, SV43 occupied a transfer orbit while it was moved from slot 5 to slot 3 in orbit plane F. The fractional frequency shift associated with this orbit adjustment was measured carefully[15] and found to be -1.85×10^{-13}. During such orbit adjustments, typically the satellite is raised or lowered in altitude by 20 km or so. Also, earth's oblateness causes a periodic fractional frequency shift with period of almost 6 hours and amplitude 0.695×10^{-14}. This means that quadrupole effects on SV clock frequencies may be important in the consideration of frequency breaks induced by orbit changes, especially since some of the recently launched Rubidium clocks show stabilities of order 8×10^{-15}. Thus, some approximate expressions for the frequency effects due to earth's oblateness, on SV clock frequencies, are needed. These effects will be discussed with the help of Lagrange's planetary perturbation equations.

Five distinct relativistic effects, discussed in Sect. 5 above, are incorporated into the System Specification Document, ICD-GPS-200.[14] These are: the effect of earth's mass on gravitational frequency shifts of atomic reference clocks fixed on the earth's surface relative to clocks at infinity; the effect of earth's oblate mass distribution on gravitational frequency shifts of atomic clocks fixed on earth's surface; second-order Doppler shifts of clocks fixed on earth's surface due to earth rotation; gravitational frequency shifts of clocks in GPS satellites due to earth's mass; and second-order Doppler shifts of clocks in GPS satellites due to their motion through an earth-centered inertial (ECI) Frame. The combination of second-order Doppler and gravitational frequency shifts given in Eq. (22) for a clock in a GPS satellite leads directly to the following expression for the fractional frequency shift of a satellite clock relative to a reference clock fixed on earth's geoid:

$$\frac{\Delta f}{f} = -\frac{1}{2}\frac{v^2}{c^2} - \frac{GM_E}{rc^2} - \frac{\Phi_0}{c^2} , \tag{39}$$

where v is the satellite speed in a local ECI reference frame, and Φ_0 is the effective gravitational potential on the earth's rotating geoid.

If the GPS satellite orbit can be approximated by a Keplerian orbit of semi-major axis a, then Eq. (27) gives

$$\frac{\Delta f}{f} = -\frac{3GM_E}{2ac^2} - \frac{\Phi_0}{c^2} + \frac{2GM_E}{c^2}\left[\frac{1}{r} - \frac{1}{a}\right] . \tag{40}$$

The first two terms in Eq. (40) give rise to the "factory frequency offset", which is supposed to be applied to GPS clocks before launch in order to make them beat at a rate equal to that of reference clocks on earth's surface. The last term in Eq. (40) is very small when the orbit eccentricity e is small; when integrated over time these terms give rise to the so-called "$e\sin E$" effect or "eccentricity effect." In most of the following discussion we shall assume that eccentricity is very small.

Clearly, from Eq. (40), if the semi-major axis should change by an amount δa due to an orbit adjustment, the satellite clock will experience a fractional frequency change

$$\frac{\delta f}{f} = +\frac{3GM_E\delta a}{2c^2a^2} . \tag{41}$$

The factor $3/2$ in this expression arises from the combined effect of second-order Doppler and gravitational frequency shifts. If the semi-major axis increases, the satellite will be higher in earth's gravitational potential and will be gravitationally blue-shifted more, while at the same time the satellite velocity will be reduced, reducing the size of the second-order Doppler shift (which is generally a red shift). The net effect would make a positive contribution to the fractional frequency shift.

Earth's quadrupole moment. If good estimates of the semi-major axis before and after an orbit adjustment were available, Eq. (41) would provide a means of calculating or predicting the frequency change. Perturbations of GPS orbits due to earth's quadrupole mass distribution are significant compared to the change in semi-major axis associated with the orbit change discussed above. For the semi-major axis, if the eccentricity is very small the dominant contribution has a period twice the orbital period and has amplitude $3J_2a_1^2\sin^2 i_0/(2a_0) \approx 1658$ m. Here $a_0 = 2.656175 \times 10^7$ m, is the SV orbit semi-major axis. This raises the question whether it is sufficiently accurate to describe GPS orbits as Keplerian while estimating such semi-major axis changes. In this section, we estimate the effect of earth's quadrupole moment on the orbital elements of a nominally circular orbit and thence on the change in frequency induced by an orbit change. Previously, such an effect on the SV clocks has been neglected, and indeed, due in part to a remarkable coincidence, it does turn out to be small. The analysis provides a method of finding the semi-major axis before and after the orbit change.

The oscillation in the semi-major axis would significantly affect calculations of the semi-major axis at any particular time. This suggests that Eq. (27) needs to be reexamined in light of the periodic perturbations on the semi-major axis. Therefore, in this section we develop an approximate description of a satellite orbit, of small eccentricity, taking into account earth's quadrupole moment to first order. Terms of order $J_2 \times e$ will be neglected.

Conservation of energy. The gravitational potential of a satellite at position (x, y, z) in equatorial ECI coordinates in the model under consideration here is

$$V(x,y,z) = -\frac{GM_E}{r}\left(1 - \frac{J_2a_1^2}{r^2}\left[\frac{3z^2}{2r^2} - \frac{1}{2}\right]\right). \qquad (42)$$

Since the force is conservative in this model (solar radiation pressure, thrust, *etc.* are not considered), the kinetic plus potential energy is conserved. Let ϵ be the energy per unit mass of an orbiting mass point. Then

$$\epsilon = \text{constant} = \frac{v^2}{2} + V(x,y,z) = \frac{v^2}{2} - \frac{GM_E}{r} + V'(x,y,z), \qquad (43)$$

where $V'(x,y,z)$ is the perturbing potential due to the earth's quadrupole potential. It is shown in textbooks[11] that, with the help of Lagrange's planetary perturbation theory, the conservation of energy condition can be put in the form

$$\epsilon = -\frac{GM_E}{2a} + V'(x,y,z), \qquad (44)$$

where a is the perturbed (osculating) semi-major axis. In other words, for the perturbed orbit,

$$\frac{v^2}{2} - \frac{GM_E}{r} = -\frac{GM_E}{2a}. \tag{45}$$

On the other hand, the net fractional frequency shift relative to a clock at rest at infinity is determined by the second-order Doppler shift (a red shift) and a gravitational red shift. The total relativistic fractional frequency shift is

$$\frac{\Delta f}{f} = -\frac{v^2}{2} - \frac{GM_E}{r} + V'(x, y, z). \tag{46}$$

The conservation of energy condition can be used to express the second-order Doppler shift in terms of the potential. Here we are interested in fractional frequency changes caused by changing the orbit, so it will make no difference if the calculations use a clock at rest at infinity as a reference rather than a clock at rest on earth's surface. From perturbation theory we need expressions for the square of the velocity, for the radius r, and for the perturbing potential. We refer to the literature[16,11] for the perturbed osculating elements. These are exactly known, to all orders in the eccentricity, and to first order in J_2. We shall keep only the leading terms in eccentricity e, and quote the resulting perturbations for the kinetic and potential energy terms.

$$\frac{v^2}{2} = \frac{GM_E}{2\bar{a}} (1 + 2e_0 \cos E) + \frac{3GM_E J_2 a_1^2}{2\bar{a}^3} \left(1 - \frac{3}{2} \sin^2 i_0 \right)$$
$$+ \frac{GM_E J_2 a_1^2}{2\bar{a}^3} \sin^2 i_0 \cos 2(\omega_0 + f). \tag{47}$$

where $\bar{a}$ is the mean perturbed osculating semi-major axis. Perturbations to the semi-major axis and eccentricity give rise to the following expression for the monopole contribution to the gravitational potential:

$$-\frac{GM_E}{r} = -\frac{GM_E}{\bar{a}} (1 + e_0 \cos E) - \frac{3GM_E J_2 a_1^2}{2\bar{a}^3} \left(1 - \frac{3}{2} \sin^2 i_0 \right)$$
$$+ \frac{GM_E J_2 a_1^2 \sin^2 i_0}{4\bar{a}^3} \cos 2(\omega_0 + f). \tag{48}$$

Thus, the quadrupole potential causes a change in the radius resulting in an important change in the contributions from the monopole portion of the potential.

Since the perturbing potential contains the small factor J_2, to leading order we may substitute unperturbed values for r and z into $V'(x, y, z)$, which yields the expression

$$V'(x, y, z) = -\frac{GM_E J_2 a_1^2}{2\bar{a}^3}\left(1 - \frac{3}{2}\sin^2 i_0\right) - \frac{3GM_E J_2 a_1^2 \sin^2 i_0}{4\bar{a}^3}\cos 2(\omega_0 + f).$$
(49)

Each of the results quoted above have contributions proportional to $(1 - 3\sin^2 i_0/2)$. Due to a fortuitous choice of inclination, $i_0 = 55°$, all such contributions are negligibly small. Because this term is negligible, numerical calculations of the total energy per unit mass provide a means of evaluating the mean perturbed semi-major axis $\bar{a}$.

Conservation of energy. It is now very easy to check conservation of energy. Adding kinetic energy per unit mass to two contributions to the potential energy then gives

$$\epsilon = \frac{v^2}{2} - \frac{GM_E}{r} + V' = -\frac{GM_E}{2\bar{a}}.$$
(50)

This verifies that the perturbation theory gives a constant energy, of the same form as that for a pure Keplerian orbit. Numerical calculations of the total energy per unit mass then yield the mean perturbed semi-major axis $\bar{a}$.

Calculation of fractional frequency shift. The fractional frequency shift calculation is very similar to the calculation of the energy, except that the second-order Doppler term contributes with a *negative* sign. The result is

$$\begin{aligned}\frac{\Delta f}{f} &= -\frac{v^2}{2c^2} - \frac{GM_E}{c^2 r} + \frac{V'}{c^2} \\ &= -\frac{GM_E}{\bar{a}c^2}\left(\frac{3}{2} + 2\cos E_0\right) - \frac{GM_E J_2 a_1^2 \sin^2 i_0}{\bar{a}^3 c^2}\cos 2(\omega_0 + f).\end{aligned}$$
(51)

The first term, when combined with the reference potential at earth's geoid, gives rise to the "factory frequency offset." The second term gives rise to the eccentricity effect. The last term has amplitude

$$\frac{GM_E J_2 a_1^2 \sin^2 i_0}{a_0^3 c^2} = 6.95 \times 10^{-15},$$
(52)

which may be large enough to consider when calculating frequency shifts produced by orbit changes. Therefore, this contribution may have to be considered in the future in the determination of the semi-major axis, but for now we neglect it.

The result suggests the following method of computing the fractional frequency shift: averaging the shift over one orbit, the periodic term will average down to a negligible value. So if one has a good estimate for the nominal semi-major axis parameter, the term $-3GM_E/2a_0c^2$ gives the average fractional frequency shift. On the other hand, the average energy per unit mass is given by $\epsilon = -GM_E/2a_0$. Therefore, the precise ephemerides, specified in an ECI frame, can be used to compute the average value for ϵ, then the average fractional frequency shift will be

$$\frac{\Delta f}{f} = 3\epsilon/c^2 \,. \tag{53}$$

When this approach is applied to the orbit change of July, 2000, the fractional frequency change is calculated to be

$$\frac{\Delta f}{f} = -1.77 \times 10^{-13} \,. \tag{54}$$

This agrees with the measured value to within about 3.3%. Applications to other orbit change events have worked so well that they are now included in the estimates of frequency changes, before the orbit adjustments occur. This results in improved GPS performance since otherwise it would take days to measure the frequency changes; during such measurements the satellites would be unusable.

Quadrupole time correction. The last periodic term in Eq. (51) is of a form similar to that which gives rise to the eccentricity correction, which is applied by GPS receivers. Considering only the last periodic term, the additional time elapsed on the orbiting clock will be given by

$$\delta t_{J_2} = \int_{path} dt \left[-\frac{GM_E J_2 a_1^2 \sin^2 i_0}{\bar{a}^3 c^2} \cos(2\omega_0 + 2nt) \right] \,, \tag{55}$$

where to a sufficient approximation we have replaced the quantity f in the integrand by $nt = \sqrt{GM_E/\bar{a}^3}t$; n is the approximate mean motion of GPS satellites. Integrating and dropping the constant of integration (assuming as usual that such constant time offsets are lumped with other contributions) gives the periodic relativistic effect on the elapsed time of the SV clock due to earth's quadrupole moment:

$$\delta t_{J_2} = -\sqrt{\frac{GM_E}{\bar{a}^3}} \frac{J_2 a_1^2 \sin^2 i_0}{2c^2} \sin(2\omega_0 + 2nt) \,. \tag{56}$$

The correction that should be applied by the receiver is the *negative* of this expression. The phase of the correction is zero when the satellite passes through earth's equatorial plane going northwards. If not accounted for, this effect on the SV clock time would give rise to a peak-to-peak periodic navigational error in position of approximately $2c \times \delta t_{J_2} = 1.43$ cm.

Summary. In the present calculation, the effect of earth's quadrupole moment on the Keplerian orbit was accounted for. It was not necessary to compute the orbit eccentricity. This approximate treatment of the orbit makes no attempt to consider perturbations that are non-gravitational in nature – *e.g.*, solar radiation pressure. As a general conclusion, the fractional frequency shift can be estimated to very good accuracy from the expression for the "factory frequency offset,"

$$\frac{\delta f}{f} = +\frac{3GM_E\delta a}{2c^2a^2} .\qquad (57)$$

10. Secondary Relativistic Effects

There are several additional significant relativistic effects that must be considered at the level of accuracy of a few cm (which corresponds to 100 picoseconds of delay). Many investigators are modeling systematic effects down to the millimeter level so these effects, which currently are not sufficiently large to affect navigation, may have to be considered in the future.

Signal Propagation Delay. The Shapiro signal propagation delay may be easily derived in the standard way from the metric, Eq. (18), which incorporates the choice of coordinate time rate expressed by the presence of the term in Φ_0/c^2. Setting $ds^2 = 0$ and solving for the increment of coordinate time along the path increment $d\sigma = \sqrt{dr^2 + r^2d\theta^2 + r^2 \sin^2\theta d\phi^2}$ gives

$$dt = \frac{1}{c}\left[1 - \frac{2V}{c^2} + \frac{\Phi_0}{c^2}\right] d\sigma .\qquad (58)$$

The time delay is sufficiently small that quadrupole contributions can be neglected. Integrating along the straight line path a distance l between the transmitter and receiver gives for the time delay

$$\Delta t_{delay} = \frac{\Phi_0}{c^2}\frac{l}{c} + \frac{2GM_E}{c^3}\ln\left[\frac{r_1 + r_2 + l}{r_1 + r_2 - l}\right] ,\qquad (59)$$

where r_1 and r_2 are the distances of transmitter and receiver from earth's center. The second term is the usual expression for the Shapiro time delay. It is modified for GPS by a term of opposite sign (Φ_0 is negative), due to the choice of coordinate time rate. This tends to cancel the logarithm term. The net effect for a satellite to earth link is less than 2 cm and for most purposes can be neglected. One must keep in mind, however, that in the main term, l/c, l is a coordinate distance and further small relativistic corrections are required to convert it to a proper distance.

Effect on Geodetic Distance. At the level of a few millimeters, spatial curvature effects should be considered. For example, using Eq. (18), the proper distance between a point at radius r_1 and another point at radius r_2 directly above the first is approximately

$$\int_{r_1}^{r_2} dr \left[1 + \frac{GM_E}{c^2 r} \right] = r_2 - r_1 + \frac{GM_E}{c^2} \ln \left(\frac{r_2}{r_1} \right) . \tag{60}$$

The difference between proper distance and coordinate distance, and between the earth's surface and the radius of GPS satellites, is approximately $4.43 \ln(4.2)$ mm ≈ 6.3 mm. Effects of this order of magnitude would enter, for example, in the comparison of laser ranging to GPS satellites, with numerical calculations of satellite orbits based on relativistic equations of motion using coordinate times and coordinate distances.

Phase Wrap-Up. Transmitted signals from GPS satellites are right circularly polarized and thus have negative helicity. For a receiver at a fixed location, the electric field vector rotates counterclockwise, when observed facing into the arriving signal. Let the angular frequency of the signal be ω in an inertial frame, and suppose the receiver spins rapidly with angular frequency Ω which is parallel to the propagation direction of the signal. The antenna and signal electric field vector rotate in opposite directions and thus the received frequency will be $\omega + \Omega$. In GPS literature this is described in terms of an accumulation of phase called "phase wrap-up." This effect has been known for a long time[17,18,19,21], and has been experimentally measured with GPS receivers spinning at rotational rates as low as 8 cps. It is similar to an additional Doppler effect; it does not affect navigation if four signals are received simultaneously by the receiver as in Eqs. (1).

Effect of Other Solar System Bodies. One set of effects that has been "rediscovered" many times are the red shifts due to other solar system bodies. The Principle of Equivalence implies that sufficiently near the earth, there can be no linear terms in the effective gravitational potential due to other solar system bodies, because the earth and its satellites are in free fall in the fields of all these other bodies. The net effect locally can only come from tidal potentials, the third terms in the Taylor expansions of such potentials about the origin of the local freely falling frame of reference. Such tidal potentials from the sun, at a distance r from earth, are of order $GM_\odot r^2 / R^3$ where R is the earth-sun distance.[22] The gravitational frequency shift of GPS satellite clocks from such potentials is a few parts in 10^{16} and is currently neglected in the GPS.

11. Applications

The number of applications of GPS has been astonishing. Accurate positioning and timing, other than for military navigation, include synchronization of power line nodes for fault detection, communications, VLBI, navigation in deep space, tests of fundamental physics, measurements on pulsars, tests of gravity theories, vehicle tracking, search and rescue, surveying, mapping, and navigation of commercial aircraft, to name a few. These are too numerous to go into in much detail here, but some applications are worth mentioning. Civilian applications have overtaken military applications to the extent that SA was turned off in May of 2000.

The Nobel-prizewinning work of Joseph Taylor and his collaborators[23] on the measurement of the rate of increase of the binary pulsar period depended on GPS receivers at the Arecibo observatory, for transferring UTC from the U.S. Naval Observatory and NIST to the local clock. Time standards around the world are compared using GPS in common-view; with this technique SA would cancel out, as well as do many sources of systematic errors such as ionospheric and tropospheric delays. Precise position information can assist in careful husbandry of natural resources, and animal and vehicle fleet tracking can result in improved efficiency. Precision agriculture makes use of GPS receivers in real-time application of pesticides or fertilizers, minimizing waste. Sunken vessels or underwater ruins with historically significant artifacts can be located using the GPS and archeologists can return again and again with precision to the same location. Monster ore trucks or earth-moving machines can be fitted with receivers and controlled remotely with minimal risk of collision or interference with other equipment. Disposable GPS receivers dropped through tropical storms transmit higher resolution measurements of temperature, humidity, pressure, and wind speed than can be obtained by any other method; these have led to improved understanding of how tropical storms intensify. Slight movements of bridges or buildings, in response to various loads, can be monitored in real time. Relative movements of remote parts of earth's crust can be accurately measured in a short time, contributing to better understanding of tectonic processes within the earth and, possibly, to future predictions of earthquakes. With the press of a button, a lost hiker can send a distress signal that includes the hikers' location.

These and many other creative applications of precise positioning and timing are leading to a rapid expansion of GPS products and services. Over 50 manufacturers produce more than 350 different GPS products for commercial, private, and military use. The number of receivers manufactured each year is in excess of two million, and different applications are continually being invented. Marketing studies predict that sales of GPS equipment and services will grow to over \$34 billion by 2006. Revenue for the European GALILEO system is projected to be 10 billion Euros per year.

12. Conclusions

The GPS is a remarkable laboratory for applications of the concepts of special and general relativity. GPS is also valuable as an outstanding source of pedagogical examples. It is deserving of more scrutiny from relativity experts. It is particularly important to confirm that the basis for synchronization is on a firm conceptual foundation.

References

1. Fliegel, H. F., and R. S. DiEsposti, "GPS and Relativity: an Engineering Overview,", Aerospace Corp. Report No. ATR-97(3389)-1, 20 Dec. (1996).
2. Malys, S., and Slater, "Maintenance and Enhancement of the World Geodetic System 1984," Proc. ION-GPS-94, Salt Lake City, UT, Sept. 20-23, 1994, pp 17-24 (1994); National Imagery and Mapping Agency Technical Report 8350.2, "World Geodetic System 1984," Third Edition, Amendment 1, NIMA Stock No. DMATR83502WGS84, NSN 7643-01-402-0347.
3. Ashby, N., and Weiss, M., "Global Positioning Receivers and Relativity," NIST Technical Note 1385, U. S. Government Printing Office, Washington, D.C., March (1999).
4. Ashby, N., "An Earth-Based Coordinate Clock Network," NBS Technical Note 659, U. S. Dept. of Commerce (1975). U. S. Government Printing Office, Washington, D.C. 20402 (S. D. Catalog # C13:46:659).
5. Allan, D.W., Ashby, N., and Weiss, M., *Science* **228**, 69–70, (1985).
6. See http://www.danof.obspn.fr/IAU_Resolutions/Resol_UAI.htm for resolutions adopted at the 20th General Assembly, Manchester, August 2000.
7. Lambeck, K., *Geophysical Geodesy*, Oxford Science Publications, The Clarendon Press, Oxford pp 13-18, (1988).
8. Ashby, N., and Spilker, J.J., Jr., "Introduction to relativistic effects on the Global Positioning System," Chapter 18 in Parkinson, B.W., and Spilker, J.J., Jr., eds., *Global Positioning System: Theory and Applications*, Vol. I, American Institute of Aeronautics and Astronautics, Inc., Washington, D.C., pp 623-697 (1996).
9. Garland, G.D., *Earth's Shape and Gravity*, Pergamon Press, New York (1965).
10. Ashby, N., and Allan, D.W., "Practical Implications of Relativity for a Global Coordinate Time Scale," *Radio Science* **14**, 649-669 (1979).
11. Fitzpatrick, P., *The Principles of Celestial Mechanics*, Academic Press, New York (1970).

12. Alley, C., "Proper time experiments in gravitational fields with atomic clocks, aircraft, and laser light pulses," in *Proc. NATO Advanced Study Institute on Quantum Optics and Experimental General Relativity*, August 1981, Bad Windesheim, Germany, eds., Meystre, P., and Scully, M. O., Plenum, New York p. 363 (1983).

13. Buisson, J.A., Easton, R.L., and McCaskill, T.B., *Proceedings of the 9th Annual Precise Time and Time Interval Applications and Planning Meeting*, pp 177-200, Technical Information and Administrative Support Division, Goddard Space Flight Center, Greenbelt, MD, (1977).

14. "NAVSTAR GPS Space Segment/Navigation User Interfaces," ICD-GPS-200, Revision C, ARINC Research Corp., Fountain Valley, CA, (1993).

15. Epstein, M., Stoll, E., and Fine, J., "Study of SVN43 clock from 06/18/00 to 12/30/00," in Proc. 33rd Annual Precise Time and Time Interval Systems and Applications Meeting, Long Beach, Nov. 2001, L. Breakiron, Ed., U.S. Naval Observatory, Washington, D.C.

16. Kozai, Y., Astronomical Journal **64**, 367 (1959).

17. Kraus, J. D., *Antennas*, 2nd Ed., McGraw-Hill (1988), reprinted by Cygnus-Quasar Books, Powell, Ohio.

18. Mashhoon, B., *Phys. Rev.* **A**47, 4498 (1993).

19. Mashhoon, B., *Phys. Lett.* **A**198, pp. 9-13 (1995).

20. Spilker, J.J., Jr., and Parkinson, B.W., "Overview of GPS Operation and Design," Chapter 2 in Parkinson, B.W., and Spilker, J.J., Jr., eds., *Global Positioning System: Theory and Applications*, Vol. I, American Institute of Aeronautics and Astronautics, Inc., Washington, D.C., p 33 (1996).

21. Tetewsky, A.K., and Mullen, F.E., "Effects of Platform Rotation on GPS with Implications for GPS Simulators," Proc. ION-GPS-96, Kansas City, MO, Sept. 17-20, pp. 1917-1925 (1996).

22. Ashby, N., and Bertotti, B., "Relativistic Effects in Local Inertial Frames," *Phys. Rev.* **D**34, 2246–2258 (1986).

23. Hulse, R.A., and Taylor, J.H., "Discovery of a pulsar in a binary system," *Astophys. J.* **195**, L51-53 (1975); Taylor, J.H., "Binary Pulsars and General Relativity," *Rev. Mod. Phys.* **66**, 711-719 (1994).

Part III

Beyond Einstein
Unifying General Relativity with Quantum Physics

...a really new field of experience will always lead to crystalliza-tion of a new system of scientific concepts and laws...when faced with essentially new intellectual challenges, we continually follow the example of Columbus who possessed the courage to leave the known world in almost the insane hope of finding land again beyond the sea.
—W. Heisenberg (Changes in the Foundation of Exact Science)

CHAPTER 11

SPACETIME IN SEMICLASSICAL GRAVITY

L. H. FORD

Institute of Cosmology
Department of Physics and Astronomy
Tufts University, Medford, MA 02155, USA
ford@cosmos.phy.tufts.edu

This article will summarize selected aspects of the semiclassical theory of gravity, which involves a classical gravitational field coupled to quantum matter fields. Among the issues which will be discussed are the role of quantum effects in black hole physics and in cosmology, the effects of quantum violations of the classical energy conditions, and inequalities which constrain the extent of such violations. We will also examine the first steps beyond semiclassical gravity, when the effects of spacetime geometry fluctuations start to appear.

1. Introduction

This article will deal with the semiclassical approximation, in which the gravitational field is classical, but is coupled to quantum matter fields. The semiclassical theory consists of two aspects: (1) Quantum field theory in curved spacetime and (2) The semiclassical Einstein equation. Quantum field theory in curved spacetime describes the effects of gravity upon the quantum fields. Here a number of nontrivial effects arise, including particle creation, negative energy densities, and black hole evaporation. The semiclassical Einstein equation describes how quantum fields act as the source of gravity. This equation is usually taken to be the classical Einstein equation, with the source as the quantum expectation value of the matter field stress tensor operator, that is

$$G_{\mu\nu} = 8\pi \langle T_{\mu\nu} \rangle. \tag{1}$$

This expectation value is only defined after suitable regularization and renormalization.

In this article, we will use units (Planck units) in which Newton's constant, the speed of light, and $\hbar$ are set to one: $G = c = \hbar = 1$. This makes all physical quantities dimensionless. Thus masses, lengths, and times are expressed as dimensionless multiples of the Planck mass, $m_P = \sqrt{\hbar c / G} = 2.2 \times 10^{-5}$g, the Planck length, $\ell_P = \sqrt{\hbar G / c^3} = 1.6 \times 10^{-33}$cm, and the Planck time, $t_P = \sqrt{\hbar G / c^5} = 5.4 \times 10^{-44}$s, respectively.

2. Renormalization of $\langle T_{\mu\nu} \rangle$

Here we will outline of the procedure for extracting a meaningful, finite part from the formally divergent expectation value of the stress tensor. More detailed accounts can be found in the books by Birrell and Davies[1] and by Fulling[2]. The first step is to introduce a formal regularization scheme, which renders the expectation value finite, but dependent upon an arbitrary regulator parameter. One possible choice is to separate the spacetime points at which the fields in $T_{\mu\nu}$ are evaluated, and then to average over the direction of separation. This leaves $\langle T_{\mu\nu} \rangle$ depending upon an invariant measure of the distance between the two points. This is conventionally chosen to be one-half of the square of the geodesic distance, denoted by σ.

The asymptotic form for the regularized expression in the limit of small σ can be shown to be

$$\langle T_{\mu\nu} \rangle \sim A \frac{g_{\mu\nu}}{\sigma^2} + B \frac{G_{\mu\nu}}{\sigma} + \left(C_1 H^{(1)}_{\mu\nu} + C_2 H^{(2)}_{\mu\nu} \right) \ln \sigma. \tag{2}$$

Here A, B, C_1, and C_2 are constants, $G_{\mu\nu}$ is the Einstein tensor, and the $H^{(1)}_{\mu\nu}$ and $H^{(2)}_{\mu\nu}$ tensors are covariantly conserved tensors which are quadratic in the Riemann tensor. Specifically, they are the functional derivatives with respect to the metric tensor of the square of the scalar curvature and of the Ricci tensor, respectively:

$$H^{(1)}_{\mu\nu} \equiv \frac{1}{\sqrt{-g}} \frac{\delta}{\delta g^{\mu\nu}} \left[\sqrt{-g} R^2 \right]$$

$$= 2 \nabla_\nu \nabla_\mu R - 2 g_{\mu\nu} \nabla_\rho \nabla^\rho R - \frac{1}{2} g_{\mu\nu} R^2 + 2 R R_{\mu\nu}, \tag{3}$$

and

$$H^{(2)}_{\mu\nu} \equiv \frac{1}{\sqrt{-g}} \frac{\delta}{\delta g^{\mu\nu}} \left[\sqrt{-g} R_{\alpha\beta} R^{\alpha\beta} \right] = 2 \nabla_\alpha \nabla_\nu R^\alpha_\mu - \nabla_\rho \nabla^\rho R_{\mu\nu}$$

$$- \frac{1}{2} g_{\mu\nu} \nabla_\rho \nabla^\rho R - \frac{1}{2} g_{\mu\nu} R_{\alpha\beta} R^{\alpha\beta} + 2 R^\rho_\mu R_{\rho\nu}. \tag{4}$$

The divergent parts of $\langle T_{\mu\nu} \rangle$ may be absorbed by renormalization of counterterms in the gravitational action. Write this action as

$$S_G = \frac{1}{16\pi G_0} \int d^4x \sqrt{-g} \left(R - 2\Lambda_0 + \alpha_0 R^2 + \beta_0 R_{\alpha\beta} R^{\alpha\beta} \right). \qquad (5)$$

We now include a matter action, S_M, and vary the total action, $S = S_G + S_M$, with respect to the metric. If we replace the classical stress tensor in the resulting equation by the quantum expectation value, $\langle T_{\mu\nu} \rangle$, we obtain the semiclassical Einstein equation including the quadratic counterterms:

$$G_{\mu\nu} + \Lambda_0 g_{\mu\nu} + \alpha_0 H_{\mu\nu}^{(1)} + \beta_0 H_{\mu\nu}^{(2)} = 8\pi G_0 \langle T_{\mu\nu} \rangle. \qquad (6)$$

We may remove the divergent parts of $\langle T_{\mu\nu} \rangle$ in redefinitions of the coupling constants G_0, Λ_0, α_0, and β_0. The renormalized values of these constants are then the physical parameters in the gravitational theory. After renormalization, G_0 is replaced by G, the renormalized Newton's constant, which is the value actually measured by the Cavendish experiment. Similarly, Λ_0 becomes the renormalized cosmological constant Λ, which must be determined by observation. This is analogous to any other renormalization in field theory, such as the renormalization of the mass and charge of the electron in quantum electrodynamics.

In any case, the renormalized value of $\langle T_{\mu\nu} \rangle$ is obtained by subtracting the terms which are divergent in the coincidence limit. However, we are free to perform additional finite renormalizations of the same form. Thus, $\langle T_{\mu\nu} \rangle_{ren}$ is defined only up to the addition of multiples of the four covariantly conserved, geometrical tensors $g_{\mu\nu}$, $G_{\mu\nu}$, $H_{\mu\nu}^{(1)}$, and $H_{\mu\nu}^{(2)}$. Apart from this ambiguity, Wald[3] has shown under very general assumptions that $\langle T_{\mu\nu} \rangle_{ren}$ is unique. Hence, at the end of the calculation, the answer is independent of the details of the regularization and renormalization procedures employed.

3. The Stability Problem in the Semiclassical Theory

The classical Einstein equation is a second order, nonlinear, differential equation for the spacetime metric tensor, because the Einstein tensor involves up to second derivatives of the metric. As a second order system of hyperbolic equations, it possesses a well-posed initial value formulation: if one specifies the metric and its first derivatives on a spacelike hypersurface, there exists a unique solution of the equations[4]. This is the usual situation in physics, where the fundamental equations can be cast as a second order system. (For example, Maxwell's equations are equivalent to a set of second order wave equations for the vector and scalar potentials.)

There is a problem with the semiclassical Einstein equation in that it is potentially a fourth-order system of equations. This arises from terms involving second derivatives of the curvature tensor, and hence fourth derivatives of the metric. This leads to the unpleasant feature that a unique solution would require specification of the metric and its first three derivatives on a spacelike hypersurface. Even worse, it can lead to instability. The situation is analogous to that in classical electrodynamics when radiation reaction in included in the equation of motion of a charged particle[5]. The Abraham-Lorentz equation, which includes the radiation reaction force for a nonrelativistic particle, is third-order in time and possesses runaway solutions. In electrodynamics, the problem is partially solved by replacing the third-order Abraham-Lorentz equation by an integrodifferential equation which is free of runaway solutions, but exhibits acausal behavior on short time scales. However, this acausality is on a time scale small compared to the Compton time of the particle. As such, it lies outside of the domain of validity of classical electrodynamics.

Several authors[6,7,8] have discussed the instability problem in semiclassical gravity theory. Some of the proposed resolutions of this problem involve reformulating the theory to eliminate unstable solutions (analogous to the integrodifferential equation in electrodynamics), or regarding the semiclassical theory as valid only for spacetimes which pass a stability criterion. These are sensible approaches to the issue. Basically, one wishes to have a theory which can approximately describe the backreaction of quantum fields on scales well above the Planck scale. It is important to keep in mind that the semiclassical theory is an approximation which must ultimately fail in situations where the quantum nature of gravity itself plays a crucial role.

4. The Hawking Effect

One of the great successes of quantum field theory in curved spacetime and of semiclassical gravity is the elegant connection between black hole physics and thermodynamics forged by the Hawking effect. Classical black hole physics suffers from Bekenstein's paradox[9]: one could throw hot objects into a black hole and apparently decrease the net entropy of the universe. This paradox can be resolved by assigning an entropy to a black hole which is proportional to the area of the event horizon. Hawking[10] carried this reasoning one step further by showing that black holes are hot objects in a literal sense and emit thermal radiation. The outgoing radiation consists of particles quantum mechanically created in a region outside of the event horizon, and carries away energy and entropy from the black hole. The resulting decrease in mass of the hole arises from a steady flux of *negative energy* into the horizon, and is consistently described by the semiclassical Einstein equation, Eq. (1), so long as the black hole's mass is well above the Planck mass.

Although the Hawking effect provides an elegant unification of thermodynamics, gravity and quantum field theory, there are still unanswered questions. One is the "information puzzle", the issue of whether information which goes into the black hole during its semiclassical phase can be recovered. Hawking[11] originally proposed that this information is irrevocably lost and that black hole evaporation is not described by a unitary evolution. This view has been disputed by several other physicists[12], who have argued that a complete quantum mechanical description of the evaporation process should be unitary. More recently, Hawking[13] has agreed with this view. However, even if the evolution is unitary, the details by which information is recovered are still unclear. One possibility is that the outgoing radiation is not exactly thermal, but contains some subtle correlations which carry the information about the details of the matter which fell into the black hole. If this suggestion is correct, it is not clear just how these correlations arise.

A second mystery raised by the Hawking effect is the "tranplanckian problem". This problem arises because the modes which will eventually become populated with the outgoing thermal radiation start out with extremely high frequencies before the black hole formed. These modes enter the collapsing body and then exit just before the horizon forms, undergoing an enormous redshift. However, as they enter and pass through the body, their frequencies are vastly higher than the Planck scale. If one postulates that full quantum gravity will impose an effective cutoff at the Planck scale, then there seems to be a conflict; a cutoff at any reasonable frequency would eliminate the modes needed for the Hawking radiation. For a black hole of mass M to evaporate, one needs to start with modes whose frequency is of order

$$\omega \approx \frac{e^{M^2}}{M}. \tag{7}$$

For a stellar mass size black hole, this corresponds to $\omega \approx 10^{10^{75}}$ g, which is vastly larger than the mass of the observable universe. One possible resolution[14,15,16] of this problem is to postulate a modified dispersion relation which allows for "mode creation", whereby the modes would appear shortly before they are needed to carry the thermal radiation. However, this solution will require new microphysics, including breaking of local Lorentz invariance.

5. Quantum Effects in the Early Universe

It is likely that there is a period in the history of the universe during which quantum effects are important, but one is sufficiently far from the Planck regime that a full theory of quantum gravity is not needed. In this case, the semiclassical theory is applicable. Among the quantum effects expected in an expanding universe is quantum particle creation[17]. Inflationary models with inflation occurring at scales below the Planck scale are plausible models for the early universe in which semiclassical gravity should hold. Indeed, such models predict that the density perturbations which later grew into galaxies had their origins as quantum fluctuations during the inflationary epoch[18,19,20,21,22]. This leads to the remarkable prediction that the large scale structure of the present day universe had its origin in quantum fluctuations of a scalar inflaton field. More precisely, quantum fluctuations of a nearly massless scalar field in deSitter spacetime translate into an approximately scale invariant spectrum of density perturbations. This picture seems to be consistent with recent observations of the cosmic microwave background radiation[23].

6. The Dark Energy Problem

There is now strong evidence that the expansion of the present day universe is accelerating. This evidence came first from observations of type Ia supernovae[24,25]. This acceleration could be due to a nonzero value for the cosmological constant, but other possibilities are consistent with the observational data. These possibilities go under the general term "dark energy", and require a negative pressure whose magnitude is approximately equal to the energy density. It has sometimes been suggested that the dark energy could be viewed as due to quantum zero point energy. However, there are some serious difficulties with this viewpoint. If we adopt the convention renormalization approach discussed in Sect. 2, then the renormalized value of the cosmological constant Λ is completely arbitrary. At this level, quantum field theory in curved spacetime can no more calculate Λ then quantum electrodynamics can calculate the mass of the electron. We could take a more radical approach and seek some physical principle which effectively fixes the value of the regulator parameter to a definite, nonzero value. However, for the first term on the right hand side of Eq. (2) to be the dark energy, we would have to take $\sigma \approx (0.01\text{cm})^2$. It is very hard to imagine what new physics would introduce a cutoff on a scale of the order of 0.01cm.

There is still a possibility that the dark energy could be due to some more complicated mechanism which involves quantum effects. One appealing idea is that there might be a mechanism for the cosmological constant to decay from a large value in the early universe to a smaller, but nonzero value today. Numerous authors [26,27,28,29,30,31,32,33,34] have discussed models for the decay of the cosmological constant, or models which otherwise attribute a quantum origin to the dark energy[35,36]. However, at the present time there is no widely accepted model which successfully links dark energy with quantum processes.

7. Negative Energy Density for Quantum Fields

One crucial feature of quantum matter fields as a source of gravity is that they do not always satisfy conditions obeyed by known forms of classical matter, such as positivity of the local energy density. Negative energy densities and fluxes arise even in flat spacetime. A simple example is the Casimir effect[37], where the vacuum state of the quantized electromagnetic field between a pair of perfectly conducting plates separated by a distance L is a state of constant negative energy density

$$\rho = \langle T_{tt} \rangle = -\frac{\pi^2}{720L^4}. \tag{8}$$

Even if the plates are not perfectly conducting, it is still possible to arrange for the energy density at the center to be negative[38].

Negative energy density can also arise as the result of quantum coherence effects. In fact, it may be shown under rather general assumptions that quantum field theories admit states for which the energy density will be negative somewhere[39,40]. In simple cases, such as a free scalar field in Minkowski spacetime, one can find states in which the energy density can become arbitrarily negative at a given point.

We can illustrate the basic phenomenon of negative energy arising from quantum coherence with a very simple example. Let the quantum state of the system be a superposition of the vacuum and a two particle state:

$$|\Psi\rangle = \frac{1}{\sqrt{1+\epsilon^2}}(|0\rangle + \epsilon|2\rangle). \qquad (9)$$

Here we take the relative amplitude ϵ to be a real number. Let the energy density operator be normal-ordered:

$$\rho =: T_{tt} :, \qquad (10)$$

so that $\langle 0|\rho|0\rangle = 0$. Then the expectation value of the energy density in the above state is

$$\langle\rho\rangle = \frac{1}{1+\epsilon^2}\left[2\epsilon\mathrm{Re}(\langle 0|\rho|2\rangle) + \epsilon^2\langle 2|\rho|2\rangle\right]. \qquad (11)$$

We may always choose ϵ to be sufficiently small that the first term on the right hand side dominates the second term. However, the former term may be either positive or negative. At any given point, we could choose the sign of ϵ so as to make $\langle\rho\rangle < 0$ at that point. This example is a limiting case of a more general class of quantum states which may exhibit negative energy densities, the squeezed states.

Note that the integral of ρ over all space is the Hamiltonian, which does have non-negative expectation values:

$$\langle H\rangle = \int d^3x\langle\rho\rangle \geq 0. \qquad (12)$$

In the above *vacuum + two particle* example, the matrix element $\langle 0|\rho|2\rangle$, which gives rise to the negative energy density, has an integral over all space which vanishes, so only $\langle 2|\rho|2\rangle$ contributes to the Hamiltonian.

8. Some Possible Consequences of Quantum Violation of Classical Energy Conditions

The existence of negative energy density can give rise to a number of effects in which the predictions of semiclassical gravity differ significantly from those of classical gravity theory.

8.1. *Singularity avoidance*

In the 1960's, several elegant theorems were proven by Penrose, Hawking, and others[41] which demonstrate the inevitability of singularity formation in gravitational collapse described by classical relativity. These singularity theorems imply that the curvature singularities found in the exact solutions for black holes or for cosmological models are generic and signal a breakdown of classical relativity theory. However, this does not tell us whether a full quantum theory of gravity is needed to give a physically consistent, that is, singularity free, picture of the end state of gravitational collapse or the origin of the universe.

A crucial feature of the proofs of the singularity theorems is the assumption of a classical energy condition. There are several such conditions that can be used, but a typical example is the weak energy condition. This states that the stress tensor $T_{\mu\nu}$ must satisfy $T_{\mu\nu} u^\mu u^\nu \geq 0$ for all timelike vectors u^μ. Thus all observers must see the local energy density being nonnegative. It is not hard to understand why there could not be a singularity theorem without an energy condition: the Einstein tensor $G_{\mu\nu}$ is a function of the metric and its first two derivatives. Thus, every twice-differentiable metric is a solution of the Einstein equation, $G_{\mu\nu} = 8\pi T_{\mu\nu}$ for some choice of $T_{\mu\nu}$. We can also understand the role which the weak energy and related conditions play. Positive energy density will generate an attractive gravitational field and cause light rays to focus. Once gravitational collapse has proceeded beyond a certain point, the formation of a singularity is inevitable as long as gravity remains attractive. The way to circumvent this conclusion is with exotic matter, such as negative energy density, which can cause repulsive gravitational effects.

Given that quantum fields can violate the classical energy conditions, there is a possibility that the semiclassical theory can produce realistic, nonsingular black hole and cosmological solutions. This is a topic which has been investigated by several authors[43,44,45]. However, it is difficult to avoid having the curvature reach Planck dimensions before saturating. In this case, the applicability of the semiclassical theory is questionable. It is possible to avoid this difficulty with a carefully selected quantum states[43], a nonminimal scalar field which violates the energy conditions at the classical level[44], or by going to models where gravity itself is quantized[45].

8.2. Creation of naked singularities

There is an opposite effect which might be caused by negative energy: the *creation* of a naked singularity. The singularities formed in gravitational collapse in classical relativity tend to be hidden from the outside universe by event horizons. Penrose[42] has made a "cosmic censorship conjecture" to the effect that this must always be the case. This implies that the breakdown of predictability caused by the singularity is limited to the region inside the horizon. It is not yet known whether this conjecture is true, even in the context of classical relativity with classical matter, obeying classical energy conditions. However, unrestricted negative energy would allow a counterexample to this conjecture. The Reissner-Nordström solution of Einstein's equation describes a black hole of mass M and electric charge Q. However, these black hole solutions have an upper limit on the electric charge in relation to the mass of $Q \leq M$ (in our units). There are solutions for which $Q > M$, but these describe a naked singularity. Simple classical mechanisms for trying to convert a charged black hole into a naked singularity fail. If we try to increase the charge of a black hole, the work needed to overcome the electrostatic repulsion causes the black hole's mass to increase at least as much as the charge and keep $Q \leq M$. However, unrestricted negative energy would offer a way to violate cosmic censorship and create a naked singularity. We could shine a beam of negative energy involving an uncharged quantum field into the black hole, decrease M without changing Q, and thereby cause a naked singularity to appear[46,47].

8.3. *Violation of the second law of thermodynamics?*

If it is possible to create unrestricted beams of negative energy, then the second law would seem to be in jeopardy. One could shine the beam of negative energy on a hot object and decrease its entropy without a compensating entropy increase elsewhere. The purest form of this experiment would involve shining the negative energy on a black hole. If the negative energy is carried by photons with wavelengths short compared to the size of the black hole, it will be completely absorbed. That is, there will be no backscattered radiation which might carry away entropy. Then the black hole's mass, and hence its entropy, will decrease in violation of the second law[48].

8.4. *Traversable wormholes and warp drive spacetimes*

As noted above, virtually any conceivable spacetime is a solution of Einstein's equation with some choice for the source. If the source violates the classical energy conditions, some bizarre possibilities arise. An example are the traversable wormholes of Morris, Thorne and Yurtsever[49]. These would function as tunnels which could connect otherwise widely separated regions of the universe by a short pathway. An essential requirement for a wormhole is exotic matter which violates the weak energy condition. The reason for this is that light rays must first enter one mouth of the wormhole, begin to converge and later diverge so as to exit the other mouth of the wormhole without coming to a focal point. In other words,the spacetime inside the wormhole must act like a diverging lens, which can only be achieved by exotic matter.

The existence of traversable wormholes would be strange enough, but they have an even more disturbing feature: they can be manipulated to create a time machine[50]. If the mouths of a wormhole move relative to one another, it is possible for the resulting spacetime to possess "closed timelike paths". On such a path, an observer could return to the same point in space and in time, and by speeding up slightly, arrive at the starting point before leaving. Needless to say, this would turn physics as we currently understand it on its head and open the door to disturbing causal paradoxes.

An equally bizarre possibility was raised by Alcubierre[51], who constructed a spacetime that functions as science fiction style "warp drive". It consist of a bubble of flat spacetime surrounded by expanding and contracting regions imbedded in an asymptotically flat spacetime. The effect of the expansion and contraction is to cause the bubble to move faster than the speed of light, as measured by a distant observer, even though locally everything moves inside the lightcone. Again, negative energy is essential for the existence of this spacetime.

9. Quantum Inequalities

It is clear that unrestrained violation of the classical energy conditions would create major problems for physics. However, it is also clear that quantum field theory does allow for some violations of these conditions. This leads us to ask if there are constraints on negative energy density in quantum field theory. The answer is yes; there are inequalities which restrict the magnitude and duration of the negative energy seen by any observer, known as *quantum inequalities* [48,52,53,54,55,56,57,58,59]. In four spacetime dimensions, a typical inequality for a massless field takes the form[53,54,58]

$$\int \rho(t)\, g(t)\, dt \geq -\frac{c}{t_0^4}\,. \tag{13}$$

Here $\rho(t)$ is the energy density measured in the frame of an inertial observer, $g(t)$ is a sampling function with characteristic width t_0, and c is a numerical constant which is typically less than one. The value of c depends upon the form of $g(t)$ (e.g. Gaussian versus Lorentzian). The sampling function and its width can be chosen arbitrarily, subject to some differentiability conditions on $g(t)$. The essential message of an inequality such Eq. (13)

is that there is an inverse relation between the duration and magnitude of negative energy density. In particular, if an observer sees a pulse of negative energy density with a magnitude of order ρ_m lasting a time of order τ, then we must have $\rho_m < 1/\tau^4$.

Furthermore, that negative energy cannot arise in isolation, but must be accompanied by compensating positive energy. This fact, plus the quantum inequalities, place very severe restrictions on the physical effects which negative energy can create. Here is a brief summary of the implications of quantum inequalities for some of the possible effects listed above.

9.1. *Violations of the second law and of cosmic censorship*

If we were to shine a pulse of negative energy onto a black hole so as to decrease its entropy and violate the second law, the entropy decrease would have to last long enough to be macroscopically observable. At a minimum, it should be sustained for a time longer than the size of the event horizon. If the negative energy is constrained by an inequality of the form of Eq. (13), then it can be shown[48] that the resulting entropy decrease is of the order of Boltzmann's constant or less. This represents an entropy change associated with about one bit of information, hardly a macroscopic violation of the second law.

The attempt to create a naked singularity by shining a pulse of negative energy on an extreme, $Q = M$, charged black hole is similarly constrained. Again, any naked singularity which is formed should last for a time long compared to M. However, it can be argued[46,47] that the resulting change in the spacetime geometry may be smaller than the natural quantum fluctuations on this time scale. Thus it seems that negative energy which obeys the quantum inequality restrictions cannot produce a clear, unambiguous violation of cosmic censorship.

9.2. *Constraints on traversable wormholes and warp drive*

The simplest quantum inequalities, such as Eq. (13) have been proven only in flat spacetime, and hence do not immediately apply to curved spacetime. There is, however, a limiting case in which they can also be used in curved spacetime. This is when the sampling time t_0, as measured in a local inertial frame, is small compared to the local radii of curvature of the spacetime in the same frame. This means that the spacetime is effectively flat on the time scale of the sampling, and the flat space inequality should also apply to curved spacetime. In the special cases where explicit curved spacetime inequalities have been derived, they are consistent with this limit. That is, they reduce to the corresponding flat space inequality in the short sampling time limit.

Even in the small t_0 limit, it is possible to put very strong restrictions on the geometry of traversable wormholes and warp drives[60]. The constraints on wormhole geometries vary from one model to another. In some cases, the throat of the wormhole is limited to be close to Planck dimensions, presumably outside of the domain of validity of semiclassical gravity. In other cases, the restrictions are slightly less severe, but still require some length scales to be much smaller than others, such as a band of negative energy no more than 10^{-13}cm thick to support a wormhole with a 1m throat. This does not quite rule out all possible wormholes based upon semiclassical gravity, but makes it hard to imagine actually constructing one. Similar, very strong restrictions are placed on warp drive spacetimes[61,62], such as the Alcubierre model.

10. Beyond Semiclassical Gravity: Fluctuations

The first extension of semiclassical gravity arises when we consider fluctuations of the gravitational field. These can be due to two causes: the quantum nature of gravity itself (active fluctuations) and quantum fluctuations of the stress tensor (passive fluctuations). The extension of the semiclassical theory to include fluctuations is sometimes called stochastic gravity[63,64,65,66]. One of the criteria for the validity of the semiclassical theory based upon Eq. (1) must be that fluctuations are small[67]. This theory can break down even well above the Planck scale if the stress tensor fluctuations are sufficiently large. A simple example is a quantum state which is a superposition of two states, each of which describe a distinct classical matter distribution (e.g. a 1000kg mass on one or the other side of a room). Equation (1) predicts a gravitational field which is an average of the fields due to the two distributions separately, (the effect of two 500kg masses on opposite sides of the room). However, an actual measurement of the gravitational field should yield that of a single 1000kg mass, but in different locations in different trials.

A treatment of small fluctuations of the gravitational field offers a window into possible extensions beyond strict semiclassical gravity. First we should be clear about the operational meaning of fluctuations of gravity. A classical gravitational field or spacetime geometry can be viewed as encoding all possible motions of test particles in that geometry. Consequently, fluctuations of spacetime imply Brownian motion of the test particles, which can be characterized by mean squared deviations from classical geodesics.

Test particles can include photons, and one of the striking consequences of gravity fluctuations can be fluctuations of the lightcone. Recall that the lightcone plays a crucial role in classical relativity theory. Events which are timelike or null separated from one another can be causally related, but those at spacelike separations cannot. Similarly, an event horizon is a null surface which separates causally disjoint regions of spacetime. This rigid separation cannot be maintained when the spacetime fluctuates. A simple way to have spacetime fluctuations is with a bath of gravitons in a nonclassical state, such as a squeezed vacuum state[69,70]. Here the mean spacetime geometry is almost flat, apart from effects of the averaged stress tensor of the gravitons, but exhibits large fluctuations around this mean. These will include lightcone fluctuations, which will manifest themselves in varying arrival times of pulses from a source. Consider a source and a detector, which are both at rest relative to the average background and separated by a proper distance D, as measured in the average metric. Then the mean flight time of pulses will be D, but some individual pulses will take a longer time, and others a shorter time. A pulse which arrives in a time less than D travels outside of the lightcone of the mean spacetime, as illustrated in Fig. 1.

As noted earlier, faster than light travel can often be used to travel backwards in time. However, there is a crucial step needed to link the two: Lorentz invariance. One must exploit the fact that one can interchange the time order of spacelike separated events by changing Lorentz frames. In the present example, Lorentz symmetry is broken by the existence of a preferred rest frame, that of the graviton bath. Thus one cannot conclude that there is any problem with causality created by these lightcone fluctuations.

Because an event horizon is a special case of a lightcone, there should be horizon fluctuations in any model with spacetime geometry fluctuations. In the case of a black hole horizon, this raises the possibility of information leaking out of the black hole, or of the horizon fluctuations drastically altering the semiclassical derivation of black hole evaporation. One estimate[71] of the magnitude of the effects of quantum horizon fluctuations concluded that they are too small to alter the Hawking radiation for black holes much larger than the Planck mass. However, other authors[72,73] have argued for a much larger effect. It has also been suggested[74] that horizon fluctuations might provide the new physics needed to gracefully solve the tranplanckian problem. This is clearly an area where more work is needed.

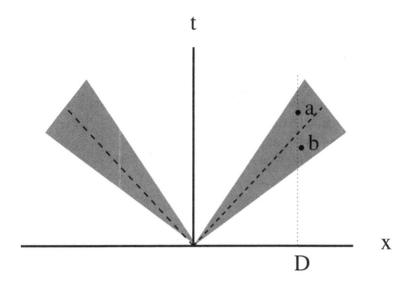

Fig. 1. The effects of lightcone fluctuations are illustrated. The dashed lines represent the average lightcone. However, pulses which are emitted at the origin can arrive at the worldline of a detector (vertical dotted line) a mean distance D away at different times. A pulse detected at point a travels slower than the mean speed of light, but one detected at point b has traveled faster than the mean speed of light, and hence outside of the mean lightcone.

The passive fluctuations of gravity driven by quantum stress tensor fluctuations are just one manifestation of stress tensor fluctuations. They are also responsible for fluctuation forces on macroscopic bodies, such as Casimir force fluctuations[75,76,77,78] and radiation pressure fluctuations[79]. This provides the possibility of an electromagnetic analog model for passive quantum gravity. The same techniques are needed to define integrals of the stress tensor correlation function in both contexts. In both cases, one needs to use a regularization method, such as dimensional regularization[80] or an integration by parts. Some of the physical effect which have recently been studied using the latter technique are angular blurring and luminosity fluctuations[81] of the image of a distant source seen through a fluctuating spacetime.

11. Summary

The semiclassical theory, with quantum matter fields and a classical gravitational field, provides a crucial link between the purely classical theory and a more complete quantum theory of gravity. Any viable candidate for a full quantum theory of gravity must reproduce the predictions of semiclassical gravity in an appropriate limit. In addition, semiclassical gravity contains a rich array of physical effects which are not found at the classical level, including black hole evaporation, cosmological particle creation, and negative energy density effects. The simplest extensions of the semiclassical theory to include spacetime fluctuations provide another array of effects, including lightcone and horizon fluctuations, which will have to be better understood in the context of a more complete theory.

Acknowledgments

I would like to thank Tom Roman for useful discussions and comments on the manuscript. This work was supported by the National Science Foundation under Grant PHY-0244898.

References

1. N.D. Birrell and P.C.W. Davies, *Quantum Fields in Curved Space*, (Cambridge University Press, 1982).
2. S.A. Fulling, *Aspects of Quantum Field Theory in Curved Space-Time*, (Cambridge University Press, 1989).
3. R.M. Wald, Commun. Math. Phys. **54**, 1 (1977).
4. The initial value problem in general relativity is actually more complicated than suggested by this sentence. This arises from the need to satisfy constraints and to have a spacetime which is globally hyperbolic. For a more detailed discussion, see, for example, R.M. Wald, *General Relativity*, (University of Chicago Press, Chicago, 1984), Chap. 10.
5. See, for example, J.D. Jackson, *Classical Electrodynamics*, 2nd ed. (Wiley, New York, 1975), Chap. 17.
6. G.T. Horowitz and R.M. Wald, Phys. Rev. D **17**, 414 (1978).
7. L. Parker and J.Z. Simon, Phys. Rev. D **47**, 1339 (1993), gr-qc/9611064.
8. P.R. Anderson, C. Molina-Paris, and E. Mottola, Phys. Rev. D **67**, 024026 (2003), gr-qc/0209075.
9. J.D. Bekenstein, Phys. Rev. D **7**, 2333 (1973)
10. S.W. Hawking, Commun. Math. Phys. **43**, 199 (1975).
11. S.W. Hawking, Phys. Rev. D **14**, 2460 (1976).
12. G. 't Hooft, Nucl. Phys. B **335**, 138 (1990).
13. S.W. Hawking, Talk presented at GR17, Dublin, July 2004.
14. T. Jacobson, Phys. Rev. D **44**, 1731 (1991).
15. W. G. Unruh, Phys. Rev. D **51**, 2827 (1995), gr-qc/9409008.
16. S. Corley and T. Jacobson, Phys. Rev. D **54**, 1568 (1996), hep-th/9601073.
17. L. Parker, Phys. Rev. **183**, 1057 (1969).
18. V. Mukhanov and G. Chibisov, JETP Lett. **33**, 532 (1981).

19. A. Guth and S-Y Pi, Phys. Rev. Lett. **49**, 1110 (1982).
20. S.W. Hawking, Phys. Lett. B **115**, 295 (1982).
21. A.A. Starobinsky, Phys. Lett. B **117**, 175 (1982).
22. J.M. Bardeen, P.J. Steinhardt and M. S. Turner, Phys. Rev. D **28**, 679 (1983).
23. For a recent review, see M. Giovannini, astro-ph/0412601.
24. B.P. Schmidt *et al*, Astrophys.J. **507**, 46 (1998), astro-ph/9805200.
25. S. Perlmutter *et al*, Astrophys.J. **517**, 565 (1999), astro-ph/9812133.
26. A.D. Dolgov, in *The Very Early Universe*, G.W. Gibbons, S.W. Hawking, and S.T.C. Siklos, eds. (Cambridge University Press, 1983).
27. L.H. Ford, Phys. Rev. D **31**, 710 (1985).
28. L.H. Ford, Phys. Rev. D **35**, 2339 (1987).
29. S. Weinberg, Rev. Mod. Phys. **61**, 1 (1989).
30. A.D. Dolgov, M.B. Einhorn, and V.I. Zakharov, Phys. Rev. D **52**, 717 (1995), gr-qc/9403056.
31. N.C. Tsamis and R.P. Woodard, Nucl. Phys. B **474**, 235 (1996), hep-ph/9602315; Ann. Phys. **253**, 1 (1997), hep-ph/9602316; Phys. Rev. D **54**, 2621 (1996), hep-ph/9602317.
32. L.R. Abramo and R.P. Woodard, Phys. Rev. D **65**, 063516 (2001), astro-ph/0109273.
33. R. Schützhold, gr-qc/0204018.
34. L.H. Ford, in *On the Nature of the Dark Energy*, Proceedings of the 18th IAP Astrophysics Colloquium, P. Braz, J. Martin, and J-P Uzan, eds. (Frontier Group, Paris, 2002), gr-qc/0210096.
35. V. Sahni and S. Habib, Phys. Rev. Lett. **81**, 1766 (1998), hep-ph/9808204.
36. L. Parker and A. Raval, Phys. Rev. D **60**, 123502 (1999), gr-qc/9908013.
37. H.B.G. Casimir, Proc. K. Ned. Akad. Wet. **B51**, 793 (1948).
38. V. Sopova and L.H. Ford, Phys. Rev. D, **66**, 045026 (2002), quant-ph/0204125.
39. H. Epstein, V. Glaser, and A. Jaffe, Nuovo Cimento **36**, 1016 (1965).
40. C.-I Kuo, Nuovo Cimento **112B**, 629 (1997), gr-qc/9611064.
41. See. for example, S.W. Hawking and G.F. R. Ellis, *The large scale structure of space-time*, (Cambridge Univ. Press, 1973).
42. R. Penrose, in *General Relativity: An Einstein Centenary Survey*, S.W. Hawking and W. Israel, eds. (Cambridge Univ. Press, 1979).
43. L. Parker and S.A. Fulling, Phys. Rev. D **7**, 2357 (1973).
44. A. Saa, E. Gunzig, L. Brenig, V. Faraoni, T.M. Rocha Filho, and A. Figueiredo, Int. J. Theor. Phys. **40**, 2295 (2001), gr-qc/0012105.
45. P. Hajicek and C. Kiefer, Int. J. Mod. Phys. D **10**, 775 (2001), gr-qc/0107102.
46. L.H. Ford and T.A. Roman, Phys. Rev. D **41**, 3662 (1990).
47. L.H. Ford and T.A. Roman, Phys. Rev. D **46**,1328 (1992).
48. L.H. Ford, Proc. R. Soc. London **A364**, 227 (1978).
49. M. Morris, K. Thorne, and U. Yurtsever, Phys. Rev. Lett. **61**, 1446 (1988).
50. See, for example, M. Visser, *Lorentzian Wormholes; From Einstein to Hawking,* (AIP, Woodbury, N. Y., 1995).

51. M. Alcubierre, Class. Quantum Grav. **11**, L73 (1994).
52. L.H. Ford, Phys. Rev. D **43**, 3972 (1991).
53. L.H. Ford and T.A. Roman, Phys. Rev. D **51**, 4277 (1995), gr-qc/9410043.
54. L.H. Ford and T.A. Roman, Phys. Rev. D **55**, 2082 (1997), gr-qc/9607003.
55. E.E. Flanagan, Phys. Rev. D, **56**, 4922 (1997), gr-qc/9706006.
56. M.J. Pfenning and L.H. Ford, Phys. Rev. D **55**, 4813 (1997), gr-qc/9608005.
57. M.J. Pfenning and L.H. Ford, Phys. Rev. D **57**, 3489 (1998), gr-qc/9710055.
58. C.J. Fewster and S.P. Eveson, Phys. Rev. D **58**, 084010 (1998), gr-qc/9805024.
59. C.J. Fewster, Class. Quantum Grav. **17**, 1897 (2000), gr-qc/9910060.
60. L.H. Ford and T.A. Roman, Phys. Rev. D **53**, 5496 (1996), gr-qc/9510071.
61. M.J. Pfenning and L.H. Ford, Class. Quantum Grav., **14**, 1743 (1997), gr-qc/9702026.
62. A.E Everett and T.A. Roman, Phys. Rev. D **56**, 2100 (1997), gr-qc/9702049.
63. J.M. Moffat, Phys. Rev. D **56**, 6264 (1997), gr-qc/9610067.
64. R. Martin and E. Verdaguer, Phys. Rev. D **60**, 084008 (1999), gr-qc/9904021.
65. B.L. Hu and E. Verdaguer, Class. Quant. Grav. **20**, R1 (2003), gr-qc/0211090.
66. B.L. Hu and E. Verdaguer, Living Rev. Rel. **7**, 3 (2004), gr-qc/0307032.
67. C.I. Kuo and L.H. Ford, Phys. Rev. D **47**, 4510 (1993), gr-qc/9304008.
68. B.L. Hu and N.G. Phillips, Int. J. Theor. Phys. **39**, 1817 (2000), gr-qc/0004006.
69. L.H. Ford, Phys. Rev. D **51**, 1692 (1995), gr-qc/9410047.
70. L.H. Ford and N.F. Svaiter, Phys. Rev. D **54**, 2640 (1996), gr-qc/9604052.
71. L.H. Ford and N.F. Svaiter, Phys. Rev. D **56**, 2226 (1997), gr-qc/9704050.
72. R.D. Sorkin, *Two Topics concerning Black Holes: Extremality of the Energy, Fractality of the Horizon*, gr-qc/9508002; *How Wrinkled is the Surface of a Black Hole?*, gr-qc/9701056.
73. A. Casher, F. Englert, N. Itzhaki, and R. Parentani, Nucl.Phys. B **484**, 419 (1997), hep-th/9606106.
74. C. Barrabes, V. Frolov, and R. Parentani, Phys. Rev. D **62**, 04020 (2000), gr-qc/0001102.
75. G. Barton, J. Phys. A **24**, 991 (1991); **24**, 5563 (1991).
76. C. Eberlein, J. Phys. A **25**, 3015 (1992); A **25**, 3039 (1992).
77. M.T. Jaekel and S. Reynaud, Quantum Opt. **4**, 39 (1992); J. Phys. I France **2**, 149 (1992); **3**, 1 (1993); **3**, 339 (1993).
78. C.-H. Wu, C.-I Kuo and L.H. Ford, Phys. Rev. A **65**, 062102 (2002), quant-ph/0112056.
79. C.-H. Wu and L. H. Ford, Phys. Rev. D **64**, 045010 (2001), quant-ph/0012144.
80. L.H. Ford and R.P. Woodard, Class. Quant. Grav. in press, gr-qc/0411003.
81. J. Borgman and L. H. Ford, Phys. Rev. D **70**, 064032 (2004), gr-qc/0307043.

CHAPTER 12

SPACE TIME IN STRING THEORY

TOM BANKS

Department of Physics and Astronomy – NHETC,
Rutgers University, Piscataway,
NJ 08540, USA
and
Department of Physics – SCIPP,
U. C. Santa Cruz, Santa Cruz,
CA 95064, USA
banks@scipp.ucsc.edu

This paper explores the holographic description of space-time in string theory. Super-Poincare invariant versions of the theory are described by moduli spaces of S-matrices. The parameters describing the S-matrix have geometrical interpretations in extreme regions of the moduli space, but the S-matrix interpolates smoothly between compact spaces of radically different topology. Asymptotically Anti-de Sitter universes are defined by quantum field theories living on the conformal boundary of AdS space. The geometry of compact factors in the space-time is encoded in internal degrees of freedom of the field theory. The cosmological constant is a discrete input parameter determining the high energy behavior of the system. Later sections discuss the description of stable or unstable de Sitter space-times, and the associated question of the breaking of Supersymmetry, as well as a conjectural holographic theory of quantum space-time geometry.

1. Introduction

The historical origin of String Theory[1a] was an attempt to construct a Poincare invariant scattering matrix for particles in Minkowski space, without relying on local field theory. Today we know that the construction really describes a theory of quantum gravity in asymptotically flat space-time[b]. The theory contains a number of stable particles, always including the graviton, and the scattering matrix computes the amplitude for a given finite collection of these particles to propagate from the infinite past and turn into a, perhaps different, finite set of particles in the infinite future. Perturbative string theory provides us with an algorithm for computing these amplitudes to all orders in a power series expansion in a dimensionless *string coupling*, g_S. Repeated attempts to introduce more localized amplitudes in string theory, all ended in failure.

Similar, but more complicated expansions exist for the scattering matrix with a variety of other choices of asymptotic space-time. These include asymptotically flat space-time in the presence of certain infinite, static p-dimensional extended objects called p-branes. If the codimension of the p-brane is ≥ 3, these expansions are also finite. In several cases, low energy limits of these scattering matrices, can be related to other kinds of space-time asymptotics, in particular Anti-de Sitter (AdS) spaces. We will discuss this case extensively below.

In many cases there are strong arguments that these perturbative expansions are asymptotic expansions of a well defined unitary, causal and Poincare invariant scattering operator[c].

The key feature of all of these expansions is that the answers depend only on the coordinates of the conformal boundary of space-time. The existence of the bulk space-time can only be inferred indirectly, and only in certain

---

[a]The string duality revolution taught us that string theory is only a theory of strings in certain asymptotic limits. For a while, I advocated the use of the term M-theory as a name for the underlying model of which all known theories were limits. Now I believe that we have found many consistent models of quantum gravity, and that the full set is unlikely to arise as limits of a single model. The name String theory is an historical artifact, which we should probably continue to use until we really understand what's going on.

[b]In four asymptotically flat dimensions the scattering matrix is not well defined, because any process emits an infinite number of soft gravitons with probability one. We expect to be able to resolve this problem along lines that are fairly well, but not completely, understood[2].

[c]In the asymptotically AdS case, the scattering matrix is replaced by the correlation functions of a conformal field theory (CFT) living on the conformal boundary of space-time.

limiting situations where a semi-classical approximation is valid. This is the first aspect of what we will call *The Holographic Principle*: gauge invariant observables in theories of quantum gravity based on string theory depend only on the boundary coordinates. Bulk physics is to be reconstructed only by some sort of "inverse holographic map", which we expect to be only approximately defined, and only applicable in certain situations.

In fact, simple intuitions about quantum gravity lead us to expect just such an answer. Consider a finite causal diamond $\mathcal{D}_{PQ}$ in a Lorentzian space-time; the intersection of $\mathcal{I}_P^-$, the interior of the forward light cone of some point P, with the interior of the backward light cone of a point $Q \in \mathcal{I}_P^-$. Ordinary intuition from quantum field theory, with a spatial ultraviolet (UV) cut off of order L_P suggests that there are a finite number of degrees of freedom associated with this region. The covariant entropy bound, conjectured by Bousso[3] on the basis of the seminal work of Fischler and Susskind[4], suggests something much more stringent: a finite entropy. The present author and Fischler[5] suggested that, since a local observer could have no preferred Hamiltonian operator, this actually requires a finite number of states[d] .

If this is the case, the physics of a finite causal diamond is intrinsically ambiguous. Precise measurements in quantum systems require us to correlate the quantum observables with *pointer observables* of a nearly classical system. One of the necessary conditions for a system to be exactly classical is that it have an infinite number of states. If all measurements inside a causal diamond must be made by machinery which has a finite number of quantum states, they will have an intrinsic uncertainty associated with an irreducible quantum fluctuation of the measuring apparatus. Consequently, the mathematical theory of such a region (its time evolution operator) cannot be defined with absolute precision.

A more physical way to state the same problem is that a large classical measuring apparatus will have a large gravitational back reaction on the system it is trying to measure. In the extreme, the measuring apparatus and the system will collapse into a huge black hole. Indeed, it has been suggested[6] that black hole production is the inevitable consequence of ordinary scattering at very high energy. Thus, although a scattering matrix may exist if black hole production and evaporation is a unitary process, no

---

[d]Finite entropy for *e.g.* a thermal density matrix for a system with an infinite number of states, depends on the grading of the states by eigenvalues of the Hamiltonian, as well as a bound on the asymptotic degeneracy of the spectrum.

observable associated with a fixed finite causal diamond can be unambiguously defined, because the entire diamond can end up inside the horizon of a black hole created in scattering.

While these intuitive arguments may not be convincing on their own, they combine with our documented inability to define localized observables in string theory to form a very strong argument for the validity of the Holographic Principle.

What then is the meaning of local space-time geometry in string theory? We will approach this question (which as yet has no complete answer) in stages. For space-times of the form (in the low energy classical approximation) $M^{(1,d-1)} \times K$ with K a compact manifold, the quantum string theory is defined by an S-matrix for particles in the asymptotically Minkowski factor space. The geometry of K appears in two ways. First, given enough supersymmetry (SUSY) there are continuous moduli spaces of S-matrices, which are related in a manner similar (but not identical) to moduli spaces of vacuum states in SUSic field theory. Some of the moduli can be identified, when all dimensions of K are large compared to the length scale defined by the string tension[e], with geometric moduli of K. As we vary the moduli into the stringy region the moduli space can change its structure. When the number of supercharges is 16 or more, such changes cannot occur. However, even in this case, the geometric picture can change radically and we can end up, deep in the Planck regime, with a different string theory compactified on a wildly different space.

The other geometric feature of K which is well defined in the quantum theory, is a set of conserved charges identified with wrapping numbers of extended objects called p-branes, around non-trivial topological cycles of K. Even when K undergoes drastic "Planckian cosmetic surgery" into a different manifold, these charges still label stable states which appear in the S-matrix. Features of K's geometry, which have an unambiguous *holographic image* in terms of S-matrix data, are exact features of the quantum theory, while the rest is meaningful only in certain extreme limits of the moduli space.

We next turn to space-times of the form $AdS_d \times K$. Here a similar picture emerges, but the holographic data is encoded in the correlation functions of a local field theory, which lives on the conformal boundary of AdS_d: $R \times S^{d-2}$. There are no longer continuous moduli spaces of so-

---

[e]Which is always larger than the Planck scale, when string perturbation theory is applicable.

lutions. This relation, the famous AdS/CFT correspondence, will enable us to explore the meaning of the cosmological constant in string theory, and the utility of the concept of effective potential. It will also teach us that black holes dominate the high energy spectrum of a model of quantum gravity.

We briefly discuss the two current approaches to asymptotically de Sitter space-times, and their implications for the way in which string theory may make contact with data about the real world.

Finally, we outline a possible holographic approach to local geometry, which takes into account the irreducible ambiguity of physics in a local region, and attempts to relate it to the gauge symmetry of diffeomorphism invariance.

A note to the reader: the charge to the writer of this article was to give a non-technical explanation of the current picture of space-time in string theory. String theory is full of technicalities, and I have tried to avoid most of them. However, in order to keep the article of reasonable length I have had to assume familiarity with the terminology of complex differential geometry. The reader dismayed by terms like "Ricci flat Kahler manifold" should consult the excellent physicist's introduction to this subject in [7], and the many references cited there. I will also occasionally use the string theory jargon, *target space*, which refers to the field space of a quantum field theory, usually the quantum field theory describing the world volume dynamics of some extended object. Finally I will use the term *moduli space of theories* to refer to a multiparameter collection of S-matrices for Super-Poincare invariant models of quantum gravity. These are quite analogous to (though not the same as) spaces of vacuum states (continuous superselection sectors) in quantum field theory.

2. Branes, Charges and BPS States

Much of our information about the non-perturbative structure of string theories came from the discovery that the theories actually contain p dimensional extended objects called p-branes, for many values of p. Some configurations of these objects are exactly stable and have properties whose dependence on parameters like the string coupling, can be computed exactly. This enables us to explore regimes inaccessible to any one perturbation expansion.

A p brane can couple to a rank $p + 1$ antisymmetric tensor gauge potential via the formula

$$S_{int} = \mu_p \int d\,\sigma^{a_1} \wedge \ldots \wedge d\,\sigma^{a_{p+1}} A_{a_1 \ldots a_{p+1}}(x(\sigma)) = \mu_p \int_{W_p} A_{p+1}.$$

The potential A_{p+1} is a $p+1$ form defined in all of space-time, but the integral is only over the brane world volume. This coupling is invariant under the gauge transformation

$$A_{p+1} \to A_{p+1} + d\omega_p,$$

where ω_p is a p form, if the brane world volume is closed or the gauge transformation vanishes on its boundary. The gauge invariant field strength $F_{p+2} = dA_{p+1}$. If we add a Maxwell term $-\frac{1}{4} \int F_{p+2} \wedge *F_{p+2}$[f] to the action, F_{p+2} satisfies

$$d * F_{p+2} = \mu_p J_{p+1},$$

where J_{p+1} is the de Rham current concentrated on the brane world volume. The p form charge $Z_{a_1 \ldots a_p}$ obtained by integrating the time component of the current over a space like surface is conserved. Notice that only the spatial components of the brane charge are non-zero. Non-vanishing brane charge indicates the existence of an infinite brane whose asymptotic world volume picks out an asymptotic Lorentz frame where the brane is at rest.

We can also introduce magnetic sources as the world volumes of $d-p-4$ branes where the Bianchi identity $dF_{p+2} = 0$ breaks down. Nepomechie and Teitelboim[8] showed that the electric and magnetic charges satisfied a generalization of the Dirac quantization condition.

The relation between p brane charges and SUSY is a consequence of the fact that the product of two spinor representations is a direct sum of anti-symmetric tensor representations. Let us elaborate in eleven dimensions, the maximal dimension in which interacting low energy supersymmetric effective field theories can exist. The spinor super-charge, Q_a, has 32 real components. The most general right hand side for the supercharge anticommutation relation is

$$[Q_a, Q_b]_+ = (\gamma^0 [\gamma^\mu P_\mu + \gamma^{\mu\nu} Z_{\mu\nu} + \gamma^{\mu_1 \ldots \mu_5} Z_{\mu_1 \ldots \mu_5}])_{ab}.$$

This suggests the existence of 2-branes and 5-branes in any supersymmetric quantum theory in eleven dimensions. It is indeed true that the unique interacting supersymmetric low energy field theory in eleven dimensions, $11D$ SUGRA, has solutions corresponding to these branes. They are called the $M2$ and $M5$ branes.

---

[f] $*$ represents the Hodge duality operator.

In supersymmetric 4 dimensional field theories, Bogolmonyi and Prasad and Sommerfield[9](BPS), found soliton solutions whose energy could be computed exactly, including all quantum corrections, by doing a classical calculation. The $M2$ and $M5$ brane tensions have exactly this property. The basic idea underlying such BPS states is easy to understand. One attempts to represent the SUSY algebra on states with fixed values of the momentum and the brane charges. It has the form (in the brane rest frame)

$$[Q_a, Q_b]_+ = (P_0 + Z)_{ab},$$

where Z is a diagonal matrix with two eigenvalues of opposite sign. Its eigenvalues are quantized by the DNT condition. This is a Clifford algebra with 32 generators unless $P_0 = ||Z||$, in which case it has only 16. Thus, BPS branes, whose tension is related to their charge by this relation, must have a tension given by this relation in any quantum theory in which the symmetry algebra is valid[g]. The classical calculation shows that the M branes satisfy the BPS property. The Q_a are Hermitian operators in a positive metric Hilbert space. The anti-commutation relations then imply that BPS states have the smallest tension for a given charge, so they are absolutely stable.

A subtle question is whether *e.g.* the M2 brane of charge 2 is a bound state of two charge 1 branes. This is a threshold bound state problem, and such problems are notoriously difficult in ordinary quantum mechanics. It turns out that the answer is different in 11 asymptotically flat dimensions (where there are no bound states of either $M2$ or $M5$ branes) and on manifolds with various numbers of compactified dimensions.

If we compactify one dimension on a circle of radius R, SUSY remains unbroken, and the algebra changes only by having the momentum in the tenth spatial dimension appear as a quantized charge coupled to a $U(1)$ gauge field in an effective description in 10 space-time dimensions. The ten dimensional low energy field theory is uniquely determined by SUSY. The radius appears as the constant mode of a ten dimensional scalar field called the dilaton, and the symmetries do not allow any potential for this field. Thus, we are led to believe in a moduli space of ten dimensional asymptotically flat theories of gravity, with type IIA SUSY, the dimensional reduction of the eleven dimensional algebra, and a single dimensionless parameter. The ten dimensional Planck length is $2\pi R(l_P^{(10)})^8 = (l_P^{(11)})^9$.

---

[g]Of course, in 11 AF dimensions, there is a unique supersymmetric quantum theory if there is any at all. However, we will introduce parameters by compactification. The dimension of the BPS representation cannot change as we vary these parameters.

However, $M2$ branes wrapped around the circle appear as ten dimensional strings. The tension of these strings is computable using the BPS formula and goes to zero linearly with the radius of the circle. Parametrically it is of order $R(l_P^{(11)})^{-3}$. We can compute the coupling between these strings and the ten dimensional gravitational field, and it goes to zero like $(R/l_P^{(11)})^{3/2}$. That is, we expect a theory of weakly coupled supersymmetric strings to arise in the limit of small R. Needless to say, this expectation is spectacularly confirmed. What is more, the resulting theory enables us to construct a systematic weak coupling expansion of the S-matrix for gravitons, which is finite and predictive to all orders. Finally, the perturbative type IIA string theory reproduces all of the BPS p-branes expected from the 11 dimensional point of view. This includes the unwrapped $M2$ and $M5$ branes, the $M5$ brane wrapped around the circle, the 0 branes which carry Kaluza-Klein charge, and their magnetically dual 6 branes. The formulas for the tensions of all of these branes can be computed either in the $11D$ SUGRA approximation or the weak coupling string theory approximation, and the results agree. Furthermore, in string theory we can show that the higher order perturbative corrections to these calculations all vanish. In perturbative string theory, most of these branes are realized as Dirichlet or D - branes: perturbation theory around configurations with some collection of infinite flat D-branes is calculated in terms of diagrams where open strings are allowed to propagate on the D-brane world volume. The rules for coupling open and closed strings are unique and follow organically from the definition of string Feynman diagrams in terms of conformal field theory in two dimensions[10]. If the codimension of the union of brane world volumes is ≥ 3 the perturbation theory is straightforward, while for co-dimension < 3 there are infrared divergences, which can sometimes, but not always be understood.

I cannot go into all the details here[11] but only want to emphasize three points. One is that the geometrical description of the tenth spatial dimension is nowhere apparent in the weak coupling limit, but that the modulus (radius) of this geometry and the generator of the asymptotic symmetry group which comes from it's isometry, do parametrize 10 dimensional scattering amplitudes. The second is that we see no evidence for a quantization of length. We can take the radius of the tenth spatial dimension to be as small as we wish. It is hard to see how such a result could be consistent with a loop quantum gravity treatment. Finally I note that, quite contrary to what we expect from KK compactification at the effective field theory level, compactification seems to have *increased* the number of degrees of freedom of the theory. This is consistent with a hint from black hole physics. The black hole entropy formula in d asymptotically flat dimensions, which we will argue gives us the true high energy density of states, gives (in Planck units), $S(E) \sim E^{\frac{d-2}{d-3}}$. The density of high energy states grows more rapidly if the number of AF dimensions is smaller.

3. $11 - 2 = 10$

Something more remarkable happens if we compactify one more dimension on a circle, or more generally, we compactify two dimensions on a torus of area A and complex structure parameter τ. This gives us a three parameter moduli space of supersymmetric theories. For finite A in eleven dimensional Planck units, we get a nine dimensional theory. As $A \to 0$ we get a sequence of low energy states by considering multiply wrapped $M2$ branes on the torus. Detailed analysis shows that in this case we do get threshold bound states, so every state in the theory carries an additional quantum number n, the $M2$ brane wrapping number. The BPS formula for multiple charges, shows that if we have a particle in 9 dimensions carrying n units of charge, then its energy is

$$E = \sqrt{\mathbf{p}^2 + n^2 (AT)^2},$$

where T is the mass per unit area of the M2 brane, which is $\propto M_{11}^3$. This looks like the formula for massless particles in 9 space dimensions, with one of them compactified on a circle of radius $R_9^{-1} = AT$. Further analysis confirms this suspicion and shows that the relevant theory is the Type IIB string theory[h]

The phenomenon of two shrinking dimensions morphing into one large dimension can be understood on another level in Matrix Theory[12]. This is a non-perturbative formulation of the theory in an approximation known as *discrete light cone quantization* (DLCQ). M-theory on a two torus is written in terms of a supersymmetric Yang Mills theory living on an auxiliary two torus whose geometry is dual (in the sense of Fourier transforms) to the original. The coordinates of objects in the two compact dimensions are $2+1$ dimensional gauge potentials A_i (we are in $A_0 = 0$ gauge). The transition from small to large two torus is a weak to strong coupling transition for the gauge theory. The small area limit is one in which the scalar field ϕ dual

---

[h]The strings are $M2$ branes wrapped on non-contractible 1-cycles of the torus. In the large complex structure limit $\mathrm{Im}\tau \to \infty$, the tension of strings wrapped around the shortest 1-cycle goes to zero and we again get a perturbative expansion.

to the gauge field strength via $F_{0i} = \epsilon_{ij}\partial_j\phi$ is approximately classical. ϕ is a periodic variable, and it's period is the circumference of the Type IIB circle. Thus, in this picture, the two coordinates of the small torus become strongly fluctuating quantum variables, and physics is better described in terms of an almost classical dual variable whose target space is the 9th spatial dimension in the space-time of the IIB theory.

When we compactify more dimensions, a variety of similar miracles occur, but we obtain no new limiting theories[11]. The entire moduli space of M-theory compactified on T^k with $k \leq 7$[i] is connected, and the $k = 7$ case has the largest number of degrees of freedom.

The properties of this moduli space are very similar to those of supersymmetric quantum field theories. The values of the moduli are superselection sectors: the scattering matrix for one value of the moduli does not include states with another value. On the other hand, effective supergravity arguments[17] indicate that an experimenter at one value of the moduli can create regions of space of arbitrarily large size and arbitrarily long lifetime, in whose interior the moduli take on any value a finite distance away in moduli space. Another property that may be shared with field theory moduli spaces is that high energy behavior is independent of the value of the moduli (as long as we do not take extreme limits which change the number of compactified dimensions). This argument depends on the conjecture [6] that in d asymptotically flat dimensions, scattering at high center of mass energy and impact parameters satisfying $b < E^{\frac{1}{d-3}}$, is dominated by black hole production. The black hole decays thermally, with a temperature that goes to zero asymptotically. For fixed d the spectrum of black holes, and of the massless states which are produced in their thermal decay, is independent of the moduli, once everything is expressed in terms of the d dimensional Planck length.

4. $11 - 4 = 10$

We now discuss models possessing only 16 real supercharges. It turns out that these fall into a number of disconnected moduli spaces. The simplest class correspond to a simple generalization of toroidal compactification. In eleven AF dimensions, M-theory is invariant under the combination of

---

[i]Even in $k = 7$, the scattering matrix has infrared divergences which must be treated by including states with finite amounts of incoming and outgoing classical gravitational radiation. For $k > 7$ the IR divergences are much worse. It is not clear whether these models exist, or what their observables are.

an orientation reversing transformation on space-time with an orientation reversing transformation on $M2$ brane world volumes. Instead of simply identifying points under a discrete translation group, we combine it with such an orientation reversal, insisting that the combined operation have no fixed points[j]. This leads to the moduli space of CHL[14] strings and its generalizations[15]. Only half of the supercharges survive the projection onto objects invariant under the discrete translation-reflection group.

Another way to find models with only 16 supercharges is to compactify on manifolds, which have a number of covariantly constant spinors equal to half the flat space maximum. These manifolds have the form $K3 \times T^k$, $k \leq 3$. The $K3$ surfaces are the 2 (complex) dimensional Ricci-flat Kahler manifolds. They are a single topological manifold, with a 19 parameter moduli space of Ricci flat metrics. The second Betti number of $K3$ is 22. The intersection form on $H_2(K3)$ has signature $(3, 19)$. It is the direct sum of three copies of a $(1, 1)$ signature lattice and two copies of the Cartan matrix of E_8.

If we shrink the volume of $K3$ to zero, we get a new low tension, string by wrapping four directions of the $M5$ brane on $K3$. This is the heterotic string[16]. For small $K3$ volume, the heterotic string model is compactified on a large three torus, whose geometry arises in much the same way as that of the IIB circle. Momentum and heterotic string winding number around the three toroidal directions are dual to membrane wrapping numbers around two cycles of $K3$, associated with the $(1, 1)^3$ lattice. Membrane wrapping numbers around the E_8^2 cycles are realized as charges of a $U(1)^{16}$ gauge group in the 7 asymptotically flat dimensions. The $U(1)$ gauge fields themselves are three form gauge fields of the eleven dimensional theory, which have the form $\sum a_I \otimes \omega_I$, where ω_I are the harmonic two forms on $K3$, dual to the cycles with E_8^2 intersection matrix. These charges are quantized in the way that would be realized in the Higgs phase of an $E_8 \times E_8$ gauge group, with Higgs field in the adjoint.

Mathematicians have long known that the singularities of $K3$ are found at places in the moduli space where the volume of some set of two cycles shrinks to zero. The shrinking cycles are associated with the Cartan matrix of some ADE Dynkin diagram. M-theory allows us to understand these singularities as the result of new massless particles ($M2$ branes wrapped on the shrinking cycle), which appear in the spectrum at these special points in

---

[j]A simple case *with* fixed points is the Horava-Witten description of the strongly coupled heterotic string[13].

moduli space. Indeed, the Wilsonian approach to the renormalization group has led us to expect that all infrared singularities in scattering amplitudes (as a function of superselection parameters) can be explained in terms of particles which become massless at some points in the moduli space. A non-singular treatment can be obtained by properly including emission and absorption of these light states in the low energy S-matrix. What is remarkable is that in M-theory, this same mechanism appears to resolve what we think of geometrically as short distance singularities, where 2-cycles shrink to zero size. The connection between short distances and low energy comes through the BPS formula for the wrapped M2 brane mass.

Readers should especially note the holographic nature of all of these arguments. When features of bulk geometry become small compared to the Planck scale[k], even in a way that is geometrically singular, we can no longer thing of them as geometry. But geometrical properties like winding numbers survive as quantum numbers in the scattering matrix of the asymptotically flat dimensions. The holographic map of the geometry remains smooth even in limits where geometrical notions lose their validity. The same quantum numbers can take on radically different geometrical meanings in different limits of moduli space, where the notion of the geometry of a compact manifold makes sense.

The smallest seven dimensional supermultiplet contains vector mesons, so the new massless particles that appear at $K3$ singularities are Yang Mills bosons, charged under the $U(1)^{16}$ gauge group, which is present for all values of the moduli. The connection of non-abelian gauge symmetry to singularities of $K3$ geometry is the basic mechanism by which Yang Mills fields appear in asymptotically flat versions of string/M-theory. Non-abelian Kaluza-Klein groups appear only in models with asymptotically AdS dimensions.

The diverse moduli spaces of 32 and 16 supercharge quantum gravity appear to be disconnected. There is no low energy gravity Lagrangian which allows one to create e.g. arbitrarily large, long lived bubbles of the $K3$ compactification model in the space-times of the $T4$ or $K2 \times T^2$ models ($K2$ is the Klein bottle). Similarly, these models have very different spectra of massless states, so if high energy amplitudes are dominated by black hole production they will not be independent of which moduli space we are on. In quantum field theory, we can have disconnected moduli spaces which are connected by a finite potential energy density barrier. We can create

---

[k]Often this happens already at the string tension scale in weakly coupled string theory.

arbitrarily large bubbles of one moduli space inside the vacuum at a point in the other. In theories of gravity this does not work[17] . The tension in the bubble wall scales like the surface area. When we try to create too large a bubble, this tension creates a black hole larger than the bubble.

The case of 8 supercharges has also been studied, although the picture there is not as complete. There are several dual descriptions of the moduli space, the simplest of which is compactification of eleven dimensions to 4 on a circle cross a Calabi-Yau three-fold: a Ricci flat Kahler manifold of complex dimension 3. For small circles this gives weakly coupled Type IIA string theory on the $CY3$. Several new features arise, the most interesting of which are stringy and quantum deformations of the classical geometrical moduli space. The Calabi-Yau moduli space consists of variations of both the complex structure and the Kahler metric. In weakly coupled string theory, one shows that the Kahler moduli space is deformed and becomes identified with the complex structure moduli space of a so-called Mirror Calabi-Yau manifold. Type IIA string theory on the original Calabi-Yau is identical to Type IIB string theory on the mirror.

Even more interesting is the quantum deformation of the moduli space, which allows smooth transitions which change the Betti numbers of the compact manifold. Again the mechanism goes through a geometrically singular manifold with shrunken cycles, and the singularity is resolved by including the massless BPS branes wrapped on these cycles[18] in the low energy description. This is the first case where SUSY allows for supermultiplets which contain neither gravitons nor gauge bosons, and the massless states include such matter multiplets. The full story of this moduli space has not been worked out. We do not know how many connected components it has, nor whether moduli spaces with less SUSY can be recovered by taking limits of moduli spaces with 8 supercharges. We do see that geometrical concepts survive in the holographic description of the compact manifold, but in a more severely distorted form.

There do exist moduli spaces of four dimensional compactifications with only four supercharges[19] but they are few and far between. In various classical limits we appear to find moduli spaces, but the massless moduli fields now live in chiral multiplets of $N = 1$, $d = 4$ SUSY. Typically, there is not enough symmetry to guarantee that the superpotential for these fields vanishes. If we try to solve the low energy field equations with a non-zero superpotential it is unusual to find asymptotically flat solutions, even if we insist on preserving SUSY. The potential in $N = 1$ SUGRA has the form:

$$V = e^{\frac{K}{M_P^2}} [K^{i\bar{j}} D_i W D_{\bar{j}} \bar{W} - \frac{3}{M_P^2} |W|^2].$$

The would be moduli space is a Kahler manifold with Kahler potential K. The superpotential, W is a section of a holomorphic line bundle over this manifold, and $D_i = \partial_i - \frac{\partial_i K}{M_P^2}$ is the covariant derivative on this bundle. Generically, there will be solutions of $D_i W = 0$, which is the condition to preserve SUSY, and also guarantees that we are at a stationary point of the potential. However, W will vanish at these points only in exceptional cases. This analysis leads us to expect supersymmetric theories with AdS asymptotics and and non-supersymmetric theories with any value of the cosmological constant, but zero seems unlikely. Indeed, attempts to break supersymmetry in supergravity and perturbative string theory sometimes preserve vanishing c.c. to one or more orders of perturbation theory, but no one has found an example with broken SUSY and exactly vanishing c.c. . This has led me to conjecture that *there are no Poincare invariant theories of quantum gravity, which are not Super-Poincare invariant.*

We will have more to say about the validity of this conjecture below. Next however, we will see what we can learn about theories of quantum gravity with AdS asymptotics.

5. The AdS/CFT Correspondence

The AdS/CFT correspondence came out of the analysis of black hole entropy in string theory. Susskind[20] was the first to suggest that the exponential degeneracies of states found in perturbative string theory were related to black hole entropy[1]. Sen pointed out that the arguments could be made more precise by considering extremal black holes which satisfied the BPS condition[22]. The idea was that BPS degeneracies can be counted in a weak coupling approximation, in which the gravitational effects that make the states into black holes are neglected. The BPS property ensures that the masses and degeneracies are independent of coupling. Sen studied black holes with zero classical horizon area. By computing instead the area of a "stretched horizon" a few string lengths from the classical horizon, he found

---

[1]This suggestion seemed a bit obscure since the powers of energy in the exponential did not match, but it was validated by the correspondence principle of Horowitz and Polchinski[21].

an entropy formula, which agreed parametrically (as a function of various charges) with the degeneracies of the states computed in weakly coupled string theory. Strominger and Vafa[23] found the first example of an extremal black hole with non-zero classical horizon area where the comparison could be made. The near horizon geometry of these black holes was $AdS_3 \times K$, with K a compact manifold. They found that the entropy formula agreed with that derived from weakly coupled string theory. In the string theory calculation the entropy is calculated from a $1 + 1$ dimensional conformal field theory describing the world volume of Dirichlet strings (D-1 branes). Even the coefficient in the entropy works. This can be traced to a special property of $1 + 1$ CFT's : the entropy is the same everywhere along a line of fixed points. Strominger and Vafa were able to compute in a soluble CFT, at regions in moduli space where the black hole horizon is tiny and gravitational effects unimportant, but get the precise extremal black hole entropy because of the BPS property and the invariance of the entropy along fixed lines.

The Strominger Vafa paper set off a flurry of activity. Entropy calculations were generalized to near extremal black branes, and calculations of gray body factors for scattering off a black hole at low energy were performed[24]. The agreement, particularly in the latter calculation, where an inclusive cross section is reproduced over a whole energy range, was spectacular. Furthermore, the successes could no longer be explained by invoking the BPS property. Maldacena[25], pondering the reason for these successes, realized that they could all be explained by a remarkable conjecture. In the perturbative string theory calculations, the black brane was described by a world volume CFT on a set of D-branes. This theory had the isometry group of $AdS_d \times K$ (with K some compact manifold) as a quantum symmetry group - it was a conformal field theory (CFT). This was the near horizon geometry of the black brane and Maldacena conjectured that the CFT was the correct quantum theory of the $AdS_d \times K$ space-time.

Maldacena's AdS/CFT conjecture was clarified some months later in work of Gubser, Klebanov and Polyakov[42] and of Witten[27]. These authors considered classical solutions of the supergravity field equations on $AdS \times K$, with boundary conditions on its conformal boundary. They calculated the action as a functional of the boundary conditions $S[\phi(b)]$. They showed in a few examples that this could be viewed as the generating functional of connected Green functions in a CFT living on the boundary of AdS space. The coordinates of the compact manifold were realized as fields in the CFT, much as in Matrix Theory. Thus, classical supergravity was shown to be an algorithm for computing the leading order, long (bulk) wavelength approximation to a boundary CFT. Note that again, the theory of quantum gravity is holographic: gauge invariant observables depend only on the boundary coordinates of the conformal boundary of an infinite space-time.

To illustrate how the AdS/CFT correspondence works, we consider a scalar field propagating in the Euclidean section of AdS space. This has the metric

$$ds^2 = d\tau^2(1 + \frac{r^2}{R^2}) + \frac{dr^2}{(1 + \frac{r^2}{R^2})} + r^2 d\Omega^2.$$

At large r the Klein-Gordan equation for the scalar is approximately

$$R^{-2}(\partial_z^2 + (d-1)\partial_z - (mR)^2)\phi = 0,$$

where $\frac{r}{R} \equiv e^z$. This has solutions $e^{\Delta_\pm z}\phi_\pm(\tau, \Omega)$, with

$$\Delta_\pm = \frac{1}{2}[1 - d \pm \sqrt{(d-1)^2 + 4m^2R^2}].$$

For most values of m^m, one of these solutions is normalizable, when considered as a wave function for a scalar particle in AdS space, while the other, Δ_+, is not normalizable. The prescription of [42][27] is to compute the effective action of the solution as a functional of the non-normalizable boundary condition ϕ_+. For free fields it is a quadratic functional. This functional is divergent, but the divergence is removed by a multiplicative renormalization. In stereographic coordinates on the boundary the renormalized two point function is just $|\mathbf{x}|^{-2\Delta_+}$. Indeed, the AdS isometry group acts as the conformal group of the boundary, so this form is fixed by symmetries and the AdS transformation properties of the scalar. This has the form of a two point function in a boundary CFT. The AdS/CFT conjecture is simply that the effective action of the bulk theory is the generating functional of correlation functions of a CFT.

There are two important aspects of this conjecture that are often overlooked by non-specialists. The first is that the set of representations of the AdS group that is used in bulk field theory in AdS, is precisely the set of highest weight unitary representations that is used in boundary CFT. This is an essential consistency condition for the validity of the conjecture. The second is that the existence of the stress tensor in boundary CFT implies that the bulk field theory contains a graviton. Stress tensor two point correlators are computed by solving the linearized gravitational field equations

---

^{m}The exceptions have been studied in Ref. 28.

in analogy to the above. A non-gravitational field theory *may* define a set of conformally invariant boundary correlators by the above procedure (one must check that the divergences can all be removed by a single multiplicative renormalization of the boundary condition) but they will not be those of a CFT because there will be no local stress tensor. Finally, I want to emphasize that the boundary IR divergences are interpreted as the standard UV divergences of CFT. This is an important aspect of the *UV/IR connection*.

The supporting evidence for the AdS/CFT correspondence is by now overwhelming[29]. It has taught us many lessons about both gravity and CFT. Here I want to review some of the most important lessons about gravity and space-time.

Perhaps the most important of all is the fact that the cosmological constant (c.c.) is not a low energy effective output of the theory, but a high energy input. The entropy of a CFT_{d-1} compactified on S^{d-2} with radius R is $c(RE)^{\frac{d-2}{d-1}}$, in the limit $ER \gg 1$. c is a measure of the total number of degrees of freedom in the theory. In $\mathcal{N} = 4$ Super Yang Mills theory with $SU(N)$ gauge group, $c \propto N^2$. Comparison with the Bekenstein-Hawking formula for AdS black holes gives $c \propto (RM_P)^{\frac{d-2}{d-1}}$, where M_P is the d dimensional Planck scale. The fact that the asymptotic structure of space-time controls the UV behavior of the theory (the UV/IR correspondence) is intuitively obvious if the UV spectrum is dominated by black holes, a conjecture I have called *Asymptotic Darkness*. Black holes of high energy, have large Schwarzchild radii and their properties depend on space-time asymptotics. Note in particular that the fact that AdS-Schwarzchild black holes have positive specific heat is crucial to the AdS/CFT correspondence. Black holes of radius larger than the AdS radius are identified with the stable canonical ensemble of the CFT. We see immediately that we cannot have a similar field theory description of asymptotically flat space.

The idea that the characteristic bulk quantum field theory connection between short space-time distances and high energy breaks down at the Planck length, is encapsulated in the conjecture that high energy, small impact parameter, scattering is dominated by black hole production. This means that there is no way to probe distances smaller than the Planck length in asymptotically flat dimensions[n]. This conjecture is supported by AdS/CFT . In that context, the statement that high energy processes enclosed in a fixed spatial region create black holes, translates

---

[n]Contrast this with our ability to shrink cycles in compact manifolds to zero size.

into the CFT statement that general high energy initial conditions lead
to thermalization.

5.1. *Potentials and domain walls*

AdS/CFT enables us to get our hands on some of the notions of effective
bulk field theory which will be important in our discussion of the applicability of string theory to the real world. In particular in gives us a new
view of the effective potential. The classical SUGRA Lagrangians relevant
for the AdS/CFT correspondence have, after dimensional reduction on the
compact manifold K, a large number of scalar fields. The Lagrangian contains a potential for these scalars, propagating in AdS space. The AdS
solution corresponds to constant scalars, sitting at a stationary point of
the potential. In the interesting cases, it is a maximum. Indeed, the formula for the correspondence between bulk mass and boundary dimension
shows that whenever the boundary operator is relevant, in the sense of the
renormalization group, then the bulk field is a tachyon. Breitenlohner and
Freedman[30] (BF), long before the invention of AdS/CFT, showed that such
tachyons were perfectly acceptable in AdS space. The curvature couplings
make the unstable solution at the top of the potential stable, for precisely
that range of tachyon masses which give real boundary dimension. The potential can have other stationary points, some of them minima, with more
negative values of the potential energy density.

In quantum field theory, when we have a potential with multiple stationary points, we can find a domain wall solution which interpolates between
the two as a function of a single spatial coordinate. In general, if we couple
the theory to gravity we find that the domain wall is no longer globally
static. The only exception which does not involve fine tuning of parameters, is the case of BPS domain walls connecting supersymmetric stationary
points. The bulk Lagrangians describing SUGRA in AdS/CFT have such
BPS solutions, and it is interesting to ask what their meaning is in the
boundary theory.

Remarkably the answer is that they give us supergravity approximations
to renormalization group flows in the boundary field theory[31]. On the side
where the domain wall solutions approach the AdS vacuum with smallest
absolute value of the c.c. their fall off is like that of a non-normalizable
rather than a normalizable mode. That is, these solutions correspond to a
change of the Lagrangian of the boundary QFT, rather than an excitation
of the original system. The change is precisely the addition of the relevant
operator corresponding to the tachyonic direction. On the other side of the

domain wall, the solution approaches the CFT dual to the AdS stationary point through one of its *irrelevant* directions. The domain wall is thus interpreted as *holographic renormalization group (HRG) flow*. The spatial direction z along which the domain wall varies, is mapped into the scale of the boundary field theory.

The holographic renormalization group interpretation of these solutions, removes a paradox, which would have disturbed us if we had tried to make a more conventional interpretation of the wall as interpolating between two vacua of the same theory. Two vacua of the same theory always have the same high energy behavior. But the CFT on the right hand side of the domain wall has a more negative cosmological constant, and therefore a smaller density of states at high energy, than that on the left hand side. In the HRG interpretation, the loss of degrees of freedom is the usual decrease of entropy we expect when we go into the infrared along a RG trajectory. Indeed, it has been shown that the holographic version of Zamolodchikov's C theorem [32] holds for these flows, whenever the bulk SUGRA theory satisfies the dominant energy condition[31].

In this context then, stationary points at negative values of the effective potential do correspond to theories of quantum gravity in AdS space (under certain conditions), but not to different states of the same theory. Rather they are different quantum field theories, connected by RG flow. This interpretation makes sense when the less negative stationary point is a maximum, and both stationary points have curvature satisfying the BF bound.

Note that in general, the functions which appear in RG equations are renormalization scheme dependent. They do not have any invariant meaning apart from the number of fixed points and the spectrum of dimensions at each fixed point. In the holographic RG, fixed points are stationary points of the bulk supergravity potential. Furthermore, the only domain wall solutions that have been given an RG interpretation in AdS/CFT, correspond to flows from a BF allowed saddle point to a stable AdS minimum or other BF allowed saddle. Other minima of the potential, particularly those with positive energy density, have not been given a reasonable interpretation in the boundary field theory.

5.2. *SUSY and large radius AdS space*

The relation between black hole and CFT entropy tells us that any CFT with large entropy would, if it were dual to quantum gravity in an AdS space, have an AdS radius much larger than the Planck length. We might be tempted to conclude that it was well described by a low energy Lagrangian involving gravity coupled to some other fields. In general we would be wrong.

The dimension $(d-1$, for $AdS_d)$ of the stress tensor in the CFT sets the scale of bulk masses in terms of the AdS radius. Any operator of dimension $(d-1)K$ corresponds to a bulk field of mass $mR = \sqrt{K^2 - K}(d-1)$. Generic CFT's which are order 1 perturbations of free field theory have a degeneracy of operators at this dimension, which grows like the exponential of a power of K. Thus, the AdS dual must have an exponentially large number of massive fields with masses that are of order the inverse AdS radius. No bulk field theory for a space-time of the form $AdS \times M$, with M a compact manifold with radius of order $\leq R$, has such a degeneracy.

Supersymmetric theories, which have a weak coupling string expansion provide an explanation for this behavior. In the canonical example of the duality between IIB string theory on $AdS_5 \times S^5$ and $\mathcal{N} = 4$ SYM theory with gauge group $SU(N)$, $N^{\frac{1}{4}}$ is the AdS radius in Planck units, while $(g^2 N)^{\frac{1}{4}}$ is the AdS radius in units of the string tension. The limit $g^2 N \ll N \to \infty$ is the limit of weak string coupling. If $g^2 N$ is of order 1 then the AdS radius is of order string scale and we do expect an exponential degeneracy of stringy states. In this regime the dimensions of most operators are indeed order one perturbations of their free field values. We expect that when $g^2 N$ is large, most of the operators get large dimension so that the corresponding masses correspond to energies of order string scale. The strongest evidence for this comes from the BMN limit of the theory[33]. There is a class of operators, called chiral operators, whose dimension is calculable and independent of $g^2 N$, but these are precisely the operators dual to the KK excitations on S^5, and they do not have an exponential degeneracy.

Thus, CFT's with the properties that correspond to space-time theories with a conventional bulk field theory description are few and far between. So far, all known examples are exactly supersymmetric, though there are non-supersymmetric, renormalizable field theories, which one can construct as relevant perturbations of a SUSic CFT with large radius dual. However, in all these cases, exact SUSY is restored as the AdS radius is taken to infinity. We have already commented above on the crucial role that SUSY plays in the stability of string theories in asymptotically flat space-time. The paucity of CFT's with a huge gap in their dimension spectrum, below which the spectrum grows only like a power law, is another argument pointing in the same direction. If we found a collection of non-supersymmetric CFT's with a tuneable (perhaps discrete) parameter which could move the exponential growth in the spectrum of operators off to infinite dimension, then we might be able to take a limit[34], which gave non-supersymmetric theories of gravity in asymptotically flat space-time. Our inability to do so thus far suggests that exact SUSY must be restored in the asymptotically flat limit.

6. The Real World: SUSY Breaking and dS Space

For much of the 15 years following the superstring revolution in 1984, the aim of the phenomenologically oriented part of the string theory community was to find a Poincare invariant, non-supersymmetric, S-matrix for a gravitational theory in 4 space-time dimensions. The idea was that we first wanted to solve particle physics, and leave the tougher problems of cosmology for later. The perturbative heterotic string had vacua with a particle spectrum (if we ignored the moduli) very close to that of the standard model (on the scale set by the string tension). Supersymmetric versions of the theory had a (generically large) number of massless moduli, but general effective field theory arguments assured us that once SUSY was broken, we would get a potential on moduli space. Efforts turned towards finding that potential, and freezing the moduli. As in any effective field theory argument, the problem of the cosmological constant seemed unsolvable, but it was a problem everyone had lived with for a long time. It is likely that most string theorists believed that some miraculous new principle would imply that the c.c. was zero.

Dine and Seiberg[35] argued that one could not find a stable minimum for the modulus that controls the string coupling, in the weak coupling approximation, *unless some other small parameter were found to control the calculation*. There were various early attempts to do this[36].

It was realized early on that simply finding a potential on moduli space was not enough. There were likely to be many minima, though at that time no one had done the proper estimates of just how many. In 1995[39] I suggested that we turn the problem of getting a reasonable value for the c.c. into a vacuum selection principle. That is, if it were difficult to find a vacuum with broken SUSY and vanishing c.c. (still an observational possibility at that time) perhaps our vacuum was singled out as the unique one with that property.

Then, observations of distant supernovae became the final straw in a pile of cosmological observations, which had been indicating a positive value of the c.c. since the mid 1980s. Although the data certainly do not prove that there is a positive c.c., it is certainly the simplest explanation of what is seen. Positive c.c. is a theoretical conundrum for the present form of string theory. Asymptotic dS spaces do not admit the kinds of boundary observables that string theory can calculate.

There have been two kinds of reaction to this in the string theory community. The first was to try to create a generalization of string theory to cope with dS space[37]. I still believe that is the correct path, although some pessimistic conclusions have appeared in the literature[38], and I will return to this below. The second was driven by the discovery of an apparent solution to the Dine-Seiberg problem. We have noted that that string theory contains many p-form gauge fields. On a compact manifold, the integral of a field strength F_p over a p-cycle is quantized, because of the DNT quantization condition. Thus, we can look for solutions of the low energy SUGRA field equations parametrized by integer values of all of these fluxes. This was proposed in [40]. Bousso and Polchinski[41] pointed out that for Calabi-Yau 3-folds with large $b_3 (\sim 100)$, this could lead to an enormous number of solutions, of order N^{b_3} with N a large integer. They pointed out that this could be used to implement an anthropic explanation of the value of the cosmological constant, along lines first suggested in [45] (see also [46]). The basic idea was that each generic choice of fluxes would stabilize the moduli and correspond to a new minimum of the effective potential, with a different energy density. Among this huge choice of minima, there are likely to be many (because the number of allowed flux configurations is exponentially larger than 10^{123}) that satisfy Weinberg's criterion for galaxy formation[44]. Galaxies, and life of our type, could only exist in those.

Giddings, Kachru and Polchinski[42] showed that in IIB supergravity compactifications on CY 3-folds, in the presence of D-branes and orientifolds, all complex structure moduli and the dilaton were stabilized at the classical level. Kachru, Kallosh, Linde and Trivedi[43] argued that the remaining moduli could be stabilized by non-perturbative effects in low energy field theory/D-brane physics, and that the inverse string coupling and CY-3 size were large at the stable minima - so that the calculations were self consistent in effective field theory. They also showed how to find meta-stable de Sitter minima, using the same techniques. A large number of refinements of these calculations have appeared.

These considerations can be criticized at both a fundamental and a phenomenological level. String theory concepts, such as D-branes and orientifolds are known to make sense only when the non-compact co-dimension of the branes is larger than 1. Here they are being used for space-filling branes. To put it another way, we have no positive evidence of the utility of the field theory notion of effective potential, as a tool for finding valid stringy models of quantum gravity. Indeed, what evidence exists[47] suggests that this is not a useful tool for this purpose.

The global space-time picture one is forced to consider if one believes in the "landscape of string theory" most likely has a Hot Big Bang in its past, and almost certainly has the fractal Penrose diagram of *eternal inflation*[48] as its future. We do not know how to talk about well defined observables in either of these contexts, so it is a bit premature to conclude that any of the meta-stable dS minima that are found by the procedures of KKLT and their followers, have anything to do with what we know as string theory. On a practical level, our lack of knowledge of what the basic setup is, makes it impossible to set up, even in principle, the computation of the next order correction to the large radius, and weak coupling approximations, which are invoked by landscape enthusiasts. We don't know what it is that these are approximations to.

If we accept the existence of the landscape, we are faced with a phenomenological dilemma. Many of the parameters in the standard model of particle physics and cosmology are much more finely tuned than would be required by anthropic considerations alone. A landscape model would have to predict that most of the parameters did not fluctuate much around their central values, and that these central values coincided with what we find in nature in order to give reasonable agreement with experiment. Only two or three parameters, notably the c.c. and the VEV of the standard model Higgs field, can be well fixed by anthropic considerations°. There is as yet no indication that the stringy landscape really makes such predictions.

A more conservative interpretation of the existence of the stringy landscape is simply that string theory gives us a lot of possible low energy theories of the world, and we should simply fit the data to determine which one we are experiencing. It is not clear whether there is any predictive power left in such a strategy. This is made even more problematic by the suggestion[P] that even given all the resources in the universe, we might not be able to build a computer capable of exploring the vast set of meta-stable states in the theory.

This subject is still unfinished and there is undoubtedly a lot left to be said about it, so I will have to leave the reader stranded in the midst of the

---

°And even this is only true if all other parameters are fixed by the theory to their real world values.

[P]Due to E. Witten.

landscape, while I go on to explore another possible route to understanding the (possibly) asymptotically de Sitter universe that we live in.

7. A Theory of Stable de Sitter Space

Quantum field theory in curved space-time indicates that de Sitter space has a finite temperature and entropy. Some time ago, Fischler and I[37] conjectured that this meant that the quantum theory of dS space had a Hilbert space with a finite number of states, determined by the c.c.[q]. This fit with the understanding of the c.c. in AdS/CFT . In both cases the c.c. is a discrete UV input parameter, determining the spectrum of the largest black holes in the theory, rather than a low energy effective parameter, subject to renormalization.

If dS space has a finite number of states, and its ground state is really the thermal ensemble described below, then it is *a fortiori stable*. It cannot have an interpretation in terms of the sort of asymptotic observables we are used to in string theory. I view the theory of stable dS space as a generalization of string theory, which approaches something like an asymptotically flat string theory (but with at most a compact moduli space) in the limit that the c.c. is taken to zero. It will become evident that this limit is crucial to the correct interpretation of physics in dS space. The theory is defined, if it is defined, as an equivalence class of Hamiltonians, which have the same asymptotic expansion (for appropriate observables) around the $\Lambda = 0$ limit.

I will quickly summarize the properties that a theory of such a stable dS space must satisfy in order to reproduce the robust predictions of field theory in curved space-time, consistent with the hypothesis of a finite dimensional Hilbert space. I will not have space to explain the arguments for these properties, which were expounded in [58]. The theory is constructed in terms of the observables in the causal patch of a given time-like observer (conveniently chosen to be geodesic). The causal patch metric is

$$ds^2 = -dt^2(1 - \frac{r^2}{R^2}) + \frac{dr^2}{(1 - \frac{r^2}{R^2})} + r^2 d\Omega^2.$$

The only part of the dS group that preserves this causal patch (we work in 4 dimensions) is $R \times SU(2)$, with R the static time translation. The spectrum of H, the generator of static time translations is bounded. There

---

[q]Strictly speaking, the dS entropy is the entropy of a thermal density matrix, but most of the eigenstates of the Hamiltonian lie below the dS temperature, so the entropy is approximately the logarithm of the number of states.

is also an approximate (super) Poincare group, which emerges in the limit of vanishing c.c.

In fact, all localized states in the causal patch are unstable, and decay to the vacuum. The vacuum is an ensemble of states with average level spacing $\frac{e^{-\pi(RM_P)^2}}{2\pi R}$, spread out between 0 and the dS temperature. The Hamiltonian for this set of states is a random matrix with this spectral bound , subject to a few constraints. The most important of these is the emergence of an approximate super-Poincare group in the $R \to \infty$ limit. The Poincare Hamiltonian P_0, in the rest frame of the static observer, satisfies

$$[H, P_0] \sim \frac{P_0}{R}.$$

This can help to explain the meta-stability (under H evolution) of low lying P_0 eigenstates, a property of both field theory in dS space, and the real world . The P_0 eigenvalues are supposed to correspond to the energies of localized objects in the static patch.

In this model, the vacuum density matrix is $\rho = e^{-2\pi RH}$, while field theory in curved space-time leads us to expect $\rho \approx e^{-2\pi RP_0}$, at least for small P_0 eigenvalues. This implies a relation between the P_0 eigenvalue and the entropy deficit (relative to the vacuum) of the corresponding eigenspace. The relation is satisfied by black holes much smaller than the dS radius, with the black hole mass parameter M identified as the P_0 eigenvalue. A simple model of this behavior can be constructed in terms of fermion creation and annihilation operators, following the rules described in the next section[58][54].

In [37] I conjectured a non-classical relation between the value of the gravitino mass and the c.c. of a stable dS space. The relation takes the scaling form $m_{3/2} \sim \Lambda^{1/4}$, and is called the hypothesis of cosmological SUSY breaking (CSB). The basic idea was that SUSY would be restored in the limit of vanishing c.c. Classical SUGRA suggests a different relation, $m_{3/2}M_P \sim \Lambda^{1/2}$. Actually, low energy effective field theory does not predict *any* relation between these two parameters. It contains a free parameter, the value of the superpotential at the minimum of the potential, which allows us to tune the c.c. to any value, no matter what the gravitino mass is. However, if we view the low energy Lagrangian parameters as subject to renormalization, then this is "unnatural fine tuning" unless the quoted relation is satisfied. By contrast, if the c.c. is a high energy input, then this fine tuning is simply required in order to make the low energy effective theory compatible with its high energy definition. The gravitino mass should then be viewed as an output parameter, and its functional form for small c.c. is to be computed. Partial evidence for the CSB scaling law has been provided in [55], but there is as yet no clean argument.

The picture that emerges from these considerations is highly constrained. The limiting $\Lambda \to 0$ theory is a super-Poincare invariant theory of gravity in four dimensions, with a compact moduli space. We have no evidence that any such theories exist, though we can easily construct low energy model Lagrangians with these properties. Phenomenological considerations related to these ideas may be found in [56].

A finite quantum theory of dS space poses certain conceptual problems related to quantum measurements[57]. Usually we view the mathematical predictions of quantum theory as the results that would appear for idealized measurements by infinite machines with precisely classical behavior. There are no such devices in a stable dS space. Indeed, our understanding of how to build pointer observables is based on quantum field theory. The tunneling amplitudes between different values of volume averaged observables in field theory are of order e^{-cS}, where c is a number of order 1 and S is the logarithm of the number of states in the volume. The entropy of field theoretic states in a causal diamond of dS space is bounded by something of order $(RM_P)^{3/2}$. Thus, the accuracy of measurements is limited to one part in $e^{(RM_P)^{3/2}}$, and the time over which a measuring apparatus is immune to quantum fluctuations is $e^{(RM_P)^{3/2}} M_P^{-1}$. These limitations are of no practical importance, but they are conceptually significant, and tell us that there will be ambiguities in mathematical constructions of the quantum theory of dS space.

The field theoretic states in a given horizon volume can be approximately described by a scattering operator, which relates incoming particle states on the past cosmological horizon to outgoing states on the future horizon. There are a variety of constraints, which make the definition of this operator ambiguous. Incoming and outgoing states must be measured at points sufficiently far from the horizon that the blue shifted dS temperature is low compared to the (Poincare) energies of the particles. Like all measurements referring to finite points in space time, these can have only an approximate meaning. The sub-energies in all scattering processes must be kept well below the mass of horizon sized black holes (*e.g.* we may try to take a limit where $R \to \infty$ with energies fixed). I would conjecture that a limiting scattering matrix exists and that it has an asymptotic expansion for $R \to \infty$ which is universal and unambiguous. I suspect the order of magnitude of the ambiguities is $e^{-c(RM_P)^{3/2}}$. The expansion coefficients of these scattering matrix elements would be the gauge invariant observables of asymptotically dS space.

8. Towards a Holographic Theory of Space-Time

The existing versions of string theory are, as we have seen, all formulated in terms of boundary correlation functions. This makes it difficult to see how local physics comes out of the formulation. At present the only clue is that in certain limits, the boundary correlators can be computed by solving bulk field equations.

The difficulties involved in a local formulation become clearer if we accept the covariant entropy bound as a fundamental axiom. A causal diamond in a Lorentzian space-time is the intersection of the causal past, $\mathcal{P}_P$ of a point P, with the causal future of a point $Q \in \mathcal{P}_P$. The maximal area $d - 2$ surface on the boundary of a causal diamond, is a holographic screen[3] for the information which passes through the boundary. For small enough causal diamonds, the area of the holographic screen is always finite. According to the covariant entropy bound, this area in Planck units is the maximal entropy that can be associated with the diamond.

Fischler and I[49] , proposed that this was the entropy of the maximally uncertain density matrix for this system, which equals the natural logarithm of the number of states in its Hilbert space. There is no other natural density matrix, which exists in all situations to which the entropy bound is supposed to apply.

If a causal diamond has only a finite number of physical states, then there cannot be a unique mathematical model which describes physics in the causal diamond. An isolated quantum system can make measurements on itself only if there are physical subsystems which behave like classical measuring devices. Subsystems can be exactly classical only if they have an infinite number of states. Thus, the model of a finite causal diamond must have an ambiguity, because its predictions can only be tested with finite precision. I believe that this ambiguity is the quantum analog of general covariance. The point of general covariance is that we cannot describe the space-time properties of a system without referring to some classical coordinate system. There are no local diffeomorphism invariant observables. In a quantum system, this problem is exacerbated by the fact that the objects defining a particular coordinate system are quantum systems. In order to make them approximately classical, we must make them large, but then they have a large gravitational interaction with the system we would like to study. The covariant entropy bound expresses this conundrum in purely quantum language.

The idea of [49] was to treat this ambiguity as the analog of coordinate invariance. A given description of a space-time in terms of the quantum mechanics of a collection of overlapping causal diamonds, is thought of as a particular physical gauge fixing of a generally covariant quantum theory of gravity. The description of small (low entropy) causal diamonds has a lot of gauge ambiguity, but large causal diamonds which can support enough degrees of freedom to make almost classical measurements have "almost gauge invariant" observables. We get mathematically gauge invariant observables only in the limit of infinite entropy causal diamonds. These are the boundary correlation functions familiar from string theory.

The first step in defining such a system is to decide on a fundamental set of observables. These are taken to be quantum pixels of the holographic screen of a causal diamond. In classical, d dimensional, Lorentzian geometry, we can identify, at each point of the holographic screen of a causal diamond, the future directed null vector N^μ normal to the screen, as well as the orientation of the screen, which is a transverse hyperplane to N^μ. The two can be packaged together into what Cartan called a pure spinor, via the Cartan-Penrose (CP) equation

$$\bar{\psi}\gamma^\mu\psi\gamma_\mu\psi = 0$$

where ψ is in the minimal dimensional Dirac spinor representation of $SO(1, d-1)$. The Dirac bilinears describe a null direction and a transverse $d-2$ plane. The equation is invariant under $\psi \to \lambda\psi$, where λ is real or complex, depending on d. Thus, this equation encodes only conformal data of the space-time.

A pure spinor has half the number of components of a general Dirac spinor. Denote these components by $S_a(\Omega)$, where Ω are coordinates on the holographic screen. These are a section of the spinor bundle over the screen, where the Riemannian structure on the screen is inherited from its embedding in space-time. It is clear that if we specify these sections for "every" causal diamond in a Lorentzian space-time, we have completely fixed its conformal structure[r]. Classically, we could fix the metric completely by specifying a compatible set of area forms, one for each screen.

---

[r]But there will be complicated consistency conditions relating the spinor sections on different holographic screens.

The Bekenstein-Hawking relation suggests however that metrical concepts originate in quantum mechanics. Thus, we quantize the pixel variables, by the unique $SO(d-2)$ invariant formula that gives a single pixel a finite number of states:

$$[S_a(n), S_b(n)]_+ = \delta_{ab}.$$

In writing $S_a(n)$ we have anticipated the fact that the whole causal diamond must have a finite number of states. We view the surface of its holographic screen as broken up into pixels, labelled by a finite set of integers n. Each pixel has area equal to the logarithm of the dimension of the minimal representation of the above Clifford algebra[s]. For 11 dimensional space-time, this dimension is 256. We see an immediate connection to $11D$ SUGRA. The $SO(d-2)$ content of this representation is exactly that of the supergraviton multiplet. Thus, in particle physics language, what we are saying is that specifying a holographic screen at each point is specifying the direction of the momentum of the massless particles, which can penetrate this pixel, as well as the possible spin states of those particles. The particle language really becomes justified only in the limit of infinite causal diamonds, in asymptotically flat space[50]. In that limit, a new quantum number, the longitudinal momentum of the massless particle, also arises, in a manner reminiscent of Matrix Theory[12].

Physically, the operators of individual pixels are independent quantum degrees of freedom and so the operators associated with different pixels should commute with each other. However, the full set of (anti)-commutation relations is invariant under a Z_2^k subgroup of the classical projective invariance of the CP equation: $S_a(n) \rightarrow (-1)^{F_n} S_a(n)$. This should be treated as a gauge symmetry, and we can do a Klein transformation to new variables that satisfy[t]

$$[S_a(n), S_b(m)]_+ = \delta_{ab}\delta_{mn}.$$

An elegant way to describe the pixelation of the holographic screen is to replace the algebra of functions on the screen by a finite dimensional associative algebra. If we want to implement space time rotation symmetries

---

[s]Our equation is written for the case where the minimal spinor representation of $SO(d-2)$ is real.

[t]For asymptotically flat space it is convenient to choose another gauge for the Z_2 symmetry. In that limit the label n encodes multiparticle states, as well as the momenta of individual particles. It is convenient to choose pixel operators for different particles to commute, in order to agree with the standard multiparticle quantum mechanics conventions.

it is probably necessary to use a non-commutative algebra, which puts us in the realm of fuzzy geometry. The spinor bundle over a non-commutative algebra is a projective module for the algebra[51]. The quantum operator algebra consists of a linear map from the spinor bundle to the algebra of operators in the Hilbert space of the causal diamond, and the $S_a(n)$ are the values of the map on a basis for the spinor bundle. The algebra of operators that we used to describe dS space in [54] is an example of this construction.

The Hilbert space of an *observer*[u] in this formalism, corresponds to a sequence of causal diamond Hilbert spaces, with each space related to the previous one by $\mathcal{H}_N = \mathcal{H}_{N-1} \otimes \mathcal{S}$, where $\mathcal{S}$ is the representation space of the single pixel Clifford algebra. The space-time picture of this sequence depends on the asymptotic boundary conditions in space-time. For example, in a Big Bang universe we would choose causal diamonds, which all begin on the Big Bang surface. Larger causal diamonds would correspond to later points on the observer's trajectory. In a space-time, like asymptotically flat or AdS space, with TCP invariance, we would choose a time reflection surface and give a sequence of causal diamonds symmetric around this surface. The observer Hilbert space is equipped with a sequence of time evolution operators $U_N(K)$ in each $\mathcal{H}_N$ describing the evolution of the system from the past to the future boundaries of each of the causal diamonds $\mathcal{H}_K$ contained in $\mathcal{H}_N$. There are consistency conditions guaranteeing that the observer's time evolution operator for a given Hilbert space $\mathcal{H}_K$ in $\mathcal{H}_N$ is consistent with its description of the same Hilbert space in $\mathcal{H}_L$ for all $K \leq L \leq N$. Note that in this formalism, one is able to describe an expanding universe, without violating unitarity. Note also that although time steps are discrete, the magnitude of the discrete steps is measured by a fixed increase in entropy. Thus, in a locally weakly curved region of space-time, the time step goes to zero for large times.

A full quantum space-time is described by an infinite set of observers[v]. A pair of nearest neighbor observers is defined by a pair of Hilbert space sequences as above, combined with an identification for each n of a common tensor factor $\mathcal{O}_n(1,2)$ in $\mathcal{H}_n(1)$ and $\mathcal{H}_n(2)$. There is a very complicated consistency condition requiring that the time evolution operators in the individual observer Hilbert spaces, agree in their operation on $\mathcal{O}_n$. More

---

[u]Observer = large quantum system with many semiclassical observables. Of course, in small causal diamonds there are no such observers, but we retain the terminology nonetheless.

[v]The case of asymptotically dS space times would be dealt with by introducing an infinite lattice of observers, each of which attains a maximal entropy. The overlap of any two maximal entropy Hilbert spaces is required to be the whole space.

generally we introduce a spatial lattice, with coordinates $\mathbf{x}$ and a collection of observer Hilbert spaces $\mathcal{H}(n, \mathbf{x})$. These are generated by operators $S_a(n, \mathbf{x})$. The overlap Hilbert space is defined by a map between subalgebras of the operator algebras at $(n, \mathbf{x})$ and $(n, \mathbf{y})$. For nearest neighbors the overlap subalgebra is missing exactly one pixel's worth of operators. The reader is urged to think of the overlap map as the quantum analog of the Lorentz parallel transport between the two lattice points.

The lattice has no *a priori* notion of distance on it, but it does impose a spatial topology, which does not change with time. This can be consistent with our observations of topology change for compactification manifolds[w] if we think of $(n, \mathbf{x})$ as describing the *non-compact* dimensions of space-time. Compactified dimensions would be an approximate notion, in which additional indices I combined with a to form a spinor of a higher dimensional space-time. The fundamental theory would have no notion of the topology of the compactification manifold, which would be an emergent phenomenon in certain limiting situations.

In general there is no reason to assume that the initial state in a causal diamond Hilbert space is pure. However, it seems reasonable to impose the Primordial Purity Postulate: the initial states in causal diamonds whose tip lies on the asymptotic past, or on the Big Bang, may be taken to be pure. A related postulate one might want to impose is the Tallness Postulate[x]: There is at least one observer whose causal past contains an entire Cauchy surface for the space-time. In the present context this means that there is a class of observers at points $\mathbf{x}_i^*$ such that asymptotically as $n \to \infty$, $\mathcal{H}(n, \mathbf{x}) \subset \mathcal{H}(n, \mathbf{x}_i^*)$. These are called the future asymptotic observers, and one might impose Primordial Purity only for the future asymptotic observer Hilbert spaces. Primordial Purity guarantees that all correlations we observe in the universe, had to develop dynamically. Many models of eternal inflation do not satisfy the Tallness principle.

The formalism described here seems to be an attractive way to discuss more or less local physics in a way that is compatible with the holographic principle. The difficulty lies in the complicated compatibility conditions

---

[w]Strictly speaking, what has been demonstrated is that in asymptotically flat string theory, we can make a smooth change of the topology of the compactification manifold by changing the moduli. It is tempting to think that the change of moduli can also occur dynamically in a cosmological space-time, but this remains to be demonstrated. The formalism we are describing could take such processes into account by the device indicated in the text.

[x]The Tallness Postulate was invented by D. Kabat and L. Susskind, quite independently of the formalism discussed here. I learned of it in private conversations with L. Susskind.

between the dynamics in different causal diamonds. It is hard to say at the moment whether this framework is compatible with string theory as we know it, or with the classical Einstein theory of gravitation. It seems almost certain that, if the dynamics allows us to construct a set of semi-classical pixel observables in the large area limit, then we should be able to reconstruct a Lorentzian geometry because we are specifying the geometry of the holographic screens of a densely spaced set of causal diamonds. Furthermore, since the covariant entropy bound is built in to our formalism, it is likely that the Einstein tensor of that space-time satisfies some form of the null energy condition. It is obvious that we need more than such plausibility arguments to verify the validity of the causal diamond formalism.

There are three partial successes in making connections between this formalism and ordinary gravitational physics. In [52] the authors constructed an explicit quantum model, satisfying all of the compatibility conditions. The model defined an emergent geometry on the lattice space-time, which was that of the flat, homogeneous isotropic universe with equation of state $p = \rho$. The Big Bang was completely non-singular in this model. It is merely an initial configuration in which causal diamonds are built from a single pixel. The scaling laws which matched to the $p = \rho$ cosmology were asymptotic relations for large entropy. This model validated the heuristic claims of [53], which postulated the $p = \rho$ FRW universe as a maximally entropic initial condition for the universe.

In [54][58] I described a model constructed by B. Fiol and myself, which reproduced the spectrum of black holes in dS space, starting from a set of fermionic pixel operators appropriate to the cosmological horizon.

Finally, in [50] I show how a large causal diamond limit of this formalism can reproduce the spectrum of asymptotic supergraviton scattering states in 11 dimensions. This paper does not touch on the question of compatibility conditions, and thus, it was impossible to get a constraint on the dynamics and compute the scattering matrix.

9. Conclusions

We have seen that the geometry of space-time in string theory is extremely mutable, and that this can be attributed to the Holographic Principle. The topology and even the dimension of compact factor manifolds can change continuously along moduli spaces of Super-Poincare invariant S-matrices. Certain quantum numbers, characterizing stable BPS particles and branes are robust indicators of the geometry, but their geometrical interpretation changes radically in different extreme regions of moduli space.

Asymptotically AdS space-times are described by correlators of a boundary quantum field theory. This description teaches us important lessons, apart from reiterating the holographic nature of quantum gravity. Most importantly it confirms the conjecture that black holes dominate the high energy spectrum of models of quantum gravity and that generic high energy processes are dominated by black hole production. A corollary of this insight is that the cosmological constant is a discrete high energy input parameter, which measures the asymptotic density of states, rather than a parameter in low energy effective theory. Low energy effective bulk field theory must be fine tuned in order to agree with the high energy input c.c.

We also learned that AdS spaces did not have continuous moduli spaces of vacuum states (the boundary theory is compactified on a sphere and so does not possess superselection sectors even when it is exactly supersymmetric). We learned that the bulk effective potential had a rather different meaning in this context than it does in non-gravitational field theories. It is related to a SUGRA approximation to the renormalization group of the boundary field theory. So far, only pairs of stationary points (the higher bulk energy density one always having at least one tachyonic direction) are identified with models of quantum gravity. The bulk field theory interpretation of a domain wall separating two vacua of the same theory is wrong. Instead the domain wall represents RG flow between two different models. Finally, we found that the only known examples where the AdS curvature is small enough that the SUGRA approximation is valid are asymptotically supersymmetric. The limit of infinite AdS radius always implies exact restoration of supersymmetry in these examples. This is consistent with our inability, so far, to find non-supersymmetric, Poincare invariant, string models.

String theorists have not yet learned to cope with space-times that have a positive cosmological constant. There are two approaches to this problem. By far the most popular is called The String Landscape. Landscape theorists have argued for the existence of a new type of string theory based on an effective potential with an exponentially large number (*e.g.* 10^{500}) of positive energy minima. These are supposed to represent meta-stable states of the new string theory, which eventually decay into a 10 or 11 dimensional non-accelerating FRW space-time. Our universe is one of the meta-stable states. The large number of minima enables one to invoke the anthropic explanation of the cosmological constant. The past asymptotics of this hypothetical space-time probably involves a Big Bang, while its future is almost certainly the fractal Penrose diagram of eternal inflation. It is alleged that the fundamental string theory observables of this system will be related to measurements done in the non-accelerating region, and that these will allow one to define the measurements done in our own universe in a rigorous, if approximate, way (much as scattering experiments define the properties of a meta-stable resonance). This proposal faces many phenomenological and fundamental challenges, as well as issues of computability associated with the huge number of meta-stable states. Work is in progress to address some of these issues.

The second approach to space-times with positive c.c. is based on the assertion that the quantum theory of dS space has a finite number, N, of states, related to the value of the cosmological constant in Planck units. As in the case of AdS space, the c.c. is a discrete high energy input. The finiteness of the state space poses problems of quantum measurement theory: such a theory can describe neither arbitrarily accurate measurements, nor measurements which remain robust over arbitrarily long time intervals. However, for the value of the c.c. indicated by observations, the limit on accuracy is about one part in $e^{10^{90}}$, and this number (in any units you care to choose) is also the size of the time over which measurements will be destroyed by quantum fluctuations. The phenomenology of this proposal is based on the idea of Cosmological SUSY Breaking. It seems to lead to a rather predictive framework for both particle physics and cosmology. Some progress has also been made in constructing a fundamental Hamiltonian description of this system. Progress on all fronts is incremental but slow.

Finally, I described a general holographic quantum theory of space-time. According to the covariant entropy bound, a causal diamond is described by a finite number of states, related to the area of its holographic screen. In the quantum theory this translates into a finite Clifford algebra of operators $S_a(n)$. The classical Cartan-Penrose equation tells us how to describe the conformal structure of the holographic screen of a causal diamond in terms of a section of the spinor bundle over the screen (viewed as a $d-2$ dimensional manifold). The $S_a(n)$ are a quantization of this spinor section. The topology of the screen is pixelated by replacing its function algebra by a finite dimensional associative (and generally non-commutative) algebra and the spinor bundle is a finite projective module over this algebra. The $S_a(n)$ are quantum operators corresponding to a basis of this module. The dimension of the irreducible representation of the Clifford algebra for fixed n determines the area of a pixel, via the Bekenstein-Hawking relation. Simple examples suggest that in asymptotically flat space-times, the S_a algebra for fixed n will generate the space of massless single particle states with fixed momentum, and that the spectrum of these states will be supersymmetric.

A full quantum space-time is described by a lattice of overlapping causal diamonds. The evolution operators in different causal diamonds are constrained by the requirement that they agree on overlaps. The topology of the non-compact part of space[y] is fixed by this lattice.

This formalism is still highly conjectural but various connections with string theory and other aspects of gravitational physics are beginning to appear.

String theory thus presents us with a collection of consistent mathematical models of quantum gravity. Extant models show us that local space-time concepts are distorted in interesting and confusing ways, and suggest the Holographic Principle as a unifying framework for understanding quantum gravity. We still have a long way to go in order to fully explore the nature of space-time in this collection of theories, and to find the version of it that fits the world we observe.

References

1. G. Veneziano, Nuovo Cimento, 57A, 190, (1968); J.A.Shapiro, Phys. Lett. 33B, 361, (1970); M.A.Virasoro, Phys. Rev. 177, 2309, (1969) ; J.H.Schwarz, (ed.) *Superstrings, the First Fifteen Years*, Vol. I,II, World Scientific, Singapore, 1985.
2. S. Weinberg, *The Quantum Theory of Fields, Vol I.*, Ch. 13, Cambridge University Press,1995.
3. R. Bousso, Rev. Mod. Phys. **74**, 825 (2002) [arXiv:hep-th/0203101].
   R. Bousso, JHEP **9907**, 004 (1999) [arXiv:hep-th/9905177].
4. W. Fischler and L. Susskind, arXiv:hep-th/9806039.
5. T. Banks and W. Fischler, arXiv:hep-th/0408076.
6. D. Amati, M. Ciafaloni and G. Veneziano, "Classical And Quantum Gravity Effects From Planckian Energy Superstring Collisions," Int. J. Mod. Phys. A **3**, 1615 (1988).
   P. C. Argyres, S. Dimopoulos and J. March-Russell, "Black holes and sub-millimeter dimensions," Phys. Lett. B **441**, 96 (1998) [arXiv:hep-th/9808138].
   H. J. Matschull, "Black hole creation in 2+1-dimensions," Class. Quant. Grav. **16**, 1069 (1999) [arXiv:gr-qc/9809087].
   T. Banks and W. Fischler, arXiv:hep-th/9906038.

---

[y]A causal diamond in dS space is considered non-compact for the purposes of this sentence.

7. M.B. Green, J.H. Schwarz, E. Witten, *Superstring Theory, Volume 2*, Cambridge University Press, 1987.

8. P. A. M. Dirac, "Quantised Singularities In The Electromagnetic Field," Proc. Roy. Soc. Lond. A **133**, 60 (1931). P. A. M. Dirac, "The Theory Of Magnetic Poles," Phys. Rev. **74**, 817 (1948). C. Teitelboim, "Gauge Invariance For Extended Objects," Phys. Lett. B **167**, 63 (1986). R. I. Nepomechie, "Magnetic Monopoles From Antisymmetric Tensor Gauge Fields," Phys. Rev. D **31**, 1921 (1985).

9. ? Bogolmonyi ; M. K. Prasad and C. M. Sommerfield, Phys. Rev. Lett. **35**, 760 (1975).

10. J. Dai, R. G. Leigh and J. Polchinski, Mod. Phys. Lett. A **4**, 2073 (1989). P. Horava, Phys. Lett. B **231**, 251 (1989).
P. Horava, Nucl. Phys. B **327**, 461 (1989). J. Polchinski, Phys. Rev. Lett. **75**, 4724 (1995) [arXiv:hep-th/9510017]. J. Polchinski, S. Chaudhuri and C. V. Johnson, arXiv:hep-th/9602052.

11. J.H. Schwarz, "Lectures on superstring and M theory dualities," Nucl. Phys. Proc. Suppl. **55B**, 1 (1997) [arXiv:hep-th/9607201]. M. J. Duff, "M theory (the theory formerly known as strings)," Int. J. Mod. Phys. A **11**, 5623 (1996) [arXiv:hep-th/9608117].

12. T. Banks, W. Fischler, S. H. Shenker and L. Susskind, "M theory as a matrix model: A conjecture," Phys. Rev. D **55**, 5112 (1997) [arXiv:hep-th/9610043]. T. Banks, Nucl. Phys. Proc. Suppl. **67**, 180 (1998) [arXiv:hep-th/9710231]. D. Bigatti and L. Susskind, arXiv:hep-th/9712072.

13. P. Horava and E. Witten, "Eleven-Dimensional Supergravity on a Manifold with Boundary," Nucl. Phys. B **475**, 94 (1996) [arXiv:hep-th/9603142].

14. S. Chaudhuri, G. Hockney and J. D. Lykken, Phys. Rev. Lett. **75**, 2264 (1995) [arXiv:hep-th/9505054].
S. Chaudhuri, S. W. Chung, G. Hockney and J. Lykken, Nucl. Phys. B **456**, 89 (1995) [arXiv:hep-ph/9501361].

15. E. Witten, JHEP **9802**, 006 (1998) [arXiv:hep-th/9712028].

16. J. A. Harvey and A. Strominger, Nucl. Phys. B **449**, 535 (1995) [Erratum-ibid. B **458**, 456 (1996)] [arXiv:hep-th/9504047]. J. H. Schwarz, Nucl. Phys. Proc. Suppl. **67**, 74 (1998).

17. T. Banks, "On isolated vacua and background independence," arXiv:hep-th/0011255.

18. A. Strominger, Nucl. Phys. B **451**, 96 (1995) [arXiv:hep-th/9504090]. B. R. Greene, D. R. Morrison and A. Strominger, Nucl. Phys. B **451**, 109 (1995) [arXiv:hep-th/9504145].

19. T. Banks and M. Dine, Phys. Rev. D **53**, 5790 (1996) [arXiv:hep-th/9508071]. E. Witten, Nucl. Phys. B **471**, 195 (1996) [arXiv:hep-th/9603150].

20. L. Susskind, arXiv:hep-th/9309145.

21. G. T. Horowitz and J. Polchinski, Phys. Rev. D **55**, 6189 (1997) [arXiv:hep-th/9612146].

22. A. Sen, "Extremal black holes and elementary string states," Mod. Phys. Lett. A **10**, 2081 (1995) [arXiv:hep-th/9504147].

23. A. Strominger and C. Vafa, "Microscopic Origin of the Bekenstein-Hawking Entropy," Phys. Lett. B **379**, 99 (1996) [arXiv:hep-th/9601029].
24. J. M. Maldacena and A. Strominger, Phys. Rev. D **55**, 861 (1997) [arXiv:hep-th/9609026]. S. S. Gubser and I. R. Klebanov, Phys. Rev. Lett. **77**, 4491 (1996) [arXiv:hep-th/9609076].
25. J. M. Maldacena, "The large N limit of superconformal field theories and supergravity," Adv. Theor. Math. Phys. **2**, 231 (1998) [Int. J. Theor. Phys. **38**, 1113 (1999)] [arXiv:hep-th/9711200].
26. S. S. Gubser, I. R. Klebanov and A. M. Polyakov, "Gauge theory correlators from non-critical string theory," Phys. Lett. B **428**, 105 (1998) [arXiv:hep-th/9802109].
27. E. Witten, "Anti-de Sitter space and holography," Adv. Theor. Math. Phys. **2**, 253 (1998) [arXiv:hep-th/9802150].
28. I. R. Klebanov and E. Witten, Nucl. Phys. B **556**, 89 (1999) [arXiv:hep-th/9905104].
29. O. Aharony, S. S. Gubser, J. M. Maldacena, H. Ooguri and Y. Oz, Phys. Rept. **323**, 183 (2000) [arXiv:hep-th/9905111].
30. P. Breitenlohner and D. Z. Freedman, Phys. Lett. B **115**, 197 (1982). P. Breitenlohner and D. Z. Freedman, Annals Phys. **144**, 249 (1982).
31. L. Girardello, M. Petrini, M. Porrati and A. Zaffaroni, "Novel local CFT and exact results on perturbations of N = 4 super Yang-Mills from AdS dynamics," JHEP **9812**, 022 (1998) [arXiv:hep-th/9810126]. D. Z. Freedman, S. S. Gubser, K. Pilch and N. P. Warner, Adv. Theor. Math. Phys. **3**, 363 (1999) [arXiv:hep-th/9904017].
32. A. B. Zamolodchikov, "'Irreversibility' Of The Flux Of The Renormalization Group In A 2-D Field Theory," JETP Lett. **43**, 730 (1986) [Pisma Zh. Eksp. Teor. Fiz. **43**, 565 (1986)].
33. D. Berenstein, J. M. Maldacena and H. Nastase, JHEP **0204**, 013 (2002) [arXiv:hep-th/0202021].
34. J. Polchinski, arXiv:hep-th/9901076.
35. M. Dine and N. Seiberg, PRINT-85-0781 (WEIZMANN) *Presented at Unified String Theories Workshop, Santa Barbara, CA, Jul 29 - Aug 16, 1985*
36. N. V. Krasnikov, "On Supersymmetry Breaking In Superstring Theories," Phys. Lett. B **193**, 37 (1987). T. Banks and M. Dine, "Coping with strongly coupled string theory," Phys. Rev. D **50**, 7454 (1994) [arXiv:hep-th/9406132].
37. T. Banks, QuantuMechanics and CosMology, Talk given at the festschrift for L.Susskind, Stanford University, May 2000; T. Banks, arXiv:hep-th/0007146. W. Fischler, *Taking de Sitter Seriously*, Talk given at The Role of Scaling Laws in Physics and Biology (Celebrating the 60th Birthday of G. West), Santa Fe, Dec. 2000, and unpublished.
38. L. Dyson, M. Kleban and L. Susskind, JHEP **0210**, 011 (2002) [arXiv:hep-th/0208013]. N. Goheer, M. Kleban and L. Susskind, JHEP **0307**, 056 (2003) [arXiv:hep-th/0212209].
39. T. Banks, arXiv:hep-th/9601151.

40. K. Becker and M. Becker, Nucl. Phys. B **477**, 155 (1996) [arXiv:hep-th/9605053]. K. Dasgupta, G. Rajesh and S. Sethi, JHEP **9908**, 023 (1999) [arXiv:hep-th/9908088].

41. R. Bousso and J. Polchinski, JHEP **0006**, 006 (2000) [arXiv:hep-th/0004134].

42. S. B. Giddings, S. Kachru and J. Polchinski, Phys. Rev. D **66**, 106006 (2002) [arXiv:hep-th/0105097].

43. S. Kachru, R. Kallosh, A. Linde and S. P. Trivedi, Phys. Rev. D **68**, 046005 (2003) [arXiv:hep-th/0301240].

44. S. Weinberg, "Anthropic Bound On The Cosmological Constant," Phys. Rev. Lett. **59**, 2607 (1987).

45. A. D. Linde, Print-83-0620 *Invited talk given at Shelter Island II Conf., Shelter Island, N.Y., Jun 1-3, 1983*
T. Banks, Phys. Rev. Lett. **52**, 1461 (1984).
T. Banks, "T C P, Quantum Gravity, The Cosmological Constant And All That..," Nucl. Phys. B **249**, 332 (1985). L. F. Abbott, "A Mechanism For Reducing The Value Of The Cosmological Constant," Phys. Lett. B **150**, 427 (1985).
T. Banks, M. Dine and N. Seiberg, "Irrational axions as a solution of the strong CP problem in an eternal universe," Phys. Lett. B **273**, 105 (1991) [arXiv:hep-th/9109040].

46. J. L. Feng, J. March-Russell, S. Sethi and F. Wilczek, Nucl. Phys. B **602**, 307 (2001) [arXiv:hep-th/0005276].

47. T. Banks,
"On isolated vacua and background independence," arXiv:hep-th/0011255.
T. Banks, "A critique of pure string theory: Heterodox opinions of diverse dimensions," arXiv:hep-th/0306074. T. Banks, "Supersymmetry, the cosmological constant and a theory of quantum gravity in our universe," Gen. Rel. Grav. **35**, 2075 (2003) [arXiv:hep-th/0305206]. T. Banks, M. Dine and E. Gorbatov, "Is there a string theory landscape?," JHEP **0408**, 058 (2004) [arXiv:hep-th/0309170]. T. Banks, "Landskepticism or why effective potentials don't count string models," arXiv:hep-th/0412129.

48. A. H. Guth, Phys. Rept. **333**, 555 (2000) [arXiv:astro-ph/0002156], and references therein. A. D. Linde, "Eternally Existing Selfreproducing Chaotic Inflationary Universe," Phys. Lett. B **175**, 395 (1986).

49. T. Banks and W. Fischler, arXiv:hep-th/0310288.

50. T. Banks, *Type II Von Neumann Algebras and M-theory*, Manuscript in preparation

51. A. Connes, *Non-Commutative Geometry*, Academic Press, 1994.

52. T. Banks, W. Fischler and L. Mannelli, arXiv:hep-th/0408076.

53. T. Banks and W. Fischler, "An holographic cosmology," arXiv:hep-th/0111142.
T. Banks and W. Fischler, "Holographic cosmology 3.0," arXiv:hep-th/0310288.
T. Banks and W. Fischler, "Holographic cosmology," arXiv:hep-th/0405200.

54. T. Banks, arXiv:astro-ph/0305037.

55. T. Banks, arXiv:hep-th/0206117. T. Banks, arXiv:hep-th/0503066. T. Banks, W. Fischler, L. Mannelli, *Infrared Divergences in dS/CFT*, Manuscript in Preparation.

56. T. Banks, arXiv:hep-ph/0203066. T. Banks, arXiv:hep-ph/0408260.

57. T. Banks, W. Fischler and S. Paban, JHEP **0212**, 062 (2002) [arXiv:hep-th/0210160].

58. T. Banks, arXiv:hep-th/0503066.

CHAPTER 13

QUANTUM GEOMETRY AND ITS RAMIFICATIONS

ABHAY ASHTEKAR

Institute for Gravitational Physics and Geometry,
Physics Department, 104 Davey, Penn State,
University Park, PA 16802, USA
Max Planck Institut für Gravitationsphysik, Albert Einstein Institut,
Am Mühlenberg 1, D14476 Golm, Germany
ashtekar@gravity.psu.edu

The goal of this article is to present a broad perspective on loop quantum gravity, a non-perturbative, background independent approach to the problem of unification of general relativity and quantum physics, based on a specific theory of quantum Riemannian geometry. The chapter is addressed to non-experts. Therefore, the emphasis is on underlying ideas, conceptual issues and the overall status of the program rather than mathematical details and associated technical subtleties. This review complements that by Martin Bojowald which focusses on applications of quantum geometry to cosmology.

1. Setting the Stage

General relativity and quantum theory are among the greatest intellectual achievements of the 20th century. Each of them has profoundly altered the conceptual fabric that underlies our understanding of the physical world. Furthermore, each has been successful in describing the physical phenomena in its own domain to an astonishing degree of accuracy. And yet, they offer us *strikingly* different pictures of physical reality. Indeed, at first one is surprised that physics could keep progressing blissfully in the face of so deep a conflict. The reason of course is the 'accidental' fact that the values of fundamental constants in our universe conspire to make the Planck length truly minute and Planck energy absolutely enormous compared to laboratory scales. Thanks to this coincidence, we can happily maintain a schizophrenic attitude and use the precise, geometric picture of reality offered by general relativity while dealing with cosmological and astrophysical phenomena, and the quantum-mechanical world of chance and intrinsic uncertainties while dealing with atomic and subatomic particles. Clearly, this strategy is quite appropriate as a practical stand. But it is highly unsatisfactory from a conceptual viewpoint. Everything in our past experience in physics tells us that the two pictures we currently use must be approximations, special cases that arise as appropriate limits of a single, universal theory. That theory must therefore represent a synthesis of general relativity and quantum mechanics. This would be the quantum theory of gravity. The burden on this theory is huge: Not only should it correctly describe all the known physical phenomena, but it should also adequately handle the Planck regime. This is the theory that we invoke when faced with phenomena, such as the big bang and the final state of black holes, where the worlds of general relativity and quantum mechanics must unavoidably meet.

The challenge of constructing a quantum gravity theory has been with us for many decades now. The long series of investigations in the ensuing years has unveiled a number of concrete problems. These come in two varieties. First, there are the issues that are 'internal' to individual programs: For example, the incorporation of physical —rather than half flat— gravitational fields in the twistor program discussed by Roger Penrose; mechanisms for breaking of supersymmetry and dimensional reduction in string theory reviewed by Tom Banks; and issues of space-time covariance in the canonical approach discussed in this chapter. The second category consists of physical and conceptual questions that underlie the whole subject. To set the stage from which one can gauge overall progress, I will now focus on the second type of issues by recalling three long standing issues that *any* satisfactory quantum theory of gravity should address.

• *Black holes:* In the early seventies, using imaginative thought experiments, Bekenstein argued that black holes must carry an entropy proportional to their area [7,19,28].[a] About the same time, Bardeen, Carter and Hawking (BCH) showed that black holes in equilibrium obey two basic laws, which have the same form as the zeroth and the first laws of thermodynamics, provided one equates the black hole surface gravity κ with some multiple of the temperature T in thermodynamics and the horizon area a_{hor} to a corresponding multiple of the entropy S.[7,19,28] However, at

---

[a]Since this article is addressed to non-experts, except in the discussion of very recent developments, I will generally refer to books and review articles which summarize the state of the art at various stages of development of quantum gravity. References to original papers can be found in these reviews.

first this similarity was thought to be only a formal analogy because the BCH analysis was based entirely on *classical* general relativity and simple dimensional considerations show that the proportionality factors must involve Planck's constant $\hbar$. Two years later, using quantum field theory on a black hole background space-time, Hawking showed that black holes in fact radiate quantum mechanically as though they are black bodies at temperature $T = \hbar\kappa/2\pi$.[7,14] Using the analogy with the first law, one can then conclude that the black hole entropy should be given by $S_{BH} = a_{hor}/4G\hbar$. This conclusion is striking and deep because it brings together the three pillars of fundamental physics — general relativity, quantum theory and statistical mechanics. However, the argument itself is a rather hodge-podge mixture of classical and semi-classical ideas, reminiscent of the Bohr theory of atom. A natural question then is: what is the analog of the more fundamental, Pauli-Schrödinger theory of the Hydrogen atom? More precisely, what is the statistical mechanical origin of black hole entropy? What is the nature of a quantum black hole and what is the interplay between the quantum degrees of freedom responsible for entropy and the exterior curved geometry? Can one derive the Hawking effect from first principles of quantum gravity? Is there an imprint of the classical singularity on the final quantum description, e.g., through 'information loss' during black hole formation and evaporation?

• *The big-bang*: It is widely believed that the prediction of a singularity, such as the big-bang of classical general relativity, is primarily a signal that the physical theory has been pushed beyond the domain of its validity. A key question to any quantum gravity theory, then, is: What replaces the big-bang? Are the classical geometry and the continuum picture only approximations, analogous to the 'mean (magnetization) field' of ferro-magnets? If so, what are the microscopic constituents? What is the space-time analog of the Heisenberg model of a ferro-magnet? How close to the singularity can we trust the continuum space-time of classical general relativity? When formulated in terms of these fundamental constituents, is the evolution of the *quantum* state of the universe free of singularities? If so, what is on the 'other side' of the big-bang? An infinite quantum foam or another, large classical space-time? Is the evolution 'across' the singularity fully determined by quantum Einstein's equations? Or, as in the pre-big-bang scenario of string theory, for example, is a new principle essential to ensure deterministic dynamics?

• *Planck scale physics and the low energy world:* In general relativity, there is no background metric, no inert stage on which dynamics unfolds.

Geometry itself is dynamical. Therefore, as indicated above, one expects that a fully satisfactory quantum gravity theory would also be free of a background space-time geometry. However, of necessity, a background independent description must use physical concepts and mathematical tools that are quite different from those of the familiar, low energy physics which takes place on a flat, background space-time. A major challenge then is to show that this low energy description does arise from the pristine, Planckian world in an appropriate sense, bridging the vast gap of some 16 orders of magnitude in the energy scale. In this 'top-down' approach, does the fundamental theory admit a 'sufficient number' of semi-classical states? Do these semi-classical sectors provide enough of a background geometry to anchor low energy physics? Can one recover the familiar description? If the answers to these questions are in the affirmative, can one pin point why the standard 'bottom-up' perturbative approach fails in the gravitational case? That is, what is the essential feature which makes the fundamental description mathematically coherent but is absent in the standard perturbative quantum gravity?

There are of course many other challenges as well. Here are a few examples. A primary goal of physics is to predict the future from the past. But if there is no space-time in the background, what does time-evolution even mean? How does one extend the measurement theory and the associated interpretation of the quantum framework when space-time geometry is itself a part of the quantum system? On a more technical level, how does one construct gauge (i.e. diffeomorphism) invariant quantum observables and introduce practical methods of computing their properties? Are there manageable ways of computing S-matrices? Of exploring the role of topology and the phenomenon of topology change? Should the structure of quantum mechanics itself be modified, e.g., through a gravity induced non-linearity? The list continues.

Every approach sets its own priorities as to which of these are more central than the others and several of these questions are discussed in articles by Banks, Dowker, Gambini and Pullin and Penrose. In loop quantum gravity described in this chapter, one adopts the view that one should first tackle squarely the three issues discussed in some detail above and then explore other questions. Indeed, these three issues are rooted in deep conceptual challenges at the interface of general relativity and quantum theory and all three have been with us longer than any of the current leading approaches.

This chapter organized as follows. Section 2 summarizes the main features of loop quantum gravity. This framework has led to a rich set of results on the first two sets of physical issues discussed above.[b] Section 3 discusses applications to black holes and complements Bojowald's review of applications to cosmology. Section 4 presents a summary and an outlook.

2. A Bird's Eye View of Loop Quantum Gravity

I will now briefly summarize the salient features and current status of the framework underlying loop quantum gravity. The emphasis is on structural and conceptual issues; detailed treatments can be found in more complete and more technical recent accounts[24,25,38] and references therein. The development of the subject can be seen by following older monographs[11,13,15] and reviews on geometrodynamics.[1-6]

2.1. Viewpoint

The central lesson of general relativity is that gravity is encoded in space-time geometry. It is this feature that is directly responsible for the most spectacular predictions of the theory discussed in the second part of this book: black holes, big bang and gravitational waves. But it also leads to its most severe limitations. Inside black holes and at the big-bang, not only do matter fields become singular, but so does geometry. Space-time simply ends. Physics of general relativity comes to an abrupt halt. *The key idea at the heart of loop quantum gravity is to retain the interplay between geometry and gravity but overcome the limitations of general relativity by replacing classical Riemannian geometry by its suitable quantum analog.* Thus, as in general relativity, there is no background metric, no passive arena on which quantum dynamics of matter is to unfold. Quantum geometry is just as physical and dynamical as quantum matter. This viewpoint is in striking contrast to approaches developed by particle physicists where one typically begins with quantum matter on a classical background geometry and uses perturbation theory to incorporate quantum effects of gravity. In

---

[b]A summary of the status of semi-classical issues can be found in some recent reviews[24,38]. Also, I will discuss spin-foams — the path integral partner of the canonical approach discussed here — only in passing. This program[20,25] has led to fascinating insights on a number of mathematical physics issues — especially the relation between quantum gravity and state sum models — and is better suited to the analysis of global issues such as topology change. However, it is yet to shed new light on conceptual and physical issues discussed above.

loop quantum gravity, there *is* a background manifold[c] but no background fields whatsoever.

In the classical domain, general relativity stands out as the best available theory of gravity, some of whose predictions have been tested to an amazing degree of accuracy, surpassing even the legendary tests of quantum electrodynamics. Therefore, it is natural to ask: *Does quantum general relativity, coupled to suitable matter or supergravity, its supersymmetric generalization, exist as consistent theories non-perturbatively?* Although its underlying quantum geometry is not rigidly tied to general relativity, much of the effort in loop quantum gravity is devoted to answering these questions.

In the particle physics circles the answer is often assumed to be in the negative, not because there is concrete evidence against non-perturbative quantum gravity, but because of the analogy to the theory of weak interactions. There, one first had a 4-point interaction model due to Fermi which works quite well at low energies but which fails to be renormalizable. Progress occurred not by looking for non-perturbative formulations of the Fermi model but by replacing the model by the Glashow-Salam-Weinberg renormalizable theory of electro-weak interactions, in which the 4-point interaction is replaced by $W^{\pm}$ and Z propagators. Therefore, it is often assumed that perturbative non-renormalizability of quantum general relativity points in a similar direction. However this argument overlooks the fact that general relativity is qualitatively different. Perturbative treatments pre-suppose that the underlying space-time can be assumed to be a continuum *at all scales* of interest to physics under consideration. This assumption is safe for weak interactions. In the gravitational case, on the other hand, the scale of interest is *the Planck length* ℓ_{Pl} and there is no physical basis to pre-suppose that the continuum picture should be valid down to that scale. The failure of the standard perturbative treatments may largely be due to this grossly incorrect assumption and a non-perturbative treatment which correctly incorporates the physical micro-structure of geometry may well be free of these inconsistencies.

Are there any situations, outside loop quantum gravity, where such expectations are borne out in detail mathematically? The answer is in the affirmative. There exist quantum field theories (such as the Gross-Neveu model in three dimensions) in which the standard perturbation expan-

---

[c]In 2+1 dimensions, although one begins in a completely analogous fashion, in the final picture one can get rid of the background manifold as well. Thus, the fundamental theory can be formulated combinatorially.[11,12] To achieve this in 3+1 dimensions, one needs more complete theory of intersecting knots in 3 dimensions.

sion is not renormalizable, although the theory is *exactly soluble!* Failure
of the standard perturbation expansion can occur because one insists on
perturbing around the trivial, Gaussian point rather than the more physical, non-trivial fixed point of the renormalization group flow. Interestingly,
thanks to recent work by Lauscher, Reuter, Percacci, Perini and others,
there is now non-trivial and growing evidence that situation may be similar in Euclidean quantum gravity. Impressive calculations have shown that
pure Einstein theory may also admit a non-trivial fixed point. Furthermore,
the requirement that the fixed point should continue to exist in presence of
matter constrains the couplings in non-trivial and interesting ways.[36] It is
therefore of considerable interest to follow up on these indications.

Finally, recall that in classical general relativity, while requirements of
background independence and general covariance do restrict the form of interactions between gravity and matter fields and among matter fields themselves. The situation is the same in loop quantum gravity; so far it does
not have a built-in principle which *determines* these interactions. Consequently, in its present form, it is not a satisfactory candidate for unification
of all known forces. Rather, the first goal of loop quantum gravity is to
construct a consistent, non-perturbative theory by elevating the interplay
between geometry and gravity to the quantum level. Since this interplay
has had profound implications in the classical domain, it is reasonable to
hope that quantum general relativity will also have qualitatively new predictions, pushing further the existing frontiers of physics. Section 3 and
Bojowald's article provide considerable support for this hope.

2.2. *Quantum Geometry*

Although loop quantum gravity does not provide a natural unification of
dynamics of all interactions, it does provide a kinematical unification. More
precisely, in this approach one begins by formulating general relativity in
the mathematical language of connections, the basic variables of gauge theories of electro-weak and strong interactions. Thus, now the configuration
variables are not metrics as in Wheeler's geometrodynamics[2,3,4,5,6] but certain *spin-connections*; the emphasis is shifted from distances and geodesics
to holonomies and Wilson loops.[11,15] Consequently, the basic kinematical
structures are the same as those used in gauge theories. A key difference,
however, is that while a background space-time metric is available and crucially used in gauge theories, there are no background fields whatsoever
now. Their absence is forced upon us by the requirement of general covariance (more precisely, diffeomorphism invariance).

Now, most of the techniques used in the familiar, Minkowskian quantum theories are deeply rooted in the availability of a flat back-ground metric. In particular, it is this structure that enables one to single out the vacuum state, perform Fourier transforms to decompose fields canonically into creation and annihilation parts, define masses and spins of particles and carry out regularizations of products of operators. Already when one passes to quantum field theory in *curved* space-times, extra work is needed to construct mathematical structures that can adequately capture underlying physics (see Ford's article in this volume). In our case, the situation is much more drastic[12]: there is no background metric whatsoever! Therefore new physical ideas and mathematical tools are now necessary. Fortunately, they were constructed by a number of researchers in the mid-nineties and have given rise to a detailed quantum theory of Riemannian geometry.[24,25,38]

Because the situation is conceptually so novel and because there are no direct experiments to guide us, reliable results require a high degree of mathematical precision to ensure that there are no hidden infinities. Achieving this precision has been a priority in the program. Thus, while one is inevitably motivated by heuristic, physical ideas and formal manipulations, generally the final results are mathematically rigorous. In particular, due care is taken in constructing function spaces, defining measures and functional integrals, regularizing products of field operators, and calculating eigenvectors and eigenvalues of geometric operators. Consequently, the final results are all free of divergences, well-defined, and respect the background independence (diffeomorphism invariance).

Let us now turn to specifics. For simplicity, I will focus on the gravitational field; matter couplings are discussed in references [11,13,25,38]. The basic gravitational configuration variable is an SU(2)-connection, A_a^i on a 3-manifold M representing 'space'. As in gauge theories, the momenta are the 'electric fields' E_i^a.[d] However, in the present gravitational context, they also acquire a *space-time* meaning: they can be naturally interpreted as orthonormal triads (with density weight 1) and determine the dynamical, Riemannian geometry of M. Thus, in contrast to Wheeler's geometrodynamics[2,3,4], the Riemannian structures, including the positive-definite metric on M, is now built from *momentum* variables.

The basic kinematic objects are: i) holonomies $h_e(A)$ of A_a^i, which dictate how spinors are parallel transported along curves or edges e; and

---

[d]Throughout, indices $a, b, ..$ will refer to the tangent space of M while the 'internal' indices $i, j, ...$ will refer to the Lie algebra of SU(2).

ii) fluxes $E_{S,t} = \int_S t_i E_i^a d^2 S_a$ of electric fields, E_i^a (smeared with test fields t_i) across a 2-surface S. The holonomies —the raison d'être of connections— serve as the 'elementary' configuration variables which are to have unambiguous quantum analogs.

The first step in quantization is to use the Poisson algebra between these configuration and momentum functions to construct an abstract (star-) algebra $\mathcal{A}$ of elementary quantum operators. This step is straightforward. The second step is to introduce a representation of this algebra by 'concrete' operators on a Hilbert space (which is to serve as the kinematic setup for the Dirac quantization program).[6,9,11] For systems with an infinite number of degrees of freedom, this step is highly non-trivial. In Minkowskian field theories, for example, the analogous kinematic algebra of canonical commutation relations admits infinitely many *inequivalent* representations even after asking for Poicaré invariance! The standard Fock representation is uniquely selected *only* when a restriction to non-interacting theories is made. The general viewpoint is that the choice of representation is dictated by (symmetries and more importantly) the dynamics of the theory under consideration. A priori this task seems daunting for general relativity. However, it turns out that the diffeomorphism invariance —dictated by 'background independence'— is enormously more powerful than Poincaré invariance. Recent results by Lewandowski, Okolow, Sahlmann and Thiemann, and by Fleischhack show that *the algebra $\mathcal{A}$ admits a unique diffeomorphism invariant state!*[29,24,30] Using it, through a standard procedure due to Gel'fand, Naimark and Segal, one can construct a unique representation of $\mathcal{A}$. Thus, remarkably, there is a unique kinematic framework for *any* diffeomorphism invariant quantum theory for which the appropriate point of departure is provided by $\mathcal{A}$, *irrespective of the details of dynamics!* This tightness adds a considerable degree of confidence in the basic framework.

Chronologically, this concrete representation was in fact introduced already in early nineties by Ashtekar, Baez, Isham and Lewandowski. It led to the detailed theory of quantum geometry that underlies loop quantum gravity. Once a rich set of results had accumulated, researchers began to analyze the issue of uniqueness of this representation and systematic improvements over several years culminated in the simple statement given above.

Let me describe the salient features of this representation.[24,38] Quantum states span a specific Hilbert space $\mathcal{H}$ consisting of wave functions of connections which are square integrable with respect to a natural, diffeomorphism invariant measure. This space is very large. However, it can be conveniently decomposed into a family of orthogonal, *finite* dimensional sub-spaces $\mathcal{H} = \oplus_{\gamma,\vec{j}} \mathcal{H}_{\gamma,\vec{j}}$, labelled by graphs γ, each edge of which itself is labelled by a spin (i.e., half-integer) j.[24,25] (The vector $\vec{j}$ stands for the collection of half-integers associated with all edges of γ.) One can think of γ as a 'floating lattice' in M —'floating' because its edges are arbitrary, rather than 'rectangular'. (Indeed, since there is no background metric on M, a rectangular lattice has no invariant meaning.) Mathematically, $\mathcal{H}_{\gamma,\vec{j}}$ can be regarded as the Hilbert space of a spin-system. These spaces are extremely simple to work with; this is why very explicit calculations are feasible. Elements of $\mathcal{H}_{\gamma,\vec{j}}$ are referred to as *spin-network states*.[17,24,25]

In the quantum theory, the fundamental excitations of geometry are most conveniently expressed in terms of holonomies.[24] They are thus *one-dimensional, polymer-like* and, in analogy with gauge theories, can be thought of as 'flux lines' of electric fields/triads. More precisely, they turn out to be *flux lines of area*, the simplest gauge invariant quantities constructed from the momenta E_i^a: an elementary flux line deposits a quantum of area on any 2-surface S it intersects. Thus, if quantum geometry were to be excited along just a few flux lines, most surfaces would have zero area and the quantum state would not at all resemble a classical geometry. This state would be analogous, in Maxwell theory, to a 'genuinely quantum mechanical state' with just a few photons. In the Maxwell case, one must superpose photons coherently to obtain a semi-classical state that can be approximated by a classical electromagnetic field. Similarly, here, semi-classical geometries can result only if a huge number of these elementary excitations are superposed in suitable dense configurations.[24,38] The state of quantum geometry around you, for example, must have so many elementary excitations that approximately $\sim 10^{68}$ of them intersect the sheet of paper you are reading. Even in such semi-classical states, the true microscopic geometry is still distributional, concentrated on the underlying elementary flux lines. But the highest energy contemporary particle accelerators can only probe distances of the order of 10^{-18} cm. Since the microstructure becomes manifest only at the Planck scale 10^{-33} cm, even these accelerators can only see a coarse-grained geometry which can be approximated by a smooth metric. This explains why the continuum picture we use in physics today works so well. However, it is only an approximation that arises from coarse graining of semi-classical states.

The basic quantum operators are the holonomies $\hat{h}_e$ along curves or edges e in M and the fluxes $\hat{E}_{S,t}$ of triads $\hat{E}_i^a$. Both are self-adjoint on

$\mathcal{H}$. Furthermore detailed work by Ashtekar, Lewandowski, Rovelli, Smolin, Thiemann and others shows that *all eigenvalues of geometric operators constructed from the fluxes of triad are discrete.*[17,24,25,38] This key property is, in essence, the origin of the fundamental discreteness of quantum geometry. For, just as the classical Riemannian geometry of M is determined by the triads E_i^a, all Riemannian geometry operators —such as the area operator $\hat{A}_S$ associated with a 2-surface S or the volume operator $\hat{V}_R$ associated with a region R— are constructed from $\hat{E}_{S,t}$. However, since even the classical quantities A_S and V_R are non-polynomial functionals of triads, the construction of the corresponding $\hat{A}_S$ and $\hat{V}_R$ is quite subtle and requires a great deal of care. But their final expressions are rather simple.[24,25,38]

In this regularization, the underlying background independence turns out to be a blessing. For, diffeomorphism invariance constrains the possible forms of the final expressions *severely* and the detailed calculations then serve essentially to fix numerical coefficients and other details. Let me illustrate this point with the example of the area operators $\hat{A}_S$. Since they are associated with 2-surfaces S while the states are 1-dimensional excitations, the diffeomorphism covariance requires that the action of $\hat{A}_S$ on a state $\Psi_{\gamma,\vec{j}}$ must be concentrated at the intersections of S with γ. The detailed expression bears out this expectation: the action of $\hat{A}_S$ on $\Psi_{\gamma,\vec{j}}$ is dictated simply by the spin labels j_I attached to those edges of γ which intersect S. For all surfaces S and 3-dimensional regions R in M, $\hat{A}_S$ and $\hat{V}_R$ are densely defined, self-adjoint operators. *All their eigenvalues are discrete.* Naively, one might expect that the eigenvalues would be uniformly spaced given by, e.g., integral multiples of the Planck area or volume. Indeed, for area, such assumptions were routinely made in the initial investigations of the origin of black hole entropy and, for space-time volume, they are made in quantum gravity approaches based on causal sets described in Dowker's chapter. In quantum Riemannian geometry, this expectation is *not* borne out; the distribution of eigenvalues is quite subtle. In particular, the eigenvalues crowd rapidly as areas and volumes increase. In the case of area operators, the complete spectrum is known in a *closed form*, and the first several hundred eigenvalues have been explicitly computed numerically. For a large eigenvalue a_n, the separation $\Delta a_n = a_{n+1} - a_n$ between consecutive eigenvalues decreases exponentially: $\Delta a_n \leq \exp{-(\sqrt{a_n}/\ell_{\mathrm{Pl}})}\,\ell_{\mathrm{Pl}}^2$! Because of such strong crowding, the continuum approximation becomes excellent quite rapidly just a few orders of magnitude above the Planck scale. At the Planck scale, however, there is a precise and very specific replacement. This is the arena of quantum geometry. The premise is that the standard perturbation theory fails because it ignores this fundamental discreteness.

There is however a further subtlety. This non-perturbative quantization has a one parameter family of ambiguities labelled by $\gamma > 0$. This γ is called the Barbero-Immirzi parameter and is rather similar to the well-known θ-parameter of QCD.[24,25,38] In QCD, a single classical theory gives rise to inequivalent sectors of quantum theory, labelled by θ. Similarly, γ is classically irrelevant but different values of γ correspond to unitarily inequivalent representations of the algebra of geometric operators. The overall mathematical structure of all these sectors is very similar; the only difference is that the eigenvalues of all geometric operators scale with γ. For example, the simplest eigenvalues of the area operator $\hat{A}_S$ in the γ quantum sector is given by[e]

$$a_{\{j\}} = 8\pi\gamma\ell_{\mathrm{Pl}}^2 \sum_I \sqrt{j_I(j_I + 1)} \tag{1}$$

where $\{j\}$ is a collection of $1/2$-integers j_I, with $I = 1, \ldots N$ for some N. Since the representations are unitarily inequivalent, as usual, one must rely on Nature to resolve this ambiguity: Just as Nature must select a specific value of θ in QCD, it must select a specific value of γ in loop quantum gravity. With one judicious experiment —e.g., measurement of the lowest eigenvalue of the area operator $\hat{A}_S$ for a 2-surface S of any given topology— we could determine the value of γ and fix the theory. Unfortunately, such experiments are hard to perform! However, we will see in Section 3.2 that the Bekenstein-Hawking formula of black hole entropy provides an indirect measurement of this lowest eigenvalue of area for the 2-sphere topology and can therefore be used to fix the value of γ.

2.3. *Quantum dynamics*

Quantum geometry provides a mathematical arena to formulate non-perturbative dynamics of a class of candidate quantum theories of gravity,

---

[e]In particular, the lowest non-zero eigenvalue of area operators is proportional to γ. This fact has led to a misunderstanding in certain particle physics circles where γ is thought of as a regulator responsible for discreteness of quantum geometry. As explained above, this is *not* the case; γ is analogous to the QCD θ and quantum geometry is discrete in *every* permissible γ-sector. Note also that, at the classical level, the theory is equivalent to general relativity only if γ is *positive*; if one sets $\gamma = 0$ by hand, one cannot recover even the kinematics of general relativity. Similarly, at the quantum level, setting $\gamma = 0$ would lead to a meaningless theory in which *all* eigenvalues of geometric operators vanish identically.

without any reference to a background classical geometry. It is not rigidly tied to general relativity. However, for reasons explained in section 2.1, so far most of the detailed work on quantum dynamics is restricted to general relativity, where it provides tools to write down quantum Einstein's equations in the Hamiltonian approach and calculate transition amplitudes in the path integral approach. Until recently, effort was focussed primarily on Hamiltonian methods. However, over the last five years or so, path integrals —called *spin foams*— have drawn a great deal of attention. This work has led to fascinating results suggesting that, thanks to the fundamental discreteness of quantum geometry, path integrals defining quantum general relativity may be finite. A summary of these developments can be found in references[20,25]. In this Section, I will summarize the status of the Hamiltonian approach. For brevity, I will focus on source-free general relativity, although there has been considerable work also on matter couplings.[24,25,38]

For simplicity, let me suppose that the 'spatial' 3-manifold M is compact. Then, in any theory without background fields, Hamiltonian dynamics is governed by constraints. Roughly this is because in these theories diffeomorphisms correspond to gauge in the sense of Dirac. Recall that, on the Maxwell phase space, gauge transformations are generated by the functional $\mathcal{D}_a E^a$ which is constrained to vanish on physical states due to Gauss law. Similarly, on phase spaces of background independent theories, diffeomorphisms are generated by Hamiltonians which are constrained to vanish on physical states.

In the case of general relativity, there are three sets of constraints. The first set consists of the three Gauss equations

$$\mathcal{G}_i := \mathcal{D}_a E^a_i = 0, \tag{2}$$

which, as in Yang-Mills theories, generates internal SU(2) rotations on the connection and the triad fields. The second set consists of a co-vector (or diffeomorphism) constraint

$$\mathcal{C}_b := E^a_i F^i_{ab} = 0, \tag{3}$$

which generates spatial diffeomorphism on M (modulo internal rotations generated by $\mathcal{G}_i$). Finally, there is the key scalar (or Hamiltonian) constraint

$$\mathcal{S} := \epsilon^{ijk} E^a_i E^b_j F_{ab\,k} + \ldots = 0 \tag{4}$$

which generates time-evolutions. (The ... are extrinsic curvature terms, expressible as Poisson brackets of the connection, the total volume constructed from triads and the first term in the expression of $\mathcal{S}$ given above. We will not need their explicit forms.) Our task in quantum theory is three-folds: i) Elevate these constraints (or their 'exponentiated versions') to well-defined operators on the kinematical Hilbert space $\mathcal{H}$; ii) Select physical states by asking that they be annihilated by these constraints; iii) introduce an inner-product on the space of solutions to obtain the final Hilbert space $\mathcal{H}_{final}$, isolate interesting observables on $\mathcal{H}_{final}$ and develop approximation schemes, truncations, etc to explore physical consequences. I would like to emphasize that, even if one begins with Einstein's equations at the classical level, non-perturbative dynamics gives rise to interesting quantum corrections. Consequently, *the effective classical equations derived from the quantum theory exhibit significant departures from classical Einstein's equations.* This fact has had important implications in quantum cosmology.

How has loop quantum gravity fared with respect to these tasks? Since the canonical transformations generated by the Gauss and the diffeomorphism constraints have a simple geometrical meaning, it has been possible to complete the three steps. For the Hamiltonian constraint, on the other hand, there are no such guiding principles whence the procedure is more involved. In particular, specific regularization choices have to be made and the final expression of the Hamiltonian constraint is not unique. A systematic discussion of ambiguities can be found in reference [24]. At the present stage of the program, such ambiguities are inevitable; one has to consider all viable candidates and analyze if they lead to sensible theories. A *key open problem* in loop quantum gravity is to show that the Hamiltonian constraint —either Thiemann's or an alternative such as the one of Gambini and Pullin— admits a 'sufficient number' of semi-classical states. Progress on this problem has been slow because the general issue of semi-classical limits is itself difficult in *any* background independent approach.[f] However, a systematic understanding has now begun to emerge and is providing the necessary 'infra-structure'.[24,38] Recent advance in quantum cosmology, described in Section 3.2, is an example of progress in this direction. The symmetry reduction simplifies the theory sufficiently so that most of the ambiguities in the definition of the Hamiltonian constraint disappear and the remaining can be removed using physical arguments. The resulting theory has rich physical consequences. The interplay between the full theory and models obtained by symmetry reduction is now providing crucial inputs to cosmology from the full theory *and* useful lessons for the full theory from cosmology.

---

[f]In the dynamical triangulation[18,27] and causal set[22] approaches, for example, a great deal of care is required to ensure that even the dimension of a typical space-time is 4.

To summarize, from the mathematical physics perspective, in the Hamiltonian approach the crux of dynamics lies in quantum constraints. The quantum Gauss and diffeomorphism constraints have been solved satisfactorily and detailed regularization schemes have been proposed for the Hamiltonian constraint. This progress is notable; for example, the analogous tasks were spelled out in geometrodynamics[2,3,4] some 35 years ago but still remain unfulfilled. In spite of this technical success, however, it is not clear if any of the proposed strategies to solve the Hamiltonian constraint incorporates the familiar low energy physics in the *full* theory, i.e., beyond symmetry reduced models. Novel ideas are being pursued to address this issue. I will list them in section 4.

3. Applications of Quantum Geometry

In this section, I will summarize two developments that answer several of the questions raised under first two bullets in section 2.1. The first application is to black holes and the second to cosmology. The two are complementary. In the discussion of black holes, one considers full theory but the main issue of interest —analysis of black hole entropy from statistical mechanical considerations— is not sensitive to the details of how the Hamiltonian constraint is solved. In quantum cosmology, on the other hand, one considers only a symmetry reduced model but the focus is on the Hamiltonian constraint which dictates quantum dynamics. Thus, as in all other approaches to quantum gravity, concrete advances can be made because there exist physically interesting problems which can be addressed without having a complete solution to the full theory.

3.1. *Black-holes*

This discussion is based on work of Ashtekar, Baez, Corichi, Domagala, Engle, Krasnov, Lewandowski, Meissner and Van den Broeck, much of which was motivated by earlier work of Krasnov, Rovelli, Smolin and others.[24,31,32]

As explained in the Introduction, since mid-seventies, a key question in the subject has been: What is the statistical mechanical origin of the entropy $S_{BH} = (a_{hor}/4\ell_{Pl}^2)$ of large black holes? What are the microscopic degrees of freedom that account for this entropy? This relation implies that a solar mass black hole must have $(\exp 10^{77})$ quantum states, a number that is *huge* even by the standards of statistical mechanics. Where do all these states reside? To answer these questions, in the early nineties Wheeler had suggested the following heuristic picture, which he christened 'It from

Bit'. Divide the black hole horizon into elementary cells, each with one Planck unit of area, ℓ_{Pl}^2 and assign to each cell two microstates, or one 'bit'. Then the total number of states $\mathcal{N}$ is given by $\mathcal{N} = 2^n$ where $n = (a_{\text{hor}}/\ell_{\text{Pl}}^2)$ is the number of elementary cells, whence entropy is given by $S = \ln \mathcal{N} \sim a_{\text{hor}}$. Thus, apart from a numerical coefficient, the entropy ('It') is accounted for by assigning two states ('Bit') to each elementary cell. This qualitative picture is simple and attractive. But can these heuristic ideas be supported by a systematic analysis from first principles? Quantum geometry has supplied such an analysis. As one would expect, while some qualitative features of this picture are borne out, the actual situation is far more subtle.

A systematic approach requires that we first specify the class of black holes of interest. Since the entropy formula is expected to hold unambiguously for black holes in equilibrium, most analyses were confined to *stationary*, eternal black holes (i.e., in 4-dimensional general relativity, to the Kerr-Newman family). From a physical viewpoint however, this assumption seems overly restrictive. After all, in statistical mechanical calculations of entropy of ordinary systems, one only has to assume that the given system is in equilibrium, not the whole world. Therefore, it should suffice for us to assume that the black hole itself is in equilibrium; the exterior geometry should not be forced to be time-independent. Furthermore, the analysis should also account for entropy of black holes whose space-time geometry cannot be described by the Kerr-Newman family. These include astrophysical black holes which are distorted by external matter rings or 'hairy' black holes of mathematical physics with non-Abelian gauge fields for which the uniqueness theorems fail. Finally, it has been known since the mid-seventies that the thermodynamical considerations apply not only to black holes but also to cosmological horizons. A natural question is: Can these diverse situations be treated in a single stroke?

Within the quantum geometry approach, the answer is in the affirmative. The entropy calculations have been carried out in the 'isolated horizons' framework which encompasses all these situations. Isolated horizons serve as 'internal boundaries' whose intrinsic geometries (and matter fields) are time-independent, although the geometry as well as matter fields in the external space-time region can be fully dynamical. The zeroth and first laws of black hole mechanics have been extended to isolated horizons.[28] Entropy associated with an isolated horizon refers to the family of observers in the exterior, for whom the isolated horizon is a physical boundary that separates the region which is accessible to them from the one which is not.

This point is especially important for cosmological horizons where, without reference to observers, one can not even define horizons. States which contribute to this entropy are the ones which can interact with the states in the exterior; in this sense, they 'reside' on the horizon.

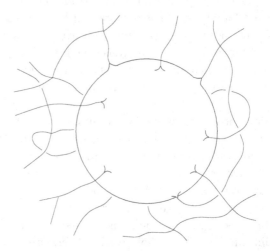

Fig. 1.   Quantum Horizon. Polymer excitations in the bulk puncture the horizon, endowing it with quantized area. Intrinsically, the horizon is flat except at punctures where it acquires a quantized deficit angle. These angles add up to endow the horizon with a 2-sphere topology.

In the detailed analysis, one considers space-times admitting an isolated horizon as inner boundary and carries out a systematic quantization. The quantum geometry framework can be naturally extended to this case. The isolated horizon boundary conditions imply that the intrinsic geometry of the quantum horizon is described by the so called U(1) Chern-Simons theory on the horizon. This is a well-developed, topological field theory. A deeply satisfying feature of the analysis is that there is a seamless matching of three otherwise independent structures: the isolated horizon boundary conditions, the quantum geometry in the bulk, and the Chern-Simons theory on the horizon. In particular, one can calculate eigenvalues of certain physically interesting operators using purely bulk quantum geometry without any knowledge of the Chern-Simons theory, or using the Chern-Simons theory without any knowledge of the bulk quantum geometry. The two theories have never heard of each other. But the isolated horizon boundary conditions require that the two infinite sets of numbers match exactly. This is a highly non-trivial requirement. But the numbers do match, thereby providing a coherent description of the quantum horizon.[24]

In this description, the polymer excitations of the bulk geometry, each labelled by a spin j_I, pierce the horizon, endowing it an elementary area a_{j_I} given by (1). The sum $\sum_I a_{j_I}$ adds up to the total horizon area a_{hor}. The intrinsic geometry of the horizon is flat except at these punctures, but at each puncture there is a *quantized* deficit angle. These add up to endow the horizon with a 2-sphere topology. For a solar mass black hole, a typical horizon state would have 10^{77} punctures, each contributing a tiny deficit angle. So, although quantum geometry *is* distributional, it can be well approximated by a smooth metric.

The counting of states can be carried out as follows. First one constructs a micro-canonical ensemble by restricting oneself only to those states for which the mass and angular momentum multipole moments lie in small intervals around fixed values $M_{hor}^{(n)}, J_{hor}^{(n)}$. (As is usual in statistical mechanics, the leading contribution to the entropy is independent of the precise choice of these small intervals.) For each set of punctures, one can compute the dimension of the surface Hilbert space, consisting of Chern-Simons states compatible with that set. One allows all possible sets of punctures (by varying both the spin labels and the number of punctures) and adds up the dimensions of the corresponding *surface* Hilbert spaces to obtain the number $\mathcal{N}$ of permissible surface states. One finds that the horizon entropy S_{hor} is given by

$$S_{hor} := \ln \mathcal{N} = \frac{\gamma_o}{\gamma} \frac{a_{hor}}{4\ell_{Pl}^2} - \frac{1}{2} \ln(\frac{a_{hor}}{\ell_{Pl}^2}) + o \ln(\frac{a_{hor}}{\ell_{Pl}^2}) \tag{5}$$

where $\gamma_o \approx 0.2735$ is a root of an algebraic equation [32,31] and $o(x)$ denote quantities for which $o(x)/x$ tends to zero as x tends to infinity. Thus, for large black holes, the leading term is indeed proportional to the horizon area. This is a non-trivial result; for example, early calculations often led to proportionality to the square-root of the area.

However, even for large black holes, one obtains agreement with the Hawking-Bekenstein formula only in the sector of quantum geometry in which the Barbero-Immirzi parameter γ takes the value $\gamma = \gamma_o$. Thus, while all γ sectors are equivalent classically, the standard quantum field theory in curved space-times is recovered in the semi-classical theory only in the γ_o sector of quantum geometry. It is quite remarkable that thermodynamic considerations involving *large* black holes can be used to fix the quantization ambiguity which dictates such Planck scale properties as eigenvalues of geometric operators. Note however that the value of γ can be fixed by demanding agreement with the semi-classical result just in one case —e.g., a spherical horizon with zero charge, or a cosmological horizon in the de Sitter space-time, or, ... Once the value of γ is fixed, the theory is completely fixed and we can ask: Does this theory yield the Hawking-Bekenstein value of entropy of *all* isolated horizons, irrespective of the values of charges, angular momentum, and cosmological constant, the amount of distortion, or hair. The answer is in the affirmative. Thus, the agreement with quantum field theory in curved space-times holds in *all* these diverse cases.[24,33]

Why does γ_o not depend on other quantities such as charges? This important property can be traced back to a key consequence of the isolated horizon boundary conditions: detailed calculations show that only the gravitational part of Poisson brackets (more precisely, symplectic structure) has a surface term at the horizon; the matter Poisson brackets only have volume terms. (Furthermore, the gravitational surface term is insensitive to the value of the cosmological constant.) Since quantization is dictated by Poisson brackets, there are no independent surface quantum states associated with matter. This provides a natural explanation of the fact that the Hawking-Bekenstein entropy depends only on the horizon area and is independent of electro-magnetic (or other) charges.

So far, all matter fields were assumed to be minimally coupled to gravity (there was no restriction on their couplings to each other). If one allows non-minimal gravitational couplings, the isolated horizon framework (as well as other methods) show that entropy should depend not just on the area *but also on the values of non-minimally coupled matter fields at the horizon*. At first, this non-geometrical nature of entropy seems to be a major challenge to approaches based on quantum geometry. However it turns out that, in presence of non-minimal couplings, the geometrical orthonormal triads E_i^a are no longer functions just of the momenta conjugate to the gravitational connection A_a^i *but depend also on matter fields*. Thus quantum Riemannian geometry —including area operators— can no longer be analyzed just in the gravitational sector of the quantum theory. The dependence of the triads and area operators on matter fields is such that the counting of surface states leads precisely to the correct expression of entropy, again for the same value of the Barbero-Immirzi parameter γ. This is a subtle and non-trivial check on the robustness of the quantum geometry approach to the statistical mechanical calculation of black hole entropy.

Finally, let us return to Wheeler's 'It from Bit'. The horizon can indeed be divided into elementary cells. But they need not have the same area; the area of a cell can be $8\pi\gamma\ell_{\mathrm{Pl}}^2\sqrt{j(j+1)}$ where j is an *arbitrary* half-integer subject only to the requirement that $8\pi\gamma\ell_{\mathrm{Pl}}^2\sqrt{j(j+1)}$ does not exceed the total horizon area a_{hor}. Wheeler assigned to each elementary cell two bits. In the quantum geometry calculation, this corresponds to focussing just on $j = 1/2$ punctures. While the corresponding surface states are already sufficiently numerous to give entropy proportional to area, other states with higher j values also contribute to the leading term in the expression of entropy.[g]

To summarize, quantum geometry naturally provides the micro-states responsible for the huge entropy associated with horizons. In this analysis, all black hole and cosmological horizons are treated in an unified fashion; there is no restriction, e.g., to near-extremal black holes. The sub-leading term has also been calculated and shown to be $-\frac{1}{2}\ln(a_{\mathrm{hor}}/\ell_{\mathrm{Pl}}^2)$.[32] Finally, in this analysis quantum Einstein's equations *are* used. In particular, had we not imposed the quantum Gauss and diffeomorphism constraints on surface states, the spurious gauge degrees of freedom would have given an infinite entropy. However, detailed considerations show that, because of the isolated horizon boundary conditions, the Hamiltonian constraint has to be imposed just in the bulk. Since in the entropy calculation one traces over bulk states, the final result is insensitive to the details of how this (or any other bulk) equation is imposed. Thus, as in other approaches to black hole entropy, the calculation is feasible because it does not require a complete knowledge of quantum dynamics.

3.2. *Big bang*

Over the last five years, quantum geometry has led to some striking results in quantum cosmology. Most of these are summarized in Bojowald's chapter. My goal is to complement his discussion with some recent results which provide a firmer foundation to the subject, validating qualitative expectations based on effective, classical descriptions. For completeness, however, I must first summarize the salient features of the underlying framework which is discussed in greater detail in Bojowald's chapter.

Traditionally, in quantum cosmology one has proceeded by first imposing spatial symmetries — such as homogeneity and isotropy — to freeze out all but a finite number of degrees of freedom *already at the classical*

---

[g]These contributions are also conceptually important for certain physical considerations — e.g. to 'explain' why the black hole radiance does not have a purely line spectrum.

level and then quantizing the reduced system. In the simplest (i.e., homogeneous, isotropic) model, the basic variables of the reduced classical system are the scale factor a and matter fields ϕ. The symmetries imply that space-time curvature goes as $\sim 1/a^n$, where $n > 0$ depends on the matter field under consideration. Einstein's equations then predict a big-bang, where the scale factor goes to zero and the curvature blows up. As indicated in Section 1, this is reminiscent of what happens to ferro-magnets at the Curie temperature: magnetization goes to zero and the susceptibility diverges. By analogy, the key question is: Do these 'pathologies' disappear if we re-examine the situation in the context of an appropriate quantum theory? In traditional quantum cosmologies, without additional input, they do not. That is, typically, to resolve the singularity one either has to use matter (or external clocks) with unphysical properties or introduce additional boundary conditions, e.g., by invoking new principles, that dictate how the universe began.

In a series of seminal papers Bojowald has shown that the situation in loop quantum cosmology is quite different: the underlying quantum geometry makes a *qualitative* difference very near the big-bang.[26,24] At first, this seems puzzling because after symmetry reduction, the system has only a *finite* number of degrees of freedom. Thus, quantum cosmology is analogous to quantum mechanics rather than quantum field theory. How then can one obtain qualitatively new predictions? Ashtekar, Bojowald and Lewandowski — with key input from Fredenhagen — have clarified the situation: if one follows the program laid out in the full theory, then even for the symmetry reduced model one is led to an inequivalent quantum theory — a new quantum mechanics!

This is still puzzling because in quantum mechanics we have the von-Neumann uniqueness theorem for the representations of the Weyl algebra generated by $U(\lambda) = \exp i\lambda x$ and $V(\mu) = \exp i\mu p$. How can then new quantum mechanics arise? Recall first that the assumptions of the von-Neumann theorem: i) the representation of the algebra be irreducible; ii)$U(\lambda)$ and $V(\mu)$ be represented by 1-parameter groups of unitary operators; and iii) $U(\lambda)$ and $V(\lambda)$ be weakly continuous in the parameters λ and μ respectively. The last requirement is a necessary and sufficient condition ensuring that the two groups are generated by self-adjoint operators x and p. In the context of quantum mechanics, all three conditions are natural. Now, in full quantum geometry, holonomies are the analogs of $U(\lambda)$ and triads are the analogs of p. The (kinematical) Hilbert space $\mathcal{H}$ carries a representation of the algebra they generate. However, a key characteristic of this representation is that, while the holonomies are well-defined operators on $\mathcal{H}$, there are no operators corresponding to connections themselves. Since loop quantum cosmology mimics the structure of the full theory, one is now led to drop the analog of the requirement that $U(\lambda)$ be continuous in λ. With this weakening of assumptions, the von Neumann uniqueness result no longer holds: Inequivalent representations of the Weyl algebra emerge even in quantum mechanics.[h] This is why the representation used in loop quantum cosmology is *inequivalent* to that used in the older, traditional quantum cosmology. And in the new representation, quantum evolution is well-defined right through the big-bang singularity.

More precisely, the situation in dynamics can be summarized as follows. Because of the underlying symmetries, the Gauss and the diffeomorphism constraints can be eliminated by gauge fixing already in the classical theory. Therefore, dynamics is dictated just by the Hamiltonian constraint. Let us consider the simplest case of homogeneous, isotropic cosmologies coupled to a scalar field. In the traditional quantum cosmology, this constraint is the celebrated Wheeler-DeWitt equation[2,3] — a second order differential equation on wave functions $\Psi(a, \phi)$ that depend on the scale factor a and the scalar field ϕ. Unfortunately, some of the coefficients of this equations diverge at $a = 0$, making it impossible to obtain an unambiguous evolution across the singularity. In loop quantum cosmology, the scale factor naturally gets replaced by μ the momentum conjugate to the connection. μ ranges over the entire real line and is related to the scale factor via $|\mu| = (\text{const } a^2;$ negative μ correspond to right handed triads, positive to left handed, and $\mu = 0$ corresponds to the degenerate triad representing the singularity. Let us expand out the quantum state as $| \Psi >= \sum \Psi(\mu, \phi) | \mu \phi >$ Then, the Hamiltonian constraint takes the form:

$$C^+(\mu)\Psi(\mu + 4\mu_o, \phi) + C^o(\mu)\Psi(\mu\,\phi) + C^-(\mu)\Psi(\mu - 4\mu_o, \phi) = \ell_{\rm Pl}^2\,\hat{H}_\phi\Psi(\mu, \phi)$$
$$(6)$$

where $C^\pm(\mu), C^o(\mu)$ are fixed functions of μ; μ_o, a constant, determined by the lowest eigenvalue of the area operator and $\hat{H}_\phi$ is the matter Hamiltonian. Again, using the analog of the Thiemann regularization from the

---

[h]In the analog of the representation used in loop quantum cosmology, the Hilbert space is not $L^2(\mathbb{R}, dx)$ but $L^2(\bar{\mathbb{R}}_{\rm Bohr}, d\mu_o)$ where $\bar{\mathbb{R}}_{\rm Bohr}$ is the Bohr compactification of the real line and $d\mu_o$ the natural Haar measure thereon. (Here, Bohr refers to the mathematician Harold Bohr, Nils' brother who developed the theory of almost periodic functions.) Now, although $U(\lambda)$ is well-defined, there is no operator corresponding to x itself. The operator p on the other hand is well-defined.

full theory, one can show that the matter Hamiltonian is a well-defined operator.

Primarily, being a constraint equation, (6) restricts the physically permissible $\Psi(\mu\,\phi)$. However, *if* we choose to interpret μ as a heuristic time variable, (6) can be interpreted as an 'evolution equation' which evolves the state through discrete time steps. The highly non-trivial result is that the coefficients $C^{\pm}(\mu), C^o(\mu)$ are such that *one can evolve right through the classical singularity*, i.e., right through $\mu = 0$. Since *all* solutions have this property, the classical singularity is resolved. However, as in the full theory, to complete the quantization program, one has to introduce the appropriate scalar product on the space of solutions to the constraint, define physically interesting operators on the resulting Hilbert space $\mathcal{H}_{\text{final}}$ and examine their expectation values and fluctuations, especially near the singularity.

All these steps have been carried out in detail in the case when ϕ is a massless scalar field[39]. Specifically, in each classical solution, ϕ is a monotonic function of time. Therefore, one can regard it as an 'internal clock' with respect to which the scale factor evolves. With this interpretation, the discrete equation (6) takes the form $\partial_t^2 \Psi = -\Theta\Psi$, where Θ is a self-adjoint operator, independent of $\phi \sim t$. This is precisely the form of the Klein-Gordon equation in static space-times. (In technical terms, this provides a satisfactory 'deparametrization' of the theory.) Therefore, one can use techniques from quantum field theory in static space-times to construct an appropriate inner product and define a complete family of ('Dirac') observables. Using the two, one can construct semi-classical states —analogs of coherent states of a harmonic oscillator— and write down explicit expressions for expectation values and fluctuations of physical observables in them. As one might expect, the evolution is well-defined across the singularity but quantum fluctuations are huge in its neighborhood.

Now that there is a well-defined theory, one can use numerical methods to evolve quantum states and compare quantum dynamics with the classical one in detail. Since we do not want to make a priori assumptions about what the quantum state was at the big-bang, it is best to start the evolution not from the big bang but from late times ('now'). Consider then wave functions which are sharply peaked at a classical trajectory at late times and evolve them backward. The first question is: how long does the state remain semi-classical? A pleasant surprise is that it does so till *very early times* —essentially till the epoch when the matter density reaches the Planck density. Now, this is precisely what one would physically expect. However, with a complicated difference equation such as (6), a priori there

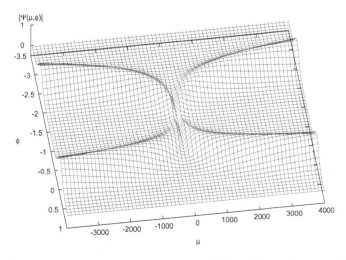

Fig. 2. Comparison between quantum and classical evolutions via plot of $|\Psi(\mu,\phi)|$. Since $\mu \to -\mu$ changes only the triad orientation, it suffices to consider just $\mu \geq 0$. Backward quantum evolution in the top half of the figure shows that Ψ follows the 'expanding branch' of the classical trajectory until it enters the Planck regime. Then quantum fluctuations become large and on the other side of the big-bang, the quantum state remerges as semi-classical, joining on to the 'contracting branch'. Thus, quantum geometry in the Planck regime bridges two vast semi-classical regions.

is no guarantee that semi-classicality would not be lost very quickly. In particular, this result provides support for the standard practice, e.g., in inflationary models, of assuming a classical continuum in the very early universe. Next, one can ask what happens to the quantum state very near and beyond the big-bang. As explained above, the state loses semi-classicality (i.e. fluctuations become large) near the big-bang. Does it then remain in a 'purely quantum regime' forever or does it again become semi-classical beyond a Planck regime on the 'other side' of the big bang? This is a question that lies entirely outside the domain of the standard Wheeler-Dewitt equation because it loses predictivity at the big-bang. In loop quantum cosmology, on the other hand, the evolution is well-defined and completely deterministic also beyond the big-bang. A priori there is no way to know what the answer would be. Space-time may well have been a 'quantum foam' till the big-bang and classicality may then have emerged only after the big-bang. Or, there may have been a classical space-time also on the 'other side'. Detailed numerical calculations show that the wave function becomes semi-classical again on the other side; *there is a 'bounce'*. Thus, loop quantum cosmology predicts that the universe did not originate at the big bang but has a long prior history. Through quantum dynamics, the universe tunnels from a contracting phase in the distant past ('before the bang') to an expanding phase in the distant future ('now') in a specific manner. Classically, of course such a transition is impossible.

To summarize, the infinities predicted by the classical theory at the big-bang are artifacts of assuming that the classical, continuum space-time approximation is valid right up to the big-bang. In the quantum theory, the state can be evolved through the big-bang without any difficulty. However, the classical, continuum completely fails near the big-bang; figuratively, the classical space-time 'dissolves'. This resolution of the singularity without any 'external' input (such as matter violating energy conditions) is dramatically different from what happens with the standard Wheeler-DeWitt equation of quantum geometrodynamics[2,3,4,5,6]. However, for large values of the scale factor, the two evolutions are close; as one would have hoped, quantum geometry effects intervene only in the 'deep Planck regime' resulting in a quantum bridge connecting two classically disconnected space-times. From this perspective, then, one is led to say that the most striking of the consequences of loop quantum gravity are not seen in standard quantum cosmology because it 'washes out' the fundamental discreteness of quantum geometry.

4. Summary and Outlook

From the historical and conceptual perspectives of section 1, loop quantum gravity has had several successes. Thanks to the systematic development of quantum geometry, several of the roadblocks encountered by quantum geometrodynamics[2,3,4,5,6] were removed. There is a framework to resolve the functional analytic issues related to the presence of an infinite number of degrees of freedom. Integrals on infinite dimensional spaces are rigorously defined and the required operators have been systematically constructed. Thanks to this high level of mathematical precision, the canonical quantization program has leaped past the 'formal' stage of development. More importantly, although some key issues related to quantum dynamics still remain, it has been possible to use the parts of the program that are already well established to extract useful and highly non-trivial physical predictions. In particular, some of the long standing issues about the nature of the big-bang and properties of quantum black holes have been resolved. In this section, I will further clarify some conceptual issues, discuss current research and outline some directions for future.

• *Quantum geometry.* From conceptual considerations, an important issue is the *physical* significance of discreteness of eigenvalues of geometric operators. Recall first that, in the classical theory, differential geometry simply provides us with formulas to compute areas of surfaces and volumes of regions in a Riemannian manifold. To turn these quantities into physical observables of general relativity, one has to define the surfaces and regions *operationally*, e.g. using matter fields. Once this is done, one can simply use the formulas supplied by differential geometry to calculate values of these observable. The situation is similar in quantum theory. For instance, the area of the isolated horizon is a Dirac observable in the classical theory and the application of the quantum geometry area formula to *this* surface leads to physical results. In 2+1 dimensions, Freidel, Noui and Perez have recently introduced point particles coupled to gravity.[34] The physical distance between these particles is again a Dirac observable. When used in this context, the spectrum of the length operator has direct physical meaning. In all these situations, the operators and their eigenvalues correspond to the 'proper' lengths, areas and volumes of physical objects, measured in the rest frames. Finally sometimes questions are raised about compatibility between discreteness of these eigenvalues and Lorentz invariance. As was recently emphasized by Rovelli, there is no tension whatsoever: it suffices to recall that discreteness of eigenvalues of the angular momentum operator $\hat{J}_z$ of non-relativistic quantum mechanics is perfectly compatible with the rotational invariance of that theory.

• *Quantum Einstein's equations.* The challenge of quantum dynamics in the full theory is to find solutions to the quantum constraint equations and endow these physical states with the structure of an appropriate Hilbert space. We saw in section 3.2 that this task can be carried to a satisfactory completion in symmetry reduced models of quantum cosmology. For the general theory, while the situation is well-understood for the Gauss and diffeomorphism constraints, it is far from being definitive for the Hamiltonian constraint. It *is* non-trivial that well-defined candidate operators representing the Hamiltonian constraint exist on the space of solutions to the Gauss and diffeomorphism constraints. However there are many ambiguities[24] and none of the candidate operators has been shown to lead to a 'sufficient number of' semi-classical states in 3+1 dimensions. A second important open issue is to find restrictions on matter fields and their couplings to gravity for which this non-perturbative quantization can be carried out to a satisfactory conclusion. As mentioned in section 2.1, the renormalization group approach has provided interesting hints. Specifically, Luscher and

Reuter have presented significant evidence for a non-trivial fixed point for pure gravity in 4 dimensions. When matter sources are included, it continues to exist only when the matter content and couplings are suitably restricted. For scalar fields, in particular, Percacci and Perini have found that polynomial couplings (beyond the quadratic term in the action) are ruled out, an intriguing result that may 'explain' the triviality of such theories in Minkowski space-times.[36] Are there similar constraints coming from loop quantum gravity? To address these core issues, at least four different avenues are being pursued: the Gambini-Pullin framework based on (Vassiliev) invariants of intersecting knots[15]; the spin-foam approach based on path integral methods[20,25]; and the discrete approach summarized in the chapter by Gambini and Pullin, and the 'master constraint program' pursued by Dittrich and Thiemann[23] and the related 'affine quantum gravity' approach of Klauder[21].

• *Quantum cosmology.* As we saw in section 3, loop quantum gravity has resolved some of the long-standing physical problems about the nature of the big-bang. In quantum cosmology, there is ongoing work by Ashtekar, Bojowald, Willis and others on obtaining 'effective field equations' which incorporate quantum corrections. Thanks to recent efforts by Pawlowski, Singh and Vandersloot, numerical loop quantum cosmology has now emerged as a new field. By a suitable combination of analytical and numerical methods, it is now feasible to analyze in detail a large class of homogeneous models with varying matter content. These models serve as a 'background' which can be perturbed. Since the dynamics of loop quantum cosmology is *deterministic* across the singularity, evolution of inhomogeneities can be studied in detail. This is in striking contrast with, say the pre-big-bang scenario or cyclic universes where new input is needed to connect the classical branches before and after the big-bang.

• *Quantum Black Holes.* As in other approaches to black hole entropy, concrete progress could be made because the analysis does not require detailed knowledge of how quantum dynamics is implemented in *full* quantum theory. Also, restriction to large black holes implies that the Hawking radiation is negligible, whence the black hole surface can be modelled by an isolated horizon. To incorporate back-reaction, one would have to extend the present analysis to *dynamical horizons.*[28] It is now known that, in the classical theory, the first law can be extended also to these time-dependent situations and the leading term in the expression of the entropy is again given by $a_{\mathrm{hor}}/4\ell_{\mathrm{Pl}}^2$. Hawking radiation will cause the horizon of a large black hole to shrink *very* slowly, whence it is reasonable to expect that the

Chern-Simons-type description of the quantum horizon geometry can be extended also to this case. The natural question then is: Can one describe in detail the black hole evaporation process and shed light on the issue of information loss?

The standard space-time diagram of the evaporating black hole is shown in figure 3. It is based on two ingredients: i) Hawking's original calculation of black hole radiance, in the framework of quantum field theory on a *fixed* background space-time; and ii) heuristics of back-reaction effects which suggest that the radius of the event horizon must shrink to zero. It is generally argued that the semi-classical process depicted in this figure should be reliable until the very late stages of evaporation when the black hole has shrunk to Planck size and quantum gravity effects become important. Since it takes a very long time for a large black hole to shrink to this size, one then argues that the quantum gravity effects during the last stages of evaporation will not be sufficient to restore the correlations that have been lost due to thermal radiation over such a long period. Thus there is loss of information. Intuitively, the lost information is 'absorbed' by the final singularity which serves as a new boundary to space-time.

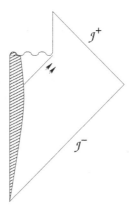

Fig. 3. The standard space-time diagram depicting black hole evaporation.

However, loop quantum gravity considerations suggest that this argument is incorrect in two respects. First, the semi-classical picture breaks down not just at the end point of evaporation but in fact *all along what is depicted as the final singularity*. Recently, using ideas from quantum cosmology, the interior of the Schwarzschild horizon was analyzed in the context of loop quantum gravity. Again, it was found that the singularity is resolved due to quantum geometry effects.[35] Thus, the space-time does *not* have a singularity as its final boundary. The second limitation of the semi-classical picture of figure 3 is its depiction of the event horizon. The notion of an event horizon is teleological and refers to the *global* structure of space-time. Resolution of the singularity introduces a domain in which there is no classical space-time, whence the notion ceases to be meaningful; it is simply 'transcended' in quantum theory. This leads to a new, possible paradigm for black hole evaporation in loop quantum gravity in which the dynamical horizons evaporate with emission of Hawking radiation, the initial pure state evolves to a final pure state and there is no information loss.[37] Furthermore, the semi-classical considerations are not simply dismissed; they turn out to be valid in certain space-time regions and under certain approximations. But for fundamental conceptual issues, they are simply inadequate. I should emphasize however that, although elements that go into the construction of this paradigm seem to be on firm footing, many details will have to be worked out before it can acquire the status of a model.

• *Semi-classical issues.* A frontier area of research is contact with low energy physics. Here, a number of fascinating challenges appear to be within reach. Fock states have been isolated in the polymer framework[24] and elements of quantum field theory on quantum geometry have been introduced.[38] These developments lead to concrete questions. For example, in quantum field theory in flat space-times, the Hamiltonian and other operators are regularized through normal ordering. For quantum field theory on quantum geometry, on the other hand, the Hamiltonians are expected to be manifestly finite.[38,24] Can one then show that, in a suitable approximation, normal ordered operators in the Minkowski continuum arise naturally from these finite operators? Can one 'explain' why the so-called Hadamard states of quantum field theory in curved space-times are special? These issues also provide valuable hints for the construction of viable semi-classical states of quantum geometry. The final and much more difficult challenge is to 'explain' why perturbative quantum general relativity fails if the theory exists non-perturbatively. As mentioned in section 1, heuristically the failure can be traced back to the insistence that the continuum space-time geometry is a good approximation even below the Planck scale. But a more detailed answer is needed. Is it because, as recent developments in Euclidean quantum gravity indicate[36], the renormalization group has a non-trivial fixed point?

• *Unification.* Finally, there is the issue of unification. At a kinematical level, there is already an unification because the quantum configuration space of general relativity is the same as in gauge theories which govern the strong and electro-weak interactions. But the non-trivial issue is that of dynamics. I will conclude with a speculation. One possibility is to use the 'emergent phenomena' scenario where new degrees of freedom or particles, which were not present in the initial Lagrangian, emerge when one considers excitations of a non-trivial vacuum. For example, one can begin with solids and arrive at phonons; start with superfluids and find rotons; consider superconductors and discover cooper pairs. In loop quantum gravity, the micro-state representing Minkowski space-time will have a highly non-trivial Planck-scale structure. The basic entities will be 1-dimensional and polymer-like. Even in absence of a detailed theory, one can tell that the fluctuations of these 1-dimensional entities will correspond not only to gravitons but also to other particles, including a spin-1 particle, a scalar and an anti-symmetric tensor. These 'emergent states' are likely to play an important role in Minkowskian physics derived from loop quantum gravity. A detailed study of these excitations may well lead to interesting dynamics that includes not only gravity but also a select family of non-gravitational fields. It may also serve as a bridge between loop quantum gravity and string theory. For, string theory has two a priori elements: unexcited strings which carry no quantum numbers and a background space-time. Loop quantum gravity suggests that both could arise from the quantum state of geometry, peaked at Minkowski (or, de Sitter) space. The polymer-like quantum threads which must be woven to create the classical ground state geometries could be interpreted as unexcited strings. Excitations of these strings, in turn, may provide interesting matter couplings for loop quantum gravity.

Acknowledgments

My understanding of quantum gravity has deepened through discussions with a large number of colleagues. Among them, I would especially like to thank John Baez, Martin Bojowald, Christian Fleischhack, Klaus Fredenhagen, Rodolfo Gambini, Jerzy Lewandowski, Don Marolf, Jose Mourão, Roger Penrose, Jorge Pullin, Carlo Rovelli, Hanno Sahlmann, Lee Smolin, and Thomas Thiemann. Tomasz Pawlowski and Param Singh kindly gave permission to quote some unpublished results. This work was supported in part by the NSF grant PHY 0090091, the Alexander von Humboldt Foundation, the Sir C.V. Raman Chair of the Indian Academy of Sciences and the Eberly research funds of The Pennsylvania State University.

References

1. Arnowitt R, Deser S and Misner C W 1962 The dynamics of general relativity, in *Gravitation: An introduction to current research* ed Witten L (John Wiley, New York)
2. Wheeler J A 1962 *Geometrodynamics*, (Academic Press, New York)
3. Wheeler J A 1964 Geometrodynamics and the issue of the final state *Relativity, Groupos and Topology* eds DeWitt C M and DeWitt B S (Gordon and Breach, New York)
4. Komar A 1970 Quantization program for general relativity, in *Relativity* Carmeli M, Fickler S. I. and Witten L (eds) (Plenum, New York)
5. Ashtekar A and Geroch R 1974 Quantum theory of gravitation, *Rep. Prog. Phys.* **37** 1211-1256
6. Isham C. J. 1975 An introduction to quantum gravity, in *Quantum Gravity, An Oxford Symposium* Isham C J, Penrose R and Sciama D W (Clarendon Press, Oxford)
7. Israel W and Hawking S W eds 1980 *General Relativity, An Einstein Centenary Survey* (Cambridge UP, Cambridge).
8. Bergmann P G and Komar A 1980 The phase space formulation of general relativity and approaches toward its canonical quantization *General Relativity and Gravitation vol 1, On Hundred Years after the Birth of Albert Einstein*, Held A ed (Plenum, New York)
9. Kuchař K 1981 Canonical methods of quantization, in *Quantum Gravity 2, A Second Oxford Symposium* Isham C J, Penrose R and Sciama D W (Clarendon Press, Oxford)
10. Isham C J 1981 Quantum gravity–An overview, in *Quantum Gravity 2, A Second Oxford Symposium* Isham C J, Penrose R and Sciama D W (Clarendon Press, Oxford)
11. Ashtekar A 1991 *Lectures on non-perturbative canonical gravity,* Notes prepared in collaboration with R. S. Tate (World Scientific, Singapore)
12. Ashtekar A Mathematical problems of non-perturbative quantum general relativity in *Gravitation and Quantizations: Proceedings of the 1992 Les Houches summer school* eds Julia B and Zinn-Justin J (Elsevier, Amsterdam); also available as gr-qc/9302024
13. Baez J and Muniain J P 1994 *Gauge fields, knots and gravity* (World Scientific, Singapore)
14. Wald R M 1994 *Quantum field theory in curved space-time and black hole thermodynamics* (Chicago UP, Chicago)
15. Gambini R and Pullin J 1996 *Loops, knots, gauge theories and quantum gravity* (Cambridge UP, Cambridge)
16. Carlip S 1998 *Quantum gravity in 2+1 dimensions* (Cambrige UP, Cambridge)
17. Rovelli C 1998 Loop quantum gravity *Living Rev. Rel.* **1** 1

18. Loll R 1998 Discrete approaches to quantum gravity in four dimensions *Living Rev. Rel.* **1** 13
19. Wald R M 2001 Black hole thermodynamics, *Living Rev. Rel.* **6**
20. Perez A 2003 Spin foam models for quantum gravity *Class. Quant. Grav.* **20** R43–R104
21. Klauder J 2003 Affine quantum gravity *Int. J. Mod. Phys.* **D12** 1769-1774
22. Sorkin R 2003 Causal sets: discrete gravity, gr-qc/0309009
23. Thiemann T 2003 The phoenix project: Master constraint programme for loop quantum gravity, gr-qc/0305080
24. Ashtekar A and Lewandowski L 2004 Background independent quantum gravity: A status report, *Class. Quant. Grav.* **21** R53–R152
25. Rovelli C 2004 *Quantum Gravity* (Cambridge University Press, Cambridge)
26. Bojowald M and Morales-Tecotl H A 2004 Cosmological applications of loop quantum gravity *Lect. Notes Phys.* **646** 421-462, also available at gr-qc/0306008
27. Ambjorn J, Jurkiewicz J and Loll R 2004 Emergence of a 4D world from causal quantum gravity hep-th/0404156
28. Ashtekar A and Krishnan B 2004 Isolated and Dynamical horizons and their properties, *Living Rev. Rel.* **7**, 10; gr-qc/0407042
29. Lewandowski J, Okołów A, Sahlmann H, Thiemann T 2004 Uniqueness of the diffeomorphism invariant state on the quantum holonomy-flux algebra, gr-qc/0504147
30. Fleischhack C 2004 Representations of the Weyl algebra in quantum geometry *Preprint*, math-ph/0407006
31. Domagala M and Lewandowski J 2004 Black hole entropy from Quantum Geometry *Class. Quant. Grav.* **21** 5233–5244
32. Meissner K A 2004 Black hole entropy in loop quantum gravity *Class. Quant. Grav.* **21** 5245–5252
33. Ashtekar A, Engle J and Van den Broeck C 2004 *Class. Quant. Grav.* **22**, L27–L34
34. Perez A and Noui K 2004 Three dimensional loop quantum gravity: coupling to point particles *Preprint* gr-qc/0402111
35. Ashtekar A and Bojowald M 2004 Non-Singular Quantum Geometry of the Schwarzschild Black Hole Interior *Preprint*
36. Perini D 2004 *The asymptotic safety scenario forgravity and matter* Ph.D. Dissertation, SISSA
37. Ashtekar A and Bojowald M 2005 Black hole evaporation: A paradigm, Class. Quant. Grav. **22** 3349–3362
38. Thiemann T 2005 *Introduction to modern canonical quantum general relativity* (Cambridge University Press, Cambridge); draft available as gr-qc/0110034
39. Ashtekar A, Pawlowski T and Singh P 2005, Quantum nature of the big bang: An analytical and numerical study, IGPG pre-print

CHAPTER 14

LOOP QUANTUM COSMOLOGY

MARTIN BOJOWALD

Max-Planck-Institute for Gravitational Physics
Albert-Einstein-Institute
Am Mühlenberg 1, 14476 Potsdam, Germany
mabo@aei.mpg.de

The expansion of our universe, when followed backward in time, implies that it emerged from a phase of huge density, the big bang. These stages are so extreme that classical general relativity combined with matter theories is not able to describe them properly, and one has to refer to quantum gravity. A complete quantization of gravity has not yet been developed, but there are many results about key properties to be expected. When applied to cosmology, a consistent picture of the early universe arises which is free of the classical pathologies and has implications for the generation of structure which are potentially observable in the near future.

1. Introduction

General relativity provides us with an extremely successful description of the structure of our universe on large scales, with many confirmations by macroscopic experiments and so far no conflict with observations. The resulting picture, when applied to early stages of cosmology, suggests that the universe had a beginning a finite time ago, at a point where space, matter, and also time itself were created. Thus, it does not even make sense to ask what was there before since "before" does not exist at all. At very early stages, space was small such that there were huge energy densities to be diluted in the later expansion of the universe that is still experienced today. In order to explain also the structure that we see in the form of galaxies in the correct statistical distribution, the universe not only needs to expand but do so in an accelerated manner, a so-called inflationary period, in its early stages. With this additional input, usually by introducing inflation with exponential acceleration[1,2,3] lasting long enough to expand the scale factor $a(t)$, the radius of the universe at a given time t, by a ratio $a_{\text{final}}/a_{\text{initial}} > e^{60}$. The resulting seeds for structure after the inflationary phase can be observed in the anisotropy spectrum of the cosmic microwave background (most recently of the WMAP satellite[4,5]), which agrees well with theoretical predictions over a large range of scales.

Nonetheless, there are problems remaining with the overall picture. The beginning was extremely violent with conditions such as diverging energy densities and tidal forces under which no theory can prevail. This is also true for general relativity itself which led to this conclusion in the first place: there are situations in the universe which, according to the singularity theorems, can be described only by solutions to general relativity which, under reasonable conditions on the form of matter, must have a singularity in the past or future.[6] There, space degenerates, e.g. to a single point in cosmology, and energy densities and tidal forces diverge. From the observed expansion of our current universe one can conclude that according to general relativity there must have been such a singularity in the past (which does not rule out further possible singularities in the future). This is exactly what is usually referred to as the "beginning" of the universe, but from the discussion it is clear that the singularity does not so much present a boundary to the universe as a boundary to the classical theory: The theory predicts conditions under which it has to break down and is thus incomplete. Here it is important that the singularity in fact lies only a finite time in the past rather than an infinite distance away, which could be acceptable. A definitive conclusion about a possible beginning can therefore be reached only if a more complete theory is found which is able to describe these very early stages meaningfully.

Physically, one can understand the inevitable presence of singularities in general relativity by the characteristic property of classical gravitation being always attractive. In the backward evolution in which the universe contracts, there is, once matter has collapsed to a certain size, simply no repulsive force strong enough to prevent the total collapse into a singularity. A similar behavior happens when not all the matter in the universe but only that in a given region collapses to a small size, leading to the formation of black holes which also are singular.

This is the main problem which has to be resolved before one can call our picture of the universe complete. Moreover, there are other problematic issues in what we described so far. Inflation has to be introduced into the picture, which currently is done by assuming a special field, the inflaton, in addition to the matter we know. In contrast to other matter, its properties must be very exotic so as to ensure accelerated expansion which with Einstein's equations is possible only if there is negative pressure. This is achieved by choosing a special potential and initial conditions for the inflaton, but there is no fundamental explanation of the nature of the inflaton and its properties. Finally, there are some details in the anisotropy spectrum which are hard to bring in agreement with theoretical models. In particular, there seems to be less structure on large scales than expected, referred to as a loss of power.

2. Classical Cosmology

In classical cosmology one usually assumes space to be homogeneous and isotropic, which is an excellent approximation on large scales today. The metric of space is then solely determined by the scale factor $a(t)$ which gives the size of the universe at any given time t. The function $a(t)$ describes the expansion or contraction of space in a way dictated by the Friedmann equation[7]

$$\left(\frac{\dot{a}}{a}\right)^2 = \frac{8\pi}{3}G\rho(a) \tag{1}$$

which is the reduction of Einstein's equations under the assumption of isotropy. In this equation, G is the gravitational constant and $\rho(a)$ the energy density of whatever matter we have in the universe. Once the matter content is chosen and $\rho(a)$ is known, one can solve the Friedmann equation in order to obtain $a(t)$.

As an example we consider the case of radiation which can be described phenomenologically by the energy density $\rho(a) \propto a^{-4}$. This is only a phenomenological description since it ignores the fundamental formulation of electrodynamics of the Maxwell field. Instead of using the Maxwell Hamiltonian in order to define the energy density, which would complicate the situation by introducing the electromagnetic fields with new field equations coupled to the Friedmann equation, one uses the fact that on large scales the energy density of radiation is diluted by the expansion and in addition red-shifted. This leads to a behavior proportional a^{-3} from dilution times a^{-1} from redshift. In this example we then solve the Friedmann equation $\dot{a} \propto a^{-1}$ by $a(t) \propto \sqrt{t - t_0}$ with a constant of integration t_0. This demonstrates the occurrence of singularities: For any solution there is a time $t = t_0$ where the size of space vanishes and the energy density $\rho(a(t_0))$ diverges. At this point not only the matter system becomes unphysical, but also the gravitational evolution breaks down: When the right hand side of (1) diverges at some time t_0, we cannot follow the evolution further by setting up an initial value problem there and integrating the equation. We can thus only learn that there is a singularity in the classical theory, but do not obtain any information as to what is happening there and beyond. These are the two related but not identical features of a singularity: energy densities diverge and the evolution breaks down.

One could think that the problem comes from too strong idealizations such as symmetry assumptions or the phenomenological description of matter. That this is not the case follows from the singularity theorems which do not depend on these assumptions. One can also illustrate the singularity problem with a field theoretic rather than phenomenological description of matter. For simplicity we now assume that matter is provided by a scalar ϕ whose energy density then follows from the Hamiltonian

$$\rho(a) = a^{-3} H(a) = a^{-3}(\tfrac{1}{2}a^{-3}p_\phi^2 + a^3 V(\phi)) \tag{2}$$

with the scalar momentum p_ϕ and potential $V(\phi)$. At small scale factors a, there still is a diverging factor a^{-3} in the kinetic term which we recognized as being responsible for the singularity before. Since this term dominates over the non-diverging potential term, we still cannot escape the singularity by using this more fundamental description of matter. This is true unless we manage to arrange the evolution of the scalar in such a way that $p_\phi \to 0$ when $a \to 0$ in just the right way for the kinetic term not to diverge. This is difficult to arrange in general, but is exactly what is attempted in slow-roll inflation (though with a different motivation, and not necessarily all the way up to the classical singularity).

For the evolution of p_ϕ we need the scalar equation of motion, which can be derived from the Hamiltonian H in (2) via $\dot\phi = \{\phi, H\}$ and $\dot p_\phi = \{p_\phi, H\}$. This results in the isotropic Klein–Gordon equation in a time-dependent background determined by $a(t)$,

$$\ddot\phi + 3\dot a a^{-1}\dot\phi + V'(\phi) = 0. \tag{3}$$

In an expanding space with positive $\dot a$ the second term implies friction such that, if we assume the potential $V'(\phi)$ to be flat enough, ϕ will change only slowly (slow-roll). Thus, $\dot\phi$ and $p_\phi = a^3\dot\phi$ are small and at least for some time we can ignore the kinetic term in the energy density. Moreover, since ϕ changes only slowly we can regard the potential $V(\phi)$ as a constant Λ which again allows us to solve the Friedmann equation with $\rho(a) = \Lambda$. The solution $a \propto \exp(\sqrt{8\pi G\Lambda/3}\, t)$ is inflationary since $\ddot{a} > 0$ and non-singular: a becomes zero only in the limit $t \to -\infty$.

Thus, we now have a mechanism to drive a phase of accelerated expansion important for observations of structure. However, this expansion must be long enough, which means that the phase of slowly rolling ϕ must be long. This can be achieved only if the potential is very flat and ϕ starts sufficiently far away from its potential minimum. Flatness means that the ratio of $V(\phi_{\text{initial}})$ and ϕ_{initial} must be of the order 10^{-10}, while ϕ_{initial} must be huge, of the order of the Planck mass.[8] These assumptions are necessary for agreement with observations, but are in need of more fundamental explanations.

Moreover, inflation alone does not solve the singularity problem.[9] The non-singular solution we just obtained was derived under the approximation that the kinetic term can be ignored when $\dot{\phi}$ is small. This is true in a certain range of a, depending on how small $\dot{\phi}$ really is, but never very close to $a = 0$. Eventually, even with slow-roll conditions, the diverging a^{-3} will dominate and lead to a singularity.

3. Quantum Gravity

For decades, quantum gravity has been expected to complete the picture which is related to well-known properties of quantum mechanics in the presence of a non-zero $\hbar$.

3.1. *Indications*

First, in analogy to the singularity problem in gravity, where everything falls into a singularity in finite time, there is the instability problem of a classical hydrogen atom, where the electron would fall into the nucleus after a brief time. From quantum mechanics we know how the instability problem is solved: There is a finite ground state energy $E_0 = -\frac{1}{2}me^4/\hbar^2$, implying that the electron cannot radiate away all its energy and not fall further once it reaches the ground state. From the expression for E_0 one can see that quantum theory with its non-zero $\hbar$ is essential for this to happen: When $\hbar \to 0$ in a classical limit, $E_0 \to -\infty$ which brings us back to the classical instability. One expects a similar role to be played by the Planck length $\ell_{\mathrm{P}} = \sqrt{8\pi G\hbar/c^3} \approx 10^{-35}$m which is tiny but non-zero in quantum theory. If, just for dimensional reasons, densities are bounded by ℓ_{P}^{-3}, this would be finite in quantum gravity but diverge in the classical limit.

Secondly, a classical treatment of black body radiation suggests the Rayleigh–Jeans law according to which the spectral density behaves as $\rho(\lambda) \propto \lambda^{-4}$ as a function of the wave length. This is unacceptable since the divergence at small wave lengths leads to an infinite total energy. Here, quantum mechanics solves the problem by cutting off the divergence with Planck's formula which has a maximum at a wave length $\lambda_{\text{max}} \sim h/kT$ and approaches zero at smaller scales. Again, in the classical limit λ_{max} becomes zero and the expression diverges.

In cosmology the situation is similar for matter in the whole universe rather than a cavity. Energy densities as a function of the scale factor behave as, e.g., a^{-3} if matter is just diluted or a^{-4} if there is an additional redshift factor. In all cases, the energy density diverges at small scales, comparable to the Rayleigh–Jeans law. Inflation already provides an indication that the behavior must be different at small scales. Indeed, inflation can only be achieved with negative pressure, while all matter whose energy falls off as a^{-k} with non-negative k has positive pressure. This can easily be seen from the thermodynamical definition of pressure as the negative change of energy with volume. Negative pressure then requires the energy to increase with the scale factor at least at small scales where inflation is required (e.g., an energy Λa^3 for exponential inflation). This could be reconciled with standard forms of matter if there is an analog to Planck's formula, which interpolates between decreasing behavior at large scales and a behavior increasing from zero at small scales, with a maximum in between.

3.2. *Early quantum cosmology*

Since the isotropic reduction of general relativity leads to a system with finitely many degrees of freedom, one can in a first attempt try quantum mechanics to quantize it. Starting with the Friedmann equation (1) and replacing $\dot{a}$ by its momentum $p_a = 3a\dot{a}/8\pi G$ gives a Hamiltonian which is quadratic in the momentum and can be quantized easily to an operator acting on a wave function depending on the gravitational variable a and possibly matter fields ϕ. The usual Schrödinger representation yields the Wheeler–DeWitt equation[10,11]

$$\frac{3}{2}\left(-\frac{1}{9}\ell_{\text{P}}^4 a^{-1}\frac{\partial}{\partial a}a^{-1}\frac{\partial}{\partial a}\right)a\psi(a,\phi) = 8\pi G\hat{H}_\phi(a)\psi(a,\phi) \qquad (4)$$

with the matter Hamiltonian $\hat{H}_\phi(a)$. This system is different from usual quantum mechanics in that there are factor ordering ambiguities in the kinetic term, and that there is no derivative with respect to coordinate time

t. The latter fact is a consequence of general covariance: the Hamiltonian is a constraint equation restricting allowed states $\psi(a, \phi)$, rather than a Hamiltonian generating evolution in coordinate time. Nevertheless, one can interpret equation (4) as an evolution equation in the scale factor a, which is then called internal time. The left hand side thus becomes a second order time derivative, and it means that the evolution of matter is measured relationally with respect to the expansion or contraction of the universe, rather than absolutely in coordinate time.

Straightforward quantization thus gives us a quantum evolution equation, and we can now check what this implies for the singularity. If we look at the equation for $a = 0$, we notice first that the matter Hamiltonian still leads to diverging energy densities. If we quantize (2), we replace p_ϕ by a derivative, but the singular dependence on a does not change; a^{-3} would simply become a multiplication operator acting on the wave function. Moreover, $a = 0$ remains a singular point of the quantum evolution equation in internal time. There is nothing from the theory which tells us what physically happens at the singularity or beyond (baring intuitive pictures which have been developed from this perspective[12,13]).

So one has to ask what went wrong with our expectations that quantizing gravity should help. The answer is that quantum theory itself did not necessarily fail, but only our simple implementation. Indeed, what we used was just quantum mechanics, while quantum gravity has many consistency conditions to be fulfilled which makes constructing it so complicated. At the time when this formalism was first applied there was in fact no corresponding full quantum theory of gravity which could have guided developments. In such a simple case as isotropic cosmology, most of these consistency conditions trivialize and one can easily overlook important issues. There are many choices in quantizing an unknown system, and tacitly making one choice can easily lead in a wrong direction.

Fortunately, the situation has changed with the development of strong candidates for quantum gravity. This then allows us to reconsider the singularity and other problems from the point of view of the full theory, making sure that also in a simpler cosmological context only those steps are undertaken that have an analog in the full theory.

3.3. Loop quantum gravity

Singularities are physically extreme and require special properties of any theory aimed at tackling them. First, there are always strong fields (classically diverging) which requires a non-perturbative treatment. Moreover, classically we expect space to degenerate at the singularity, for instance a single point in a closed isotropic model. This means that we cannot take the presence of a classical geometry to measure distances for granted, which is technically expressed as background independence. A non-perturbative and background independent quantization of gravity is available in the form of loop quantum gravity,[14,15,16] which by now is understood well enough in order to be applicable in physically interesting situations.

Here, we only mention salient features of the theory which will turn out to be important for cosmology; for further details see [17]. The first one is the kind of basic variables used, which are the Ashtekar connection[18,19] describing the curvature of space and a triad (with density weight) describing the metric by a collection of three orthonormal vectors in each point. These variables are important since they allow a background independent representation of the theory, where the connection A_a^i is integrated to holonomies

$$h_e(A) = \mathcal{P} \exp \int_e A_a^i \tau_i \dot{e}^a \mathrm{d}t \tag{5}$$

along curves e in space and the triad E_i^a to fluxes

$$F_S(E) = \int_S E_i^a \tau^i n_a \mathrm{d}^2 y \tag{6}$$

along surfaces S. (In these expressions, $\dot{e}^a$ denotes the tangent vector to a curve and n_a the co-normal to a surface, both of which are defined without reference to a background metric. Moreover, $\tau_j = -\frac{1}{2}i\sigma_j$ in terms of Pauli matrices). While usual quantum field theory techniques rest on the presence of a background metric, for instance in order to decompose a field in its Fourier modes and define a vacuum state and particles, this is no longer available in quantum gravity where the metric itself must be turned into an operator. On the other hand, some integration is necessary since the fields themselves are distributional in quantum field theory and do not allow a well-defined representation. This "smearing" with respect to a background metric has to be replaced by some other integration sufficient for resulting in honest operators.[20,21] This is achieved by the integrations in (5) and (6), which similarly lead to a well-defined quantum representation. Usual Fock spaces in perturbative quantum field theory are thereby replaced by the loop representation, where an orthonormal basis is given by spin network states.[22]

This shows that choosing basic variables for a theory to quantize has implications for the resulting representation. Connections and triads can naturally be smeared along curves and surfaces without using a background metric and then represented on a Hilbert space. Requiring diffeomorphism invariance, which means that a background independent theory must not change under deformations of space (which can be interpreted as changes of coordinates), even selects a unique representation.[23,24,25,26,27] These are basic properties of loop quantum gravity, recognized as important requirements for a background independent quantization. Already here we can see differences to the Wheeler–DeWitt quantization, where the metric is used as a basic variable and then quantized as in quantum mechanics. This is possible in the model but not in a full theory, and in fact we will see later that a loop quantization will give a representation inequivalent to the Wheeler–DeWitt quantization.

The basic properties of the representation have further consequences. Holonomies and fluxes act as well-defined operators, and fluxes have discrete spectra. Since spatial geometry is determined by the triad, spatial geometry is discrete, too, with discrete spectra for, e.g., the area and volume operator.[28,29,30] The geometry of space-time is more complicated to understand since this is a dynamical situation which requires solving the Hamiltonian constraint. This is the analog of the Wheeler–DeWitt equation in the full theory and is the quantization of Einstein's dynamical equations. There are candidates for such operators,[31] well-defined even in the presence of matter[32] which in usual quantum field theory would contribute divergent matter Hamiltonians. Not surprisingly, the full situation is hard to analyze, which is already the case classically, without assuming simplifications from symmetries. We will thus return to symmetric, in particular isotropic models, but with the new perspective provided by the full theory of loop quantum gravity.

4. Quantum Cosmology

Symmetries can be introduced in loop quantum gravity at the level of states and basic operators,[33,34,35] such that it is not necessary to reduce the classical theory first and then quantize as in the Wheeler–DeWitt quantization. Instead, one can view the procedure as quantizing first and then introducing symmetries which ensures that consistency conditions of quantum gravity are observed in the first step before one considers treatable situations. In particular, the quantum representation derives from symmetric states and basic operators, while the Hamiltonian constraint can be obtained with constructions analogous to those in the full theory. Between the dynamics of models and the full theory there is thus no complete link yet and not all ingredients of models have been derived so far. But the formulation of quantum gravity in a background independent manner implies characteristic properties which are also realized in models. This allows us to reconsider the singularity problem, now with methods from full quantum gravity. In fact, symmetric models present a class of systems which can often be treated explicitly while still being representative for general phenomena. For instance, the prime examples of singular situations in gravity, and some of the most widely studied physical applications, are already obtained in isotropic or spherically symmetric systems, which allow access to cosmology and black holes.

4.1. Representation

Before discussing the quantum level we reformulate isotropic cosmology in connection and triad variables instead of a. The role of the scale factor is now played by the triad component p with $|p| = a^2$ whose canonical momentum is the isotropic connection component $c = -\frac{1}{2}\dot{a}$ with $\{c, p\} = 8\pi G/3$. The main difference to metric variables is the fact that p, unlike a, can take both signs with $\mathrm{sgn}\, p$ being the orientation of space. This is a consequence of having to use triad variables which not only know about the size of space but also its orientation (depending on whether the set of orthonormal vectors is left or right handed).

States in the full theory are usually written in the connection representation as functions of holonomies. Following the reduction procedure for an isotropic symmetry group leads to orthonormal states which are functions of the isotropic connection component c and given by[36]

$$\langle c | \mu \rangle = e^{i\mu c/2} \qquad \mu \in \mathbb{R}. \tag{7}$$

On these states the basic variables p and c are represented by

$$\hat{p} | \mu \rangle = \tfrac{1}{6} \ell_{\mathrm{P}}^2 \mu | \mu \rangle \tag{8}$$

$$\widehat{e^{i\mu' c/2}} | \mu \rangle = | \mu + \mu' \rangle \tag{9}$$

with the properties:

(i) $[\widehat{e^{i\mu' c/2}}, \hat{p}] = -\frac{1}{6}\ell_{\mathrm{P}}^2 \mu' \widehat{e^{-i\mu' c/2}} = i\hbar(\{e^{i\mu' c/2}, p\})^{\wedge}$,

(ii) $\hat{p}$ has a discrete spectrum and

(iii) only exponentials $e^{i\mu'c/2}$ of c are represented, not c directly.

These statements deserve further explanation: First, the classical Poisson relations between the basic variables are indeed represented correctly, turning the Poisson brackets into commutators divided by $i\hbar$. On this representation, the set of eigenvalues of $\hat{p}$ is the full real line since μ can take arbitrary real values. Nevertheless, the spectrum of $\hat{p}$ is discrete in the technical sense that eigenstates of $\hat{p}$ are normalizable. This is indeed the case in this non-separable Hilbert space where (7) defines an orthonormal basis. The last property follows since the exponentials are not continuous in the label μ', for otherwise one could simply take the derivative with respect to μ' at $\mu' = 0$ and obtain an operator for c. The discontinuity can be seen, e.g., from

$$\langle\mu|\widehat{e^{i\mu'c/2}}|\mu\rangle = \delta_{0,\mu'}$$

which is not continuous.

These properties are quite unfamiliar from quantum mechanics, and indeed the representation is inequivalent to the Schrödinger representation (the discontinuity of the c-exponential evading the Stone–von Neumann theorem which usually implies uniqueness of the representation). In fact, the loop representation is inequivalent to the Wheeler–DeWitt quantization which just assumed a Schrödinger like quantization. In view of the fact that the phase space of our system is spanned by c and p with $\{c,p\} \propto 1$ just as in classical mechanics, the question arises how such a difference in the quantum formulation arises.

As a mathematical problem the basic step of quantization occurs as follows: given the classical Poisson algebra of observables Q and P with $\{Q,P\} = 1$, how can we define a representation of the observables on a Hilbert space such that the Poisson relations become commutator relations and complex conjugation, meaning that Q and P are real, becomes adjointness? The problem is mathematically much better defined if one uses the bounded expressions e^{isQ} and $e^{it\hbar^{-1}P}$ instead of the unbounded Q and P, which still allows us to distinguish any two points in the whole phase space. The basic objects e^{isQ} and $e^{it\hbar^{-1}P}$ upon quantization will then not commute but fulfill the commutation relation (Weyl algebra)

$$e^{isQ}e^{it\hbar^{-1}P} = e^{ist}e^{it\hbar^{-1}P}e^{isQ} \tag{10}$$

as unitary operators on a Hilbert space.

In the Schrödinger representation this is done by using a Hilbert space $L^2(\mathbb{R}, \mathrm{d}q)$ of square integrable functions $\psi(q)$ with $\int_{\mathbb{R}} \mathrm{d}q|\psi(q)|^2$ finite. The representation of basic operators is

$$e^{isQ}\psi(q) = e^{isq}\psi(q)$$
$$e^{it\hbar^{-1}P}\psi(q) = \psi(q + t)$$

which indeed are unitary and fulfill the required commutation relation. Moreover, the operator families as functions of s and t are continuous and we can take the derivatives in $s = 0$ and $t = 0$, respectively:

$$-i\left.\frac{\mathrm{d}}{\mathrm{d}s}\right|_{s=0} e^{isQ} = q$$

$$-i\hbar\left.\frac{\mathrm{d}}{\mathrm{d}t}\right|_{t=0} e^{it\hbar^{-1}P} = \hat{p} = -i\hbar\frac{\mathrm{d}}{\mathrm{d}q}.$$

This is the familiar representation of quantum mechanics which, according to the Stone–von Neumann theorem is unique under the condition that e^{isQ} and $e^{it\hbar^{-1}P}$ are indeed continuous in both s and t.

The latter condition is commonly taken for granted in quantum mechanics, but in general there is no underlying physical or mathematical reason. It is easy to define representations violating continuity in s or t, for instance if we use a Hilbert space $\ell^2(\mathbb{R})$ where states are again maps ψ_q from the real line to complex numbers but with norm $\sum_q |\psi_q|^2$ which implies that normalizable ψ_q can be non-zero for at most countably many q. We obtain a representation with basic operators

$$e^{isQ}\psi_q = e^{isq}\psi_q$$
$$e^{it\hbar^{-1}P}\psi_q = \psi_{q+t}$$

which is of the same form as before. However, due to the different Hilbert space the second operator $e^{it\hbar^{-1}P}$ is no longer continuous in t which can be checked as in the case of $e^{i\mu c/2}$. In fact, the representation for Q and P is isomorphic to that of p and c used before, where a general state $|\psi\rangle = \sum_\mu \psi_\mu|\mu\rangle$ has coefficients ψ_μ in $\ell^2(\mathbb{R})$.

This explains mathematically why different, inequivalent representations are possible, but what are the physical reasons for using different representations in quantum mechanics and quantum cosmology? In quantum mechanics it turns out that the choice of representation is not that important and is mostly being done for reasons of familiarity with the standard choice. Physical differences between inequivalent representations can only occur at very high energies[37] which are not probed by available experiments and do not affect characteristic quantum effects related to the ground state or excited states. Thus, quantum mechanics as we know it can well be formulated in an inequivalent representation, and also in quantum field theory this can be done and even be useful.[38]

In quantum cosmology we have a different situation where it is the high energies which are essential. We do not have direct observations of this regime, but from conceptual considerations such as the singularity issue we have learned which problems we have to face. The classical singularity leads to the highest energies one can imagine, and it is here where the question of which representation to choose becomes essential. As shown by the failure of the Wheeler–DeWitt quantization in trying to remove the singularity, the Schrödinger representation is inappropriate for quantum cosmology. The representation underlying loop quantum cosmology, on the other hand, implies very different properties which become important at high energies and can shed new light on the singularity problem.

Moreover, by design of symmetric models as obtained from the full theory, we have the same basic properties of a loop representation in cosmological models and the full situation where they were recognized as being important for a background independent quantization: discrete fluxes $\hat{F}_S(E)$ or $\hat{p}$ and a representation only of holonomies $h_e(A)$ or $e^{i\mu c/2}$ but not of connection components A_a^i or c. These basic properties have far-reaching consequences as discussed in what follows:[39]

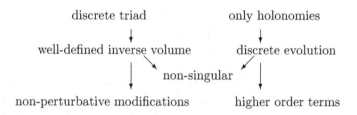

By this reliable quantization of representative and physically interesting models with a known relation to full quantum gravity we are finally able to resolve long-standing issues such as the singularity problem.

4.2. *Quantum evolution*

We will first look at the quantum evolution equation which we obtain as the quantized Friedmann equation. This is modeled on the Hamiltonian constraint of the full theory such that we can also draw some conclusions for the viability of the full constraint.

4.2.1. *Difference equation*

The constraint equation will be imposed on states of the form $|\psi\rangle = \sum_\mu \psi_\mu |\mu\rangle$ with summation over countably many values of μ. Since the states $|\mu\rangle$ are eigenstates of the triad operator, the coefficients ψ_μ which can also depend on matter fields such as a scalar ϕ represent the state in the triad representation, analogous to $\psi(a, \phi)$ before. For the constraint operator we again need operators for the conjugate of p, related to $\dot{a}$ in the Friedmann equation. Since this is now the exponential of c, which on basis states acts by shifting the label, it translates to a finite shift in the labels of coefficients $\psi_\mu(\phi)$. Plugging together all ingredients for a quantization of (1) along the lines of the constraint in the full theory leads to the difference equation[40,36]

$$(V_{\mu+5} - V_{\mu+3})\psi_{\mu+4}(\phi) - 2(V_{\mu+1} - V_{\mu-1})\psi_\mu(\phi) \tag{11}$$

$$+(V_{\mu-3} - V_{\mu-5})\psi_{\mu-4}(\phi) = -\tfrac{4}{3}\pi G\ell_P^2 \hat{H}_{\text{matter}}(\mu)\psi_\mu(\phi)$$

with volume eigenvalues $V_\mu = (\ell_P^2 |\mu|/6)^{3/2}$ obtained from the volume operator $\hat{V} = \hat{p}^{3/2}$, and the matter Hamiltonian $\hat{H}_{\text{matter}}(\mu)$.

We again have a constraint equation which does not generate evolution in coordinate time but can be seen as evolution in internal time. Instead of the continuous variable a we now have the label μ which only jumps in discrete steps. As for the singularity issue, there is a further difference to the Wheeler–DeWitt equation since now the classical singularity is located at $p = 0$ which is in the interior rather than at the boundary of the configuration space. Nevertheless, the classical evolution in the variable p breaks down at $p = 0$ and there is still a singularity. In quantum theory, however, the situation is very different: while the Wheeler–DeWitt equation does not solve the singularity problem, the difference equation (11) uniquely evolves a wave function from some initial values at positive μ, say, to negative μ.[41,36] Thus, the evolution does not break down at the classical singularity and can rather be continued beyond it. Quantum gravity is thus a theory which is more complete than classical general relativity and is free of limitations set by classical singularities.

An intuitive picture of what replaces the classical singularity can be obtained from considering evolution in μ as before. For negative μ, the volume V_μ decreases with increasing μ while V_μ increases for positive μ. This leads to the picture of a collapsing universe before it reaches the classical big bang singularity and re-expands. While at large scales the classical description is good,[42] when the universe is small close to the classical singularity it starts to break down and has to be replaced by discrete quantum geometry. The resulting quantum evolution does not break down, in contrast to the classical space-time picture which dissolves. Using the fact that the sign of μ, which defines the orientation of space, changes during the transition through the classical singularity one can conclude that the universe turns its inside out during the process. This can have consequences for realistic matter Hamiltonians which violate parity symmetry.

4.2.2. Meaning of the wave function

An important issue in quantum gravity which is still outstanding is the interpretation of the wave function and its relation to the problem of time. In the usual interpretation of quantum mechanics the wave function determines probabilities for measurements made by an observer outside the quantum system. Quantum gravity and cosmology, however, are thought of as theories for the quantum behavior of a whole universe such that, by definition, there cannot be an observer outside the quantum system. Accordingly, the question of how to interpret the wave function in quantum cosmology is more complicated. One can avoid the separation into a classical and quantum part of the problem in quantum mechanics by the theory of decoherence which can explain how a world perceived as classical emerges from the fundamental quantum description.[43] The degree of "classicality" is related to the number of degrees of freedom which do not contribute significantly to the evolution but interact with the system nonetheless. Averaging over those degrees of freedom, provided there are enough of them, then leads to a classical picture. This demonstrates why macroscopic bodies under usual circumstances are perceived as classical while in the microscopic world, where a small number of degrees of freedom is sufficient to capture crucial properties of a system, quantum mechanics prevails. This idea has been adapted to cosmology, where a large universe comes with many degrees of freedom such as small inhomogeneities which are not of much relevance for the overall evolution. This is different, however, in a small universe where quantum behavior becomes dominant.

Thus, one can avoid the presence of an observer outside the quantum system. The quantum system is described by its wave function, and in some circumstances one can approximate the evolution by a quantum part being looked at by classical observers within the same system. Properties are then encoded in a relational way: the wave function of the whole system contains information about everything including possible observers. Now, the question has shifted from a conceptual one — how to describe the system if no outside observers can be available — to a technical one. One needs to understand how information can be extracted from the wave function and used to develop schemes for intuitive pictures or potentially observable effects. This is particularly pressing in the very early universe where everything including what we usually know as space and time are quantum and no familiar background to lean on is left.

One lesson is that evolution should be thought of as relational by determining probabilities for one degree of freedom under the condition that another degree of freedom has a certain value. If the reference degree of freedom (such as the direction of the hand of a clock) plays a distinguished role for determining the evolution of others, it is called internal time: it is not an absolute time outside the quantum system as in quantum mechanics, and not a coordinate time as in general relativity which could be changed by coordinate transformations. Rather, it is one of the physical degrees of freedom whose evolution is determined by the dynamical laws and which shows how other degrees of freedom change by interacting with them. From this picture it is clear that no external observer is necessary to read off the clock or other measurement devices, such that it is ideally suited to cosmology. What is also clear is that now internal time depends on what we choose it to be, and different questions require different choices. For a lab experiment the hand of a clock would be a good internal time and, when the clock is sufficiently isolated from the physical fields used in the experiment and other outside influence, will not be different from an absolute time except that it is mathematically more complicated to describe. The same clock, on the other hand, will not be good for describing the universe when we imagine to approach a classical singularity. It will simply not withstand the extreme physical conditions, dissolve, and its parts will behave in a complicated irregular manner ill-suited for the description of evolution. Instead, one has to use more global objects which depend on what is going on in the whole universe.

Close to a classical singularity, where one expects monotonic expansion or contraction, the spatial volume of the universe is just the right quantity as internal time. A wave function then determines relationally how matter fields or other gravitational degrees of freedom change with respect to the expansion or contraction of the universe. In our case, this is encoded in the wave function $\psi_\mu(\phi)$ depending on internal time μ (which through the volume defines the size of the universe but also spatial orientation) and matter fields ϕ. By showing that it is subject to a difference equation in μ

which does not stop at the classical singularity $\mu = 0$ we have seen that relational probabilities are defined for all internal times without breaking down anywhere. This shows the absence of singularities and allows developing intuitive pictures, but does not make detailed predictions before relational probabilities are indeed computed and shown how to be observable at least in principle.

Here, we encounter the main issue in the role of the wave function: we have a relational scheme to understand what the wave function should mean but the probability measure to be used, called the physical inner product, is not known so far. We already used a Hilbert space which we needed to define the basic operators and the quantized Hamiltonian constraint, where wave functions ψ_μ, which by definition are non-zero for at most countably many values $\mu \in \mathbb{R}$, have the inner product $\langle \psi | \psi' \rangle = \sum_\mu \bar{\psi}_\mu \psi'_\mu$. This is called the kinematical inner product which is used for setting up the quantum theory. But unlike in quantum mechanics where the kinematical inner product is also used as physical inner product for the probability interpretation of the wave function, in quantum gravity the physical inner product must be expected to be different. This occurs because the quantum evolution equation (11) in internal time is a constraint equation rather than an evolution equation in an external absolute time parameter. Solutions to this constraint in general are not normalizable in the kinematical inner product such that a new physical inner product on the solution space has to be found. There are detailed schemes for a derivation which have been applied in different models.[44,45,46] In particular the recent Ref. [46] computes the inner product for a model with remaining physical degrees of freedom, which confirms and solidifies expectations obtained from the difference equation alone. This also strengthens the link to a simplified, effective route to extract physical statements which will be discussed in Sec. 4.4 together with the main results.

A related issue, which is also of relevance for the classical limit of the theory is that of oscillations on small scales of the wave function. Being subject to a difference equation means that ψ_μ is not necessarily smooth but can change rapidly when μ changes by a small amount even when the volume is large. In such a regime one expects classical behavior, but small scale oscillations imply that the wave function is sensitive to the Planck scale. There are also other issues related to the fact that now a difference rather than differential equation provides the fundamental law.[47] For a complete understanding of the solution space it is worthwhile to study the mathematical problem of if and when solutions with suppressed oscillations exist. This is easy to answer in the affirmative for isotropic models subject to (11) where in some cases one even obtains a unique wave function.[48,49] However, already in other homogeneous but anisotropic models the issue is much more complicated to analyze.[50,51]

In a more general situation than homogeneous cosmology there is an additional complication. In general, it is very difficult to find an internal time to capture the evolution of a complicated quantum system, which is called the problem of time in general relativity. In cosmology, the volume is a good internal time to understand the singularity, but it would not be good for the whole history if the universe experiences a recollapse where the volume would not be monotonic. This is even more complicated in inhomogeneous situations such as the collapse of matter into a black hole. Since we used internal time μ to show how quantum geometry evolves through the classical singularity, it seems that the singularity problem in general cannot be solved before the problem of time is understood. Fortunately, while the availability of an internal time simplifies the analysis, requirements on a good choice can be relaxed for the singularity problem. An internal time provides us with an interpretation of the constraint equation as an evolution equation, but the singularity problem can be phrased independently of this as the problem to extend the wave function on the space of metrics or triads. This implies weaker requirements and also situations can be analyzed where no internal time is known. The task then is to find conditions which characterize a classical singularity, analogous to $p = 0$ in isotropic cosmology, and find an evolution parameter which at least in individual parts of an inhomogeneous singularity allows to see how the system can move through it. This has been established in models with spherical symmetry such as non-rotating black holes.[52] These inhomogeneous cases are now under study but only partially understood so far, such that in the next section we return to isotropic cosmology.

4.3. *Densities*

In the previous discussion we have not yet mentioned the matter Hamiltonian on the right hand side, which diverges classically and in the Wheeler–DeWitt quantization when we reach the singularity. If this were the case here, the discrete quantum evolution would break down, too. However, as we will see now the matter Hamiltonian does not diverge, which is again a consequence of the loop representation.

4.3.1. *Quantization*

For the matter Hamiltonian we need to quantize the matter field and in quantum gravity also coefficients such as a^{-3} in the kinetic term which now become operators, too. In the Wheeler–DeWitt quantization where a is a multiplication operator, a^{-3} is unbounded and diverges at the classical singularity. In loop quantum cosmology we have the basic operator $\hat{p}$ which one can use to construct a quantization of a^{-3}. However, a straightforward quantization fails since, as one of the basic properties, $\hat{p}$ has a discrete spectrum containing zero. In this case, there is no densely defined inverse operator which one could use. This seems to indicate that the situation is even worse: an operator for the kinetic term would not only be unbounded but not even be well-defined. The situation is much better, however, when one tries other quantizations which are more indirect. For non-basic operators such as a^{-3} there are usually many ways to quantize, all starting from the same classical expression. What we can do here, suggested by constructions in the full theory,[32] is to rewrite a^{-3} in a classically equivalent way as

$$a^{-3} = (\pi^{-1}G^{-1}\mathrm{tr}\tau_3 e^{c\tau_3}\{e^{-c\tau_3}, \sqrt{V}\})^6$$

where we only need a readily available positive power of $\hat{p}$. Moreover, exponentials of c are basic operators, where we just used su(2) notation $e^{c\tau_3} = \cos\frac{1}{2}c + 2\tau_3\sin\frac{1}{2}c$ in order to bring the expression closer to what one would have in the full theory, and the Poisson bracket will become a commutator in quantum theory.

This procedure, after taking the trace, leads to a densely defined operator for a^{-3} despite the nonexistence of an inverse of $\hat{p}$:[53]

$$\widehat{a^{-3}} = \left(8i\ell_{\mathrm{P}}^{-2}(\sin\tfrac{1}{2}c\sqrt{\hat{V}}\cos\tfrac{1}{2}c - \cos\tfrac{1}{2}c\sqrt{\hat{V}}\sin\tfrac{1}{2}c)\right)^6. \tag{12}$$

That this operator is indeed finite can be seen from its action on states $|\mu\rangle$ which follows from that of the basic operators:

$$\widehat{a^{-3}}|\mu\rangle = \left(4\ell_{\mathrm{P}}^{-2}(\sqrt{V_{\mu+1}} - \sqrt{V_{\mu-1}})\right)^6|\mu\rangle \tag{13}$$

immediately showing the eigenvalues which are all finite. In particular, at $\mu = 0$ where we would have the classical singularity the density operator does not diverge but is zero.

This finiteness of densities finally confirms the non-singular evolution since the matter Hamiltonian

$$\hat{H}_{\mathrm{matter}} = \tfrac{1}{2}\widehat{a^{-3}}\hat{p}_\phi^2 + \hat{V}V(\phi) \tag{14}$$

in the example of a scalar is well-defined even on the classically singular state $|0\rangle$. The same argument applies for other matter Hamiltonians since only the general structure of kinetic and potential terms is used.

4.3.2. *Confirmation of indications*

The finiteness of the operator is a consequence of the loop representation which forced us to take a detour in quantizing inverse powers of the scale factor. A more physical understanding can be obtained by exploiting the fact that there are quantization ambiguities in this non-basic operator. This comes from the rewriting procedure which is possible in many classically equivalent ways, which all lead to different operators. Important properties such as the finiteness and the approach to the classical limit at large volume are robust under the ambiguities, but finer details can change. The most important choices one can make are selecting the representation j of SU(2) holonomies before taking the trace[54,55] and the power l of $|p|$ in the Poisson bracket.[56] These values are restricted by the requirement that j is a half-integer ($j = 1/2$ in the above choice) and $0 < l < 1$ to obtain a well-defined inverse power of a ($l = 3/4$ above). The resulting eigenvalues can be computed explicitly and be approximated by the formula[55,56]

$$(a^{-3})_{\text{eff}} = a^{-3} p_l (a^2/a_{\text{max}}^2)^{3/(2-2l)} \tag{15}$$

where $a_{\text{max}} = \sqrt{j/3}\,\ell_{\text{P}}$ depends on the first ambiguity parameter and the function

$$p_l(q) = \frac{3}{2l} q^{1-l} \left((l+2)^{-1} \left((q+1)^{l+2} - |q-1|^{l+2} \right) \right. \tag{16}$$
$$\left. - (l+1)^{-1} q \left((q+1)^{l+1} - \text{sgn}(q-1)|q-1|^{l+1} \right) \right) .$$

on the second.

The function $p_l(q)$, shown in Fig. 1, approaches one for $q \gg 1$, has a maximum close to $q = 1$ and drops off as q^{2-l} for $q \ll 1$. This shows that $(a^{-3})_{\text{eff}}$ approaches the classical behavior a^{-3} at large scales $a \gg a_{\text{max}}$, has a maximum around a_{max} and falls off like $(a^{-3})_{\text{eff}} \sim a^{3/(1-l)}$ for $a \ll a_{\text{max}}$. The peak value can be approximated, e.g. for $j = 1/2$, by $(a^{-3})_{\text{eff}}(a_{\text{max}}) \sim 3l^{-1} 2^{-l} (1 - 3^{-l})^{3/(2-2l)} \ell_{\text{P}}^{-3}$ which indeed shows that densities are bounded by inverse powers of the Planck length such that they are finite in quantum gravity but diverge in the classical limit. This confirms our qualitative expectations from the hydrogen atom, while details of the coefficients depend on the quantization.

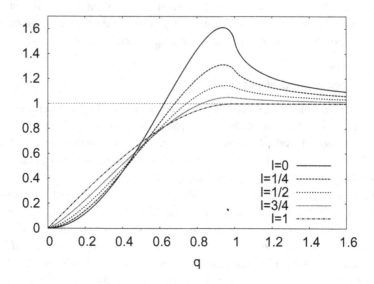

Fig. 1. The function $p_l(q)$ in (16) for some values of l, including the limiting cases $l = 0$ and $l = 1$.

Similarly, densities are seen to have a peak at $a_{\max}$ whose position is given by the Planck length (and an ambiguity parameter). Above the peak we have the classical behavior of an inverse power, while below the peak the density increases from zero. As suggested by the behavior of radiation in a cavity whose spectral energy density

$$\rho_T(\lambda) = 8\pi h\lambda^{-5}(e^{h/kT\lambda} - 1)^{-1} = h\lambda^{-5}f(\lambda/\lambda_{\max})$$

can, analogously to (15), be expressed as the diverging behavior λ^{-5} multiplied with a cut-off function $f(y) = 8\pi/((5/(5 - x))^{1/y} - 1)$ with $x = 5 + W(-5e^{-5})$ (in terms of the Lambert function $W(x)$, the inverse function of xe^x) and $\lambda_{\max} = h/xkT$, we obtain an interpolation between increasing behavior at small scales and decreasing behavior at large scales in such a way that the classical divergence is cut off.

We thus have an interpolation between increasing behavior necessary for negative pressure and inflation and the classical decreasing behavior (Fig. 2). Any matter density turns to increasing behavior at sufficiently small scales without the need to introduce an inflaton field with tailor-made properties. In the following section we will see the implications for cosmological evolution by studying effective classical equations incorporating this characteristic loop effect of modified densities at small scales.

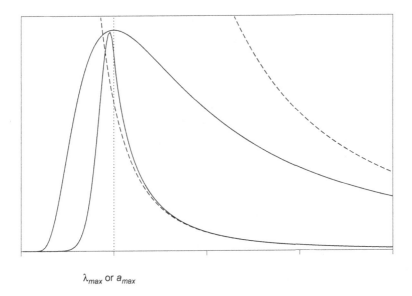

λ_{max} or a_{max}

Fig. 2. Comparison between the spectral energy density of black body radiation (wide curve) and an effective geometrical density with their large scale approximations (dashed).

4.4. Phenomenology

The quantum difference equation (11) is rather complicated to study in particular in the presence of matter fields and, as discussed in Sec. 4.2.2, difficult to interpret in a fully quantum regime. It is thus helpful to work with effective equations, comparable conceptually to effective actions in field theories, which are easier to handle and more familiar to interpret but still show important quantum effects. This can be done systematically,[36,57,58] starting with the Hamiltonian constraint operator, resulting in different types of correction terms whose significance in given regimes can be estimated or studied numerically.[59] There are perturbative corrections to the Friedmann equation of higher order form in $\dot{a}$, or of higher derivative, in the gravitational part on the left hand side, but also modifications in the matter Hamiltonian since the density in its kinetic term behaves differently at small scales. The latter corrections are mainly non-perturbative since the full expression for $(a^{-3})_{\text{eff}}$ depends on the inverse Planck length, and their range can be extended if the parameter j is rather large. For these reasons, those corrections are most important and we focus on them from now on.

The effective Friedmann equation then takes the form

$$a\dot{a}^2 = \tfrac{8\pi}{3}G\left(\tfrac{1}{2}(a^{-3})_{\text{eff}}\,p_\phi^2 + a^3 V(\phi)\right) \tag{17}$$

with $(a^{-3})_{\text{eff}}$ as in (15) with a choice of ambiguity parameters. Since the matter Hamiltonian does not just act as a source for the gravitational field on the right hand side of the Friedmann equation, but also generates Hamiltonian equations of motion, the modification entails further changes in the matter equations of motion. The Klein–Gordon equation (3) then takes the effective form

$$\ddot{\phi} = \dot{\phi}\,\dot{a}\,\frac{\mathrm{d}\log(a^{-3})_{\text{eff}}}{\mathrm{d}a} - a^3(a^{-3})_{\text{eff}}V'(\phi) \tag{18}$$

and finally there is the Raychaudhuri equation

$$\frac{\ddot{a}}{a} = -\frac{8\pi G}{3}\left(a^{-3}d(a)_{\text{eff}}^{-1}\dot{\phi}^2\left(1 - \tfrac{1}{4}a\frac{\mathrm{d}\log(a^3 d(a)_{\text{eff}})}{\mathrm{d}a}\right) - V(\phi)\right) \tag{19}$$

which follows from the above equation and the continuity equation of matter.

4.4.1. Bounces

The resulting equations can be studied numerically or with qualitative analytic techniques. We first note that the right hand side of (17) behaves differently at small scales since it increases with a at fixed ϕ and p_ϕ. Viewing this equation as analogous to a constant energy equation in classical mechanics with kinetic term $\dot{a}^2$ and potential term $\mathcal{V}(a) := -\tfrac{8\pi}{3}Ga^{-1}\left(\tfrac{1}{2}(a^{-3})_{\text{eff}}\,p_\phi^2 + a^3 V(\phi)\right)$ illustrates the classically attractive nature of gravity: The dominant part of this potential behaves like $-a^{-4}$ which is increasing. Treating the scale factor analogously to the position of a classical particle shows that a will be driven toward smaller values, implying attraction of matter and energy in the universe. This changes when we approach smaller scales and take into account the quantum modification. Below the peak of the effective density the classical potential $\mathcal{V}(a)$ will now decrease, $-\mathcal{V}(a)$ behaving like a positive power of a. This implies that the scale factor will be repelled away from $a = 0$ such that there is now a small-scale repulsive component to the gravitational force if we allow for quantum effects. The collapse of matter can then be prevented if repulsion is taken into account, which indeed can be observed in some models where the effective classical equations alone are sufficient to demonstrate singularity-free evolution.

This happens by the occurrence of bounces where a turns around from contracting to expanding behavior. Thus, $\dot{a} = 0$ and $\ddot{a} > 0$. The first condition is not always realizable, as follows from the Friedmann equation (1). In particular, when the scalar potential is non-negative there is no bounce, which is not changed by the effective density. There are then two possibilities for bounces in isotropic models, the first one if space has positive curvature rather than being flat as assumed here,[60,61] the second one with a scalar potential which can become negative.[62,63] Both cases allow $\dot{a} = 0$ even in the classical case, but this always corresponds to a maximum rather than minimum. This can easily be seen for the case of negative potential from the Raychaudhuri equation (19) which in the classical case implies negative $\ddot{a}$. With the modification, however, the additional term in the equation provides a positive contribution which can become large enough for $\ddot{a}$ to become positive at a value of $\dot{a} = 0$ such that there is a bounce.

This provides intuitive explanations for the absence of singularities from quantum gravity, but not a general one. The generic presence of bounces depends on details of the model such as its matter content or which correction terms are being used,[64,65] and even with the effective modifications there are always models which classically remain singular. Thus, the only general argument for absence of singularities remains the quantum one based on the difference equation (11), where the conclusion is model independent[41] and which also confirms bounce pictures qualitatively.[59,46]

4.4.2. *Inflation*

A repulsive contribution to the gravitational force can not only explain the absence of singularities, but also enhances the expansion of the universe on scales close to the classical singularity. Thus, as seen also in Fig. 3 the universe accelerates just from quantum effects, providing a mechanism for inflation without choosing special matter.

Via the generation of structure, inflationary phases of the universe can have an imprint on the observable cosmic microwave background. Observations imply that the predicted power spectrum of anisotropies must be nearly independent of the scale on which the anisotropies are probed, which implies that the inflationary phase responsible for structure formation must be close to exponential acceleration. This is true for slow-roll inflation, but also for the inflationary phase obtained from the effective density once a non-zero scalar potential is taken into account.[66] For more detailed comparisons between theory and observations one needs to consider how inhomoge-

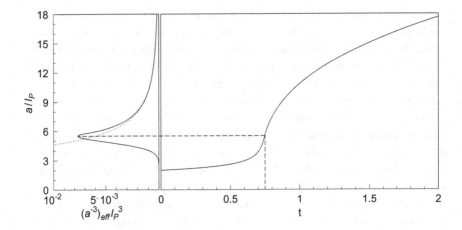

Fig. 3. Numerical solution to the effective Friedmann equation (17) with a vanishing scalar potential. While the modification in the density on the left is active the expansion is accelerated, which stops automatically once the universe expands to a size above the peak in the effective density.

neous fields evolve, which already requires us to relax the strong symmetry assumption of homogeneity. The necessary methods are not well-developed at the current stage (see Refs. 35, 67, 68 for the basic formulation), but preliminary calculations of implications on the power spectrum have been undertaken nonetheless. Ref. 69 indicates that loop inflation can be distinguished from simple inflaton models because the power depends differently on scales.

It turns out that this loop phase alone can provide a sufficient amount of inflation only for unnatural choices of parameters (such as extremely large j), and those cases are even ruled out by observations already. At this point, the modified matter dynamics of (18) and its $\dot{\phi}$-term becomes important. Classically, it is a friction term which is used for slow-roll inflation. But in the modified regime at small scales the sign of the term changes once $(a^{-3})_{\text{eff}}$ is increasing. Thus, at those small scales friction turns into antifriction and matter is driven up its potential if it has a non-zero initial momentum (even a tiny one, e.g., from quantum fluctuations). After the loop phase matter fields slow down and roll back toward their minima, driving additional inflation. The potentials need not be very special since structure formation in the first phase and providing a large universe happen by different mechanisms. When matter fields reach their minima they start to oscillate and usual re-heating to obtain hot matter can commence.

Loop quantum cosmology thus presents a viable alternative to usual inflaton models which is not in conflict with current observations but can be verified or ruled out with future experiments such as the Planck satellite. Its attractive feature is that it does not require the introduction of an inflaton field with its special properties, but provides a fundamental explanation of acceleration from quantum gravity. This scenario is thus encouraging, but so far there are still several open questions not just about details. Compared to inflaton models, which have been studied extensively in all possible details, many opportunities for potential new effects but also the need for crucial viability tests still remain.

Even if we assume the presence of an inflaton field, its properties are less special than in the purely classical treatment. We still need to assume a potential which is sufficiently flat, but there is now an explanation of initial values far away from the minimum. For this we again use the effective Klein–Gordon equation and the fact that ϕ is driven up its potential. One can then check that for usual inflaton potentials the value of typical initial conditions, as a function of chosen ambiguity parameters and initial fluctuations of the scalar, is just what one needs for sufficient inflation in a wide range.[70,71] After the modifications in the density subside, the inflaton keeps moving up the potential from its initial push, but is now slowed down by the friction term. Eventually, it will stop and turn around, entering a slow roll phase in its approach to the potential minimum. Thus, the whole history of the expansion is described by a consistent model as illustrated in Fig. 4, not just the slow roll phase after the inflaton has already obtained its large initial values.

One may think that such a second phase of slow-roll inflation washes away potential quantum gravity effects from the early expansion. That this is not necessarily the case has been shown in Ref. 70, based on the fact that around the turning point of the inflaton the slow-roll conditions are violated. In this scenario, structure we see today on the largest scales was created at the earliest stages of the second inflationary phase since it was enlarged by the full inflationary phase. If the second inflationary regime did not last too long, these scales are just observable today where in fact the observed loss of power can be taken as an indication of early violations of slow-roll expansion. Thus, loop quantum cosmology can provide an explanation, among others, for the suppression of power on large scales.

There are diverse scenarios since different phases of inflation can be combined, and eventually the right one has to be determined from observations. One can also combine bounces and inflationary regimes in order

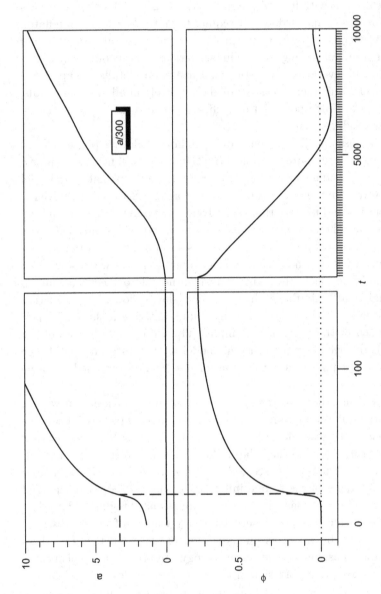

Fig. 4. History of the scale factor (top) and inflaton (bottom) with the left hand side in slow motion. (Tics on the right horizontal axis mark increments in t by 100.) The upper right data are rescaled so as to fit on the same plot. Units for a and ϕ are Planck units, and parameters are not necessarily realistic but chosen for plotting purposes. Dashed lines mark the time and scale factor where classical behavior of $(a^{-3})_{\mathrm{eff}}$ starts.

to obtain cyclic universes which eventually reach a long phase of accelerated expansion.[72] In particular, this allows the conctruction of models which start close to a simple, static initial state and, after a series of cycles, automatically reach values of the scalar to start inflation. In this way, a semiclassical non-singular inflationary model[73,74,75] is formulated which evades the singularity theorem of Ref. 9.

Current observations are already beginning to rule out certain, very large values of the ambiguity parameter j such that from future data one can expect much tighter limits. In all these scenarios the non-perturbative modification of the density is important, which is a characteristic feature of loop quantum cosmology. At larger scales above the peak there are also perturbative corrections which imply small changes in the cosmological expansion and the evolution of field modes. This has recently been investigated[76] with the conclusion that potential effects on the power spectrum would be too small to be noticed by the next generation satellites. The best candidates for observable effects from quantum gravity thus remain the non-perturbative modifications in effective densities.

5. Conclusions and Outlook

What we have described is a consistent picture of the universe which is not only observationally viable but also mathematically well-defined and non-singular. There are instances where quantum gravity is essential, and others where it is helpful in achieving important effects. The background independent quantization employed here is very efficient: There are a few basic properties, such as the discreteness of spatial geometry and the representation only of exponentials of curvature, which are behind a variety of applications. Throughout all the developments, those properties have been known to be essential for mathematical consistency before they were recognized as being responsible for physical phenomena.

For instance, for the singularity issue the basic properties are all needed in the way they turn out to be realized. First, the theory had to be based on triad variables which now not only provides us with the sign of orientation, and thus two sides of the classical singularity, but also in more complicated models positions the classical singularity in phase space such that it becomes accessible by quantum evolution. Then, the discreteness of spatial geometry encoded in triad operators and the representation of exponentials of curvature play together in the right way to remove divergences in densities and extend the quantum evolution through the classical singularity. These features allow general results about the absence of singularities without any new or artificial ingredients, and lead to a natural solution of a long-standing problem which has eluded previous attempts for decades. Symmetry assumptions are still important in order to be able to perform the calculations, but they can now be weakened considerably and are not responsible for physical implications. The essential step is to base the symmetry reduction on a candidate for full quantum gravity which is background independent and therefore allows one to study the essence of quantum geometry.

Absence of singularities in this context is a rather general statement about the possibility to extend a quantum wave function through a regime which classically would appear as a singularity. More explicit questions, such as what kind of new region one is evolving to and whether it again becomes classical or retains traces of the evolution through a quantum regime, depend on details of the relevant constraint operators. This includes, for instance, quantization ambiguities and the question whether a symmetric operator has to be used. The latter aspect is also important for technical concepts such as a physical inner product.

Here we discussed only isotropic models which are classically described solely by the scale factor determining the size of space. But a more realistic situation has to take into account also the shape of space, and changes of the distribution of geometry and matter between different points of space. The methods we used have been extended to homogeneous models, allowing for anisotropic spaces, and recently to some inhomogeneous ones, defined by spherical symmetry and some forms of cylindrical symmetry. In all cases, essential aspects of the general mechanism for removing classical singularities which has first been seen only in the simple isotropic models are known to be realized.[a] Moreover, in the more complicated systems it is acting much more non-trivially, again with the right ingredients provided by the background independent quantization. Nevertheless, since the inhomogeneous constraints are much more complicated to analyze, absence of singularities for them has not yet been proven completely. The inhomogeneous systems now also allow access to black hole and gravitational wave models such that their quantum geometry can be studied, too.

Effective equations are a useful tool to study quantum effects in a more

---

[a]This does not refer to the boundedness of densities or curvature components for *all geometries*, which is known not to be present in anisotropic models[77,78] or even on some degenerate configurations in the full theory.[79] What is relevant is the behavior on configurations seen along the dynamical evolution.

familiar setting given by classical equations of motion. They show diverse effects whose usefulness in cosmological phenomenology is often surprising. Also here, the effects were known to occur from the quantization and the transfer into effective classical equations, before they turned out to be helpful. In addition to inflationary scenarios and bounces which one can see in isotropic cosmologies, modified densities have more implications in less symmetric models. The anisotropic Bianchi IX model, for instance, is classically chaotic which is assumed to play a role in the complicated approach to a classical singularity.[80,81] With the effective modifications the dynamics changes and simplifies, removing the classical chaos.[82] This has implications for the effective approach to a classical singularity and can provide a more consistent picture of general singularities.[83] Effective classical equations can also be used to study the collapse of matter to a black hole, with modifications in the development of classical singularities and horizons.[84] This can now also be studied with inhomogeneous quantum models which allow new applications for black holes and cosmological phenomenology where the evolution of inhomogeneities is of interest in the context of structure formation.

With these models there will be new effects not just in cosmology but also for black holes and other systems which further check the overall consistency of the theory. Moreover, a better understanding of inhomogeneities evolving in a cosmological background will give us a much better computational handle on signatures in the cosmic microwave or even gravitational wave background, which may soon be testable with a new generation of observations. One may wonder how it can be possible to observe quantum gravity effects, given that the Planck scale is so many orders of magnitude away from scales accessible by today's experiments. The difference in scales, however, does not preclude the observation of indirect effects even though direct measurements on the discreteness scale are impossible, as illustrated by a well-known example: Brownian motion allows to draw conclusions about the atomic structure of matter and its size by observations on much larger scales.[85] Similarly, cosmological observations can carry information on quantum gravity effects which otherwise would manifest themselves only at the Planck scale.

References

1. A. H. Guth, *Phys. Rev. D* **23**, 347–356 (1981).
2. A. D. Linde, *Phys. Lett. B* **108**, 389–393 (1982).
3. A. Albrecht and P. J. Steinhardt, *Phys. Rev. Lett.* **48**, 1220–1223 (1982).

4. D. N. Spergel et al. *Astrophys. J. Suppl.* **148**, 175 (2003), [astro-ph/0302209].
5. T. Padmanabhan, this volume.
6. S. W. Hawking and R. Penrose, *Proc. Roy. Soc. Lond.* A **314**, 529–548 (1970).
7. A. Friedmann, *Z. Phys.* **10**, 377–386 (1922).
8. A. D. Linde, *Phys. Lett.* B **129**, 177–181 (1983).
9. A. Borde, A. H. Guth and A. Vilenkin, *Phys. Rev. Lett.* **90**, 151301 (2003), [gr-qc/0110012].
10. B. S. DeWitt, *Phys. Rev.* **160**, 1113–1148 (1967).
11. D. L. Wiltshire, in *Cosmology: The Physics of the Universe*, Ed. B. Robson, N. Visvanathan and W. S. Woolcock (World Scientific, Singapore, 1996), p. 473–531, [gr-qc/0101003].
12. A. Vilenkin, *Phys. Rev.* D **30**, 509–511 (1984).
13. J. B. Hartle and S. W. Hawking, *Phys. Rev.* D **28**, 2960–2975 (1983).
14. C. Rovelli, *Living Reviews in Relativity* **1**, 1 (1998), [gr-qc/9710008], http://www.livingreviews.org/Articles/Volume1/1998-1rovelli.
15. T. Thiemann, Introduction to Modern Canonical Quantum General Relativity, [gr-qc/0110034].
16. A. Ashtekar and J. Lewandowski, *Class. Quantum Grav.* **21**, R53–R152 (2004), [gr-qc/0404018].
17. A. Ashtekar, this volume.
18. A. Ashtekar, *Phys. Rev.* D **36**, 1587–1602 (1987).
19. J. F. Barbero G., *Phys. Rev.* D **51**, 5507–5510 (1995), [gr-qc/9410014].
20. C. Rovelli and L. Smolin, *Nucl. Phys.* B **331**, 80–152 (1990).
21. A. Ashtekar, J. Lewandowski, D. Marolf, J. Mourão and T. Thiemann, *J. Math. Phys.* **36**, 6456–6493 (1995), [gr-qc/9504018].
22. C. Rovelli and L. Smolin, *Phys. Rev.* D **52**, 5743–5759 (1995).
23. H. Sahlmann, Some Comments on the Representation Theory of the Algebra Underlying Loop Quantum Gravity, [gr-qc/0207111].
24. H. Sahlmann, When Do Measures on the Space of Connections Support the Triad Operators of Loop Quantum Gravity?, [gr-qc/0207112].
25. A. Okolow and J. Lewandowski, *Class. Quantum Grav.* **20**, 3543–3568 (2003), [gr-qc/0302059].
26. H. Sahlmann and T. Thiemann, On the superselection theory of the Weyl algebra for diffeomorphism invariant quantum gauge theories, [gr-qc/0302090].
27. C. Fleischhack, Representations of the Weyl Algebra in Quantum Geometry, [math-ph/0407006].
28. C. Rovelli and L. Smolin, *Nucl. Phys.* B **442**, 593–619 (1995), [gr-qc/9411005], Erratum: *Nucl. Phys.* B 456 (1995) 753.
29. A. Ashtekar and J. Lewandowski, *Class. Quantum Grav.* **14**, A55–A82 (1997), [gr-qc/9602046].
30. A. Ashtekar and J. Lewandowski, *Adv. Theor. Math. Phys.* **1**, 388–429 (1997), [gr-qc/9711031].
31. T. Thiemann, *Class. Quantum Grav.* **15**, 839–873 (1998), [gr-qc/9606089].
32. T. Thiemann, *Class. Quantum Grav.* **15**, 1281–1314 (1998), [gr-qc/9705019].
33. M. Bojowald, *Quantum Geometry and Symmetry*, PhD thesis, RWTH Aachen, 2000, published by Shaker-Verlag, Aachen.

34. M. Bojowald and H. A. Kastrup, *Class. Quantum Grav.* **17**, 3009–3043 (2000), [hep-th/9907042].
35. M. Bojowald, *Class. Quantum Grav.* **21**, 3733–3753 (2004), [gr-qc/0407017].
36. A. Ashtekar, M. Bojowald and J. Lewandowski, *Adv. Theor. Math. Phys.* **7**, 233–268 (2003), [gr-qc/0304074].
37. A. Ashtekar, S. Fairhurst and J. Willis, *Class. Quant. Grav.* **20**, 1031–1062 (2003), [gr-qc/0207106].
38. W. Thirring and H. Narnhofer, *Rev. Math. Phys.* **SI1**, 197–211 (1992).
39. M. Bojowald and H. A. Morales-Técotl, in *Proceedings of the Fifth Mexican School (DGFM): The Early Universe and Observational Cosmology*, Lect. Notes Phys. 646, Springer-Verlag, 2004, p. 421–462, [gr-qc/0306008].
40. M. Bojowald, *Class. Quantum Grav.* **19**, 2717–2741 (2002), [gr-qc/0202077].
41. M. Bojowald, *Phys. Rev. Lett.* **86**, 5227–5230 (2001), [gr-qc/0102069].
42. M. Bojowald, *Class. Quantum Grav.* **18**, L109–L116 (2001), [gr-qc/0105113].
43. D. Giulini, C. Kiefer, E. Joos, J. Kupsch, I. O. Stamatescu and H. D. Zeh, *Decoherence and the Appearance of a Classical World in Quantum Theory*, Springer, Berlin, Germany, 1996.
44. G. M. Hossain, *Class. Quantum Grav.* **21**, 179–196 (2004), [gr-qc/0308014].
45. K. Noui, A. Perez and K. Vandersloot, *Phys. Rev. D* **71**, 044025 (2005), [gr-qc/0411039].
46. A. Ashtekar, T. Pawlowski and P. Singh, Quantum nature of the big bang: An analytical and numerical study, in preparation.
47. M. Bojowald and G. Date, *Class. Quantum Grav.* **21**, 121–143 (2004), [gr-qc/0307083].
48. M. Bojowald, *Phys. Rev. Lett.* **87**, 121301 (2001), [gr-qc/0104072].
49. M. Bojowald, *Gen. Rel. Grav.* **35**, 1877–1883 (2003), [gr-qc/0305069].
50. D. Cartin, G. Khanna and M. Bojowald, *Class. Quantum Grav.* **21**, 4495–4509 (2004), [gr-qc/0405126].
51. D. Cartin and G. Khanna, *Phys. Rev. Lett.* **94**, 111302 (2005), [gr-qc/0501016].
52. M. Bojowald, *Phys. Rev. Lett.* **95**, 061301 (2005), [gr-qc/0506128].
53. M. Bojowald, *Phys. Rev. D* **64**, 084018 (2001), [gr-qc/0105067].
54. M. Gaul and C. Rovelli, *Class. Quantum Grav.* **18**, 1593–1624 (2001), [gr-qc/0011106].
55. M. Bojowald, *Class. Quantum Grav.* **19**, 5113–5130 (2002), [gr-qc/0206053].
56. M. Bojowald, in *Proceedings of the International Conference on Gravitation and Cosmology (ICGC 2004), Cochin, India*, Vol. 63, Pramana, 2004, p. 765–776, [gr-qc/0402053].
57. A. Ashtekar, M. Bojowald and J. Willis, in preparation.
58. J. Willis, *On the Low-Energy Ramifications and a Mathematical Extension of Loop Quantum Gravity*, PhD thesis, The Pennsylvania State University, 2004.
59. M. Bojowald, P. Singh and A. Skirzewski, *Phys. Rev. D* **70**, 124022 (2004), [gr-qc/0408094].
60. P. Singh and A. Toporensky, *Phys. Rev. D* **69**, 104008 (2004), [gr-qc/0312110].

61. G. V. Vereshchagin, *JCAP* **07**, 013 (2004), [gr-qc/0406108].
62. J. E. Lidsey, D. J. Mulryne, N. J. Nunes and R. Tavakol, *Phys. Rev. D* **70**, 063521 (2004), [gr-qc/0406042].
63. M. Bojowald, R. Maartens and P. Singh, *Phys. Rev. D* **70**, 083517 (2004), [hep-th/0407115].
64. G. Date and G. M. Hossain, *Class. Quantum Grav.* **21**, 4941–4953 (2004), [gr-qc/0407073].
65. G. Date and G. M. Hossain, *Phys. Rev. Lett.* **94**, 011302 (2005), [gr-qc/0407074].
66. G. Date and G. M. Hossain, *Phys. Rev. Lett.* **94**, 011301 (2005), [gr-qc/0407069].
67. M. Bojowald and R. Swiderski, *Phys. Rev. D* **71**, 081501 (2005), [gr-qc/0410147].
68. M. Bojowald and R. Swiderski, Spherically Symmetric Quantum Geometry: Hamiltonian Constraint, in preparation.
69. G. M. Hossain, *Class. Quant. Grav.* **22**, 2511–2532 (2005), [gr-qc/0411012].
70. S. Tsujikawa, P. Singh and R. Maartens, *Class. Quantum Grav.* **21**, 5767–5775 (2004), [astro-ph/0311015].
71. M. Bojowald, J. E. Lidsey, D. J. Mulryne, P. Singh and R. Tavakol, *Phys. Rev. D* **70**, 043530 (2004), [gr-qc/0403106].
72. D. J. Mulryne, N. J. Nunes, R. Tavakol and J. Lidsey, *Int. J. Mod. Phys. A* **20**, 2347-2357 (2005), [gr-qc/0411125].
73. G. F. R. Ellis and R. Maartens, *Class. Quant. Grav.* **21**, 223–232 (2004), [gr-qc/0211082].
74. G. F. R. Ellis, J. Murugan and C. G. Tsagas, *Class. Quant. Grav.* **21**, 233–250 (2004), [gr-qc/0307112].
75. D. J. Mulryne, R. Tavakol, J. E. Lidsey and G. F. R. Ellis, *Phys. Rev. D* **71**, 123512 (2005), [astro-ph/0502589].
76. S. Hofmann and O. Winkler, The Spectrum of Fluctuations in Inflationary Quantum Cosmology, [astro-ph/0411124].
77. M. Bojowald, *Class. Quantum Grav.* **20**, 2595–2615 (2003), [gr-qc/0303073].
78. M. Bojowald, G. Date, and K. Vandersloot, *Class. Quantum Grav.* **21**, 1253–1278 (2004), [gr-qc/0311004].
79. J. Brunnemann and T. Thiemann, Unboundedness of Triad-Like Operators in Loop Quantum Gravity, [gr-qc/0505033]
80. V. A. Belinskii, I. M. Khalatnikov and E. M. Lifschitz, *Adv. Phys.* **13**, 639–667 (1982).
81. A. D. Rendall, this volume.
82. M. Bojowald and G. Date, *Phys. Rev. Lett.* **92**, 071302 (2004), [gr-qc/0311003].
83. M. Bojowald, G. Date and G. M. Hossain, *Class. Quantum Grav.* **21**, 3541–3569 (2004), [gr-qc/0404039].
84. M. Bojowald, R. Goswami, R. Maartens and P. Singh, A black hole mass threshold from non-singular quantum gravitational collapse, *Phys. Rev. Lett.* **95**, 091302 (2005), [gr-qc/0503041].
85. A. Einstein, *Annalen Phys.* **17**, 549–560 (1905).

CHAPTER 15

CONSISTENT DISCRETE SPACE-TIME

RODOLFO GAMBINI

Instituto de Física, Facultad de Ciencias,
Iguá 4225, Montevideo, Uruguay

JORGE PULLIN

Department of Physics and Astronomy,
Louisiana State University,
Baton Rouge, LA 70803, USA

We review recent efforts to construct gravitational theories on discrete space-times, usually referred to as the "consistent discretization" approach. The resulting theories are free of constraints at the canonical level and therefore allow to tackle many problems that cannot be currently addressed in continuum quantum gravity. In particular the theories imply a natural method for resolving the big bang (and other types) of singularities and predict a fundamental mechanism for decoherence of quantum states that might be relevant to the black hole information paradox. At a classical level, the theories may provide an attractive new path for the exploration of issues in numerical relativity. Finally, the theories can make direct contact with several kinematical results of continuum loop quantum gravity. We review in broad terms several of these results and present in detail as an illustration the classical treatment with this technique of the simple yet conceptually challenging model of two oscillators with constant energy sum.

1. Introduction

The idea that space-time might be discrete has arisen at various levels in gravitational physics. On one hand, some approaches hypothesize that at a fundamental level a discrete structure underlies space-time. Other approaches start with a continuum theory but upon quantization discrete structures associated with space-time emerge. Finally, discretizations are widely used in physics, and in gravity in particular, as a calculational tool at two levels: a) at the time of numerically computing predictions of the theory (classical and quantum mechanically) and b) as a regularization tool for quantum calculations.

Whatever the point of view that may lead to the consideration of a discrete space-time, the formulation of gravitational theories on such structures presents significant challenges. At the most immediate level, the presence of discrete structures can conflict with diffeomorphism invariance, a desirable property of gravitational theories. This manifests itself in various ways. For instance, if one simply proceeds to discretize the equations of motion of general relativity, as is common in numerical relativity applications, one finds that the resulting equations are inconsistent: the evolution equations do not preserve the constraints, as they do in the continuum. In another example, if one considers the discretization of the constraints of canonical quantum gravity, the resulting discrete constraints fail to close an algebra, which can be understood as another manifestation of the inconsistency faced in numerical relativity.

A new viewpoint has recently emerged towards the treatment of theories on discrete space-times. At the most basic level, the viewpoint advocates discretizing the action of the theory and working out the resulting equations of motion rather than discretizing the equations of motion directly. The resulting equations of motion stemming from the discrete action are generically guaranteed to be consistent. So immediately the problem of consistency is solved. This has led to the approach being called the "consistent discretization" approach. This approach has been pursued in the past in numerical approaches to unconstrained theories and is known as "variational integrators" (see Lew et al.[1] for a review) and it appears to have several desirable properties. Constrained systems have only been considered if they are holonomic (i.e. they only depend on the configuration variables), although there are some recent results in the mathematical literature for anholonomic constraints[2].

In spite of these positive prospects, the resulting theories have features that at first sight may appear undesirable. For instance, quantities that in the usual continuum treatment are Lagrange multipliers of first class constrained systems (and therefore freely specifiable), become determined by the equations of motion. The equations that determine the Lagrange multipliers in general may have undesirable complex solutions.

On the other hand, the approach has potentially very attractive features: equations that in the continuum are constraints among the dynamical variables become evolution equations that are automatically solved by the scheme. Having no constraints in the theory profoundly simplifies things at the time of quantization. The conceptually hard problems of canonical quantum gravity are almost entirely sidestepped by this approach. For instance, one can introduce a relational description in the theory and therefore solve the "problem of time" that created so much trouble in canonical quantum gravity. The resulting relational description naturally implies a loss of unitarity that may have implications for the black hole information puzzle. The discrete theories also have a tendency to avoid singularities, since the latter usually do not fall on the computational grid. At a quantum level this implies that singularities have zero probability. This provides a singularity   · avoidance mechanism that is distinct from the one usually advocated in loop quantum cosmology. From the point of view of numerical relativity, the resulting evolution schemes preserve the constraints of the continuum theory to a great degree of accuracy, at least for solutions that approximate the continuum limit. This is different from usual "free evolution" schemes which can converge to continuum solutions that violate the constraints. As we will see, we are still somewhat away from being able to advocate that the resulting schemes can be competitive in numerical relativity. But given that enforcing the constraints has been identified by some researchers as the main obstacle to numerical relativity, the proposed schemes deserve some consideration.

An aspect of presupposing a discrete structure for space-time that may also appear undesirable is that in loop quantum gravity the discrete structures only emerge after quantization. The initial formulation of the theory is in the continuum. Therefore there is the risk that the new viewpoint may not be able to make contact with the many attractive kinematical results of loop quantum gravity. We will see that a connection is in fact possible, and it retains the attractive aspects of both approaches.

In this paper we would like to review the "consistent discretization" approach to gravitational theories. In Section 2 we will outline the basics of the strategy. In Section 3 we concretely apply the method to a simple yet important model problem so the reader can get a flavor of what to expect from the method classically. In Section 4 we discuss cosmological models and in Section 5 the introduction of a relational time and the black hole information puzzle. In Section 6 we discuss connections with continuum loop quantum gravity. We end with a discussion.

2. Consistent Discretizations

We start by considering a continuum theory representing a mechanical system. Although our ultimate interest is in field theories, the latter become mechanical systems when discretized. Its Lagrangian will be denoted by $\hat{L}(q^a, \dot{q}^a)$, $a = 1 \dots M$. This setting is general enough to accommodate, for instance, totally constrained systems. In such case $\dot{q}$ will be the derivative of the canonical variables with respect to the evolution parameter. It is also general enough to include the systems that result from formulating on a discrete space-time lattice a continuum field theory.

We discretize the evolution parameter in intervals (possibly varying upon evolution) $t_{n+1} - t_n = \epsilon_n$ and we label the generalized coordinates evaluated at t_n as q_n. We define the discretized Lagrangian as

$$L(n, n+1) \equiv L(q_n^a, q_{n+1}^a) \equiv \epsilon_n \hat{L}(q^a, \dot{q}^a) \tag{1}$$

where

$$q^a = q_n^a \quad \text{and} \quad \dot{q}^a \equiv \frac{q_{n+1}^a - q_n^a}{\epsilon_n}. \tag{2}$$

Of course, one could have chosen to discretize things in a different fashion, for instance using a different approximation for the derivative, or by choosing to write the continuum Lagrangian in terms of different variables. The resulting discrete theories generically will be different and will approximate the continuum theory in different ways. However, given a discrete theory, the treatment we outline in this paper is unique.

The action can then be written as

$$S = \sum_{n=0}^{N} L(n, n+1). \tag{3}$$

It should be noted that in previous treatments[3,4] we have written the Lagrangian in first order form, i.e. $L = \int dt \, (p\dot{q} - H(p,q))$. It should be emphasized that this is contained as a particular case in the treatment we are presenting in this paper. In this case one takes both q and p to be configuration variables, and one is faced with a Lagrangian that involves q_n, p_n and q_{n+1} as variables, being independent of p_{n+1}. The reason we frequently resort to first order formulations in the various concrete examples we discuss is that the Ashtekar formulation is naturally a first order one and we usually tend to frame things in a closely related way. But again, there is no obstruction to using either first or second order formulations with our framework, they are both contained as particular cases.

If the continuum theory is invariant under reparameterizations of the evolution parameter, one can show that the information about the intervals ϵ_n may be absorbed in the Lagrange multipliers. In the case of standard mechanical systems it is simpler to use an invariant interval $\epsilon_n = \epsilon$.

The Lagrange equations of motion are obtained by requiring the action to be stationary under variations of the configuration variables q^a fixed at the endpoints of the evolution interval $n = 0, n = N + 1$,

$$\frac{\partial L(n, n+1)}{\partial q_n^a} + \frac{\partial L(n-1, n)}{\partial q_n^a} = 0. \tag{4}$$

We introduce the following definition of the canonically conjugate momenta of the configuration variables,

$$p_{n+1}^a \equiv \frac{\partial L(n, n+1)}{\partial q_{n+1}^a} \tag{5}$$

$$p_n^a \equiv \frac{\partial L(n-1, n)}{\partial q_n^a} = -\frac{\partial L(n, n+1)}{\partial q_n^a} \tag{6}$$

Where we have used Eq. (4). The equations (5) and (6) define a canonical transformation for the variables q_n, p_n to q_{n+1}, p_{n+1} with a the type 1 generating function $F_1 = -L(q_n^a, q_{n+1}^a)$. Notice that the evolution scheme is implicit, one can use the bottom equation (since we are in the non-singular case) to give an expression for q_{n+1} in terms of q_n, p_n, which in turn can be substituted in the top equation to get an equation for p_{n+1} purely in terms of q_n, p_n.

It should be noted that there are several other possible choices, when going from the set of equations (5,6) to an explicit evolution scheme (see Di Bartolo *et al.*[5] for further details.)

The canonical transformation we introduced becomes singular as an evolution scheme if $\left| \dfrac{\partial^2 L(n, n+1)}{\partial q_{n+1}^a \partial q_n^b} \right|$ vanishes. If the rank of the matrix of second partial derivatives is K the system will have $2(M - K)$ constraints of the form,

$$\Phi_A(q_n^a, p_n^a) = 0 \tag{7}$$

$$\Psi_A(q_{n+1}^a, p_{n+1}^a) = 0. \tag{8}$$

And these constraints need to be enforced during evolution, which may lead to new constraints. We refer the reader for the detailed Dirac analysis to Di Bartolo *et al.*[5].

To clarify ideas, let us consider an example. The model consists of a parameterized free particle in a two dimensional space-time under the influence of a linear potential. The discrete Lagrangian is given by,

$$L_n \equiv L(q_n^a, \pi_n^a, N_n, q_{n+1}^a, \pi_{n+1}^a, N_{n+1}) \tag{9}$$

$$= \pi_n^a(q_{n+1}^a - q_n^a) - N_n[\pi_n^0 + \frac{1}{2}(\pi_n^1)^2 + \alpha q_n^1].$$

We have chosen a first order formulation for the particle (otherwise there are no constraints and the example is trivial). However, this Lagrangian is of the type we considered in this paper, one simply needs to consider all variables, q^a, π^a, N as configuration variables. The system is clearly singular since the $\pi's$ and N only appear at level n (or in the continuum Lagrangian, their time derivatives are absent). When considered as a Type I generating function, the above Lagrangian leads to the equations

$$p_{\pi, n+1}^a = \frac{\partial L_n}{\partial \pi_{n+1}^a} = 0, \tag{10}$$

$$p_{q, n+1}^a = \frac{\partial L_n}{\partial q_{n+1}^a} = \pi_n^a, \tag{11}$$

$$p_{N, n+1} = \frac{\partial L_n}{\partial N_{n+1}} = 0, \tag{12}$$

and

$$p_{\pi, n}^a = -\frac{\partial L_n}{\partial \pi_n^a} = -(q_{n+1}^a - q_n^a) + \pi_n^1 N_n \delta_1^a + N_n \delta_0^a, \tag{13}$$

$$p_{q, n}^a = -\frac{\partial L_n}{\partial q_n^a} = \pi_n^a + \delta_1^a \alpha N_n, \tag{14}$$

$$p_{N, n} = -\frac{\partial L_n}{\partial N_n} = \pi_n^0 + \frac{1}{2}(\pi_n^1)^2 + \alpha q_n^1. \tag{15}$$

The constraints (10,12,14,15) can be imposed strongly to eliminate the π's and the N's and obtain an explicit evolution scheme for the q's and the p_q's,

$$q_n^0 = q_{n+1}^0 - \frac{C_{n+1}}{\alpha p_{q, n+1}^1}, \tag{16}$$

$$q_n^1 = q_{n+1}^1 - \frac{C_{n+1}}{\alpha}, \tag{17}$$

$$p_{q, n}^0 = p_{q, n+1}^0, \tag{18}$$

$$p_{q, n}^1 = p_{q, n+1}^1 + \frac{C_{n+1}}{p_{q, n+1}^1}, \tag{19}$$

and the Lagrange multipliers get determined to be,

$$N_n = \frac{C_{n+1}}{\alpha p^1_{q,\,n+1}}, \tag{20}$$

where $C_{n+1} = p^0_{q,\,n+1} + (p^1_{q,\,n+1})^2/2 + \alpha q^1_{n+1}$. The evolution scheme runs backwards, one can construct a scheme that runs forward by solving for N and π at instant n when imposing the constraints strongly. The two methods yield evolution schemes of different functional form since one propagates "forward" in time and the other "backward". The inequivalence in the functional form stems from the fact that the discretization of the time derivatives chosen in the Lagrangian is not centered. It should be emphasized that if one starts from given initial data and propagates forwards with the first system of equations and then backwards using the second, one will return to the same initial data.

So we see in the example how the mechanism works. It yields evolution equations that usually are implicit as evolution schemes. The equations are consistent. The Lagrange multipliers get determined by the scheme and there are no constraints left on the canonical variables. The evolution is implemented by a (non-singular) canonical transformation. The number of degrees of freedom is larger than those in the continuum. There will exist different sets of initial data that lead to different solutions for the discrete theory but nevertheless will just correspond to different discrete approximations and parameterizations of a single evolution of the continuum theory.

3. The Rovelli Model

To analyze the method in a simple —yet challenging— model we consider the model analyzed by Rovelli[6] in the context of the problem of time in canonical quantum gravity: two harmonic oscillators with constant energy sum. The intention of this section is to illustrate how the method works and some of the expectations one can hold when applying the method to more complex situations. The model itself can obviously be treated with more straightforward discretization techniques given its simplicity. The fact that the method works well for the model should not be construed as proof that it will be successful in other numerical applications. Current efforts suggest that the technique is successfully applicable to Gowdy cosmologies.

The model has canonical coordinates q^1, q^2, p^1, p^2 with the standard Poisson brackets and a constraint given by,

$$C = \frac{1}{2}\left((p^1)^2 + (p^2)^2 + (q^1)^2 + (q^2)^2\right) - M = 0, \tag{21}$$

with M a constant. No Hamiltonian system can correspond to this dynamical system since the presymplectic space is compact and therefore cannot contain any $S \times R$ structure. Nevertheless, we will see that the consistent discretization approach does yield sensible results. This helps dispel certain myths about the consistent discretization scheme. Since it determines Lagrange multipliers a lot of people tend to associate the scheme as some sort of "gauge fixing". For this model however, a gauge fixing solution would be unsatisfactory, since it would only cover a portion of phase space. We will see this is not the case in the consistent discretization scheme. We will also see that the evolution scheme is useful numerically in practice.

We start by writing a discrete Lagrangian for the model,

$$L(n, n+1) = p_n^1 \left(q_{n+1}^1 - q_n^1 \right) + p_n^2 \left(q_{n+1}^2 - q_n^2 \right) \tag{22}$$

$$- \frac{N_n}{2} \left((p_n^1)^2 + (p_n^2)^2 + (q_n^1)^2 + (q_n^2)^2 - 2M \right),$$

and working out the canonical momenta for all the variables, i.e., $P_q^1, P_q^2, P_p^1, P_p^2$. One then eliminates the $p^{1,2}$ and the $P_p^{1,2}$ and is left with evolution equations for the canonical pairs,

$$q_{n+1}^1 = q_n^1 + N_n \left(P_{q,n}^1 - 2q_n^1 \right) \tag{23}$$

$$q_{n+1}^2 = q_n^2 + N_n \left(P_{q,n}^2 - 2q_n^2 \right) \tag{24}$$

$$P_{q,n+1}^1 = P_{q,n}^1 - N_n q_n^1 \tag{25}$$

$$P_{q,n+1}^2 = P_{q,n}^2 - N_n q_n^2. \tag{26}$$

The Lagrange multiplier gets determined by the solution(s) of a quadratic equation,

$$\left((q_n^1)^2 + (q_n^2)^2 \right) (N_n)^2 - 2 \left(P_{q,n}^1 q_n^1 + P_{q,n}^2 q_n^2 \right) N_n +$$

$$+ \left(P_{q,n}^1 \right)^2 + \left(P_{q,n}^2 \right)^2 + \left(q_n^1 \right)^2 + \left(q_n^2 \right)^2 - 2M = 0. \tag{27}$$

We would like to use this evolution scheme to follow numerically the trajectory of the system. For this, we need to give initial data. Notice that if one gives initial data that satisfy the constraint identically at level n, the quadratic equation for the lapse has a vanishing independent term and therefore the solution is that the lapse N vanishes (the nonvanishing root will be large and far away from the continuum generically). To construct initial data one therefore considers a set for which the constraint vanishes and introduces a small perturbation on one (or more) of the variables. Then one will have evolution. Notice that one can make the perturbation as small as desired. The smaller the perturbation, the smaller the lapse and the closer the solution will be to the continuum.

For concreteness, we choose the following initial values for the variables, $M = 2$,

$$q_0^1 = 0, \tag{28}$$

$$q_0^2 = (\sqrt{3} - \Delta)\sin(\frac{\pi}{4}), \tag{29}$$

$$P_{q,0}^1 = 1, \tag{30}$$

$$P_{q,0}^1 = (\sqrt{3} - \Delta)\cos(\frac{\pi}{4}), \tag{31}$$

We choose the parameter Δ to be the perturbation, i.e., $\Delta = 0$ corresponds to an exact solution of the constraint, for which the observable $A = 1/2$ (see below for its definition). The evolution scheme can easily be implemented using a computer algebra program like Maple or Mathematica.

Before we show results of the evolution, we need to discuss in some detail how the method determines the lapse. As we mentioned, it is obtained by solving the quadratic equation (27). This implies that generically there will be two possible solutions and in some situations they could be negative or complex. One can choose any of the two solutions at each point during the evolution. It is natural numerically to choose one "branch" of the solution and keep with it. However, if one encounters that the roots become complex, we have observed that it is possible to backtrack to he previous point in the iteration, choose the alternate root to the one that had been used up to that point and continue with the evolution. Similar procedure could be followed when the lapse becomes negative. It should be noted that negative lapses are not a problem per se, it is just that the evolution will be retraced backwards. We have not attempted to correct such retracings, i.e. in the evolutions shown we have only "switched branches" whenever the lapse becomes complex. This occurs when the discriminant in the quadratic equation (27) vanishes.

Figure 1 shows the evolution of q_1 as a function of n. The figure looks choppy since the "rate of advance" (magnitude of the lapse) varies during the evolution.

Figure 1 also exhibits that in a reparameterization invariant theory like this one it is not too useful to plot parameterization dependent quantities. One would have to exhibit relational quantities that are true observables of the theory to obtain physically relevant information. In this particular model, Rovelli has given an explicit expression for a relational ("evolving") observable

$$q^2(q^1) = \sqrt{M/A - 1}\left[q^1\cos\phi \pm \sqrt{2A - (q^1)^2}\sin\phi\right], \tag{32}$$

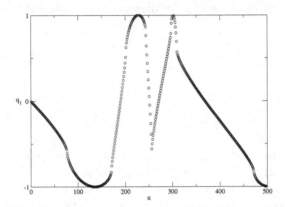

Fig. 1.   The evolution of one of the variables of the oscillator model as a function of the
discretization parameter n. The irregular nature of the curve is due to the fact that the
lapse is a dynamical variable and therefore the rate of advance changes during evolution.
The sharp feature around $n = 250$ is due to the fact that the lapse becomes negative
and the evolution runs backwards for a while, until about $n = 310$.

where A and ϕ are constants of the motion ("perennials"), whose expression
in terms of the coordinates is,

$$4A = 2M + (p^1)^2 - (p^2)^2 + (q^1)^2 - (q^2)^2, \tan\phi = \frac{p^1 q^2 - p^2 q^1}{p^1 p^2 + q^2 q^1}. \quad (33)$$

The relational observable gives an idea of the trajectory in configuration
space in a manner that is invariant under reparameterizations. In figure 2
we show the error in the evaluation of the relational observable with respect
to the exact expression of the continuum in our model.

This model has two independent "perennials" that can be used to construct relational observables like the one we just discussed. The first one of
these perennials happens to be an exact conserved quantity of the discretized theory. The relation between perennials in the continuum and
conserved quantities of the discrete theory was further discussed in[7]. The
perennial in question is,

$$O_1 = p^1 q^2 - p^2 q^1. \quad (34)$$

Another perennial is given by

$$O_2 = (p^1)^2 - (p^2)^2 + (q^1)^2 - (q^2)^2. \quad (35)$$

This quantity is not an exact conserved quantity of the discrete model, it is
conserved approximately, as we can see in figure 3. We see that the discrete

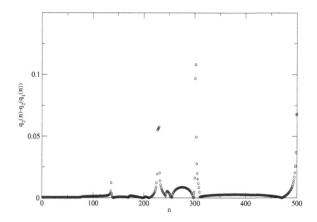

Fig. 2. The error in the evaluation of the relational observable in the discretized evolution, compared with respect to the exact continuum expression. We show the absolute
error, but since the quantities are of order one it can also be understood as the relative
error. The peaks in the error are due to the functional form of the observable involving
the square root of $2A - q_1^2$. This vanishes when $q_1 = \pm 1$ (see previous plot) and this
magnifies the error in the observable.

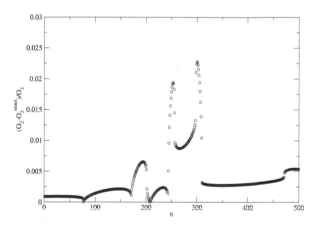

Fig. 3. The model has two "perennials". One of them is an exact conserved quantity
of the discrete theory, so we do not present a plot for it. The second perennial (O_2)
is approximately conserved. The figure shows the relative error in its computation in
the discrete theory. It is worthwhile noticing that, unlike what is usual in free evolution
schemes, errors do not accumulate, they may grow for a while but later they might
diminish.

R. Gambini and J. Pullin

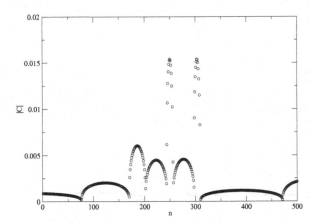

Fig. 4.   Absolute value of the constraint of the continuum theory, evaluated in the discrete theory. The plot shows that the constraint of the continuum theory does not increase in the discrete theory as a function of evolution (a major desired goal in numerical relativity). The value of the constraint is also a measure of the error in the quantities of the discrete theory with respect to those of the continuum one (compare with the error in the observable in figure 3, for example) that can be used to independently assess the accuracy of the theory in convergence studies.

theory conserves the perennial quite well in relative error and even though in intermediate steps of the evolution the error grows, it decreases later.

In figure 4 we depict the absolute value of the constraint of the continuum theory as a function of discrete time n. It is interesting to observe that in the discrete theory the variables approximate the ones of the continuum with an error that is proportional to the value of the constraint. Therefore the value of the constraint can be taken as an indicator of how accurately one is mirroring the continuum theory. It is a nice feature to have such an error indicator that is independent of the knowledge of the exact solution. This is clearly exhibited by contrasting (4) with (3) and seeing how the value of the constraint mirrors the error in the perennial. It should be noted that the proportionality factor between the value of the constraint and the error is a function of the dynamical variables and therefore the value of the constraint should only be taken as an indicator, not an exact measure of the error. However, it is a good indicator when one carries out convergence studies, since there the dynamical variables do not change in value significantly from one resolution to the next and the constraint diminishes as one converges to the continuum.

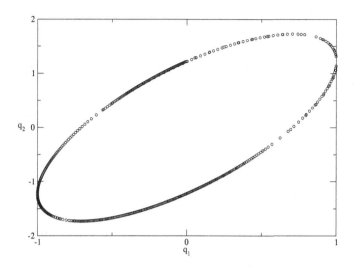

Fig. 5. The orbit in configuration space. As it is readily seen, the consistent discrete approach covers the entire available configuration space. This clearly exhibits that the approach is not a "gauge fixing". Gauge fixed approaches cannot cover the entire configuration space due to its compact nature. The dynamical changes in the value of the lapse can be seen implicitly through the density of points in the various regions of the trajectory. Also apparent is that the trajectory is traced on more than one occasion in various regions. Deviation from the continuum trajectory is not noticeable in the scales of the plot.

Figure 5 shows the trajectory in configuration space. As we see, the complete trajectory is covered by the discretized approach. This is important since many people tend to perceive the consistent discretization approach as "some sort of gauge fixing". This belief stems from the fact that when one gauge fixes a theory, the multipliers get determined. In spite of this superficial analogy, there are many things that are different from a gauge fixing. For instance, as we discussed before, the number of degrees of freedom changes. In addition to this, this example demonstrates another difference. If one indeed had gauge fixed this model, one would fail to cover the entire available configuration space, given its compact nature.

An issue of interest in any numerical method is the concept of convergence. Any reasonable numerical scheme should be such that one has control on the approximation of the exact solution. Figure 6 shows the convergence in the error of the estimation of the observable O_2. We see that one has convergence in the traditional sense, i.e. making the discretization

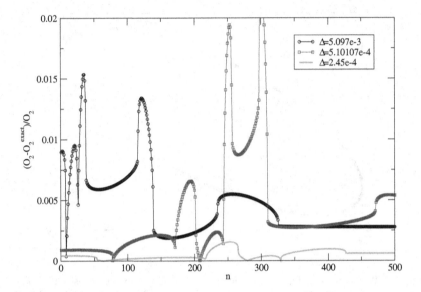

Fig. 6.   Convergence of the method with increasing resolution. We display the relative
error in one of the perennials of the theory as a function of n. The two runs with the
coarser resolutions are shown with one point out of every ten displayed. The finer run
is shown with one point displayed out of every thirty. The range of n displayed corresponds to a full trajectory along the ellipse in configuration space. So the improvement
in accuracy is throughout the entire evolution with different levels of improvement at
different points. See the text as to why we fine tune the evolution steps for the various
runs in the convergence study.

step smaller lowers the errors. However, one notes differences with the usual
type of convergence in the sense that here we have that it is not uniform as
a function of the evolution time. One notes, for example, that at isolated
points some of the coarser runs have very low errors. To understand this
one needs to recall that in our approach discrete expressions differ from the
continuum ones as,

$$O = O_{\text{continuum}} + f(p, q)C \qquad (36)$$

with C the constraints. It could happen that at a particular point in a coarse
evolution the constraint chances to take a value very close to zero. Slightly
finer resolution runs may not land on top of that particular point and
therefore will apparently have larger errors in the region. Eventually, if one

increases the resolution enough, one will be able to achieve better accuracy than with the coarser run. The reader may ponder why the parameter we have chosen to characterize the discretization is listed with several digits of precision in the convergence runs. The reason is the following: our algorithm exhibits some sensitivity on the parameter in the following sense. If one chooses slightly different values of Δ, the way in which the scheme will usually cover the phase space will be different. For instance, one choice may cover the configuration space ellipse in one continuous sweep, another close choice may reverse course several times before covering the ellipse. From a physical point of view this is no problem, but from the point of view of comparing runs as a function of n one wishes to compare runs that behave in the same way. Therefore when one "halves the stepsize" one may need to fine tune by trial an error to make sure one is comparing evolutions that behave similarly in terms of the (unphysical) parameter n.

4. Applications in Classical and Quantum Cosmology

To try to seek an application more connected with gravitational physics, yet simple enough that we can solve things analytically, let us consider a cosmological model. The model in question will be a Friedmann model with a cosmological constant and a scalar field. To make the solution of the model analytically tractable we will consider a very massive scalar field (i.e. we will ignore the kinetic term of the field in the action). We have actually solved this and other models without this approximation numerically and obtained results that are conceptually similar to the ones we present here.

The Lagrangian for the model, written in terms of Ashtekar's new variables (there is no impediment in treating the model with traditional variables if one wishes) is,

$$L = E\dot{A} + \pi\dot{\phi} - NE^2(-A^2 + (\Lambda + m^2\phi^2)|E|) \tag{37}$$

where Λ is the cosmological constant, m is the mass of the scalar field ϕ, π is its canonically conjugate momentum and N is the lapse with density weight minus one. Here E and A are the functions of time which are what remains of the triad and connection for the homogeneous case. The appearance of $|E|$ in the Lagrangian is due to the fact that the term cubic in E is supposed to represent the spatial volume and therefore should be positive definite. In terms of the ordinary lapse α we have $\alpha = N|E|^{3/2}$. The equations of motion and the only remaining constraint (Hamiltonian) are

$$\dot{A} = N(\Lambda + m^2\phi^2)\text{sgn}(E)E^2 \tag{38}$$

$$\dot{E} = 2NE^2A \tag{39}$$

$$\dot{\phi} = 0 \tag{40}$$

$$\dot{\pi} = -2N|E|^3m^2\phi \tag{41}$$

$$A^2 = (\Lambda + m^2\phi^2)|E| \tag{42}$$

It immediately follows from the large mass approximation that $\phi = $ constant. To solve for the rest of the variables, we need to distinguish four cases, depending on the signs of E and A. Let us call $\epsilon = \text{sgn}(E)$ and $\chi = \text{sgn}(A)$. Then the solution (with the choice of lapse $\alpha = 1$) is,

$$A = \chi \exp\left(\chi\epsilon t\sqrt{\Lambda + m^2\phi^2}\right) \tag{43}$$

$$E = \epsilon\frac{\exp\left(2\chi\epsilon t\sqrt{\Lambda + m^2\phi^2}\right)}{\Lambda + m^2\phi^2} \tag{44}$$

There are four possibilities according to the signs ϵ, χ. If $\epsilon = \chi = 1$ or $\epsilon = \chi = -1$ we have a universe that expands. If both have different signs, the universe contracts. This just reflects that the Lagrangian is invariant if one changes the sign of either A or E and the sign of time. It is also invariant if one changes simultaneously the sign of both A and E.

Let us turn to the observables of the theory (quantities that have vanishing Poisson brackets with the constraint (42) and therefore are constants of the motion). The theory has four phase space degrees of freedom with one constraint, therefore there should be two independent observables. Immediately one can construct an observable $O_1 = \phi$, since the latter is conserved due to the large mass approximation. To construct the second observable we write the equation for the trajectory,

$$\frac{d\pi}{dA} = \frac{-2Em^2\phi}{\Lambda + m^2\phi^2} = -\frac{2A^2m^2\phi}{(\Lambda + m^2\phi^2)^2}\text{sgn}E \tag{45}$$

where in the latter identity we have used the constraint. Integrating, we get the observable,

$$O_2 = \pi + \frac{2}{3}\frac{m^2\phi}{(\Lambda + m^2\phi^2)^2}A^3\text{sgn}E \tag{46}$$

and using the constraint again we can rewrite it,

$$O_2 = \pi + \frac{2}{3} \frac{m^2 \phi}{\Lambda + m^2 \phi^2} AE. \tag{47}$$

Although the last two expressions are equivalent, we will see that upon discretization only one of them becomes an exact observable of the discrete theory.

We consider the evolution parameter to be a discrete variable. Then the Lagrangian becomes

$$L(n, n+1) = E_n(A_{n+1} - A_n) + \pi_n(\phi_{n+1} - \phi_n) \tag{48}$$
$$- N_n E_n^2 (-A_n^2 + (\Lambda + m^2 \phi_n^2)|E_n|)$$

The discrete time evolution is generated by a canonical transformation of type 1 whose generating function is given by $-L$, viewed as a function of the configuration variables at instants n and $n+1$. The canonical transformation defines the canonically conjugate momenta to all variables. The transformation is such that the symplectic structure so defined is preserved under evolution. The configuration variables will be $(A_n, E_n, \pi_n, \phi_n, N_n)$ with canonical momenta $(P_n^A, P_n^E, P_n^\phi, P_n^\pi, P_n^N)$ defined in the traditional fashion by functional differentiation of the action with respect to the canonical variables. We do not reproduce their explicit expression here for reasons of brevity, the reader can consult them in reference[7]. The definitions of the momenta can be combined in such a way as to yield a simpler evolution system,

$$A_{n+1} - A_n = N_n (P_{n+1}^A)_n^2 (\Lambda + m^2 \phi_n^2) \mathrm{sgn} P_{n+1}^A, \tag{49}$$

$$P_{n+1}^A - P_n^A = 2N_n A_n (P_{n+1}^A)^2 \tag{50}$$

$$\phi_{n+1} - \phi_n = 0 \tag{51}$$

$$P_{n+1}^\phi - P_n^\phi = -2N_n (P_{n+1}^A)^2 m^2 \phi_n |P_{n+1}^A|, \tag{52}$$

$$0 = -A_n^2 + (\Lambda + m^2 \phi_n^2)|P_{n+1}^A|, \tag{53}$$

and the phase space is now spanned by $A_n, P_n^A, \phi_n, P_n^\phi$.

From (50) we determine,

$$P_{n+1}^A = \frac{1 + \xi \sqrt{1 - 8 P_n^A A_n N_n}}{4 A_n N_n}, \tag{54}$$

where $\xi = \pm 1$ and we will see the final solution is independent of ξ. Substituting in (53) and solving for the lapse we get,

$$N_n = \frac{\left[-P_n^A \left(\Lambda + m^2 \phi_n^2 \right) + A_n^2 \mathrm{sgn} P_n^A \right] \left(\Lambda + m^2 \phi_n^2 \right)}{2 A_n^5}. \tag{55}$$

Let us summarize how the evolution scheme presented actually operates. Let us assume that some initial data $A^{(0)}, P^A_{(0)}$, satisfying the constraints of the continuum theory, are to be evolved. The recipe will consist of assigning $A_0 = A^{(0)}$ and $P^A_1 = P^A_{(0)}$. Notice that this will automatically satisfy (53). In order for the scheme to be complete we need to specify P^A_0. This is a free parameter. Once it is specified, then the evolution equations will determine all the variables of the problem, including the lapse. Notice that if one chooses P^A_0 such that, together with the value of A_0 they satisfy the constraint, then the right hand side of the equation for the lapse (55) would vanish and no evolution takes place. It is clear that one can choose P^A_0 in such a way as to make the evolution step as small as desired.

The equation for the lapse (55) implies that the lapse is a real number for any real initial data. But it does not immediately imply that the lapse is positive. However, it can be shown that the sign of the lapse, once it is determined by the initial configuration, does not change under evolution. The proof is tedious since one has to separately consider the various possible signs of ϵ and χ. This is an important result. In spite of the simplicity of the model, there was no a priori reason to expect that the construction would yield a real lapse. Or that upon evolution the equation determining the lapse could not become singular or imply a change in the sign of the lapse, therefore not allowing a complete coverage of the classical orbits in the discrete theory.

In general the discrete theory, having more degrees of freedom than the continuum theory, will have more constants of the motion than observables in the continuum theory. In this example, the discrete theory has four degrees of freedom. One can find four constants of the motion. One of them we already discussed. The other one is ϕ. The two other constants of the motion can in principle be worked out. One of them is a measure of how well the discrete theory approximates the continuum theory and is only a function of the canonical variables (it does not depend explicitly on n). The constant of the motion is associated with the canonical transformation that performs the evolution in n. It is analogous to the Hamiltonian of the discrete theory. The expression can be worked out as a power series involving the discrete expression of the constraint of the continuum theory. This constant of motion therefore vanishes in the continuum limit. The other constant of the motion also vanishes in the continuum limit.

That is, we have two of the constants of the motion that reduce to the observables of the continuum theory in the continuum limit and two others that vanish in such limit. The discrete theory therefore clearly has a remnant of the symmetries of the continuum theory. The canonical transformations associated with the constants of the motion which have non-vanishing continuum limit map dynamical trajectories to other trajectories that can be viewed as different choices of lapse in the discrete theory. This is the discrete analog of the reparameterization invariance of the continuum theory. As we will see soon the lapse in the discrete theory is determined up to two constants. The choice of these two constants is the remnant of the reparameterization invariance of the continuum theory that is present in the discrete theory.

Figure (7) shows the comparison of the discrete evolution of the model with the exact solution of the continuum theory. As we see the discrete theory approximates the continuum theory very well.

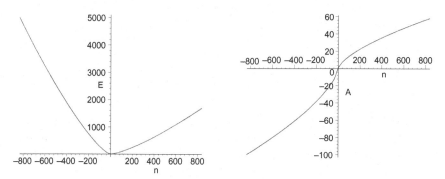

Fig. 7. The discrete evolution of the triad E and the connection A as a function of the discrete evolution parameter n. We have chosen initial conditions that produce a positive branch of A for $n > 0$ and a negative branch for $n < 0$. For the triad we chose both branches positive.

Figure (8) blows up the region of the evolution close to the singularity. As it can be seen the discrete theory evolves through the singularity. Emerging on the other side the evolution has a different value for the lapse and therefore a different time-step. This could be used to implement the proposal of Smolin that physical constants change when one tunnels through the singularity (in lattice theories the lattice spacing determines the "dressed" value of physical constants). See Smolin[8] for further discussion of this point.

To quantize the theory one has to implement the equations of evolution via a unitary transformation. Details of the derivation can be found in[7].

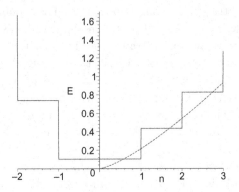

Fig. 8. The approach to the singularity in the discrete and continuum solutions. The discrete theory has a small but non-vanishing triad at $n = 0$ and the singularity is therefore avoided.

The result is

$$< A_1, \phi_1, n || A_2, \phi_2, n + 1 > = \delta(\phi_1 - \phi_2) \tag{56}$$

$$\times \exp\left(i \frac{\text{sgn}(A_1)A_2^2(A_1 - A_2)}{\Theta} \right) \sqrt{\frac{|A_1|}{\pi\Theta}}.$$

It should be noted that not all canonical transformations correspond to unitary evolutions at the quantum level. If the canonical transformation defines an isomorphism between the phase spaces at levels n and $n+1$, then one can show that the canonical transformation can be implemented by an isometry at a quantum level. If the isomorphism is an automorphism then the canonical transformation can be implemented as a unitary transformation. A good discussion of canonical transformations in quantum mechanics can be found in Anderson[9].

With the unitary transformation introduced above one can answer any question about evolution in the Heisenberg picture for the model in question. One could also choose to work in the Schrödinger picture, evolving the states. Notice that the wavefunctions admit a genuine probabilistic interpretation. This is in distinct contrast with the usual "naive Schrödinger interpretation" of quantum cosmology which attempted to ascribe probabilistic value to the square of a solution of the Wheeler–DeWitt equation (see Kuchař[10] for a detailed discussion of the problems associated with the naive interpretation).

An interesting aspect of this quantization is that for any square-integrable wavefunction and for any value of the parameter n, the expectation value of $(P_n^A)^2$, and therefore that of E^2 is non-vanishing, and so are the metric and the volume of the slice. Therefore quantum mechanically one never sees a singularity. The mechanism for elimination of the singularity is quite distinct from the one encountered in loop quantum cosmology[11]. The reader may ask how generic is our mechanism. Are singularities always avoided? Singularities are avoided (simply because the a lattice point generically will not coincide with them) provided that they do not occur at a boundary of the phase space. If they occur at a boundary, it is guaranteed that a lattice point will coincide with the singularity and therefore it will not be avoided. In the example we showed the singularity does not occur on the boundary. However, it appears this is a special feature of the large mass approximation in the scalar field. For other models we have studied, at least worked out in the Ashtekar variables, the connection diverges at the singularity and therefore the singularity is at one boundary of the phase space. This can be remedied by changing variables before discretizing so that the singularity is not on a boundary, but it is not what is the direct outcome of discretizing the theory in the Ashtekar variables. An alternative that also eliminates the singularity is to rewrite things in terms of holonomic variables as is done in loop quantum cosmology.

As we will discuss in the next section, it is more desirable to introduce a quantization that is relational in nature. This is because the evolution parameter n cannot be observed physically and therefore is not a good choice of time to have a quantization that is physical. Having implemented the evolution equations as a unitary transformation and having a probabilistic interpretation for the wavefunctions is all that is needed to work out a relational quantization in detail without any of the conceptual problems that were encountered in the past (see Page and Wootters[12] and the critique of their work by Kuchař[10]).

To define a time we therefore introduce the conditional probabilities, "probability that a given variable have a certain value when the variable chosen as time takes a given value". For instance, taking A as our time variable, let us work out first the probability that the scalar field conjugate momentum be in the range $\Delta P^\phi = [P^\phi_{(1)}, P^\phi_{(2)}]$ and "time" is in the range $\Delta A = [A_{(1)}, A_{(2)}]$ (the need to work with ranges is because we are dealing with continuous variables). We go back to the naive quantization and recall that the wavefunction $\Psi[A, \phi, n]$ in the Schrödinger representation admits a probabilistic interpretation. One can also define the amplitude $\Psi[A, P^\phi, n]$ by taking the Fourier transform. Therefore the probability of simultaneous measurement is,

$$P_{\text{sim}}(\Delta P^\phi, \Delta A) = \lim_{N \to \infty} \frac{1}{N} \sum_{n=0}^{N} \int_{P^\phi_{(1)}, A_{(1)}}^{P^\phi_{(2)}, A_{(2)}} \Psi^2[A, P^\phi, n] dP^\phi dA. \qquad (57)$$

We have summed over n since there is no information about the "level" of the discrete theory at which the measurement is performed, since n is just a parameter with no physical meaning. With the normalizations chosen if the integral in P^ϕ and A were in the range $(-\infty, \infty)$, P_{sim} would be equal to one.

To get the conditional probability $P_{\text{cond}}(\Delta P^\phi | \Delta A)$, that is, the probability that having observed A in ΔA we also observe P^ϕ in ΔP^ϕ, we use the standard probabilistic identity

$$P_{\text{sim}}(\Delta P^\phi, \Delta A) = P(\Delta A) P_{\text{cond}}(\Delta P^\phi | \Delta A) \qquad (58)$$

where $P(\Delta A)$ is obtained from expression (57) taking the integral on P^ϕ from $(-\infty, \infty)$. We therefore get

$$P_{\text{cond}}(\Delta P^\phi | \Delta A) = \frac{\lim_{N \to \infty} \frac{1}{N} \sum_{n=0}^{N} \int_{P^\phi_{(1)}, A_{(1)}}^{P^\phi_{(2)}, A_{(2)}} \Psi^2[A, P^\phi, n] dP^\phi dA}{\lim_{N \to \infty} \frac{1}{N} \sum_{n=0}^{N} \int_{-\infty, A_{(1)}}^{\infty, A_{(2)}} \Psi^2[A, P^\phi, n] dP^\phi dA}. \qquad (59)$$

Notice that all the integrals are well defined and the resulting quantity behaves as a probability in the sense that integrating from $(-\infty, \infty)$ in P^ϕ one gets unity.

Introducing probabilities is not enough to claim to have completed a quantization. One needs to be able to specify what happens to the state of the system as a measurement takes place. The most natural reduction postulate is that,

$$|\psi> \to \frac{\Pi_{P^\phi, A} |\phi>}{\sqrt{| < \phi | \Pi_{P^\phi, A} |\phi > |}}, \qquad (60)$$

where

$$\Pi_{P^\phi, A} = \sum_{n=0}^{N} |P^\phi, A, n> < P^\phi, A, n|. \qquad (61)$$

The model considered is too simple to test too much of the framework, however, we have shown that one can work out in detail the discrete treatment at a classical an quantum mechanical level without conceptual obstructions. It is a big leap to claim that because everything worked well for such a simple model these ideas will succeed in full GR. Currently we are working in detail the discretization of Gowdy cosmologies, which have field theoretic degrees of freedom. Recently achieved success[13] in such models greatly enhances our confidence in the ability of this scheme to discretize general relativity.

For now, we would like to take a glimpse at some possibilities that the framework will introduce in the full theory, we do so in the next section.

5. Fundamental Decoherence and Other Quantum Applications

Having made the case that one can approximate general relativity by a discrete theory that is constraint-free allows us to make progress in many aspects of quantum gravity. Most of the hard conceptual problems that one faces in canonical quantum gravity are related to the presence of constraints in the theory. For an unconstrained theory, most obstructions are eliminated. One of the first obstructions we can deal with is the "problem of time". To a certain extent we have shown that this is possible in Rovelli's example (which has a "problem of time") and in the cosmological model. But actually progress is possible in a more generic sense. One can, for instance, implement the relational time that was proposed by Page and Wootters[12] in the full theory. The idea consists in quantizing the theory by promoting all variables to quantum operators, unlike the usual Schrödinger quantization in which a variable called "time" is kept classical. One then picks from among the quantum operators one that we will call "clock". It should be emphasized that it is a quantum operator and therefore will have an expectation value, fluctuations, etc. One then computes conditional probabilities for the other variables in the theory to take certain values when the "clock" variable has a given value.

The resulting conditional probabilities do not evolve according to a Schrödinger equation. If one chooses as "clock" a variable that behaves close to classicality (small fluctuations) then one can show that the conditional probabilities evolve approximately by a Schrödinger evolution. But as the fluctuations in the clock variable increase there appear corrective terms in the evolution equations. The fist kind of corrective terms were evaluated in reference[14] and have the form of a Lindblad type of evolution. A feature of the evolution implied by the use of "real clocks" as we are considering is that it is not unitary. In general pure states evolve into mixed states. This is easy to understand in our discrete approach. There we saw that evolution in terms of the discrete parameter n was implemented by a canonical transformation. Upon quantization evolution is implemented by a unitary operator. However, the real clock variable we choose will in general have a probability distribution with a certain spread in n. If evolution is unitary in n then it cannot be unitary in the clock variable since in general for a given value of the clock variable one will have a superposition of states with different values of n.

This lack of unitarity could in principle be experimentally observable. We have estimated its magnitude[15]. In order to make an estimate one needs to make a model of what is the "most classical" clock one can construct. To do this we borrowed from ideas of Ng and Van Dam, Amelino-Camelia and more recently Lloyd and collaborators[16]. They start with the observation of Salecker and Wigner[17] that the accuracy of a clock is limited by its mass $\delta T > \sqrt{T/M}$ (in units of $\hbar = c = 1$) where T is the time to be measured. Then they argue that the maximum amount of mass one can concentrate can be achieved in a black hole. If the clock is a black hole its accuracy is given by its quasinormal frequencies, $\delta T > T_P^2 M$. where T_P is Planck's time. Combining the two inequalities we get, $\delta T \sim T_P \sqrt[3]{T/T_P}$

If one now considers a model system consisting of two quantum mechanical levels and estimates the lack of coherence induced by the use of the relational time one finds that it that the elements of the density matrix that are off-diagonal in the energy basis decay exponentially. The exponent is given by $t_P^{(4/3)} T^{(2/3)} \omega_{12}^2$ where T is the time we observe the system and ω_{12} is the Bohr frequency between both levels. Given the presence of the t_P factor, this effect is unobservable for almost all experimental situations. An exception could be "macroscopic" quantum states, like the ones that are starting to become available through Bose-Einstein condensates. Current technology does not produce states that are "macroscopic enough" for the effect to be observable, and it remains to be seen if future technologies could produce such states (and keep them free of ambiental decoherence effects) for this effect to be observable (for an independent discussion see Simon and Jaksch[18]).

One place where the fundamental decoherence could play a role is in the black hole information puzzle. Black holes take a very long time to evaporate and therefore there is a chance for the decoherence to build up. The question is: does it build up enough to wipe out all coherence from the states before the black hole would have done the same via Hawking evaporation? We recently estimated the effect by considering a very naive model of a black hole consisting of two energy levels separated by kT with k the Boltzmann constant and T the temperature predicted by Hawking for the black hole. We found out[19] that

$$|\rho_{12}(T_{\max})| \sim |\rho_{12}(0)| \left(\frac{M_{\text{Planck}}}{M_{\text{BH}}}\right)^{\frac{2}{3}} \tag{62}$$

where ρ_{12} is an off-diagonal element of the density matrix of the state of interest at the time the black hole would have evaporated. For a Solar sized black hole the magnitude of the off diagonal element is 10^{-28}, that is, it would have de facto become a mixed state even before invoking the Hawking effect. The information paradox is therefore unobservable in practice. It should be noted that this estimate is an optimistic one. In reality clocks fare much worse than the estimate we worked out and the fundamental decoherence will operate even faster than what we consider here.

6. Connections with Continuum Loop Quantum Gravity

In spite of the possibilities raised by the discrete approach, some readers may feel that it forces us to give up too much from the outset. This was best perhaps captured by Thiemann[20], who said "While being a fascinating possibility, such a procedure would be a rather drastic step in the sense that it would render most results of LQG obtained so far obsolete". Indeed, the kinematical structure built in loop quantum gravity, with a rigorous integration measure and the natural basis of spin foam states appear as very attractive tools to build theories of quantum gravity. We would like to discuss how to recover these structures in our discrete approach.

To make contact with the traditional kinematics of loop quantum gravity, we consider general relativity and discretize time but keep space continuous, and we proceed as in the consistent discretization approach, that is, discretizing the action and working out the equations of motion. We start by considering the action written in terms of Ashtekar variables[21],

$$S = \int dt d^3x \left(\tilde{P}_i^a F_{0a}^i - N^a C_a - NC\right) \tag{63}$$

where N, N^a are Lagrange multipliers, $\tilde{P}_i^a$ are densitized triads, and the diffeomorphism and Hamiltonian constraints are given by, $C^a = \tilde{P}_i^a F_{ab}^i$, $C = \frac{\tilde{P}_i^a \tilde{P}_j^b}{\sqrt{\det q}} \left(\epsilon^{ijk} F_{ab}^i - (1+\beta^2) K_{[a}^i K_{b]}^j\right)$ where $\beta K_a^i \equiv \Gamma_a^i - A_a^i$ and Γ_a^i is the spin connection compatible with the triad, and q is the three metric. We now proceed to discretize time. The action now reads,

$$S = \int dt d^3x \left[\text{Tr}\left(\tilde{P}^a \left(A_a(x) - V(x)A_{n+1,a}(x)V^{-1}(x) + \partial_a(V(x))V^{-1}(x)\right)\right)\right.$$
$$\left. - N^a C_a - NC + \mu\sqrt{\det q}\text{Tr}\left(V(x)V^\dagger(x) - 1\right)\right] \tag{64}$$

In the above expression $V(x) = V_I T^I$ is the parallel transport matrix along a time-like direction and F_{0a} is approximated by the holonomy along a plaquette that is finite in the "time-like" direction and infinitesimal in the "space-like" direction and $T^0 = 1/\sqrt{2}$ and $T^a = -i\sigma^a/\sqrt{2}, a = 1..3$ and σ's are the Pauli matrices and the coefficients V_I are real. We have omitted the subscript n to simplify the notation and kept it in the quantities that are evaluated at $n + 1$. The last term involves a Lagrange multiplier μ and is present in order to enforce the fact that the parallel transport matrices are unitary. We notice that the $SU(2)$ gauge invariance is preserved in the semi-discrete theory. This in turn implies that Gauss' law for the momentum canonically conjugate to the connection, $\tilde{E}_{n+1}^a \equiv V^{-1}\tilde{P}^a V$ is preserved automatically upon evolution. Remarkably, although the theory does not have a diffeomorphism constraint, one can show that the quantity that in the continuum would correspond to the diffeomorphism constraint is a conserved quantity (to intuitively see this, notice that the action is invariant under (time independent) spatial diffeomorphisms and n and $n + 1$). One can then impose that this quantity vanish and this requirement is consistent with evolution. One would therefore have a theory with the same constraints as kinematical loop quantum gravity and with an explicit evolution scheme as is usual in consistently discretized theories.

We have actually worked[22] out the procedure in detail for the case of $2+1$ dimensions, treating gravity as a BF theory. The procedure reproduces the physical space for the theory of traditional quantizations.

Summarizing, one can apply the consistent discretization approach by discretizing only time. The resulting theory has no constraints, but the diffeomorphism constraint can be introduced additionally, so the resulting theory has the same kinematics as loop quantum gravity. The dynamics is implemented via a unitary operator. As in the full discrete approach, one can envision bypassing many of the hard conceptual questions of canonical quantum gravity, for instance introducing a relational notion of time. This can actually be seen as a concrete framework to implement loop quantum gravity numerically, since it is not expected that one will be able to work out things analytically in the full case.

7. Discussion and Frequently Asked Questions

It is instructive at this point to revise a list of questions that were posed about the formalism when it was beginning to be developed[7]:

(i) Solubility of the multiplier equations: Solving the constraints by choosing the Lagrange multipliers produces a theory that is constraint free. This is analogous to what happens when one gauge fixes a theory. It is well known that gauge fixing is not a cure for the problems of general relativity since gauge fixings usually become problematic. In the same sense, it could happen that the algebraic equations that determine the Lagrange multipliers (the lapse and the shift in the case of general relativity) in our approach develop problems in their solutions (for instance, negative lapses, or complex solutions). As we argued in the text, it is clear that the formalism is not a gauge fixing nor does it share its pathologies, as we saw in the Rovelli model. We also have seen in several examples, now ranging from mechanics to Gowdy cosmologies, that the equations determining the multipliers can be solved.

(ii) Performing meaningful comparisons: When comparing a discrete theory with a continuum theory, one needs to choose the quantities that are to be compared. In particular, the continuum theory has "observables" ("perennials"), that is, quantities that have vanishing Poisson brackets with the constraints while the discrete theory is constraint-free. The Rovelli model shows how to deal with these issues. In particular the relationship between perennials and conserved quantities in the discrete theories.

(iii) The continuum limit: In continuum constrained theories with first class constraints, the Lagrange multipliers are free functions. Yet in our discrete construction, the Lagrange multipliers are determined by the initial conditions. If one wishes to take a naive continuum limit, the discrete equations that determine the Lagrange multipliers must become ill-defined. To extract meaningful information about the continuum theories, one needs to proceed differently. We have seen in the Rovelli example how one can reach better and better continuum approximations by tuning the initial data of the discrete theory.

(iv) Singularities: Discrete theories in principle have the possibility of evolving through a Big Bang singularity, since generically the singularity will not lie on a point on the lattice. We have seen this implemented concretely in cosmological models.

(v) Problem of time: We discussed evolution in the discrete by using a relational approach in terms of the observables of these theories. This shows how the problem of time can be solved in these theories.

(vi) Discretization ambiguities: An important element is to note that the Lagrange multipliers get determined by this construction only if the constraint is both a function of q and p. If the constraint is only a function

of q or of p then the constraints are automatically preserved in evolution without fixing the Lagrange multipliers. This raises a conceptual question. For certain theories in the continuum, one can make a canonical transformation to a new set of variables such that the constraints depend only on q or on p. The resulting discrete theories will therefore be very different in nature, but will have the same continuum limit. From the point of view of using discrete theories to quantize gravity, we believe this ambiguity should receive the same treatment that quantization ambiguities (choice of canonical variables, factor orderings, etc.) get: they are decided experimentally. Generically there will be different discrete theories upon which to base the quantization and some will be better at approximating nature than others in given regimes. Many of them may allow to recover the continuum limit, however they may have different discrete properties when one is far from the semiclassical regime. A full discussion of the continuum limit in the quantum case is still lacking and is one of the main open issues of the framework.

8. Conclusions

The consistent discretization approach is emerging as an attractive technique to handle discrete general relativity, both at a classical and at a quantum mechanical level. In quantum gravity, since it does away with the constraints it solves in one sweep some of the hardest conceptual issues of canonical quantization. In classical general relativity there is a growing body of evidence that suggests that the discretizations work numerically and approximate the continuum theory (including its constraints) in a convergent and stable way. Having discrete theories with these properties classically is a desirable point of view for any effort towards quantization. The discrete approach also yields directly computable evolutions quantum mechanically opening the possibility for concrete numerical quantum gravity calculations.

Acknowledgments

This work was supported in part by grants NSF-PHY0244335, NASA-NAG5-13430 and by funds of the Horace C. Hearne Jr. Institute for Theoretical Physics and CCT-LSU.

References

1. A. Lew, J. Marsden, M. Ortiz, M. West, in "Finite element methods: 1970's and beyond", L. Franca, T. Tezduyar, A. Masud, editors, CIMNE, Barcelona (2004).
2. R. McLachlan, M. Perlmutter "Integrators for nonholonomic mechanical systems" preprint (2005).
3. C. Di Bartolo, R. Gambini, J. Pullin, Class. Quan. Grav. 19, 5275 (2002).
4. R. Gambini and J. Pullin, Phys. Rev. Lett. 90, 021301, (2003).
5. C. Di Bartolo, R. Gambini, R. Porto and J. Pullin, J. Math. Phys. **46**, 012901 (2005) [arXiv:gr-qc/0405131].
6. C. Rovelli, Phys. Rev. D **42**, 2638 (1990).
7. R. Gambini and J. Pullin, Class. Quant. Grav. **20**, 3341 (2003) [arXiv:gr-qc/0212033].
8. R. Gambini and J. Pullin, Int. J. Mod. Phys. D **12**, 1775 (2003) [arXiv:gr-qc/0306095].
9. A. Anderson, Annals Phys. **232**,292-331 (1994) [arXiv:hep-th/9305054].
10. K. Kuchař, "Time and interpretations of quantum gravity", in "Proceedings of the 4th Canadian conference on general relativity and relativistic astrophysics", G. Kunstatter, D. Vincent, J. Williams (editors), World Scientific, Singapore (1992). Available online at http://www.phys.lsu.edu/faculty/pullin/kvk.pdf
11. M. Bojowald, Pramana **63**, 765 (2004)
12. D. N. Page and W. K. Wootters, Phys. Rev. D **27**, 2885 (1983); W. Wootters, Int. J. Theor. Phys. **23**, 701 (1984); D. N. Page, "Clock time and entropy" in "Physical origins of time asymmetry", J. Halliwell, J. Perez-Mercader, W. Zurek (editors), Cambridge University Press, Cambridge UK, (1992).
13. R. Gambini, M. Ponce, J. Pullin, "Consistent discretizations of Gowdy cosmologies" [gr-qc/0505043].
14. R. Gambini, R. Porto and J. Pullin, New J. Phys. **6**, 45 (2004) [arXiv:gr-qc/0402118].
15. R. Gambini, R. A. Porto and J. Pullin, Phys. Rev. Lett. **93**, 240401 (2004) [arXiv:hep-th/0406260].
16. G. Amelino-Camelia, Mod. Phys. Lett. A **9**, 3415 (1994) [arXiv:gr-qc/9603014]; Y. J. Ng and H. van Dam, Annals N. Y. Acad. Sci. **755**, 579 (1995) [arXiv:hep-th/9406110]; Mod. Phys. Lett. A **9**, 335 (1994); V. Giovanetti, S. Lloyd, L. Maccone, Science **306**, 1330 (2004); S. Lloyd, Y. J. Ng, Scientific American, November 2004.
17. E. Wigner, Rev. Mod. Phys. **29**, 255 (1957); H. Salecker, E. Wigner, Phys. Rev. **109**, 571 (1958).
18. C. Simon and D. Jaksch, Phys. Rev. **A70**, 052104 (2004) [arXiv:quant-ph/0406007].
19. R. A. Gambini, R. Porto and J. Pullin, "Fundamental decoherence in quantum gravity,", to appear in the proceedings of 2nd International Workshop DICE2004: From Decoherence and Emergent Classicality to Emergent Quan-

tum Mechanics, Castello di Piombino, Tuscany, Italy, 1-4 Sep 2004, H. T. Elze, editor [arXiv:gr-qc/0501027].

20. T. Thiemann, "The Phoenix project: Master constraint programme for loop quantum gravity," arXiv:gr-qc/0305080.

21. A. Ashtekar, J. Lewandowski Class. Quant. Grav. **21**, R53 (2004) [arXiv:gr-qc/0404018].

22. R. Gambini and J. Pullin, Phys. Rev. Lett. **94**, 101302 (2005) [arXiv:gr-qc/0409057].

CHAPTER 16

CAUSAL SETS AND
THE DEEP STRUCTURE OF SPACETIME

FAY DOWKER

Blackett Laboratory, Imperial College,
Prince Consort Road, London SW7 2AZ, UK

The causal set approach to quantum gravity embodies the concepts of causality and discreteness. This article explores some foundational and conceptual issues within causal set theory.

1. Introduction

The problem of quantum gravity in its narrow sense is the problem of finding a theory that incorporates both Quantum Mechanics and General Relativity. A broader vision is that a theory of quantum gravity would restore to physics a unified framework, a framework in which there is no fundamental division in principle between observer and observed nor between matter and spacetime. One reading of Einstein's writings on Quantum Mechanics and Unification is that he viewed success in the broad quest as a prerequisite to a solution of the narrow problem.

Every approach to quantum gravity in this broad conception must embody answers to two fundamental questions: "What is quantum mechanics?" and "What is the deep structure of spacetime?" This article will touch on the former question and focus on the latter and the answer to it provided by the approach known as causal set theory which marries the two concepts of discreteness (or atomicity) and causality.

The view that causality is a more fundamental organising principle, even than space and time, is an ancient tradition of thought. Corresponding to this view, momentary events have a better claim to be basic than objects extended in time like particles, the latter being understood as persistent patterns of events rather than enduring "substances". Within Relativity, the recognition that almost all of the geometrical properties of Minkowski space could be reduced to order theoretic relationships among point events came very early.[1]

The concept of atomicity also has a long history as do philosophical challenges to the antithetical notion of a physical continuum. Of course Quantum Mechanics itself is named for the discreteness of atomic and subatomic phenomena. In more recent times, the increasing importance of computing with its discrete algorithms and digital processing has had a pervasive influence on intellectual culture. Whatever the roots of the concept, it is certainly now the case that many workers believe that a fundamentally discrete structure to reality is required to reconcile spacetime with the quantum. Einstein himself adumbrated this view and it is impossible to resist the temptation of quoting him here:

"But you have correctly grasped the drawback that the continuum brings. If the molecular view of matter is the correct (appropriate) one, *i.e.*, if a part of the universe is to be represented by a finite number of moving points, then the continuum of the present theory contains too great a manifold of possibilities. I also believe that this too great is responsible for the fact that our present means of description miscarry with the quantum theory. The problem seems to me how one can formulate statements about a discontinuum without calling upon a continuum (space-time) as an aid; the latter should be banned from the theory as a supplementary construction not justified by the essence of the problem, which corresponds to nothing "real". But we still lack the mathematical structure unfortunately. How much have I already plagued myself in this way!"
A. Einstein in a letter to Walter Dällenbach, November 1916, translated and cited by John Stachel.[2]

Causal set theory[3,4,5,6,7] arises by combining discreteness and causality to create a substance that can be the basis of a theory of quantum gravity. Spacetime is thereby replaced by a vast assembly of discrete "elements" organised by means of "relations" between them into a "partially ordered set" or "poset" for short. None of the continuum attributes of spacetime, neither metric, topology nor differentiable structure, are retained, but emerge it is hoped as approximate concepts at macroscopic scales.

Amongst current approaches, causal set theory can claim to lie at the extreme end of the granularity scale: it is maximally discrete. No continuum concept is required to specify the underlying reality which is purely combinatorial. The elements have no internal structure; they are the fundamental units of reality. All one can do is count: count elements, count relations, count certain patterns of elements and relations. The slogan might be coined, "Real numbers are not real in a causal set."

The hypothesis that the deep structure of spacetime is a discrete poset characterises causal set theory at the *kinematical* level; that is, it is a proposition about what substance is the subject of the theory. However, kinematics needs to be completed by *dynamics*, or rules about how the substance behaves, if one is to have a complete theory. In this article I will explore some foundational issues within these two categories in a non-technical way. I have not attempted to give anything approaching a review of causal set theory. Reference 8, by Rafael Sorkin who has done more than anyone to further the causal set approach, gives more details of current developments.

2. Kinematics

2.1. *The causal set*

Mathematically, a causal set is defined to be a locally finite partially ordered set, or in other words a set C together with a relation $\prec$, called "precedes", which satisfy the following axioms:

(1) if $x \prec y$ and $y \prec z$ then $x \prec z$, $\forall x, y, z \in C$ (transitivity);
(2) if $x \prec y$ and $y \prec x$ then $x = y$ $\forall x, y \in C$ (non-circularity);
(3) for any pair of fixed elements x and z of C, the set $\{y | x \prec y \prec z\}$ of elements lying between x and z is finite.

Of these axioms, the first two say that C is a partially ordered set or poset and the third expresses local finiteness. The idea is that in the deep quantum regime of very small distances, spacetime is no longer described by a tensor field, the metric, on a differentiable manifold, but by a causal set. The discrete elements of the causal set are related to each other only by the partial ordering that corresponds to a microscopic notion of before and after, and the continuum notions of length and time arise only as approximations at large scales.

The richness of the structure of partial orders is reflected in the many different sets of terminologies used by mathematicians and physicists who study them. One of the most useful and suggestive for the purposes of quantum gravity is the "genealogical" jargon whereby one thinks of a causal set as a family tree. An element x is an *ancestor* of an element y if $x \prec y$, and y is then a *descendant* of x.

To arrive at this structure as the kinematical basis for quantum gravity, one can start by conjecturing discreteness and see how this leads to the addition of causal structure, or vice versa. The scientific arguments for the two interweave each other and it's an artificial choice. Indeed, it is the fact that the two concepts "complete" each other in the context of a proposed substructure for spacetime that is one of the strongest motivations for the causal set programme. For the purposes of the current exposition let's choose to begin by postulating a discrete spacetime.

2.2. *An analogy: discrete matter*

In quantum gravity, we know what the continuum description of spacetime is at macroscopic scales – it is a Lorentzian manifold – and are trying to discover the discrete substratum. It is useful to imagine ourselves in the analogous situation for a simple physical system. Consider a quantity of a material substance in a box in flat space for which we have a continuum description – a mass density, say – but which we suspect is fundamentally discrete. The question is, "What could the discrete state be that gives rise to this continuum approximation?" and a good first response is to try to *discretise* the continuum. We then try to give the discrete object so created "a life of its own" as an entity independent of the continuum from which it was derived. We then ask whether we can believe in this discrete object as fundamental and whether and how we can recover from it a continuum approximation which may not be exactly the original continuum from which we started but which must be "close" to it.

In the case of the material substance, let us postulate that it is made of identical atoms. The varying mass density then can only be due to differing number densities of atoms in space. We discretise by somehow distributing the atoms in the box so that the number of atoms in each sufficiently large region is approximately proportional to the mass density integrated over that region. We may or may not have reasons to suspect what the atomic mass actually is, but in any case it must be small enough so that the spacing between the atoms is much smaller than the scales on which the density varies.

Each method of discretising will produce distributions of atoms that have different properties and which type is more favourable as a fundamental state depends on which features of the continuum theory we wish to preserve and how fruitful each turns out to be when we come to propose dynamics for the discrete structure itself. Suppose for example that the continuum theory of the material is invariant under Euclidean transformations (rotations and translations) at least locally in small enough regions over which the mass density is approximately constant and ignoring edge effects. There are ways to discretise which do not respect the invariance under Euclidean transformations. For example, we can divide the box into cubes small enough so the density is approximately constant in each. In each cube, place an atom at every vertex of a Cartesian lattice with lattice spacing chosen inversely proportional to the mass density in the cube. We can produce in this way an atomic state from which we can recover approximately the correct continuum mass density but it is not invariant under the Euclidean group – there are preferred directions – and if a fully fledged fundamental discrete theory is based on such lattice-like atomic states, this will show up in deviations from exact Euclidean invariance in the continuum approximation to this full underlying theory.

If a discretisation that respects the invariance is desired, there must be no preferred directions and this can be achieved by taking a *random* distribution of atoms, in other words atomic positions chosen according to a Poisson distribution so that the expected number of atoms in a given region is the mass in that region (in atomic mass units). This will produce, with high probability, an atomic configuration that does not distinguish any direction.

Having placed the atoms down, we kick away the prop of the continuum description and ask if the distribution of atoms itself could be the underlying reality; in particular how do we start with it and recover, approximately, a continuum description? To answer this question, we can use the discretisation in reverse: a continuum mass density is a good approximation to an atomic state if that atomic state could be a discretisation of the mass density. In the case of the atomic states arising from random discretisations, this is modified to: a mass density is a good approximation if the atomic state could have arisen with relatively high probability from amongst the possible discretisations. Then it must be checked that if two continua are good approximations to the same atomic state, they must be close to each other – this is crucial if this relationship between discrete states and continuum states is to be consistent.

Finally, we propose the atomic state as the underlying reality, reinterpret *mass* as a measure of the *number* of atoms in a region, ask whether there are atomic states which have no continuum approximation and what their meaning is, start working on a dynamics for the atoms *etc.*

2.3. *Discrete spacetime*

These steps are straightforward in this simple case, and can be taken in analogous fashion in the discretisation of a continuum spacetime given by a Lorentzian metric tensor on a differentiable manifold. Let us tread the same path.

What plays the role of the mass density? A good candidate for the measure of the sheer quantity of spacetime in a region is the *volume* of that region. It is calculated by integrating over the region the volume density given by the square root of minus the determinant of the spacetime metric, $\int_{region} \sqrt{-g}\, d^4x$ and is a covariant quantity: it is independent of the coordinates used to calculate it.

In the case of quantum gravity we have independent evidence, from the entropy of black holes for example,[9] that the scale of the discreteness is of the order of the Planck scale formed from the three fundamental constants, G, $\hbar$ and c. If the Planck length is defined to be $l_p = \sqrt{\kappa\hbar}$ where $\kappa = 8\pi G$ and we have set $c = 1$, then the fundamental unit of volume will be $V_f \equiv \nu l_p^4$ where ν is a number of order one and yet to be determined. In order to discretise a spacetime, we distribute the atoms of spacetime, which we will call simply "elements", in the spacetime in such a way that the number of them in any sufficiently large region is approximately the volume of that region in fundamental volume units.

Analogous to the question of Euclidean invariance in the material example, we must ask: do we want to preserve Lorentz invariance or not? We might think of laying down a grid of coordinates on spacetime marked off at $\nu^{1/4}l_p$ intervals and placing an element at every grid vertex. The problem that immediately arises is that it is not a covariant prescription and this manifests itself dramatically in frames highly boosted with respect to the frame defined by the coordinates, where the distribution of elements will not be uniform but will contain large void regions in which there are no elements at all (see *e.g.* reference 10 for a picture). Such a coordinate-dependent discretisation will therefore violate Lorentz invariance. This is a much more serious matter than the breaking of Euclidean invariance for a coordinate-based discretisation of the material. The breaking of Lorentz invariance manifests itself by a failure of the distribution of elements to be uniform *at all* in highly boosted frames. If such a discrete entity were to be proposed as fundamental, it would have to be concluded that in highly boosted frames there can be no manifold description at all.

There is as yet no evidence that Lorentz invariance is violated so let us try to maintain it in quantum gravity. In seeking a Lorentz invariant discretisation process the crucial insight is, again, that the discretisation should be random.[5] It is performed via a process of "sprinkling" which is the Poisson process of choosing countable subsets of the spacetime for which the expected number of points chosen from any given region of spacetime is equal to its volume in fundamental units. That this process is exactly Lorentz invariant in Minkowski spacetime is a consequence of the fact that the Minkowskian volume element is equal to the Euclidean volume element on $\mathbb{R}^n$ and the fact that the Poisson process in $\mathbb{R}^n$ is invariant under any volume preserving map (see *e.g.* reference 11).

The sprinkling process results in a uniform distribution of elements in the spacetime but this set of discrete elements cannot, by itself, provide a possible fundamental description of spacetime. Here, our analogy with a simple mass density in a box breaks down. When, having constructed a distribution of atoms in the box, we take away the continuum mass density, like whisking away the tablecloth from under the crockery, the atoms retain their positions in space because Euclidean space is an assumed background – the table – for the whole setup. But in the case of quantum gravity, the sprinkled elements are meant to *be* the spacetime and if we whisk spacetime away from the elements we have sprinkled into it – removing the table not just the tablecloth – the elements just collapse into a heap of unstructured dust. The sprinkled elements must be endowed with some extra structure.

We already know that by counting elements we can recover volume information in the continuum approximation. Powerful theorems in causal analysis[12,13,14] show that what is needed in the continuum to complete volume information to give the full spacetime geometry is the *causal structure* of a spacetime.[a]

The causal structure of a spacetime is the totality of the information about which events can causally influence which other events. For each point p of the spacetime, we define the set $J^-(p)$ $(J^+(p))$ the causal past (future) of p, to be the set of points, q, in the spacetime for which there is a future (past) directed causal curve – a curve with an everywhere non-spacelike, future (past) pointing tangent vector – from q to p. The collection of all these causal past and future sets is the causal structure of the spacetime. This is often colloquially called the "light cone structure" because the boundaries of these sets from a point p are the past and future lightcones from p and the sets themselves are the lightcones and their interiors.

---

[a]The theorems apply to spacetimes that satisfy a certain global causality condition – *past and future distinguishability* – which means that distinct points have distinct causal pasts and futures and which we will assume for every continuum spacetime referred to here. We can say that the theorems imply that causal set theory predicts spacetime must satisfy this condition because only such spacetimes will be able to approximate a causal set.

Let us therefore, in our discretisation procedure, endow the elements sprinkled into the spacetime with the order given by the spacetime causal structure: elements $e_i \prec e_j$ if they are sprinkled at points p_i and p_j respectively in the continuum such that $p_i \in J^-(p_j)$. The set of sprinkled elements with this induced order is a causal set satisfying the axioms given above.

Now we give the causal set independence from the spacetime. To recover from it approximately the continuum we discretised, we follow the guidance of the example of the substance in a box: a spacetime M is a good approximation to a causal set C if C could have arisen from M by the discretisation process we have described (sprinkling and endowing with induced order) with relatively high probability.

As mentioned in the material-in-a-box example, consistency requires that if two continua are good approximations to the same discretum, they should be close to each other. This is a central conjecture, the *Hauptvermutung*, of causal set theory and it is surprisingly hard even to formulate it precisely due to the difficulty of defining a notion of distance on the space of Lorentzian manifolds. We use the intuitive idea constantly – how else would it make sense to talk of one spacetime being a small perturbation of another – but only recently has progress been made in this direction.[15,16] This progress has been inspired by causal sets, in particular by utilising and comparing the different probability distributions on the space of causal sets given by the sprinkling processes into different Lorentzian manifolds. If it is the case that using sprinklings is the only way covariantly to say what we mean by two Lorentzian spacetimes being close, it would be further evidence for causal sets as the deep structure.

2.4. *Reassessed in a quantal light*

It may be argued that the above steps leading to the proposal of causal sets for quantum gravity kinematics have been taken under the assumption that the continuum is an approximation to one single discrete spacetime whereas Quantum Mechanics would suggest that a continuum spacetime is better characterised as corresponding to a coarse grained set of many discreta. This point has validity and in addressing it, we are drawn into the realm of dynamics and the question of what form a quantum dynamics for causal sets might take, the subject of the next section. Certainly, the statement that the number and ordering of causal set elements correspond to continuum volume and causal structure, respectively, will have to be judged and interpreted in the light of the full dynamical theory. In the meantime, however, Quantum Mechanics need make us no more squeamish about the statement, "Spacetime is a causal set", than about the statement, "Things are made of atoms."

Even if it turns out that only a whole bunch of discreta can properly be said to correspond to a continuum spacetime, we can still make the claim that discrete data can give rise, in a Lorentz invariant manner, to a continuum spacetime if they are organised as a causal set. In this case, we would say that the data common to, or shared by, each member of the bunch of discreta – a coarse graining of them all – is a causet. The question would then be, what discretum can be coarse grained to give a causet?

The answer is, another "finer" causet! Indeed causets admit a notion of coarse graining that is consistent with the inverse procedures of discretisation and continuum approximation given above because it is itself a random process. To perform a $\frac{1}{q}$: 1 coarse graining of a causet, go through it and throw away each element with fixed probability $p = 1 - q$. A causet a is a coarse graining of a causet b if a could have resulted with relatively high probability from the process of coarse graining b.

On this view, a spacetime M could correspond dynamically to a set of "microscopic" states which are causets with no continuum approximation at all, but which have a common coarse graining to which M *is* a good approximation.

It should be mentioned that this process of coarse graining allows the notion of scale dependent topology to be realised in causal set theory. Many quantum gravity workers have the intuitive picture that at scales close to the Planck scale, the topology of spacetime will be complicated by many wormholes and other nontrivial topological excitations, but at every day scales the topology is trivial. As attractive as this idea is, I know of no way in the continuum to make it concrete. If spacetime is a causet, however, coarse graining it at different scales (*i.e.* with different deletion probabilities p) gives rise to different causets which may have continuum approximations with different topologies, including different dimensions.

3. Dynamics

If we imagine a bag containing all the causets with N elements, where N is some immensely large number, then drawing out of the bag a causet uniformly at random will result, with probability approaching 1 as N tends to infinity, in a causet with a very specific and rather surprising structure. This structure is of three "levels": the first level is of elements with no ancestors and has roughly $N/4$ members, the second is of elements with ancestors in level 1 and has roughly $N/2$ members and the remaining elements are in level 3 with ancestors in level 2.[17]

These 3-level causets have no continuum approximation – they are universes that "last" only a couple of Planck times. If the fundamental reality is a causal set then we have to explain why the causet that actually occurs is a manifold-like one and not one of the, vastly more numerous, 3-level causets. This is the causal set version of a problem that is common to all discrete approaches to quantum gravity: the sample space of discreta is always dominated in sheer numbers by the non-manifold-like ones and a uniform distribution over the sample space will render these bad ones overwhelmingly more likely. We need a dynamics to cancel out this "entropic" effect and to produce a measure over the sample space that is peaked on the manifold-like entities.

As mentioned in the introduction, the broad goal of quantum gravity includes the unification of observer with observed (or rather the elimination of the observer as fundamental). Thus, finding a dynamics for quantum gravity, involves a resolution of the vexed problem of the "Interpretation of Quantum Mechanics." There is no consensus on how this is to be achieved and this difficulty could be seen as a severe obstacle in the quest for quantum gravity. Turning the problems around, however, the requirements of quantum gravity, for example general covariance, might be taken as a guide for how to approach Quantum Mechanics in general. Indeed, quantum gravity points to an approach to Quantum Mechanics that is based on the histories of a system.

3.1. A histories framework for quantum causal sets

The reader may already be aware that the terms "quantum state" and "Hilbert space" *etc.* have been conspicuous by their absence in the present account. I have avoided them for reasons bound up with our broad goals in quantum gravity. The standard approach to Quantum Mechanics based on quantum states, *i.e.* elements of a Hilbert space, is tied up with the Copenhagen interpretation with its emphasis on "observables" and "observers". The quantum state is a state "at a moment of time" (in the Schrödinger picture) and to define it requires a foliation of spacetime by spatial hypersurfaces. Taking the principle of general covariance fully to heart suggests that our goals are better served by instead maintaining the fully space-time character of reality inherent in General Relativity, which points to the framework of the "sum-over-histories" for quantum gravity.[18,19,9]

The bare bones structure of a histories quantum theory is a sample space, Ω of possible histories of the system in question – for us, causal sets – and a dynamics for the system expressed in terms of a "quantum measure" μ, on Ω. I will say more about the quantum measure below. For now, note that were μ a probability measure, this would be the familiar case of a classical stochastic theory and indeed a histories quantum theory can be thought of as a generalisation of such a theory. That being so, we can prepare for our task of finding a quantum dynamics by studying classical stochastic causets.

3.2. A classical warm up

Just as a probability measure on the space of all paths on the integers – a random walk – can be given by all the "transition probabilities" from each incomplete path, γ, to γ plus an extra step, so a measure on the space of all possible past finite[b] causets can be specified by giving all the transition probabilities from each n-element causet, c, to all possible causets formed from c by adding a single new element to it. The ancestors of the newly born element are chosen from the elements of c according to the distribution of the transition probabilities. This is a process of stochastic "sequential growth" of a causet.[20]

Without any restrictions, the number of such "laws of growth" is so huge that the class of them is not very useful to study. Imposing on the dynamics two physically motivated conditions severely narrows down the class, however. These conditions are *discrete general covariance* and *Bell causality*. The first condition states that the probability of growing a particular finite partial causet does not depend on the order in which the elements are born. This is recognisably a "label independence" condition and analogous to the independence of the gravitational action from a choice of coordinates. The second condition is the closest possible analogue, in this setting of causet growth, of the condition that gives rise to the Bell Inequalities in an ordinary stochastic theory in a background with fixed causal structure. It is meant to imply that a birth taking place in one region of the causet cannot be influenced by any birth in a region spacelike to the first.

---

[b]Past finite means that every element has finitely many ancestors.

The solution of these two conditions is the class of "Rideout-Sorkin" models. Each model is specified by a sequence of non-negative real numbers, $t_0 = 1, t_1, t_2, \ldots$ and the transition probabilities can be expressed in terms of these "coupling constants." The models are also known as "generalised percolations" since they generalise the dynamics called "transitive percolation" in which a newly born element independently chooses each of the already existing elements to be its ancestor with probability p. In generalised percolation a newly born element chooses a set of elements of cardinality k to be its ancestors with relative probability t_k.[c]

This family of models has proved to be a fruitful test bed for various issues that will arise in the quantum theory, most notably that of general covariance.

3.3. The problem of general covariance

Part of the problem of general covariance is to identify exactly how the principle manifests itself in quantum gravity and how the general covariance of General Relativity arises from it. The answer will surely vary from approach to approach but some overarching comments can be made. Part of what general covariance means is that physical quantities and physical statements should be independent of arbitrary labels. It is needed as a principle when the dynamics of the system is not expressible (or not obviously expressible) directly in terms of the label invariant quantities, but can only be stated as a rule governing the labelled system. To atone for the sin of working with these meaningless labels, the principle of general covariance is invoked and physical statements must be purged of these unphysical markers.

This is a difficult thing to do even in classical General Relativity. In quantum gravity, the issue is even more fraught because it is caught up in the quantum interpretational argy bargy. But the rough form of the problem can be understood in the following way. When the dynamics is expressed in terms of labelled quantities, we can satisfy general covariance by formally "summing over all possible labellings" to form covariant classes of things (histories, operators, whatever). But in doing so, the physical meaning of those classes is lost, in other words the relation of these classes to label-independent data is obscure. Work must then be done to recover this physical meaning. As an example, consider, in flat spacetime, the set

---

[c]It would be more accurate to say "proto-ancestors" because the set has to be completed by adding all its own ancestors.

of non-abelian gauge fields which are zero in a fixed spacetime region and all fields which are gauge equivalent to them. This is a gauge invariant set of fields, but its physical meaning is obscure. Uncovering a gauge invariant characterisation of elements of the set takes work.

In causal set theory, progress has been made on this issue.[21,22] A Rideout-Sorkin dynamics naturally gives a measure on the space of *labelled* causets, $\tilde{\Omega}$ – the labelling is the order in which the elements are born. This labelling is unphysical, the only physical structure possessed by a causet is its partial order. General covariance requires that only the covariant sub-sets of $\tilde{\Omega}$ have physical meaning where a covariant subset is one which if it contains a labelled causet, also contains all other labellings of it. Such subsets can be identified with subsets of the unlabelled sample space Ω. For each covariant, measureable set, A, the dynamics provides a number $\mu(A)$, the probability that the realised causet is an element of A. But what is a physical, *i.e.* label-independent characterisation of the elements of such a set A?

We now know that given almost any causet c in Ω we can determine whether or not it is in A by asking countable logical combinations of questions of the form, "Is the finite poset b a *stem*[d] in c?" The only case in which this is not possible is when c is a so-called rogue causet for which there exists a non-isomorphic causet, c' with exactly the same stems as c. The set of rogues, however, has measure zero and so they almost surely do not occur and we can simply eliminate them from the sample space.

Which finite posets are stems in c is manifestly label-invariant data and the result implies that asking these sorts of questions (countably combined using "not", "and" and "or") is sufficient to exhaust all that the dynamics can tell us.

This result depends crucially on the dynamics. It would not hold for other sequential growth models in which the special type of causet which threatens to spoil the result does not have measure zero. So, though we can claim to have solved the "problem of covariance" for the Rideout-Sorkin models, when we have a candidate quantum dynamics, we will have to check whether the "stem questions" are still the whole story.

3.4. *The problem of Now*

Nothing seems more fundamental to our experience of the world than that we have those experiences "in time." We feel strongly that the moment

---

[d]A stem of a causet is a finite subcauset which contains all its own ancestors.

of Now is special but are also aware that it cannot be pinned down: it is elusive and constantly slips away as time flows on. As powerful as these feelings are, they seem to be contradicted by our best current theory about time, General Relativity. There is no scientific contradiction, no prediction of General Relativity that can be shown to be wrong. But, the general covariance of the theory implies that the proper way to think of spacetime is "timelessly" as a single entity, laid out once and for all like the whole reel of a movie. There's no physical place for any "Now" in spacetime and this seems at odds with our perceptions.

In the Rideout-Sorkin sequential growth models we see, if not a resolution, then at least an easing of this tension. The models are covariant, but nevertheless, the dynamics is specified as a sequential growth. An element is born, another one is born. There is growth and change. Things happen! But the general covariance means that the physical order in which they happen is a *partial* order, not a total order. This doesn't give any physical significance to a *universal* Now, but rather to events, to a Here-and-Now.

I am not claiming that this picture of accumulating events (which will have to be reassessed in the quantum theory) would *explain* why we experience time passing, but it is more compatible with our experience than the Block Universe view.

One might ask whether such a point of view, that events happen in a partial order, could be held within General Relativity itself. Certain consequences have to be grappled with: if one event, x has occurred, then before another event, y can occur to its future, infinitely many events must occur in between and this is true no matter how close in proper time x and y are. Perhaps it is possible to make this coherent but, to my mind, the discreteness of causal sets makes it easier to make sense of this picture of events occurring in a partial order.

Another thing left unexplained by General Relativity is the observational fact of the inexorable nature of time: it will not stop. Tomorrow will come. We have not so far encountered an "edge" in spacetime where time simply comes to an end (nor, for that matter, an edge in space). Spacetimes with boundaries, for example a finite portion of Minkowski space, are just as much solutions of the Einstein equations as those without and General Relativity cannot of itself account for the lack of boundaries and holes in spacetime. In a Rideout-Sorkin universe, one can prove that time cannot stop. A sequential growth must be run to infinity if it is to be generally covariant and Joe Henson has proved that in a Rideout-Sorkin model every element has a descendant and therefore infinitely many descendants.[23] If such a result continues to hold true in the quantum case, not only would it prove that tomorrow will always come but also would imply that in situations where we expect there to be no continuum approximation at all, such as a Big Crunch or the singularity of a black hole, the causal set will continue to grow afterwards.

The Rideout-Sorkin models give us a tantalising glimpse of the sorts of fundamental questions that may find their answers in quantum gravity when we have it.

3.5. *The quantum case*

The mathematical structure of ordinary unitary Quantum Mechanics in its sum-over-histories formulation is a generalisation of a classical stochastic theory in which instead of a probability measure there is a "quantum measure" on the sample space.[e] The quantum measure of a subset, A, of Ω is calculated in the familiar way by summing and then squaring the amplitudes of the fine grained elements of A. A quantum measure differs from a classical probability measure in the phenomenon of interference between histories which leads to it being non-additive. A familar example is the double slit experiment: the quantum measure of the set of histories which go through the upper slit plus the quantum measure of the set of histories which go through the lower slit is not equal to the quantum measure of the set of all histories which end up on the screen.

After the breakthrough of the Rideout-Sorkin models, it seemed that it would be relatively straightforward to generalise the derivation to obtain a quantum measure for causets. Roughly speaking, instead of transition probabilities there would be transition amplitudes and the quantum measure would be constructed from them via a generalisation of "sum and square" appropriate for a non-unitary dynamics.[25] General covariance and the appropriate quantum version of the Bell Causality condition could then be solved to find the form that the transition amplitudes must take. However, it is proving difficult to find the required Quantum Bell Causality condition, not least because the condition is not known even in the case of ordinary unitary Quantum Mechanics in a background with a fixed causal structure though we do now have at least a candidate for it.[26]

---

[e]Indeed Quantum Mechanics is the first of an infinite series of generalisations resulting in a heirarchy of measure theories with increasingly complex patterns of "interference" between histories.[24]

Even if we had in hand a covariant, causal quantum measure, there would still remain the problem of interpreting it. The interference between histories and consequent non-additive nature of the quantum measure mean that we are exploring new territory here. Reference 27 is a first attempt at a realist interpretation for quantum measure theory. It relies on the adequacy of predictions such as, "Event X is very very unlikely to occur," to cover all the predictions we want to make, which should include all the predictions we can make in standard quantum mechanics with its Copenhagen Interpretation. This adequacy is explicitly denied by Kent[28] and I have tended to agree with this judgement. However, the quantum measure is the result of taking a conservative approach to Quantum Mechanics (no new dynamics) whilst making histories primary, maintaining fundamental spacetime covariance and taking a completely realist perspective. As such it deserves to be persevered with.

4. Conclusions

The belabouring, in section 2, of the correspondence between the inverse processes of discretisation and continuum approximation makes manifest a certain conservatism of causal set theory: the steps are familiar, we've been down similar roads before. Moreover, spacetime, albeit discrete, is still considered to be real. There is no replacement of spacetime by a substance of a completely different ilk, such as a collection of $D0$-branes in 11-dimensional flat spacetime[29] to choose an example at random, as the underlying ontology. The radical kinematical input is the postulate of fundamental discreteness.

However, no matter how smooth one can make such arguments for causal sets, no scientific theory can be arrived at by pure philosophical introspection. For a start, hard scientific labour is already contained in the proof of the essential result that the causal structure fixes the spacetime metric up to local volume information. This has been strengthened by results that show that topological and geometrical information can indeed be "read off" from a causet which is a sprinkling into Minkowski spacetime. More importantly, we need to do a great deal of further work. For example, within kinematics the Hauptvermutung needs to be given a formal mathematical statement and more evidence provided for it and we need to have more results on how to read off geometrical information from a causet especially in the case of curved spacetime.

In the area of dynamics the Rideout-Sorkin models, though only classically stochastic, are proving invaluable for exploring issues such as the problem of general covariance. It is possible that a more or less direct generalisation of the derivation of these models will lead to the desired quantum theory. Finding a quantum dynamics is the central challenge for workers in causal set theory. I have given a somewhat sketchy account of some of the conceptual hurdles that need to be overcome before this can be achieved.

One thing that has not been explored in this article is how far causal set theory has come in the area of *phenomenology*, in other words in the deriving of observable consequences of the theory. It will be hard for any approach to quantum gravity to come to be universally accepted without experimental and observational evidence in the form of predictions made and verified. In this regard, causal set theory already has the advantage of a long-standing prediction of the current order of magnitude of the cosmological constant, or "dark energy density"[6,7,9] that has apparently now been verified. The argument leading to this prediction is heuristic – it depends on certain expectations about the quantum theory – and can only be made rigorous with the advent of the full quantum causal set dynamics. However, the sheer unexpectedness of the observational result amongst the wider community of theorists – some cosmologists have called it preposterous[30] – is great encouragement to investigate the arguments further. Numerical simulations of stochastic cosmologies based on the arguments bear out the conclusion that the envelope of the fluctuations of the dark energy density tends to "track" the energy density of matter [31]. Improvements would include models which allow spatial inhomogeneities in the dark energy density.

Moreover, there is great promise for further phenomenological model building. The unambiguous kinematics of the causal set approach means that there is an obvious way to try to make phenomenological models: create a plausible dynamics for matter (particles or fields, say) on the background of a causal set that is well approximated by the classical spacetime that we actually observe: Minkowski spacetime or Friedmann-Robertson-Walker spacetime, depending on the physical context. The dynamics of the matter might be classical or quantum. The limitation of such model building is that it doesn't take into account the quantum dynamics of the causal set itself, nor any back-reaction, but these models could be a first step in deriving observable effects of the underlying discreteness on phenomena such as the propagation of matter over long distances. An example of exactly this form is a model of point particle motion on a causet which leads to a prediction of a Lorentz invariant diffusion in particle momentum and therefore energy.[10] A naive application to cosmic protons doesn't work as a universal acceleration mechanism that might explain high cosmic energy rays but a quantum version might do better and the idea could be applied to other particles like neutrinos.

In causal set theory, we now have the mathematical structure that Einstein lacked, giving us a framework for a fundamentally discrete theory of spacetime which does not rely on any continuum concept as a aid. How successful it will be in realising the unification that Einstein hoped for, will be for the future to decide. But let me end by musing on a unification even beyond that of quantum gravity: the unification of kinematics and dynamics. In causal set theory as currently conceived, the subject of the theory and the laws by which it is governed are different in kind. This is apparent in the Rideout-Sorkin models for example. The law of growth is given by a sequence of non-negative numbers. This law is not part of physical reality which is the causal set. To a materialist like myself, it would be more satisfying if the laws themselves were, somehow, physically real; then the physical universe, meaning everything that exists, would be "self-governing" and not subject to laws imposed on it from outside. Should these nebulous ideas find concrete expression it would represent perhaps the ultimate unity of physics.

Acknowledgments

I gratefully acknowledge the influence of Rafael Sorkin on this article. Anyone familiar with his work will realise that I have plagiarised his writings shamelessly. I thank him and my collaborators on causal sets, Graham Brightwell, Raquel Garcia, Joe Henson, Isabelle Herbauts and Sumati Surya for their insights and expertise.

References

1. A. A. Robb, *Geometry of Time and Space.* (Cambridge University Press, Cambridge (UK), 1936). A revised version of "A theory of Time and Space" (C.U.P. 1914).
2. J. Stachel. Einstein and the quantum: Fifty years of struggle. In ed. R. Colodny, *From Quarks to Quasars, Philosophical Problems of Modern Physics*, p. 379. U. Pittsburgh Press, (1986).
3. G. 't Hooft. Quantum gravity: a fundamental problem and some radical ideas. In eds. M. Levy and S. Deser, *Recent Developments in Gravitation (Proceedings of the 1978 Cargese Summer Institute).* Plenum, (1979).
4. J. Myrheim, (1978). CERN preprint TH-2538.

5. L. Bombelli, J.-H. Lee, D. Meyer, and R. Sorkin, Space-time as a causal set, *Phys. Rev. Lett.* **59**, 521, (1987).

6. R. D. Sorkin. First steps with causal sets. In eds. R. Cianci, R. de Ritis, M. Francaviglia, G. Marmo, C. Rubano, and P. Scudellaro, *Proceedings of the ninth Italian Conference on General Relativity and Gravitational Physics, Capri, Italy, September 1990*, pp. 68–90. World Scientific, Singapore, (1991).

7. R. D. Sorkin. Space-time and causal sets. In eds. J. C. D'Olivo, E. Nahmad-Achar, M. Rosenbaum, M. P. Ryan, L. F. Urrutia, and F. Zertuche, *Relativity and Gravitation: Classical and Quantum, Proceedings of the SILARG VII Conference, Cocoyoc, Mexico, December 1990*, pp. 150–173. World Scientific, Singapore, (1991).

8. R. D. Sorkin. Causal sets: Discrete gravity (notes for the valdivia summer school). In eds. A. Gomberoff and D. Marolf, *Lectures on Quantum Gravity, Proceedings of the Valdivia Summer School, Valdivia, Chile, January 2002*. Plenum, (2005).

9. R. D. Sorkin, Forks in the road, on the way to quantum gravity, *Int. J. Theor. Phys.* **36**, 2759–2781, (1997).

10. F. Dowker, J. Henson, and R. D. Sorkin, Quantum gravity phenomenology, lorentz invariance and discreteness, *Mod. Phys. Lett.* **A19**, 1829–1840, (2004).

11. D. Stoyan, W. Kendall, and J. Mecke, In *Stochastic geometry and its applications*, chapter 2. Wiley, 2 edition, (1995).

12. S. W. Hawking, A. R. King, and P. J. McCarthy, A new topology for curved space-time which incorporates the causal, differential, and conformal structures, *J. Math. Phys.* **17**, 174–181, (1976).

13. D. B. Malament, The class of continuous timelike curves determines the topology of spacetime, *J. Math. Phys.* **18**, 1399–1404, (1977).

14. A. V. Levichev, Prescribing the conformal geometry of a lorentz manifold by means of its causal structure, *Soviet Math. Dokl.* **35**, 452–455, (1987).

15. L. Bombelli, Statistical lorentzian geometry and the closeness of lorentzian manifolds, *J. Math. Phys.* **41**, 6944–6958, (2000).

16. J. Noldus, A lorentzian lipschitz, gromov-hausdoff notion of distance, *Class. Quant. Grav.* **21**, 839–850, (2004).

17. D. Kleitman and B. Rothschild, Asymptotic enumeration of partial orders on a finite set, *Trans. Amer. Math. Society.* **205**, 205–220, (1975).

18. S. W. Hawking, Quantum gravity and path integrals, *Phys. Rev.* **D18**, 1747–1753, (1978).

19. J. B. Hartle. Space-time quantum mechanics and the quantum mechanics of space-time. In eds. J. Zinn-Justin and B. Julia, *Proceedings of the Les Houches Summer School on Gravitation and Quantizations, Les Houches, France, 6 Jul - 1 Aug 1992*. North-Holland, (1995).

20. D. P. Rideout and R. D. Sorkin, A classical sequential growth dynamics for causal sets, *Phys. Rev.* **D61**, 024002, (2000).

21. G. Brightwell, H. F. Dowker, R. S. Garcia, J. Henson, and R. D. Sorkin. General covariance and the 'problem of time' in a discrete cosmology. In ed. K. Bowden, *Correlations: Proceedings of the ANPA 23 conference, August*

16-21, 2001, Cambridge, England, pp. 1–17. Alternative Natural Philosophy Association, (2002).

22. G. Brightwell, H. F. Dowker, R. S. Garcia, J. Henson, and R. D. Sorkin, 'observables' in causal set cosmology, *Phys. Rev.* **D67**, 084031, (2003).

23. J. Henson, (2002). Unpublished result.

24. R. D. Sorkin, Quantum mechanics as quantum measure theory, *Mod. Phys. Lett.* **A9**, 3119–3128, (1994).

25. X. Martin, D. O'Connor, and R. D. Sorkin, The random walk in generalized quantum theory, *Phys. Rev.* **D71**, 024029, (2005).

26. D. Craig, F. Dowker, J. Henson, S. Major, D. Rideout, and R. Sorkin. A bell inequality analog in quantum measure theory, (2005). In preparation.

27. R. D. Sorkin. Quantum measure theory and its interpretation. In eds. D. Feng and B.-L. Hu, *Quantum Classical Correspondence: Proceedings of 4th Drexel Symposium on Quantum Nonintegrability, September 8-11 1994, Philadelphia, PA*, pp. 229–251. International Press, Cambridge, Mass., (1997).

28. A. Kent, Against many worlds interpretations, *Int. J. Mod. Phys.* **A5**, 1745, (1990).

29. T. Banks, W. Fischler, S. H. Shenker, and L. Susskind, M theory as a matrix model: A conjecture, *Phys. Rev.* **D55**, 5112–5128, (1997).

30. S. M. Carroll, Dark energy and the preposterous universe. (2001).

31. M. Ahmed, S. Dodelson, P. B. Greene, and R. Sorkin, Everpresent lambda, *Phys. Rev.* **D69**, 103523, (2004).

CHAPTER 17

THE TWISTOR APPROACH TO
SPACE-TIME STRUCTURES

ROGER PENROSE

Mathematical Institute, Oxford University,
24-29 St. Giles, Oxford OX1 3LB, UK
Institute for Gravitational Physics and Geometry, Penn State,
University Park, PA 16802-6300, USA

An outline of twistor theory is presented. Initial motivations (from 1963) are given for this type of non-local geometry, as an intended scheme for unifying quantum theory and space-time structure. Basic twistor geometry and algebra is exhibited, and it is shown that this provides a complex-manifold description of classical (spinning) massless particles. Simple quantum commutation rules lead to a concise representation of massless particle wavefunctions, in terms of contour integrals or (more profoundly) holomorphic 1st cohomology. Non-linear versions give elegant representations of anti-self-dual Einstein (or Yang-Mills) fields, describing left-handed non-linear gravitons (or Yang-Mills particles). A brief outline of the current status of the 'googly problem' is provided, whereby the right-handed particles would also be incorporated.

1. Early Motivations and Fundamental Basis of Twistor Theory

Twistor theory's original motivations, prior to December 1963 (which marks its initiation, as a physical theory[a]), came from several different directions; but in general terms, the intention was for a theory that would represent some kind of scheme for unifying basic principles coming from both quantum mechanics and relativity. Yet, the theory did not arise out of an attempt, on my part, to "quantize" space-time structure in any conventional sense. A good measure of my own reasons for not adopting this more conventional "quantum-gravity" stance came from a suspicion that the very rules of quantum mechanics might well have to be changed in such a unification, in order that its disturbing paradoxes (basically, the various forms of the *measurement* paradox) might perhaps be satisfactorily resolved as part of the proposed unification. For, in my own view, a resolution of these paradoxes would *necessarily* involve an actual (though presumably subtle) modification of the underlying quantum-mechanical rules. Nevertheless, it is clear that the rules of quantum mechanics must apply very precisely to physical systems that are, in an appropriate sense, "small". Most particularly, I had always been profoundly impressed by the physical role that quantum theory had found for the *complex number field*. This is manifested most particularly in the quantum linear superposition rule. But since quantum linearity can be regarded as the main "culprit" with regard to the measurement paradox one may expect that some kind of (complex?) *non*-linearity might begin to show itself when systems get large. For *small* systems the complex-*linear* superposition rule is extraordinarily precise.

We must, however, keep in mind that the notion of "small" that is of relevance here does not refer (simply) to small *distances*. We now know from experimental findings with EPR (Einstein–Rosen–Podolsky[b]) systems, that quantum entanglements can stretch unattenuated to at least some 30 kilometers.[c] Moreover, as an "objectivist" who had been aware of the profound puzzles that EPR phenomena would present for a consistent space-time picture of "quantum reality", I had considered these phenomena to be indicative of a need for some kind of objective "non-local geometry". Even at the more elementary level of a wavefunction for a single particle, there is an essential non-locality, as the measurement of the particle at one place forbids its detection at some distant place, even thought the spread of the wavefunction may have to encompass both locations in order for possible quantum interference to be accommodated. We should bear in mind that all such non-localities refer not just to small distances. When a notion of "smallness" is relevant, it is more likely that its appropriate measure might refer to *mass* displacements between two components of a superposition[d] (e.g. when a physical detector becomes entangled with the system), and general relativity tells us that mass displacements refer to *space-time curvature* differences.

Some years previously,[e] I had initiated *spin-network* theory. This theory had close ties with EPR-Bohm situations, these being entanglements which are manifested in *spin* correlations between widely separated events. The original form of spin-network theory had provided a kind of discrete *quantum geometry* for 3-dimensional Euclidean space, where spatial notions are taken to be *derived* rather than built initially into the theory, and it could be said to provide a non-local geometry of this nature. But I had recognized the limitations inherent in the essentially non-relativistic nature of that theory, and in its inability to describe spatial displacements. Spin networks arose from a study of the representation theory of the rotation group SO(3), and a logical route to follow might seem to be to replace this group by the Poincaré group. However, that approach did not particularly appeal to me, partly owing to the Poincaré group's non-semi-simple nature,[f] and generalizing further to the *conformal group* of Minkowski space — essentially to SO(2,4) — had, for various reasons seemed to me to be a possibility more in line with what I had in mind.

The representations of SO(3) are described in terms of 2-spinors, these providing the fundamental representation space of the *spin group* SU(2), of SO(3), and I had been very struck by the precise geometrical association between (projective SU(2)) 2-spinors and actual directions in 3-dimensional *physical space*.[g] In the case of the twistor group SO(2,4), there turned out to be an even more remarkable geometrical space-time association. Being led to consider representations of SO(2,4), instead of SO(3), we study its spin group SU(2,2), in place of SU(2). The fundamental representation space of SU(2,2) — the *reduced* (or "half") spin space for SO(2,4) — is the complex 4-dimensional vector space which I refer to as *twistor space* $\mathbb{T}$. The full (unreduced) spin space for SO(2,4) is the direct sum of twistor space $\mathbb{T}$ with its dual space $\mathbb{T}^*$. By virtue of $\mathbb{T}$'s $(+ + --)$ Hermitian structure, $\mathbb{T}^*$ can also be identified as the *complex conjugate* of $\mathbb{T}$. This Hermitian structure tells us that twistor space $\mathbb{T}$ has a 7-real-dimensional subspace $\mathbb{N}$, consisting of twistors whose (squared) "norm" $\|\mathbf{Z}\|$, given by this structure, *vanishes*. The elements of $\mathbb{N}$ are called *null twistors*, and we find that the *projective* null twistors (elements of the projective subspace $\mathbb{PN}$ of the projective twistor space $\mathbb{PT}$) are in precise geometrical correspondence with *light rays* in Minkowski space $\mathbb{M}$, i.e. null geodesics. Here we must include the limiting light rays that are the generators of the light cone at *null infinity* $\mathscr{I}$ for the conformally compactified Minkowski space $\mathbb{M}^\sharp$, these being described by the elements of a complex line $\mathbb{PI}$ lying in $\mathbb{PT}$ (where $\mathbf{I}$ is a certain complex 2-dimensional subspace of $\mathbb{T}$ representing space-time "infinity").

This geometrical fact is remarkable enough, but the relation between twistors and important physical quantities goes much farther than this. In the first place, the *real* scaling of a null twistor $\mathbf{Z}$ has a direct physical interpretation, assigning an actual (future-pointing) *4-momentum* (equivalently a frequency) to a massless particle with world-line defined by $\mathbb{PZ}$ (taking

$\mathbf{Z} \in \mathbb{N} - \mathbf{I}$). There remains a phase freedom

$$\mathbf{Z} \mapsto e^{i\theta} \mathbf{Z} \quad (\theta \text{ real}),$$

not affecting this 4-momentum. More strikingly, it turns out that *every* element $\mathbf{Z}$ of $\mathbb{T} - \mathbf{I}$ (not just those in $\mathbb{N}$), up to this same phase freedom, has an interpretation describing the *kinematics* of a massless particle which can have a *non-zero spin*. The helicity s of this particle (whose modulus is the spin) turns out to be simply

$$s = \frac{1}{2} \|\mathbf{Z}\|$$

(taking units with $\hbar = 1$). When $s \neq 0$ there is no actual "world-line" defined (in a Poincaré-invariant way), and the particle is, to some extent, *non-localized*.

What this demonstrates, since $\mathbb{T}$ is a *complex* vector space, is that twistor theory reveals a hidden Poincaré-invariant *holomorphic* (i.e. complex-analytic) structure to the kinematics of a massless particles (this kinematics being extended by the above phase freedom). The more primitive fact that the *celestial sphere* of an observer, according to relativity theory, can naturally regarded as a *Riemann sphere*[h] (a complex 1-manifold) is a particular aspect of this holomorphic structure. It had long struck me as particularly pertinent fact that only in the $1 + 3$ dimensions of our observed universe can the space of light rays through a point — i.e. the celestial sphere — be regarded as a complex manifold, and that the (non-reflective) Lorentz group can then be regarded as the group of holomorphic self-transformations of this sphere. On this view, it might be possible to view the celestial sphere as a kind of "quantum spread" of two directions, somewhat similarly to the way in which The Riemann sphere of possible "directions of spin" for a massive spin-$\frac{1}{2}$ particle can be thought of as a "quantum spread" of two independent directions (say, "up" and "down").

Nevertheless, all this twistor structure is, as yet, fully classical, so twistor theory is actually revealing a hidden role for complex-number structure which is present already at the classical level of (special) relativistic physics. This is very much in line with the driving force behind twistor theory. Rather than trying to "quantize" geometry, in some conventional sense, one seeks out strands of connection between the underlying mathematics of the quantum formalism and that of space-time geometry and kinematics. In relation to what has just been said, it may be pointed out that although this complex structure is an immediate feature only of *massless* particles, it is now a conventional standpoint to regard massless particles as being in some sense primary, with mass entering at a secondary stage, via some specific (e.g. Higgs) mechanism. Although twistor theory is perfectly capable of handling massive particles (and is neutral with respect to the specific Higgs mechanism), there indeed is a special "primitive" role for massless particles in that theory.

Something analogous could said to apply also to the "primitive" nature of Minkowski space $\mathbb{M}$, since it is here that these holomorphic twistor-related structures are most evident. Things get considerably more complicated when the effects of mass (or energy) begin to play their roles, and we are driven to consider *general relativity*. Here, the correct twistorial procedures are still only partially understood, but there is, nevertheless, a tantalizing relation between holomorphic structure and Einstein's vacuum equations, as revealed particularly in the "non-linear graviton" construction[i] of 1975/6, and its extensions. Moreover, many years previously, in the years before 1963, there had been another significant driving force behind the origination of twistor theory, which had relevance to this. I had been impressed by hints of a hidden complex structure revealed in the roles that complex (holomorphic) functions play in numerous exact solutions of the Einstein (vacuum) equations (e.g. plane-fronted waves, Robinson–Trautman solutions, stationary axi-symmetric solutions, including, as I learned later, the Kerr solution). The image of an ice-berg had come to mind, where all that we normally perceive of this hidden complex structure represents but a tiny part of it. This suggested that some novel way of looking at (possibly curved) space-time geometry might reveal some kind of hidden complex structure, even at the classical level. I had felt that an understanding of this could be a pointer to understanding how curved space-time structure might somehow become intertwined with quantum-mechanical principles (and particularly quantum theory's complex structure) in some fundamental way.

There is one further (but interrelated) motivation that may be mentioned here, namely the idea that the sought-for complex geometry should in some way automatically incorporate the notion — fundamental to quantum field theory — of the splitting field amplitudes into their positive and negative frequency parts. This procedure is neatly expressed, for a function of a single (real) variable, by the holomorphic extendibility of that function into the "top half" S^+ or "bottom half" S^- of the Riemann sphere S^2, where S^2 is divided into these two hemispheres by the equator, this representing the real line $\mathbb{R}$ (compactified to a circle by adjoining to it a "point at infinity"). In view of the importance of this frequency splitting to quantum field theory, I had imagined that our sought-for complex geometry should somehow reflect this frequency splitting in a natural but more global way, applying now all at once to entire fields, as might be globally defined on $\mathbb{M}$. As we shall be seeing in §5, this motivation is indeed satisfied in twistor theory but, as it turned out, in a much more subtle way than I had ever imagined.

2. Basic Twistor Geometry and Algebra

Choose standard Minkowski coordinates r^0, r^1, r^2, r^3 for $\mathbb{M}$ (with r^0 as the time coordinate, with $c = 1$). These are to be related to the standard complex coordinates Z^0, Z^1, Z^2, Z^3 for the vector space $\mathbb{T}$, via the *incidence relation*

$$\begin{pmatrix} Z^0 \\ Z^1 \end{pmatrix} = \frac{i}{\sqrt{2}} \begin{pmatrix} r^0 + r^3 & r^1 + ir^2 \\ r^1 - ir^2 & r^0 - r^3 \end{pmatrix} \begin{pmatrix} Z^2 \\ Z^3 \end{pmatrix}.$$

To find what locus $\mathbf{R}$ in $\mathbb{T}$ corresponds to a fixed point R in $\mathbb{M}$, we hold the coordinates r^a of R fixed and vary Z^α while maintaining the incidence relation. We have two homogeneous linear equations in the Z^α, giving us a linear subspace $\mathbf{R}$ of $\mathbb{T}$ of dimension 2 to represent R. In terms of the *projective* space $\mathbb{PT}$, we find a projective straight line $\mathbb{PR}$ (which is a Riemann sphere) to represent R. Conversely, to find the locus in $\mathbb{M}$ corresponding to a particular Z^α, we hold Z^α fixed and ask for the family of r^a which satisfy the incidence relation. Now if we require the r^a to be *real* — as indeed we should, if we are properly concerned with Minkowski space $\mathbb{M}$, rather than its complexification $\mathbb{CM}$ — then we find that the condition

$$\bar{Z}_\alpha Z^\alpha = 0$$

for a *null* twistor must hold, where (noting the order of $\mathbf{Z}$'s components, in what follows)

$$\bar{Z}_\alpha = (\bar{Z}_0, \bar{Z}_1, \bar{Z}_2, \bar{Z}_3) = complex\ conjugate\ of\ (Z^2, Z^3, Z^0, Z^1).$$

In fact, the quantity

$$\|\mathbf{Z}\| = \bar{Z}_\alpha Z^\alpha$$

$$= \bar{Z}_0 Z^0 + \bar{Z}_1 Z^1 + \bar{Z}_2 Z^2 + \bar{Z}_3 Z^3$$

$$= \frac{1}{2} (|Z^0 + Z^2|^2 + |Z^1 + Z^3|^2 - |Z^0 - Z^2|^2 - |Z^1 - Z^3|^2)$$

is the twistor *norm* referred to in §1, this being a Hermitian form of signature $(+ + - -)$ (because of the swapping of the first two components with the second in the definition of twistor complex conjugation). Assuming that $\|\mathbf{Z}\| = 0$, so $\mathbf{Z}$ is a null twistor, we find that the locus z of points in $\mathbb{M}$ which are incident with $\mathbf{Z}$ is indeed a *light ray* (null geodesic), in accordance with what was asserted in §1. At least this is strictly the case provided that Z^2 and Z^3 do not both vanish. If they do both vanish, then we can interpret the light ray z in conformally compactified Minkowski space $\mathbb{M}^\sharp$, as a generator of the light cone $\mathscr{I}$ at infinity.[j]

Note that 'the index positioning on $\bar{Z}_\alpha$ is consistent with $\bar{\mathbf{Z}}$ being a *dual* twistor (element of $\mathbb{T}^*$). Generally, the operation of twistor complex conjugation interchanges the spaces $\mathbb{T}$ and $\mathbb{T}^*$, so the up/down nature of twistor indices are reversed under complex conjugation. Sometimes I shall use the script letter $\mathcal{C}$ to denote twistor complex conjugation, as applied to (abstract-)indexed twistor (or, later, 2-spinor) quantities. Thus, we have, in particular, $\mathcal{C}Z^\alpha = \bar{Z}_\alpha$ and $\mathcal{C}\bar{Z}_\alpha = Z^\alpha$.

The geometrical correspondence between $\mathbb{M}^\sharp$ and $\mathbb{PN}$, and also between the complexification $\mathbb{CM}^\sharp$ and $\mathbb{PT}$, has many intriguing features. I mention only a very few of these here. We have seen that points of $\mathbb{M}^\sharp$ correspond to projective lines lying in $\mathbb{PN}$, and that points of $\mathbb{PN}$ correspond to light rays in $\mathbb{M}^\sharp$. In $\mathbb{M}^\sharp$, incidence is represented by a point lying on a light ray; in $\mathbb{PN}$, incidence is represented, correspondingly, by a projective line passing through a point. A *general* projective line in $\mathbb{PT}$ (*not* necessarily restricted to lie in $\mathbb{PN}$) corresponds to a *complex* space-time point, i.e. a point of $\mathbb{CM}^\sharp$. (This is a classical correspondence of 19th century geometry, often referred to as the *Klein* correspondence, $\mathbb{CM}^\sharp$ being understood as a complex 4-quadric.[k])

A general point $\mathbb{P}\mathbf{Z}$ of $\mathbb{PT}$ corresponds to a 2-complex-dimensional locus Z referred to as an *α-plane* (a standard classical terminology), this being a "self-dual" 2-plane on which the complex metric (induced from $\mathbb{CM}^\sharp$) vanishes identically. There is another type of complex 2-surface on $\mathbb{CM}^\sharp$ whose metric vanishes identically, which is "anti-self-dual", called a *β-plane*. The twistor correspondence represents β-planes in $\mathbb{CM}^\sharp$ by *complex projective 2-planes* in $\mathbb{PT}$.

The Hermitian relationship $\mathbf{Z} \leftrightarrow \bar{\mathbf{Z}}$ provides a *duality* transformation of $\mathbb{PT}$ in which points go to complex 2-planes, and *vice versa*, so this complex conjugation interchanges α-planes with β-planes on $\mathbb{CM}^\sharp$. In general, the point $\mathbb{P}\mathbf{Z}$ of $\mathbb{PT}$ will not lie on its corresponding plane $\mathbb{P}\bar{\mathbf{Z}}$ in $\mathbb{PT}$, the condition for it to do so being $\|\mathbf{Z}\| = 0$. In terms of $\mathbb{CM}^\sharp$: an α-plane Z will not generally meet its complex conjugate β-plane $\bar{Z}$, but the condition for it to do so is $\|\mathbf{Z}\| = 0$. When they do meet, their intersection is the complexification $\mathbb{C}z$ of the light ray z that we obtained earlier (the part of $\mathbb{C}z$ lying in $\mathbb{M}^\sharp$ being the light ray z itself).

Finally, we note that the projective-space structure of $\mathbb{PT}$ completely fixes the complex-*conformal* structure of $\mathbb{CM}^\sharp$ in a very direct way. By the term "complex-conformal" structure, I simply mean the structure defined by the complex null cones, this being equivalent to a locally defined complex metric, up to general conformal rescalings. Two points R, S of $\mathbb{CM}^\sharp$ are *null separated* if and only if the corresponding lines $\mathbb{P}R$ and $\mathbb{P}S$ of $\mathbb{PT}$ *intersect*, whence, the *light cone* of a point R in $\mathbb{CM}^\sharp$ is represented in $\mathbb{PT}$ by the family of lines which meet $\mathbb{P}R$. More generally, we can obtain the Minkowskian squared interval between R and S as $-4\mathbf{R}{:}\mathbf{S}(\mathbf{R}{:}\mathbf{I})^{-1}(\mathbf{S}{:}\mathbf{I})^{-1}$, where $\mathbf{R}{:}\mathbf{S}$ stands for $\frac{1}{2}\varepsilon_{\alpha\beta\rho\sigma}\mathsf{R}^{\alpha\beta}\mathsf{S}^{\rho\sigma}$, etc., and where each of $\mathsf{R}^{\alpha\beta}$, etc. is antisymmetrical and simple; i.e. has the form $\mathsf{R}^{\alpha\beta} = \mathsf{X}^{[\alpha}\mathsf{Y}^{\beta]}$, etc. Here, I have made use of the particular *infinity twistors* $\mathsf{I}^{\alpha\beta}$ and $\mathsf{I}_{\alpha\beta}$, representing the 2-dimensional "infinity" subspace I of $\mathbb{T}$ (referred to in §1) or line $\mathbb{P}\mathsf{I}$ of $\mathbb{PT}$, subject to

$$\mathsf{I}_{\alpha\beta} = \tfrac{1}{2}\varepsilon_{\alpha\beta\rho\sigma}\mathsf{I}^{\rho\sigma} = \mathcal{C}\mathsf{I}^{\alpha\beta}\,, \quad \mathsf{I}^{\alpha\beta} = \tfrac{1}{2}\varepsilon^{\alpha\beta\rho\sigma}\mathsf{I}_{\rho\sigma} = \mathcal{C}\mathsf{I}_{\alpha\beta}\,,$$

$$\mathsf{I}_{[\alpha\beta}\mathsf{I}_{\rho]\sigma} = 0\,, \quad \mathsf{I}^{[\alpha\beta}\mathsf{I}^{\rho]\sigma} = 0\,, \quad \mathsf{I}^{\alpha\beta}\mathsf{I}_{\beta\gamma} = 0\,.$$

(Here $\varepsilon_{\alpha\beta\rho\sigma}$ and $\varepsilon^{\alpha\beta\rho\sigma}$ are skew-symmetrical Levi-Civita twistors, satisfying $\varepsilon_{\alpha\beta\rho\sigma}\varepsilon^{\alpha\beta\rho\sigma} = 24$ and $\mathcal{C}\varepsilon^{\alpha\beta\rho\sigma} = \varepsilon_{\alpha\beta\rho\sigma}$ and $\mathcal{C}\varepsilon_{\alpha\beta\rho\sigma} = \varepsilon^{\alpha\beta\rho\sigma}$). Conformal-invariance breaking can be achieved by the incorporation of $\mathsf{I}^{\alpha\beta}$ or $\mathsf{I}_{\alpha\beta}$ into expressions, as desired.

3. Momentum and Angular Momentum for Massless Particles

It is convenient to use a 2-spinor notation[1] for much of twistor theory. The components of a twistor naturally fall into two pairs, where the first two are the components of an upper unprimed 2-spinor $\boldsymbol{\omega}$ and the second two, the components of a lower primed spinor $\boldsymbol{\pi}$:

$$\omega^0 = \mathsf{Z}^0\,, \quad \omega^1 = \mathsf{Z}^1\,, \quad \pi_{0'} = \mathsf{Z}^2\,, \quad \pi_{1'} = \mathsf{Z}^3\,.$$

So we can write

$$\mathsf{Z}^\alpha = (\omega^A, \pi_{A'})$$

and

$$\bar{\mathsf{Z}}_\alpha = (\bar{\pi}_A, \bar{\omega}^{A'})\,,$$

so that

$$\bar{Z}_\alpha Z^\alpha = \bar{\pi}_A \omega^A + \bar{\omega}^{A'} \pi_{A'}$$

(bearing in mind that the primed spin-space is the complex conjugate of the unprimed spin-space). Then the incidence relation becomes

$$\omega^A = i\, r^{AA'} \pi_{A'}$$

where a standard 2-spinor representation of 4-vectors is being used:

$$r^{AA'} = \frac{1}{\sqrt{2}} \begin{pmatrix} r^{00'} & r^{01'} \\ r^{10'} & r^{11'} \end{pmatrix}.$$

On change of origin, from the original origin O to a new origin Q whose position vector relative to O is q^a (i.e. $q^{AA'}$), we find

$$\omega^A \leadsto \omega^A - i\, q^{AA'} \pi_{A'} \quad \text{and} \quad \pi_{A'} \leadsto \pi_{A'}$$

This turns out to be consistent with the physical interpretation of a twistor $Z^\alpha = (\omega^A, \pi_{A'})$, as providing the 4-momentum/6-angular momentum kinematics for a massless particle. From the twistor Z^α we can construct quantities (with $\bar{\pi}_A = \mathcal{C}\pi_{A'}$ and $\bar{\omega}^{A'} = \mathcal{C}\omega^A$)

$$p_a = \bar{\pi}_A \pi_{A'} \quad \text{and} \quad M^{ab} = i\, \omega^{(A} \bar{\pi}^{B)} \varepsilon^{A'B'} - i\varepsilon^{AB} \bar{\omega}^{(A'} \pi^{B')},$$

(where in expressions such as these, I am adopting an abstract-index viewpoint,[m] which allows me to *equate* the vector/tensor (abstract) index "a" with pair of spinor (abstract) indices "AA'", etc.) and we find that they have the correct behaviour ($p_a \leadsto p_a$, $M^{ab} \leadsto M^{ab} - q^a p^b + q^b p^a$) under change of origin, for p_a to be the 4-momentum and M^{ab} the 6-angular momentum for a relativistic system. Moreover, the required conditions, for a *massless* particle, that p_a be null and future pointing, and that the Pauli-Lubanski spin vector S_a constructed from p_a and M^{ab} be *proportional* to p_a

$$S_a = \frac{1}{2} e_{abcd} p^b M^{cd} = s\, p_a$$

with s being the *helicity* (which is the standard requirement), are now automatically satisfied by the above twistorial definitions, where the only restriction on Z^α is that $\pi_{A'} \neq 0$, in order that the 4-momentum be non-zero. Conversely, given a 4-momentum p_a and a 6-angular momentum M^{ab}, subject to these "massless-particle" conditions, we find that such a Z^α always exists, uniquely up to the phase freedom

$$Z^\alpha \mapsto e^{i\theta} Z^\alpha.$$

Furthermore, it turns out that the helicity s is simply given by

$$s = \frac{1}{2}\bar{Z}_\alpha Z^\alpha = \frac{1}{2}\|\mathsf{Z}\|$$

as was asserted in §1.

All this has been entirely classical, and we have seen that twistor space, being a complex space, indeed provides us with a natural (Poincaré-invariant) complex structure associated with the classical kinematics of a massless particle. (More explicitly, owing to the above phase freedom, the complex space that we have exhibited — the twistor space $\mathbb{T}$ with the sub-space I removed — is a circle bundle over the space of classical kinematics for a massless particle.) But what about the *quantum* kinematics? It turns out that all we need is to impose standard canonical commutation rules between Z^α and $\bar{\mathsf{Z}}_\alpha$:

$$[\mathsf{Z}^\alpha, \mathsf{Z}^\beta] = 0, \quad [\bar{\mathsf{Z}}_\alpha, \bar{\mathsf{Z}}_\beta] = 0, \quad [\mathsf{Z}^\alpha, \mathsf{Z}_\beta] = \delta^\alpha_\beta,$$

and then the standard commutators for p_a and M^{ab} (as Poincaré group generators, namely $[p_a, p_b] = 0$, $[p_a, M^{bc}] = 2ig_a{}^{[b}p^{c]}$, $[M^{ab}, M_{cd}] = 4ig^{[b}{}_{[c}M^{a]}{}_{d]}$) follow unambiguously. There are no factor ordering problems here (because of the symmetry brackets in the twistor expression for M^{ab}), but we must be slightly careful in the case of the helicity operator s, which nevertheless comes out unambiguously as

$$s = \frac{1}{4}(\mathsf{Z}^\alpha \bar{\mathsf{Z}}_\alpha + \bar{\mathsf{Z}}_\alpha \mathsf{Z}^\alpha).$$

We can now consider the notion of a *twistor wavefunction* which, since Z^α and $\bar{\mathsf{Z}}_\alpha$ are conjugate variables, should be a function *either* of Z^α *or* of $\bar{\mathsf{Z}}_\alpha$, but not of both (which is analogous to the rules for ordinary position x^a and momentum p_a, where a wavefunction can be a function of x^a or of momentum p_a, but not both). But what does it mean for a function $f(\mathsf{Z}^\alpha)$ to be independent of $\bar{\mathsf{Z}}_\alpha$? The condition is $\partial f/\partial \bar{\mathsf{Z}}_\alpha = 0$, which provides us with the Cauchy–Riemann equations for f, asserting that f is *holomorphic* in Z^α. We could, alternatively, choose the conjugate (or dual) representation, where our wavefunction g is taken to be holomorphic in $\bar{\mathsf{Z}}_\alpha$, which means *anti*-holomorphic in Z^α. It is more convenient, with this representation, to use a dual twistor variable $\mathsf{W}_\alpha(= \bar{\mathsf{Z}}_\alpha)$ and to consider $g(\mathbf{W})$ simply as holomorphic in W_α.

To be definite, I shall tend phrase my arguments in terms of the Z^α-representation (and then the results using the W_α-representation follow essentially by symmetry). In the Z^α-representation, we can interpret the quantum operator $\bar{Z}_\alpha$, according to

$$\bar{Z}_\alpha = -\frac{\partial}{\partial Z^\alpha}.$$

Note that this enables us to "re-instate holomorphicity" in expressions which may, classically, have to be described *non*-holomorphically. For example, the classical expression for *helicity* is given in terms of the distinctly non-holomorphic quantity $\bar{Z}_\alpha Z^\alpha$, whereas quantum-mechanically, we have the entirely holomorphic operator

$$s = \frac{1}{4}(Z^\alpha \bar{Z}_\alpha + \bar{Z}_\alpha Z^\alpha)$$

$$= \frac{1}{4}(2Z^\alpha \bar{Z}_\alpha - \delta^\alpha_\alpha)$$

$$= \frac{1}{2}(-Z^\alpha \frac{\partial}{\partial Z}^\alpha - 2).$$

We note that the operator

$$\Upsilon = Z^\alpha \frac{\partial}{\partial Z^\alpha}$$

is Euler's *homogeneity* operator, whose eigenfunctions are homogeneous functions in Z^α with eigenvalue the degree of homogeneity. Thus, if we wish to describe a massless particle whose helicity takes the specific value $n/2$, we can use a twistor function $f(Z^\alpha)$ which is homogeneous of degree $-n - 2$. For a photon, for example, we would use a function of degree 0 for the left-handed part and of degree -4 for the right-handed part.

4. Massless Fields and their Twistor Contour Integrals

What is the relation between such a twistor wavefunction for a particle of a specific helicity and the ordinary space-time description of such a particle? I shall use 2-spinor notation (where $\Box = \nabla_a \nabla^a = \nabla_{AA'} \nabla^{AA'}$, etc.). For helicity 0 we have

$$\Box \varphi = 0\,;$$

for negative helicity $n/2$ (< 0)

$$\nabla^{AA'} \phi_{AB...L} = 0\,;$$

and for positive helicity $n/2$ (> 0)

$$\nabla^{AA'} \chi_{A'B'...L'} = 0\,.$$

Here $\phi_{AB...L}$ has $-n/2$ indices and $\chi_{A'B'...L'}$ has $+n/2$ indices (a positive number in each case), and each is totally symmetric

$$\phi_{(AB...L)} = \phi_{AB...L}, \quad \chi_{(A'B'...L')} = \chi_{A'B'...L'}.$$

In the case $s = \pm 1$ (spin 1), we can relate these equations to the more familiar source-free Maxwell equations

$$\nabla^a F_{ab} = 0, \quad \nabla_{[a} F_{bc]} = 0,$$

given by $F_{ab}(= -F_{ba})$ defined by

$$F_{ab} = F_{AA'BB'} = \phi_{AB}\varepsilon_{A'B'} + \varepsilon_{AB}\chi_{A'B'}.$$

(Recall the abstract-index conventions noted in §3; also incorporated are the basic anti-symmetrical 2-spinor Levi-Civita quantities, used for raising or lowering 2-spinor indices, ε^{AB}, $\varepsilon^{A'B'}$, ε_{AB}, and $\varepsilon_{A'B'}$, where $g_{ab} = \varepsilon_{AB}\varepsilon_{A'B'}$.) In the case $s = \pm 2$, we obtain the linearized Einstein vacuum equations,[n] expressed in terms of the linearized curvature tensor K_{abcd}, with symmetry relations

$$K_{abcd} = K_{[cd][ab]}, \quad K_{[abc]d} = 0$$

and vacuum condition

$$K_{abc}{}^a = 0,$$

all these being automatic consequences of

$$K_{abcd} = \phi_{ABCD}\varepsilon_{A'B'}\varepsilon_{C'D'} + \varepsilon_{AB}\varepsilon_{CD}\chi_{A'B'C'D'}$$

(abstract indices!), the Bianchi identity equation

$$\nabla_{[a} K_{bc]de} = 0,$$

now re-expressing the spin-2 massless free-field equation above.

I have given the above expressions for Maxwell and linear Einstein fields in the general case of *complex* fields, which is appropriate for the description of wavefunctions, but if we wish to describe a real field, we restrict to the case where the χ and ϕ fields are complex conjugates of one another:

$$\chi_{A'B'} = \bar{\phi}_{A'B'} = \mathcal{C}\phi_{AB}, \quad \chi_{A'B'C'D'} = \bar{\phi}_{A'B'C'D'} = \mathcal{C}\phi_{ABCD}.$$

In the case of complex fields (wavefunctions), we can specialize to

$$\phi_{AB...L} = 0$$

which gives a *self-dual* field (for integer spin), describing a *positive*-helicity (right-handed) massless particle, or to

$$\chi_{A'B'\ldots L'} = 0$$

which gives an *anti-self-dual* field, describing a *negative*-helicity (left-handed) massless particle.

We ask for the relation between these massless field equations and a twistor wavefunction of homogeneity degree $-n - 2$. The answer is largely expressed in the contour-integral expressions°

$$\varphi(x) = c_0 \oint f(\omega, \pi)\delta Z$$

for the case $n = 0$,

$$\chi_{A'B'\ldots L'}(x) = c_n \oint \pi_{A'}\pi_{B'}\ldots\pi_{L'} f(\omega, \pi)\delta Z$$

for $n > 0$, and

$$\phi_{AB\ldots L}(x) = c_n \oint \frac{\partial}{\partial\omega^A} \frac{\partial}{\partial\omega^B} \cdots \frac{\partial}{\partial\omega^L} f(\omega, \pi)\delta Z$$

for $n < 0$. (The constants c_n are here left undetermined, their most appropriate values to be perhaps fixed at some later date.) In each case, ω is first to be eliminated by means of the *incidence relation*

$$\omega = i x \pi$$

before the integration is performed, the quantity δZ being defined by either of the following definitions

homogeneous case ($\oint$ with 1-dimensional real contour): $\delta Z = \pi_{A'} d\pi^{A'}$

inhomogeneous case ($\oint$ with 2-dimensional real contour): $\delta Z = \frac{1}{2} d\pi_{A'} \wedge d\pi^{A'}$.

In either case, the integration removes the π-dependence, and we are left with a function solely of x. Moreover, it is a direct matter to verify that the appropriate massless field equation is indeed satisfied in each case.

In the homogeneous case we get a genuine contour integral — in the sense that the answer does not change under continuous deformations of the (closed) contour, within regions where f remains holomorphic — provided that the entire integrand (including the 1-form δZ) has homogeneity degree zero, that being the condition that its exterior derivative vanishes. This condition is ensured by the nature of the 1-form δZ and the balancing of the homogeneities of the various terms, using the homogeneity prescription for f given above. Here the contour is just a one-dimensional real curve, so the geometry is normally very simple. The inhomogeneous case, on the other hand, involves a contour which is a two-dimensional real surface. This sometimes gives additional freedom, but the geometry is often not so transparent as in the homogeneous case. However, there is a more compelling reason to expect that the inhomogeneous case provides a broader-ranging viewpoint. This arises from the fact that we get a genuine contour integral even when the homogeneities do not balance, the exterior derivative of the integrand vanishing merely by virtue of its holomorphicity. The role the homogeneity balancing to zero is now simply that the integral will now *vanish* without this balancing.

There are many reasons why the inhomogeneous prescription gives a more powerful viewpoint, but perhaps the most transparent of these is that we can now describe the wavefunction of, say, a plane-polarized photon, which is not in an eigenstate of helicity — whereas this cannot be done directly in the homogeneous case — helicities $+1$ and -1 being now involved simultaneously. Using the inhomogeneous form, we can simply add together a twistor function of homogeneity degree -4 (to describe the right-handed part of the photon) and a twistor function of homogeneity degree 0 (to describe the left-handed part). This would be acted upon by the appropriate combination $\pi_{A'}\pi_{B'} + \partial^2/\partial\omega^{A'}\partial\omega^{B'}$. The cross-terms, for which the homogeneity does not balance to zero, simply disappear upon integration, and we are left with the appropriate space-time sum of a self-dual and an anti-self-dual complex Maxwell field.

In order to get a feeling for the nature of the contour-integral expressions for massless fields generally, let us consider the homogeneous case, so we just have a 1-dimensional contour, and take $n = 0$ with a very simple twistor function (homogeneity degree -2):

$$f(\mathbf{Z}) = \frac{1}{(A_\alpha Z^\alpha)(B_\beta Z^\beta)} \, .$$

The singularities of this function are simple poles, represented in $\mathbb{PT}$ as lying along two planes $\mathbb{PA}$ and $\mathbb{PB}$, corresponding to the vanishing of the respective terms $A_\alpha Z^\alpha$ and $B_\beta Z^\beta$. The point X in $\mathbb{CM}$ whose position vector is $\boldsymbol{x}$ is represented as a line $\mathbb{PX}$ in $\mathbb{PT}$ (and this lies in $\mathbb{PN}$ whenever $\boldsymbol{x}$ is real). Recall that the line $\mathbb{PX}$ is a Riemann sphere, and the function f has, as its singularities, a simple pole at each of the two points where $\mathbb{PX}$ meets the planes $\mathbb{PA}$ and $\mathbb{PB}$. We choose a contour on this sphere which loops once between these poles, separating them, so that it cannot be shrunk away continuously without passing across one pole or the other. In this case the integration is easily performed, and we obtain a field $\varphi(\boldsymbol{x})$ that is a constant multiple of $\{(x_a - q_a)(x^a - q^a)\}^{-1}$, where q^a is the position vector of the point Q whose representation in $\mathbb{PT}$ is the line of intersection $\mathbb{P}\mathbf{Q}$ of the two planes $\mathbb{P}\mathbf{A}$ and $\mathbb{P}\mathbf{B}$. Notice that the field is singular only on the light cone of Q, which is when X and Q are null-separated, i.e. when the lines $\mathbb{P}\mathbf{X}$ and $\mathbb{P}\mathbf{Q}$ meet. This singularity arises when the poles on the Riemann sphere $\mathbb{P}\mathbf{X}$ "pinch" together, and the contour cannot pass between them.

For the wavefunction of a free particle, we require a condition of *positive frequency*. This is achieved by demanding that the space-time field extend holomorphically into the region of $\mathbb{CM}$ referred to as the *forward tube*, which consists of points whose (complex) position vectors have imaginary parts which are *past-timelike*. We find that this corresponds precisely to the family of lines lying in the upper region $\mathbb{PT}^+$ of projective twistor space. We see that this positive-frequency condition is easily satisfied for the particular scalar field φ just considered, if we arrange for $\mathbb{P}\mathbf{Q}$ to lie entirely in $\mathbb{PT}^-$, for then no line in $\mathbb{PT}^+$ can meet it.[P] It is clear, also, that something very similar will hold for any twistor function f whose singularities, in $\mathbb{PT}^+$, lie within two disjoint closed sets $\mathcal{A}$ and $\mathcal{B}$. For then, supposing that we get a non-zero field at all, with a contour on the Riemann sphere $\mathbb{P}\mathbf{X}$ separating $\mathbb{P}\mathbf{X} \cap \mathcal{A}$ from $\mathbb{P}\mathbf{X} \cap \mathcal{B}$, it follows from the fact that $\mathcal{A}$ and $\mathcal{B}$ are disjoint closed sets in $\mathbb{PT}^+$ that the contours can never get pinched. This situation clearly has considerable generality. It applies, for example, equally well to fields of arbitrary helicity and not simply to the case $n = 0$, and many types of positive-frequency wavefunctions are thereby obtained.

5. Twistor Sheaf Cohomology

We see, by means of this contour-integral representation, that twistor theory provides a powerful method of generating wavefunctions for massless fields, where the massless field equations seem to "dissolve" into pure complex analyticity. There is, however, a curious difficulty with this twistor representation, the resolution of which will lead us to a more sophisticated and fruitful point of view. One of the properties of wavefunctions that is evident in the conventional space-time representation is that such fields are Poincaré covariant (and in fact conformally covariant), and that they form a complex *linear space*. This would be clear also for the twistor representation (perhaps even more so, especially for conformal invariance), were it not for the awkward fact that we seem to need to assign some fixed region in which the singularities are to reside, and that any such assignment would destroy manifest Poincaré covariance. The reason for seeming to need to fix the singularity region is that if we add two twistor functions, the singularities of the sum are likely to occupy the union of the singularity regions of the two twistor functions individually. If these singularity regions are allowed too much freedom, then with extensive (and perhaps continuous) linear combinations, the resulting regions may not merely be complicated; in certain cases they could even preclude the finding of any appropriate contour whatsoever.

Compensating this difficulty, there is evidently some mobility in the singularity regions themselves. In the case of the particular twistor function f, considered in the previous section, we could replace B_α by adding a multiple of A_α

$$B_\alpha \rightsquigarrow B_\alpha + \lambda A_\alpha \,,$$

where λ is any fixed complex number, thereby *moving* the singularity region **B**. It is not hard to see that this gives us a completely equivalent twistor function to the one that we had before, in the sense that if we subtract one from the other (while retaining a contour that works for both), then this "difference" twistor function finds its singularities all on one side of the contour, necessarily giving *zero* for the contour integral because the contour "slips off" on the other side. The same would apply if we choose to move **A**, correspondingly, rather than **B**. This indicates that a deeper perspective on twistor functions is needed.

As a pointer to this deeper perspective, let us consider the more general situation of the two disjoint closed sets of singularities $\mathcal{A}$ and $\mathcal{B}$ in $\mathbb{PT}^+$, as introduced above (and we should also bear in mind that we may want to generalize $\mathbb{PT}^+$ itself to some other appropriate region of interest lying within $\mathbb{PT}$, or perhaps even to some other complex manifold). We may re-express the conditions on f, $\mathcal{A}$, and $\mathcal{B}$ in the following curious-looking way: the region on which f is assigned to be holomorphic is the *intersection* $\mathcal{U}_1 \cap \mathcal{U}_2$ of two open sets

$$\mathcal{U}_1 = \mathbb{PT}^+ - \mathcal{A} \quad \text{and} \quad \mathcal{U}_2 = \mathbb{PT}^+ - \mathcal{B} \,,$$

where the *union* $\mathcal{U}_1 \cup \mathcal{U}_2$ is the entire region $\mathbb{PT}^+$ that we are interested in. What is the purpose of this? We shall see in a moment. First, this situation extends to more complicated open coverings $\{\mathcal{U}_i\}$ of $\mathbb{PT}^+$, which we shall require to be *locally finite* (i.e. only a finite number of the $\mathcal{U}_i$ contain any given point of $\mathbb{PT}^+$), though the sets $\mathcal{U}_1, \mathcal{U}_2, \mathcal{U}_3, \ldots$ need not actually be finite in number. Since there are now liable to be many different intersections of pairs of these sets, our "twistor function" is now defined in terms of a *collection* $\{f_{ij}\}$ of various holomorphic functions f_{ij}, with

$$f_{ij} \text{ holomorphic on } \mathcal{U}_i \cap \mathcal{U}_j$$

subject to

$$f_{ij} = -f_{ji}$$

on intersecting pairs $\mathcal{U}_i \cap \mathcal{U}_j$, and

$$f_{ij} + f_{jk} = f_{ik}$$

on intersecting triples $\mathcal{U}_i \cap \mathcal{U}_j \cap \mathcal{U}_k$. Such a collection $\{f_{ij}\}$ is called a *1-cochain* (of holomorphic sheaf cohomology), with respect to the covering $\{\mathcal{U}_i\}$ of $\mathbb{PT}^+$. If all the members of this cochain can simultaneously be expressed in the form

$$f_{ij} = h_i - h_j \,, \quad \text{with each } h_k \text{ holomorphic on } \mathcal{U}_k$$

then we say that the collection $\{f_{ij}\}$ is a *1-coboundary*. (We may check that the particular freedom that we encountered, with the replacement $B_\alpha \rightsquigarrow B_\alpha + \lambda A_\alpha$ in our example above, is an example of adding a coboundary to f.) The elements of *1st cohomology* (here *holomorphic* 1st cohomology)^q, with respect to the covering $\{\mathcal{U}_i\}$, are the 1-cochains *factored out* by the 1-coboundaries (so two 1-cochains are deemed equivalent, as elements of 1st cohomology, if their difference is a 1-coboundary).

Now, we would like to get rid of this reference to a particular covering of $\mathbb{PT}^+$ and to refer just to $\mathbb{PT}^+$ itself. The general procedure for doing this is to take what is called a "direct limit" of finer and finer coverings, but this is complicated and non-intuitive. Fortunately, one of the miracles of complex analysis now comes to our rescue, which tells us that if the sets $\mathcal{U}_i$ are what are called *Stein* manifolds, then we are already finished, and we do not need to take the direct limit! The cohomology relative to any Stein covering $\{\mathcal{U}_i\}$ of $\mathbb{PT}^+$ is *independent* of the choice of Stein covering, and therefore refers simply to $\mathbb{PT}^+$ itself, as a whole!

But what is a Stein manifold? The definition refers simply to its intrinsic complex-manifold structure, and it does not depend on any particular imbedding of it in a larger complex manifold (such as $\mathbb{PT}^+$). It is easiest not to rely on a full definition a Stein manifold here, but merely give some examples of certain widespread classes of Stein manifolds, so that we can see how easy it is to ensure that we do have a Stein covering. In the first place, any region of $\mathbb{C}^n$ that is delineated by the vanishing of a family of (holomorphic) polynomial equations is Stein (an "affine variety"). Secondly, any region of $\mathbb{CP}^n$ obtainable by *excluding* the zero locus of a single homogeneous polynomial will be Stein. Thirdly, any region of $\mathbb{C}^n$ which has a smooth boundary satisfying a certain "convexity" condition — referred to as *holomorphic pseudo-convexity* (specified by a positive-definiteness criterion on a certain Hermitian form defined by the equation of the boundary, called the Levy form[r]). This last property tells us that we can always find small Stein-manifold open neighbourhoods of any point of a complex manifold (e.g. a small spherical ball).

An important additional property is that the *intersection* of Stein manifolds is again always Stein. We can use this fact to compare 1st cohomology elements defined with respect to two different Stein coverings $\{\mathcal{U}_i\}$ and $\{\mathcal{V}_I\}$. All we need to do is find the *common refinement* of the two coverings, which is a covering $\{\mathcal{W}_{iJ}\}$ each of whose members is the intersection of one set from each of the two coverings:

$$\mathcal{W}_{iJ} = \mathcal{U}_i \cap \mathcal{V}_J$$

(where we may ignore the empty intersections) and this will also be a Stein covering. Thus, if we have cochains $\{f_{ij}\}$ with respect to $\{\mathcal{U}_i\}$ and $\{g_{IJ}\}$ with respect to $\{\mathcal{V}_I\}$, then we can form the sum "$f+g$", with respect to the common refinement $\{\mathcal{W}_{iJ}\}$ as the cochain $\{f_{ij}|_{IJ} + g_{IJ}|_{ij}\}$, where $f_{ij}|_{IJ}$ is f_{ij} on $\mathcal{U}_i \cap \mathcal{U}_j$, restricted to its the intersection with $\mathcal{V}_I \cap \mathcal{V}_J$, and $g_{IJ}|_{ij}$ is g_{IJ} on $\mathcal{V}_I \cap \mathcal{V}_J$, restricted to its the intersection with $\mathcal{U}_i \cap \mathcal{U}_j$. It is a direct matter to show that this leads to a sum of the corresponding cohomology elements. Hence, the notion of a 1st cohomology element of $\mathbb{PT}^+$ is *intrinsic* to $\mathbb{PT}^+$, and does not really care about how we have chosen to cover it with a collection of open (Stein) sets.[s] This now gets us out of our difficulty with the twistor representation of massless fields. But it leads us to a greater degree of sophistication in the mathematics needed for the description of physical fields than we might have expected.

One aspect of this sophistication is that our twistor description is already a *non-local one*, even for the description of certain things which, in ordinary space-time, are perfectly local, like a physical field. But we should bear in mind that this twistor description is primarily for *wavefunctions*, and we recall from §1 that the wavefunction of a single particle is *already* something with puzzling non-local features (since the detection of the particle in one place immediately forbids its detection at some distant place). Accordingly, twistor theory's essentially non-local description of wavefunctions is actually something rather closer to Nature than the conventional picture in terms of a space-time "field". Indeed, it is frequently pointed out that this *holistic* character of a wavefunction distinguishes it fundamentally from the kind of *local* behaviour that is exhibited by ordinary physical fields or wavelike disturbances, this distinction contributing to the common viewpoint that a wavefunction is not to be attributed any actual "physical reality". However, we see that the cohomological character of the *twistor* formulation of a wavefunction gives it precisely the kind of holistic (non-local) nature that wavefunctions actually posses, and I would contend that the twistor formulation of a wavefunction assigns just the right kind of mathematical "reality" to a physical wavefunction.

To emphasize the essential non-locality of the concept of cohomology, we may take note of the fact that the 1st (and higher) holomorphic cohomology of any Stein manifold always *vanishes*. (I have not defined cohomology higher than the 1st here; the basic difference is that for n-cochains, we need functions defined on $(n+1)$-ple intersections, with a consistency condition on $(n+2)$-ple intersections, the coboundaries being defined in terms of "h"s on n-ple intersections.[t]) It is important, therefore, that $\mathbb{PT}^+$ is *not* Stein (it is, indeed, not pseudo-convex at its boundary $\mathbb{PN}$). It is this that allows non-trivial holomorphic 1st cohomology elements to exist. However, from what has just been said, we see that 1st cohomology of an open region always vanishes *locally*, in the sense that it vanishes if we restrict it down to a small Stein set containing any chosen point. First (and higher) cohomology, for an open complex manifold, is indeed an essentially *non-local* notion.

To end this section, we take note of the remarkable fact that the *positive-frequency* condition for a wavefunction is now neatly taken care of by the fact that we are referring simply to the holomorphic (1st) cohomology of $\mathbb{PT}^+$. Correspondingly, *negative*-frequency complex massless fields would be those which are described by the holomorphic cohomology of $\mathbb{PT}^-$. This provides a very close analogy to the way in which the positive/negative frequency splitting of a (complex) function defined on the real line — compactified into a circle S^1 — can be described in terms of holomorphic extendibility into the northern or southern hemispheres S^+, S^- of a Riemann sphere (a $\mathbb{CP}^1$) whose equator represents this S^1 (and where I am arranging things so that the "north pole" is the point $-i$, with the "south pole" at $+i$). A complex function defined on S^1 can be split into a component which extends holomorphically into S^+, namely the positive-frequency part, and a component which extends holomorphically into S^-, the negative frequency part. This splitting is very closely analogous to the splitting of a complex 1st cohomology element defined on the "equator" $\mathbb{PN}$ of $\mathbb{PT}$ (a $\mathbb{CP}^3$) into its "positive frequency part", which extends holomorphically into $\mathbb{PT}^+$, and its "negative frequency part", extending holomorphically into $\mathbb{PT}^-$. A complex function can be thought of as an element of "0th cohomology", and the whole "splitting" procedure applies also to nth cohomology, defined on an analogue of $\mathbb{PN}$, dividing $\mathbb{CP}^{2n+1}$ into two halves analogous to $\mathbb{PT}^+$ and $\mathbb{PT}^-$.

In each case, we can consider that these functions may be "twisted" to a certain degree, which refers to fixing a particular homogeneity for the functions (cohomology elements), as defined on the non-projective space $\mathbb{C}^{2n+2}$. There is, however, a more serious subtlety if we wish these cohomology elements to form a *Hilbert space*, so that there is a (positive definite) norm, or Hermitian scalar product defined. This requires an appropriate notion of "fall-off" as the cohomology element approaches the boundary. We can ensure that this scalar product exists, however, if we demand *analyticity* at S^1, or at $\mathbb{PN}$ (or at the higher-dimensional analogue of $\mathbb{PN}$, for higher cohomology), which is adequate for most purposes in twistor theory.[u]

Some readers might be disturbed by the dual role that $\mathbb{PT}^+$ seems to be playing in this discussion. On the one hand, we have seen in §3 that it is associated with *positive helicity* but, on the other, we now see that it is associated with positive *frequency*, which means positive *energy*. However, this association between the signs of helicity and energy may be regarded as a consequence of the (arbitrary) choice that we have made to express things in terms of $\mathbb{T}$ (the Z^α-description) rather than $\mathbb{T}^*$ (the W_α-description). Had we used $\mathbb{T}^*$, we would find that the forward tube is represented in terms of projective lines in $\mathbb{T}^*$ (the space of dual twistors W_α for which the norm $\|W\| = \bar{W}^\alpha W_\alpha$ is negative), whereas it would now be the massless kinematics for *negative* helicity which is represented by this space, and the association would be between the signs of *minus* the helicity and of energy.

6. The Non-Linear Graviton

There is a particular quality possessed by 1st cohomology that seems to provide a strong pointer to the future development of twistor theory. This is the existence of certain non-linear generalizations of 1st cohomology that have important (but as yet incomplete) relevance to the twistor descriptions of the known interactions of Nature: the Einstein gravitational interaction and those forces (electro-weak and strong) described by Yang–Mills theory. Recall that in 1st cohomology we have functions defined on overlaps of *pairs* of open sets, with a consistency relation on *triple* overlaps of open sets. This is closely analogous to the procedure for building a (perhaps "curved") manifold out of overlapping coordinate neighbourhoods. Here there are *transition functions* defined on overlaps of pairs, with a consistency condition on triple overlaps, and there is a condition of non-triviality for the resulting manifold that is analogous to the coboundary condition of cohomology.

The analogy can be made much more precise in the case of *small deformations* of some given complex manifold $\mathcal{X}$. Here we take a (locally finite) covering $\{\mathcal{U}_i\}$ of $\mathcal{X}$. We consider a cochain $\{F_{ij}\}$, where each F_{ij} is a holomorphic *vector field* on $\mathcal{U}_i \cap \mathcal{U}_j$. We are to think of the sets $\mathcal{U}_i$ as infinitesimally "sliding" over one another, as directed by this vector field. In fact, any non-trivial continuous deformation of $\mathcal{X}$ to a new complex manifold can be generated (infinitesimally) by such means, for some non-trivial 1st cohomology element defined by such a cochain. The converse is not quite true, as unless a certain 2nd cohomology element vanishes — which usually seems to be the case — we cannot guarantee that the given 1st cohomology element actually "exponentiates" consistently to give a finite deformation.

We shall first consider projective twistor space and, to be specific, let us take $\mathcal{X}$ to be either $\mathbb{PT}^+$ or an appropriate neighbourhood of some line $\mathbb{PR}$ in $\mathbb{PT}$. The latter situation refers to the *local* space-time neighbouring some space-time point R. We imagine some point Q, near R, moving around, so as to sweep out some open neighbourhood $\mathcal{V}$ of R in $\mathbb{CM}$; then the corresponding line $\mathbb{PQ}$ in $\mathbb{PT}$ will sweep out some small open region $\mathbb{PR}$ in $\mathbb{PT}$, called a *tubular neighbourhood* of $\mathbb{PR}$.

We are going to try to *deform* twistor space, so that it might, in some way, encode the structure of a *curved* space-time. The reason for considering $\mathbb{PT}^+$ or $\mathbb{PR}$ is that it turns out that we cannot deform the *whole* of $\mathbb{PT}$, continuously, so as to obtain a distinct complex manifold. For there are rigorous theorems which tell us that there is no complex manifold with the same topology as $\mathbb{CP}^3$ whose complex structure actually differs from that of $\mathbb{CP}^3$. But if we restrict attention to $\mathbb{PT}^+$, or to our tubular neighbourhood $\mathbb{PR}$ of the line $\mathbb{PR}$, then many such deformations are possible. For definiteness in what follows, let us work with the tubular neighbourhood case $\mathbb{PR}$.

In fact, we can use a twistor function f, homogeneous of degree $+2$ (this homogeneity corresponding to helicity $s = -2$, as is appropriate for a left-handed graviton), to generate the required deformation — where I am now assuming, for simplicity, that there are just two open sets $\mathcal{U}_1$ and $\mathcal{U}_2$ covering $\mathbb{P}\mathcal{R}$ (where we can take these sets to be the intersections of two Stein manifolds with $\mathbb{P}\mathcal{R}$, if we like, but $\mathcal{U}_1$ and $\mathcal{U}_2$ will not themselves be Stein), f being defined on $\mathcal{U}_1 \cap \mathcal{U}_2$. We may take it that the Riemann sphere $\mathbb{P}\mathbf{R}$ is divided, by its intersections with $\mathcal{U}_1$ and $\mathcal{U}_2$, into two slightly extended hemispherical open sets which overlap in an annular region, these two being "thickened out" to give us the two 3-complex-dimensional open regions $\mathcal{U}_1$ and $\mathcal{U}_2$. We shall generate our deformation by means of the vector field $\mathbf{F}$ on $\mathcal{U}_1 \cap \mathcal{U}_2$, defined (see §2) by

$$\mathbf{F} = \mathsf{I}^{\alpha\beta} \frac{\partial f}{\partial Z^\alpha} \frac{\partial}{\partial Z^\beta}$$

which we can rewrite as

$$\mathbf{F} = \varepsilon^{AB} \frac{\partial f}{\partial \omega^A} \frac{\partial}{\partial \omega^B}.$$

This vector field has homogeneity 0. In fact, in what follows, we shall first allow $\mathbf{F}$ to be a quite *general* holomorphic vector field on $\mathcal{U}_1 \cap \mathcal{U}_2$, homogeneous of degree 0. The special significance of the particular form given above will emerge a little later.

If we imagine sliding $\mathcal{U}_1$ over $\mathcal{U}_2$ by an infinitesimal amount, according to $\mathbf{F}$, then (by virtue of this homogeneity) we get an infinitesimal deformation of the complex structure of the projective space $\mathbb{P}\mathcal{R}$. In fact, we can now envisage *exponentiating* $\mathbf{F}$, so as to get a *finite* deformation $\mathbb{P}\tilde{\mathcal{R}}$ of $\mathbb{P}\mathcal{R}$. (Had we chosen a covering of $\mathbb{P}\mathcal{R}$ with more than two sets, then this might have been problematic with regard to getting the triple-overlap condition to behave consistently, and this is the reason for restricting attention the case of a two-set covering.) Roughly, this amounts to "breaking" $\mathbb{P}\mathcal{R}$ into two (overlapping) pieces and then re-gluing the pieces in a slightly displaced way, so as to obtain $\mathbb{P}\tilde{\mathcal{R}}$.

How are we to make use of $\mathbb{P}\tilde{\mathcal{R}}$ as a new kind of (projective) twistor space? We would like to have some "lines" in $\mathbb{P}\tilde{\mathcal{R}}$ that can be interpreted as the points of some sort of "space-time". We can't use the same lines as the $\mathbb{P}\mathbf{Q}$s that we had before, because these will have now become "broken" by this procedure. But fortunately, some theorems of Kodaira (1962, 1963) and Kodaira and Spencer (1958) now come to our rescue, telling us that (assuming that this deformation — though finite — is in an appropriate sense "not too big") there will always be a 4-complex-parameter family of holomorphic curves (Riemann spheres) lying in $\mathbb{P}\tilde{\mathcal{R}}$, characterized by the fact that they can be obtained *continuously* from the line $\mathbb{P}\mathbf{R}$ that we started with, lying in $\mathbb{P}\mathbb{T}$ (as the deformation continuously proceeds). Each curve $\mathbb{P}\tilde{\mathbf{Q}}$ of this family — which I shall refer to as a *line* in $\mathbb{P}\tilde{\mathcal{R}}$ — is to be represented by a point $\tilde{Q}$ of a *new* complex manifold $\tilde{\mathcal{V}}$. The complex 4-manifold $\tilde{\mathcal{V}}$ is, indeed, *defined* as the space parameterizing these lines in $\mathbb{P}\tilde{\mathcal{R}}$.

In fact, the manifold $\tilde{\mathcal{V}}$ automatically acquires a complex *conformal* structure, purely from the incidence properties of $\mathbb{P}\tilde{\mathcal{R}}$. Recall from §2 that this was the case with the relationship between $\mathbb{C}\mathbb{M}$ and $\mathbb{P}\mathbb{T}$, since null separation between two points Q and S, in $\mathbb{C}\mathbb{M}$, is represented, in $\mathbb{P}\mathbb{T}$, simply by the meeting of the corresponding lines $\mathbb{P}\mathbf{Q}$ and $\mathbb{P}\mathbf{S}$. We adopt precisely the same procedure here, so the *light cone* of a point $\tilde{Q}$ in $\tilde{\mathcal{V}}$ is defined simply as the locus of points $\tilde{S}$ whose corresponding line $\mathbb{P}\tilde{\mathbf{S}}$ meets $\mathbb{P}\tilde{\mathbf{Q}}$. Again, it follows from general theorems that this light cone's vertex is an ordinary *quadratic* (i.e. not Finsler) one arising from a complex quadratic metric (which is thereby defined, locally, up to proportionality). We can now proceed to piece together several small regions $\tilde{\mathcal{V}}$, obtained from these "local" tubular neighbourhoods $\mathbb{P}\tilde{\mathcal{R}}$, so as to obtain a more extended complex conformal manifold $\mathcal{M}$, obtained from a more extended (projective) twistor space $\mathbb{P}\mathcal{T}$, generalizing complex Minkowski space $\mathbb{C}\mathbb{M}$ and its relation to $\mathbb{P}\mathbb{T}$.

Does $\mathcal{M}$ have any particular properties, by virtue of this construction? Indeed it does. Most significantly, the very existence of *points* in $\mathbb{P}\mathcal{T}$, leads to $\mathcal{M}$ being *anti-self dual* (where the "anti" part of this terminology is purely conventional, chosen here to fit in with standard twistor conventions and use of the Z^α-representation). Why is this? The lines in $\mathbb{P}\mathcal{T}$ constitute a 4-parameter family, so those passing through some given point $\mathbb{P}\mathbf{Z}$, of $\mathbb{P}\mathcal{T}$, (2 analytic conditions) will constitute a 2-parameter family. The points of $\mathcal{M}$ representing this 2-parameter family of lines will represent a complex 2-surface, which will be called the α-*surface* Z in $\mathcal{M}$. Now since each pair of lines through $\mathbb{P}\mathbf{Z}$ must intersect (at $\mathbb{P}\mathbf{Z}$), it follows that each pair of points in the α-surface Z must be *null separated*, so that, as in $\mathbb{M}$, any α-surface must be *totally null* (vanishing induced conformal metric), and therefore is either self dual or anti-self dual. Conventionally, we call the α-surfaces *self dual*.

Now we ask: what is the condition on the Weyl curvature tensor $C_{abc}{}^d$ (which is well defined, for a conformal manifold) for the existence of a 3-parameter family of (self-dual) α-surfaces? It is a straight-forward calculation to show that the self-dual part of the Weyl curvature must consequently vanish and, conversely, that the vanishing of the self-dual Weyl tensor of a conformal (complex) 4-manifold $\mathcal{M}$ is (locally) the condition for the existence (locally) of a 3-parameter family of α-surfaces. The existence of such a family enables us to construct $\mathcal{M}$'s (projective) *twistor space* $\mathbb{P}T$, each point of $\mathbb{P}T$ representing an α-surface in $\mathcal{M}$. The above procedure thus provides us with a direct way of constructing (any) conformally anti-self-dual (complex) 4-manifold from a suitable (but "generic") complex 3-manifold $\mathbb{P}T$, the only technical difficulty, in this construction being actually *finding*[v] such a family of lines in $\mathbb{P}T$. Their existence can be ensured by the procedure to follow.

The particular form of the vector field $\mathbf{F}$, as given at the beginning of this section in terms of a twistor function f, provides us only a restricted class of such projective twistor spaces $\mathbb{P}T$, whose particular significance we shall be seeing in a moment. For the *general* projective twistor space $\mathbb{P}T$, from which an anti-self-dual conformal manifold can be constructed, we simply generate our deformation using a *general* holomorphic vector field[w] $\mathbf{F}$ on $\mathcal{U}_1 \cap \mathcal{U}_2$, homogeneous of degree 0, in place of one of the above restricted form. By use of this kind of deformation — for deformations that are "not too big" — we are assured that $\mathbb{P}T$ has the right form for it to have an appropriate 4-parameter family of lines enabling a generic anti-self-dual $\mathcal{M}$ to be constructed.

We take note of the fact that this construction provides us with a *complex* anti-self-dual conformal manifold $\mathcal{M}$. Such an $\mathcal{M}$ can be the complexification of a real Lorentzian conformal manifold only in the (relatively) uninteresting case of conformal flatness. For in the complex case, the Weyl conformal curvature tensor C_{abcd} can be written in a 2-spinor (abstract-index) form in just the same way as for the tensor K_{abcd} of §4

$$C_{abcd} = \Phi_{ABCD}\varepsilon_{A'B'}\varepsilon_{C'D'} + \varepsilon_{AB}\varepsilon_{CD}X_{A'B'C'D'} \,,$$

where Φ_{ABCD} and $X_{A'B'C'D'}$ are each totally symmetrical. In the conformally self dual case we have

$$\Phi_{ABCD} = 0$$

and in the conformally anti-self-dual case we have

$$X_{A'B'C'D'} = 0 \,,$$

but in the Lorentzian case we must have

$$X_{A'B'C'D'} = \mathcal{C}\Phi_{ABCD} \,,$$

so if one vanishes so must the other, whence C_{abcd} as a whole must vanish — the condition for (local) conformal flatness. It may be remarked, however, that in the positive-definite case (signature$+ + ++$)[x] and the split-signature case (signature $+ + --$)[y] there is a large family of conformally (anti-)self-dual 4-manifolds. These spaces, and their twistor spaces, have a considerable pure-mathematical interest,[z] there being (different) "reality conditions" on each of the independent quantities Φ_{ABCD} and $X_{A'B'C'D'}$ in these two cases.

Yet, *complex* (anti-)self-spaces are by no means devoid of *physical* interest, especially those which can arise from deformations of (parts of) $\mathbb{T}$ for which $\mathbf{F}$ has the special form $\mathbf{F} = \varepsilon^{AB}\partial f/\partial\omega^A\partial/\partial\omega^B$, as given at the beginning of this section. We note that, in these cases, the operator $\partial/\partial\pi_{A'}$ does not appear, and it follows that its infinitesimal action on $Z^\alpha = (\omega^A, \pi_{A'})$ leaves $\pi_{A'}$ unaffected. Thus, $\mathbf{F}$ generates a deformation of (part of) $\mathbb{T}$ that preserves the projection

$$\mathbf{F} : \mathbb{T} \to \bar{\mathbb{S}}^*\,,$$

(where $\bar{\mathbb{S}}^*$ is the complex conjugate of the dual $\mathbb{S}^*$ of the spin space $\mathbb{S}$; note that $\mathbb{S}$ is the space of 2-spinors like ω^A, and $\bar{\mathbb{S}}^*$ is the space of those like $\pi_{A'}$). In each coordinate patch $\mathcal{U}_i$, with standard twistor coordinates, this projection takes the form

$$(\omega^A, \pi_{A'}) \mapsto \pi_{A'}\,,$$

and this now extends to a projection $\mathbf{F}$ that applies to the whole (non-projective) curved twistor space

$$\mathbf{F} : \mathcal{T} \to \bar{\mathbb{S}}^*$$

The inverse $\mathbf{F}^{-1}$ of this projection is a *fibration* of $\mathbb{P}\mathcal{T}$, each fibre being the entire complex 2-surface in $\mathcal{T}$ which projects down to a particular π in $\bar{\mathbb{S}}^*$. In this case, the *lines* in $\mathbb{P}\mathcal{T}$ can be neatly characterized as the projective versions of the holomorphic *cross-sections* of this fibration. These are the results of maps $\mathbf{R} : \bar{\mathbb{S}}^* \to \mathcal{T}$ (whose composition $\mathbf{F} \circ \mathbf{R}$ with $\mathbf{F}$ is the identity on $\bar{\mathbb{S}}^*$) which lift $\bar{\mathbb{S}}^*$ back into $\mathbb{P}\mathcal{T}$, and they generalize what in the canonical flat case would be expressed as $\pi_{A'} \mapsto (\mathrm{i}r^{AA'}\pi_{A'}, \pi_{A'})$, where r^a is the position vector of the point in $\mathbb{CM}$ that this cross-section defines. (Note the appearance of the basic incidence relation of §2.)

The 2-surfaces of this fibration have tangent directions annihilated by a simple closed 2-form τ which is ($\frac{1}{2}$×) the exterior derivative of a 1-form ι (of homogeneity degree 2):

$$2\tau = \mathrm{d}\iota\,, \quad \iota \wedge \tau = 0$$

(the latter condition ensuring simplicity of ι, since $\tau \wedge \tau = 0$). The forms ι and τ are part of the structure of $\mathcal{T}$ that is unchanged by the deformation generated by our special kind of $\mathbf{F}$ defined from f as given above; and in the flat case $\mathbb{T}$, these forms are the two alternative versions of the (Poincaré-invariant) "$\delta\mathbf{Z}$", used in the contour integrals of §4:

$$\iota = \mathsf{I}_{\alpha\beta} Z^\alpha \mathrm{d}Z^\beta = \varepsilon^{A'B'} \pi_{A'} \mathrm{d}\pi_{B'}, \quad \tau = \frac{1}{2}\mathsf{I}_{\alpha\beta} \mathrm{d}Z^\alpha \wedge \mathrm{d}Z^\beta = \frac{1}{2}\varepsilon^{A'B'} \mathrm{d}\pi_{A'} \wedge \mathrm{d}\pi_{B'}.$$

In addition, $\mathcal{T}$ has a ("volume") 4-form ϕ which is ($\frac{1}{4}\times$) the exterior derivative of a ("projective-volume") 3-form θ (each of homogeneity degree 4), also unchanged by the deformation, where in the flat case $\mathbb{T}$,

$$\theta = \frac{1}{6}\varepsilon_{\alpha\beta\rho\sigma} Z^\alpha \mathrm{d}Z^\beta \wedge \mathrm{d}Z^\rho \wedge \mathrm{d}Z^\sigma, \quad \phi = \frac{1}{24}\varepsilon_{\alpha\beta\rho\sigma} \mathrm{d}Z^\alpha \wedge \mathrm{d}Z^\beta \wedge \mathrm{d}Z^\rho \wedge \mathrm{d}Z^\sigma,$$

and we have

$$4\phi = \mathrm{d}\theta, \quad \iota \wedge \theta = 0$$

This local structure possessed by $\mathcal{T}$ enables $\mathcal{M}$ to be assigned a metric $g_{ab}(= \varepsilon_{AB}\varepsilon_{A'B'})$, where the "$\varepsilon^{A'B'}$" in $g^{ab} = \varepsilon^{AB}\varepsilon^{A'B'}$ comes from τ (the area form in the structure of $\bar{\mathbb{S}}^*$) and "ε_{AB}" comes from "$\phi \div \tau$" (the area form in the fibres of $\mathbb{F}^{-1}$). This metric determines a connection, and because the projection $\mathbb{F}: \mathcal{T} \to \mathbb{S}^*$ is preserved in our special deformation, it follows that there is a global parallelism for elements of $\bar{\mathbb{S}}^*$ (the space of $\pi_{A'}$-spinors). The self-dual part of the Weyl curvature being zero ($X_{A'B'C'D'} = 0$), this turns out to imply the *vanishing of the Ricci tensor* ($R_{ab} = 0$). In fact, this argument reverses, and we find that any complex-Riemannian 4-manifold $\mathcal{M}$, which is both Ricci flat and conformally anti-self dual, has (locally) a twistor space $\mathcal{T}$ with the structure just provided above, from which $\mathcal{M}$ can be reconstructed by the foregoing procedure (in which the points of $\mathcal{M}$ are identified as holomorphic cross-sections of the fibration $\mathbb{F}^{-1}$).

How are we to interpret such a complex "space-time" $\mathcal{M}$ physically? The first place where such spaces were encountered, in a physical context, was with the $\mathcal{H}$-space construction of Ezra T. Newman (1976, 1979). Here, one considers the complexified future conformal infinity $\mathbb{C}\mathcal{J}^+$, of an asymptotically flat space-time (assumed analytic), and examines cross-sections ("cuts") of $\mathbb{C}\mathcal{J}^+$, which satisfy a condition of "vanishing complex asymptotic shear", this being a generalization to a curved space-time of a procedure which would locate the intersections, with $\mathbb{C}\mathcal{J}^+$, of the future light cones of points in $\mathbb{CM}$. In the case where $\mathcal{M} = \mathbb{CM}$, this procedure indeed enables $\mathbb{CM}$ to be reconstructed from the geometrical structure of its $\mathbb{C}\,\mathscr{I}^+$. However, in the general case of a gravitationally radiating (analytic) asymptotically flat space-time, one finds that this procedure does not reproduce the original space-time (complexified) but, instead, produces a Ricci-flat, conformally anti-self-dual complex 4-space called $\mathcal{H}$-space, which may be thought of as the "space-time" reconstructed from the anti-self-dual part of its outgoing gravitational field.[aa]

Subsequently it was proposed[bb] that such anti-self-dual Ricci-flat complex spaces, if subjected to an appropriate condition of "positive frequency" could be viewed as representing a *non-linear* description of the *wavefunction* of a left-handed graviton. Indeed this kind of interpretation is very much in line with the aims of twistor theory, as put forward in §1. In the standard perturbative viewpoint, a single "graviton" would be described by a solution of *linearized* general relativity, and it is an entirely flat-space quantity, where curvature and non-trivial causality structure does not arise. These geometrical properties, characteristic of general relativity proper, only occur when contributions involving an indefinitely large number of such "linear" gravitons are involved. But a non-linear graviton, as described by the above twistorial construction, is already a curved-space entity, and we may take the view that, in that description, each graviton carries its own measure of actual curvature. Moreover, being a non-linear entity, the concept of such a graviton moves us away from the standard linear structure of quantum mechanics, leading to a hope that eventually some non-linear quantum mechanics might arise, according to which the measurement paradox might eventually be resolved.

7. The Googly Problem; Further Developments

As yet, however, there is no clear relation between this kind of non-linearity and that which might be relevant to state reduction. Any such development would seem to require, as an initial step, a more complete description of the gravitational field than that which was outlined in the previous section. The "non-linear graviton" of §6 is, after all, only "half a graviton" in the sense that it restricts our consideration to only one of the two helicities that should be available to a graviton. Of course one could repeat the entire argument in terms of the *dual* twistor description — or W_α-representation — and then we should have a description of a right-handed non-linear graviton. But this is of no use to us if we wish for a comprehensive formalism in which, for example, plane-polarized gravitons might be described. In such a formalism, it would have to be possible also to describe a right-handed (i.e. *self*-dual) graviton while still using the Z^α-representation (or, equivalently, to describe a left-handed graviton using the W_α--representation). This would mean finding the appropriate non-linear version of a twistor function homogeneous of degree -6 (which, in the Z^α-representation, means helicity $s = +2$). This problem is referred to as the *googly problem* (a reference to the subtle bowling of a cricket ball with a right-handed spin about the direction of motion, using a bowling action which would appear to be imparting a left-handed spin).

I have referred to the googly problem only in the gravitational case, but there are analogues for electromagnetism and other Yang–Mills fields also. For left-handed photons (or the left-handed high-energy massless limit of W- or Z-bosons, or "gluons" of strong interactions) one would anticipate some sort of twistor deformation that is obtainable from the "exponentiation" of vector field obtainable from a twistor function homogeneous of degree 0. Indeed, there is such a construction, due to Richard Ward (1977) which provides the general anti-self-dual solution of the Yang–Mills equations, for a given specific group, in terms of such a (generally non-linear) twistor construction. Basically, what is required is to produce a (locally unconstrained) holomorphic vector bundle (for this group), over some suitable region $\mathcal{X}$ of $\mathbb{PT}$ where, as in §6, we may take $\mathcal{X}$ to be either $\mathbb{PT}^+$ or an appropriate tubular neighbourhood of some line $\mathbb{PR}$ in $\mathbb{PT}$. Then, noting that holomorphicity basically fixes the bundle to be *constant over any line* in $\mathcal{X}$, we obtain a fibre over that point in $\mathbb{CM}$ which corresponds to this line. Thus, we can transfer the whole bundle over $\mathcal{X}$ to a bundle over the corresponding region $\mathcal{V}$, in $\mathbb{CM}$. From the fact that there is a single fibre over any specific point $\mathbb{PZ}$ in $\mathcal{X}$, we find a natural connection defined on the bundle over $\mathcal{X}$ that is necessarily constant over the corresponding α-plane Z in $\mathcal{V}$. This makes the connection an *anti-self-dual* one, as required, and we find that this construction is reversible, showing the essential equivalence between anti-self-dual Yang–Mills fields on (appropriate) regions of $\mathbb{CM}$ and holomorphic bundles on the corresponding regions in $\mathbb{PT}$.

This is the geometrical essence of the Ward construction. It has found many applications, especially in the theory of integrable systems.[cc] It is clear, however, that for a full application of these ideas in basic physical theory, a satisfactory solution to the Yang–Mills googly problem is needed also. In view of its importance, it is perhaps remarkable how little interest this topic has aroused so far in the physics community, although the matter has recently received some attention in the twistor-string literature, which

I shall refer to briefly at the end of this section.

My own approach has been to concentrate attention more thoroughly on the gravitational googly problem, in the hope that the seemingly more severe restrictions on what types of construction are likely to be appropriate, in the gravitational case, may act as a guide to the correct approach. On the whole, however, there has been much frustration (for over 25 years!), and it is still not altogether clear whether the correct approach has been found. Accordingly, I shall only rather briefly outline what seem to be the three most promising modes of attack on this problem, concentrating mainly on the first. These may be classified crudely as follows:

- Geometric
- Functorial
- Twistor-string related.

It should be mentioned that a satisfactory solution of the googly problem should also inform us how the left and right helicities of the graviton are to *combine*, so we anticipate a twistorial ("Z^{α}"-)representation of solutions of the full (vacuum) Einstein equations.

The geometric approach[dd] is the most fully developed, and it has turned out to be possible to encode the information of a vacuum (i.e. Ricci-flat), analytic and appropriately asymptotically flat, complexified, space-time $\mathcal{M}$ in the structure of a deformed twistor space $\mathcal{T}$. The construction of $\mathcal{T}$ from $\mathcal{M}$ is completely explicit, and the data determining the structure of $\mathcal{M}$ seems to be given freely (i.e. without differential equations or awkward boundary conditions having to be solved). However the re-construction of $\mathcal{M}$ from $\mathcal{T}$ remains conjectural and somewhat problematic.

The local structure of this proposed complex 4-dimensional twistor space $\mathcal{T}$ can be defined in terms of the (non-vanishing) holomorphic forms ι and θ (a 1-form and a 3-form, respectively) that we had in §6, where

$$\iota \wedge d\iota = 0, \quad \iota \wedge \theta = 0,$$

as before, from which we can infer that $\mathcal{T}$ is foliated, locally, by a family of holomorphic curves — which I shall refer to as *Euler curves* defined by θ, and of holomorphic 3-surfaces, defined by ι, each of these 3-surfaces being foliated by Euler curves, but now we are to specify each of ι and θ only up to proportionality. There are, however, some restrictions on how these forms can be jointly rescaled, which can be stated as the requirement that the following two quantities Π and Σ are to be invariant:

$$\Pi = d\theta \otimes d\theta \otimes \theta, \quad \Sigma = d\theta \otimes \iota = -2\theta \oslash d\iota,$$

where we demand that the two given expressions for Σ are to be equal, the bilinear operator "$\varnothing$" being defined by

$$\eta\varnothing(\rho \wedge \sigma) = (\eta \wedge \rho) \otimes \sigma - (\eta \wedge \sigma) \otimes \rho$$

as applied to any r-form η and 1-forms ρ and σ. We can express this $\varnothing$ in "index form" as twice the anti-symmetrization of the final index of η with the two indices of the 2-form which follows the $\varnothing$ symbol. (This generalizes to a $\varnothing$-operation between an r-form and a t-form, where we take $t\times$ the antisymmetrization of the final index of the r-form with all indices of the t-form.) The preservation of the two quantities Π and Σ is really just asserting that on the overlap of two open sets $\mathcal{U}'$ and $\mathcal{U}$, the quantities ι and θ must scale according to the (somewhat strange) rules

$$\iota' = \kappa\iota, \quad \theta' = \kappa^2\theta, \quad \text{and} \quad \mathrm{d}\theta' = \kappa^{-1}\mathrm{d}\theta,$$

for some scalar function κ. The *Euler homogeneity operator* Υ (a vector field — see beginning of §3), which points along the Euler curves, can be defined, formally, by

$$\Upsilon = \theta \div \phi$$

where we recall that the 4-form ϕ of §6 is defined from θ by $4\phi = \mathrm{d}\theta$. More precisely, we can define Υ by

$$\mathrm{d}\xi \wedge \theta = \Upsilon(\xi)\phi$$

for any scalar field ξ. We find, on the overlap between open regions $\mathcal{U}'$ and $\mathcal{U}$ (primed quantities κ' and Υ' referring to $\mathcal{U}'$), that

$$\kappa' = \kappa^{-1}$$
$$\Upsilon' = \kappa^3\Upsilon$$

and, consequently,

$$\Upsilon(\kappa) = 2\kappa^{-2} - 2\kappa$$

and, equivalently $\Upsilon(\kappa^{-1}) = 2\kappa^2 - 2\kappa^{-1}$. We can deduce from all this that, on overlaps, κ^3 takes the form

$$\kappa^3 = 1 - f_{-6}(Z^\alpha)$$

in standard flat-space terms, so we obtain the encoding of a twistor function f_{-6}, homogeneous of degree -6 (in a fully cohomological way).

This geometrical means of encoding a twistor function of the required "googly" homogeneity -6 may seem somewhat strange, where the information is stored in a curious non-linear deformation of the *scaling* of the Euler curves (the curves which collapse down to points in the passage from $\mathcal{T}$ to $\mathbb{P}\mathcal{T}$). This deformation destroys the clear notion of the homogeneity degree of a function defined on $\mathcal{T}$. On the other hand, no other procedure has yet emerged for the encoding of this self-dual curvature information in a deformation of twistor space. Moreover, this curious non-linear scaling of the Euler curves can actually be seen to arise in an well-defined construction of $\mathcal{T}$ in terms of the space-time geometry of $\mathcal{M}$, where the points of $\mathcal{T}$ are defined as solutions of an explicit differential equation defined at $\mathbb{C}\mathscr{I}^{+}$, in which the *self-dual* (as opposed to anti-self-dual) Weyl curvature appears as a coefficient in the equation determining this scaling.[ee] There also appears to be a clear algebraic role for these particular scalings in certain relevant expressions.

Nonetheless, to be fully confident that such procedures are really following "correct" lines, one would like to have a clear-cut way of seeing that the resulting gravitational theory is really left/right symmetric, despite the extreme lop-sidedness that seems to be involved in this geometry. One might imagine that this could be understood at a formal algebraic level; for the algebra generated by commuting quantities Z^0, Z^1, Z^2, Z^3 is formally identical with that generated by $\partial/\partial Z^0$, $\partial/\partial Z^1$, $\partial/\partial Z^2$, $\partial/\partial Z^3$. But, can we see, for example, that if we translate (in some formal sense) a pure *left*-helicity ($s = -2$) "non-linear graviton" — as given by the prescriptions of §6 — from a construction in terms of ordinary "Z^α-coordinates" to one in terms of "$\partial/\partial Z^\alpha$-coordinates", then this now behaves (being now pure right-handed, $s = +2$) as though it were a complex manifold, as constructed in the present section, with a *flat* $\mathbb{P}\mathcal{T}$ (i.e. for which $\mathbb{P}\mathcal{T}$, is a portion of flat projective twistor space $\mathbb{P}\mathbb{T}$) so that a pure *right*-helicity graviton is now being described?

The required notions can be at least partially formulated in terms of *category theory*,[ff] where one finds a certain "functorial" relation between the multiplicative action of a quantity X and the derivative action of $\partial/\partial\mathsf{X}$, but where the latter is *dual* to the former, in the sense that the functorial arrows are reversed.[gg] One would hope to find that the above deformations generated by twistor functions of homogeneity degree $+2$ and -6 to be related in a similar way. As yet, this is not very clear, the issues being complicated by a basic obscurity about how one is to "dualize" the procedures that apply to building a manifold out of open coordinate patches. These matters are tied up with the issue of the "radius of convergence" of an analytic function defined by a power series in Z^α (so that the function has an appropriate open set on which it is defined) and whatever the corresponding notion should be for a "power series" in $\partial/\partial Z^\alpha$.

It would appear to be probable that some insights into the appropriate "quantum twistor geometry" are to be obtained from the procedures of *non-commutative geometry*[hh] applied to the original twistor space $\mathbb{T}$, since here we have basic "coordinates" Z^α and $\bar{Z}_\alpha$ which do not commute, these behaving formally like Z^α and $-\partial/\partial Z^\alpha$ I am not aware of any detailed work in this direction, however. In the absence of this, I wish to make some pertinent comments that seem to address this kind of issue from a somewhat different angle.

There is at least one way in which the replacements

$$\bar{Z}_\alpha \rightsquigarrow -\frac{\partial}{\partial Z^\alpha} \quad \text{and} \quad \frac{\partial}{\partial \bar{Z}_\alpha} \rightsquigarrow Z^\alpha$$

do find a clear mathematical representation in important twistor expressions. This is in the (positive definite) Hermitian *scalar product* $\langle f|g \rangle$ between positive-frequency twistor functions (1st cohomology elements) f and g, each of a given homogeneity degree r. Let us choose another such twistor function h, but now of homogeneity $r - 1$. Then we find the relations

$$\langle \frac{\partial f}{\partial Z^\alpha}|h \rangle = -\langle f|Z^\alpha h \rangle \,,$$

$$\langle Z^\alpha h|f \rangle = -\langle h|\frac{\partial f}{\partial Z^\alpha} \rangle \,,$$

which is consistent with the above replacements, where we must bear in mind that the quantities in the "$\langle \ldots |$" actually appear in complex-conjugate form and that in the first of these relations a minus sign comes about when the action of "$\partial/\partial \bar{Z}_\alpha$" is transferred from leftward to rightward.

To see how to ensure that these relations are satisfied, we need the general form of the scalar product, but where (for the moment) I restrict attention to cases for which $r > -4$. We find that this scalar product takes the form

$$\langle f|g \rangle = c \oint \bar{f}(W_\alpha)[W_\alpha Z^\alpha]_{-r-4} g(Z^\alpha) \mathrm{d}^4\mathbf{W} \wedge \mathrm{d}^4\mathbf{Z}$$

where c is some constant, independent of r, where $\mathrm{d}^4\mathbf{W} = \frac{1}{24}\varepsilon^{\alpha\beta\rho\sigma}\mathrm{d}W_\alpha \wedge \mathrm{d}W_\beta \wedge \mathrm{d}W_\beta \wedge \mathrm{d}W_\beta$ and correspondingly for $\mathrm{d}^4\mathbf{Z}$ (which is the 4-form ϕ above), and where (with $n > 0$)

$$[x]_{-n} = -(-x)^{-n}(n-1)!$$

so that

$$[x]_{-1} = x^{-1} \quad \text{and} \quad \frac{\mathrm{d}[x]_{-n}}{\mathrm{d}x} = [x]_{-n-1} \,.$$

The scalar product $\langle \ldots \mid \ldots \rangle$ then satisfies the required relations, above (for $r > -4$).

Now, bearing in mind what was said earlier, towards the end of §4, about the need to consider twistor functions that are not necessarily homogeneous (this being reinforced by the discussion of the googly problem, earlier in this section, in which inhomogeneous expressions arise), we find that we shall need to replace the "$[W_\alpha Z^\alpha]_{-r-4}$" term in the scalar product by a *sum* of such terms in which different values of r are involved (taking note of the fact that the "cross-terms", where the integrand has homogeneity other than zero, must vanish). If we are concerned with only a finite number of these terms, then we have no problem, but for an infinite number of terms (again restricting to $r > -4$, for the moment), then we appear to be presented with the seriously divergent series

$$F(x) = x^{-1} - 1!x^{-2} + 2!x^{-3} - 3!x^{-4} + 4!x^{-5} - 5!x^{-6} + \ldots .$$

Yet, in the 18th century, Euler had already shown that — in a formal sense at least — this series can be equated to the *convergent* expression[ii]

$$E(x) = (-\gamma - \log x) + x(1 - \gamma - \log x) + \frac{x^2}{2!}(1 + \frac{1}{2} - \gamma - \log x)$$

$$+ \frac{x^3}{3!}(1 + \frac{1}{2} + \frac{1}{3} - \gamma - \log x)$$

$$+ \frac{x^4}{4!}(1 + \frac{1}{2} + \frac{1}{3} + \frac{1}{4} - \gamma - \log x) + \ldots ,$$

where γ is *Euler's constant*. The function $E(x)$ can be defined in various equivalent ways, for example by the (equivalent) integral formulae

$$E(x) = e^x \int_x^\infty \frac{e^{-u}}{u} \, du$$

$$= \int_0^\infty \frac{e^{-u}}{(u+x)} du$$

where we *formally* have

$$F(x) - E(x) = 0 .$$

This suggests that we might use $E(W_\alpha Z^\alpha)$ in place of our divergent $F(W_\alpha Z^\alpha)$, in the contour integral expression for $\langle f \mid g \rangle$ (and that this might also serve to extend the definition of $\langle \ldots \mid \ldots \rangle$ to values of r with $r \leq -4$).

However, this is not such a simple matter because the singularity structure of the terms in the series for $E(x)$ involve logarithms, and the standard contours surrounding the *poles* that occur in the formal series for $F(x)$ run into branch cuts, and cannot be directly used. The issues arising here appear to be somewhat subtle and complicated, and they are not fully resolved at present. Various approaches for handling such logarithmic terms have been used in the past, including regarding $\log x$, in a contour integral, as arising from a limit[jj] of $x^\nu \Gamma(\nu)$ as $\nu \to 0$. But the most fruitful and satisfactory procedure appears to be to regard such terms as being treated as though the contour has a *boundary* in place of the branch cut in $\log x$, the "$\log x$" term itself being replaced, more or less, by $2\pi i$. In fact, according to a detailed study of twistor diagrams (the twistor analogue of Feynman diagrams) by Andrew Hodges (1985, 1998), this boundary should be taken at $x = k$, where k is some non-zero constant. Hodges also finds that the natural value of k appears to be given by $k = e^{-\gamma}$ (or possibly at $k = -e^{-\gamma}$), where γ is Euler's constant. This choice has to do with the requirement that infra-red divergences be regularized (as indeed they are with this prescription), and also that a certain idempotency requirement of twistor diagrams be satisfied.

It is interesting that this choice of boundary at $x = e^{-\gamma}$ corresponds to a requirement that the leading term "$(-\gamma - \log x)$", in the expansion of Euler's function $E(x)$ above, should vanish. Indeed, we appear to resolve a difficulty with the contour topology, in the definition of $\langle \ldots \mid \ldots \rangle$, if we allow a "blow down" in the twistor space $\mathbb{T}$ where our (8-dimensional) contour, in $\mathbb{T} \times \mathbb{T}^*$, encounters $W_\alpha Z^\alpha = e^{-\gamma}$, a place where $-\gamma - \log(W_\alpha Z^\alpha)$ *vanishes* on one branch of the logarithm, which seems to be a requirement for the consistency of this procedure. However the full significance of all this is not yet clear, and requires further understanding. This is work presently in progress, but there is significant hope that the procedures that Hodges has successfully developed in the theory of twistor diagrams may serve to illuminate, and to be illuminated by, this study of quantum twistor geometry.

Finally, some remarks concerning the recent *twistor string theory* are appropriate here. In December 2003, Edward Witten introduced the basis of this new body of ideas.[kk] Here, many of the procedures of string theory are united with those of twistor theory to provide some great simplifications and new insights in the theory of Yang–Mills scattering processes. These involve multiple "gluon" processes (in the massless limit), where in -and out- gluon states are taken to be pure helicity states. To some considerable extent, twistor theory is well set up to handle such situations (helicity states for massless particles being the natural building blocks of the physical interpretation of twistor theory). So it is perhaps not surprising that twistor theory can offer considerable simplifications in the description of such processes.

But it is likely that the string-theory perspective can also offer some new insights into the basis of twistor theory also. As has been emphasized at several places above, the information in a twistor wave-function is stored non-locally (in the form of 1st cohomology or in the non-linear construction of a deformed twistor space). There is no local information in these constructions. The situation is reminiscent of what happens in a topological quantum field theory, where again there is no local information, but a Lagrangian formalism can nevertheless be introduced. (An oft-cited example of this is "(2 + 1)-dimensional general relativity", where the "vacuum" is treated as a Ricci-flat region, as in standard 4-dimensional general relativity, but where in this 3-dimensional case the Weyl curvature also vanishes — automatically. There is now no local field information and no dynamics for this "gravitational field", yet a Lagrangian formalism can still be used.) Witten (1988) has shown that such a topological quantum field theory can be treated using string-theoretic procedures and non-trivial results thereby obtained (such as in the theory of knots and links in 3-dimensional space). The fact that there is no local dynamics both in the case of twistor theory and in the situations of a topological field theory suggests that there could be a link between the two. Indeed, Witten proposes such a link (this being more strictly a "holomorphic" than a topological theory[11]), and there is considerable hope that this may open up new prospects for twistor theory, where up to now there has been little in the way of a Lagrangian basis for developing a comprehensive "twistor dynamics" leading to a genuine approach to a twistorial theory of physics that can stand on its own.

As for how this might relate to a full solution to the googly problem, no serious attempt seems to have been made, so far, to tackle the issues that arise in gravitational theory. But some developments in the case of Yang–Mills theory have been suggested. These may be regarded as taking the Ward construction as encompassing the anti-self-dual part of the Yang–Mills field, but then perturbing away from this so as to provide a full description in which both self-dual and anti-self-duel parts of the Yang–Mills field are described. Although this procedure does not yet provide a full resolution of the googly problem in the Yang–Mills case (let alone the

Einstein case) it seems to indicate some new directions of procedure which could open up promising lines of new development.

Acknowledgments

I am grateful to various people for illuminating conversations, most particularly Andrew Hodges and Lionel Mason. I am grateful also to NSF for support under PHY00-90091.

Notes

a. See Penrose 1987, pp. 350, 359.

b. See Einstein, Podolsky, and Rosen (1935); Bohm (1951) Ch. 22, §§15–19; Bell (1987); Baggott (2004).

c. For results of this kind, see Tittel *et al.* (1998).

d. See Károlyházy, F. (1966); Diósi (1989); Penrose (1996, 2000, 2004).

e. This was around 1955, but only published later; see Penrose (1971, 1975, 2004). It should be mentioned that a version of spin-network theory is also used in the loop-variable approach to quantum gravity; see Ashtekar and Lewandowski (2004).

f. However, John Moussouris (1983) has had some success in pursuing this approach, in his (unpublished) Oxford D.Phil. thesis.

g. This is a well-known correspondence; see, for example, Penrose (2004), §22.9, Fig. 22.10.

h. See Terrell (1959); Penrose (1959, 2004 §18.5).

i. Penrose (1976).

j. See Penrose and Rindler (1986), Chapter 9; Penrose (2004) §§33.3, 5.

k. Penrose and Rindler (1986) §§9.2,3; Penrose (2004).

l. Penrose and Rindler (1984).

m. Penrose and Rindler (1984), Chapter 2.

n. Fierz and Pauli (1939); Fierz (1940); Penrose and Rindler (1984).

o. See Penrose (1968, 1969); Hughston (1979); Penrose and Rindler (1986); versions of these expressions can be traced back to Whittaker (1903) and Bateman (1904, 1944).

p. This type of non-singular field, termed an "elementary state", being of finite norm and positive frequency, plays an important role in twistor scattering theory (see Hodges 1985, 1998). These fields appear to have been first studied by C. Lanczos.

q. The more conventional term to use here, rather than "holomorphic cohomology" is "sheaf cohomology", with a "coherent analytic sheaf"; see Gunning and Rossi (1965), Wells (1991).

r. If the (smooth) boundary is defined by $L = 0$, where L is a smooth real-valued function of the holomorphic coordinates z_i and their complex conjugates $\bar{z}_i$, and whose gradient is non-vanishing at $L = 0$, then the Levy form is defined by the matrix of mixed partial derivatives $\partial^2 L / \partial z_i \partial \bar{z}_j$ restricted to the holomorphic tangent directions of the boundary $L = 0$. See Gunning and Rossi (1965); Wells (1991).

s. It may be remarked that a full Stein covering of $\mathbb{PT}^+$ must always involve an infinite number of open sets, because $\mathbb{PT}^+$ is not holomorphically pseudo-convex at its boundary $\mathbb{PN}$. In practice, however, one normally gets away with just a 2-set covering, encompassing a more extended region $\mathbb{PT}$ of than just $\mathbb{PT}^+$.

t. The descriptions of sheaf cohomology that I am providing here are being given only in the form of what is called Čech cohomology. This turns out to be by far the simplest for explicit representations. But there are other equivalent forms which are useful in various different contexts, most notably the Dolbeault (or $\bar{\partial}$) cohomology and that defined by "extensions" of exact sequences; see Wells (1991), Ward and Wells (1989).

u. For further details on these matters, see Eastwood, Penrose, and Wells (1981), Bailey, Ehrenpreis, and Wells (1982).

v. And checking the normal-bundle condition of the next note 23.

w. If we wish to exhibit $\mathbb{P}\mathcal{T}$ directly, rather than generating it in this way, we need to demand the existence of "lines" whose *normal bundle* is of the right holomorphic class. See Ward and Wells (1989).

x. Atiyah, Hitchin, and Singer (1978).

y. Dunajski (2002).

z. See, for example, Hitchin (1979, 1982); LeBrun (1990, 1998).

aa. See Penrose (1992).

bb. Penrose (1976).

cc. See, particularly, Mason and Woodhouse (1996) for an overview of these matters.

dd. Penrose (2001a); Frauendiener and Penrose (2001).

ee. Penrose (2001a).

ff. Eilenberg and Mac Lane (1945); Mac Lane (1988).

gg. See Penrose (2001b).

hh. Connes and Berberian (1995).

ii. See Hardy (1949).

jj. See Penrose (1968), although the needed Pochhammer-type contours were not understood at that time.

kk. Witten (2003); this was based partly on earlier results due to Parke and Taylor (1986) and by Berends and Giele (1988) on Gluon scatterings, and on twistor-related ideas of Nair (1988).

ll. See also Penrose (1988).

References

1. A. Ashtekar and J. Lewandowski, Background independent quantum gravity: a status report,*Class. Quant. Grav***21** R53-R152 (2004).
2. M. F. Atiyah, N. J. Hitchin and I. M. Singer, Self-duality in four-dimensional Riemannian geometry, *Proc. Roy. Soc. (London)* **A362**, 425–61 (1978).
3. J. Baggott, *Beyond Measure: Modern Physics, Philosophy and the Meaning of Quantum Theory* (Oxford University Press, Oxford, 2004).
4. T. N. Bailey, L. Ehrenpreis and R. O. Wells, Jr., Weak solutions of the massless field equations, *Proc. Roy. Soc. London* **A384**, 403–425 (1982).
5. H. Bateman, The solution of partial differential equations by means of definite integrals, *Proc. Lond. Math. Soc.* (2) **1**, 451–8 (1904).
6. H. Bateman, *Partial Differential Equations of Mathematical Physics* (Dover, New York, 1944).
7. J. S. Bell, *Speakable and Unspeakable in Quantum Mechanics* (Cambridge University Press, 1987).
8. F. A. Berends and W. T. Giele, Recursive calculations for processes with N gluons, *Nucl. Phys.* **B306**, 759 (1988).
9. D. Bohm, *Quantum Theory* (Prentice Hall, New York, 1951).
10. A. Connes and S. K. Berberian, *Noncommutative Geometry* (Academic Press, 1995).
11. L. Diósi, Models for universal reduction of macroscopic quantum fluctuations, *Phys. Rev.* **A40**, 1165–74 (1989).
12. M. Dunajski, Anti-self-dual four-manifolds with a parallel real spinor, *Proc. Roy. Soc. (London)* **A458(2021)**, 1205–22 (2002) .
13. M. G. Eastwood, R. Penrose and R. O. Wells, Jr., Cohomology and massless fields, *Comm. Math. Phys.* **78**, 305–51 (1981).
14. S. Eilenberg and S. Mac Lane, General theory of natural equivalences, *Trans. Am. Math. Soc.* **58**, 231–94 (1945).
15. E. Einstein, B. Podolsky and N. Rosen, Can quantum-mechanical description of physical reality be considered complete? *Phys. Rev.* **47**, 777–780 (1935).
16. M. Fierz and W. Pauli, On relativistic wave equations for particles of arbitrary spin in an electromagnetic field, *Proc. Roy. Soc. London* **A173**, 211–32 (1939).
17. M. Fierz, Uber den Drehimpuls von Teichlen mit Ruhemasse null und beliebigem Spin, *Helv. Phys. Acta* **13**, 45–60 (1940).
18. J. Frauendiener and R. Penrose, Twistors and general relativity, in *Mathematics Unlimited — 2001 and Beyond; Part 1*, eds. B. Enquist and W. Schmid (Springer-Verlag, Berlin, 2001), 479–505.
19. R. C. Gunning and R. Rossi, *Analytic Functons of Several Complex Variables* (Prentice-Hall, Englewood Cliffs, New Jersey, 1965).

20. G. H. Hardy, *Divergent Series* (Oxford University Press, New York, 1949).

21. N. J. Hitchin, Polygons and gravitons, *Math. Proc. Camb. Phil. Soc.* **85**, 465–76 (1979).

22. N. J. Hitchin, Complex manifolds and Einstein's equations, in *Twistor Geometry and Non-linear Systems*, ed. H. D. Doebner and T. D. Palev (Springer-Verlag, Berlin, 1982), pp. 73–99.

23. A. P. Hodges, A twistor approach to the regularization of divergences, *Proc. Roy. Soc. London* **A397**, 341–74; Mass eigenstatates in twistor theory, *ibid* 375–96 (1985).

24. A. P. Hodges, The twistor diagram programme, in *The Geometric Universe; Science, Geometry, and the Work of Roger Penrose*, eds. S. A. Huggett, L. J. Mason, K. P. Tod, S. T. Tsou and N. M. J. Woodhouse (Oxford Univ. Press, Oxford, 1998).

25. L. P. Hughston, *Twistors and Particles*, Lecture Notes in Physics No. 97 (Springer-Verlag, Berlin, 1979).

26. F. Károlyházy, Gravitation and quantum mechanics of macroscopic bodies, *Nuovo Cim.* **A42**, 390 (1966).

27. K. Kodaira, A theorem of completeness of characteristic systems for analytic submanifolds of a complex manifold, *Ann. Math.* (2) **75**, 146–162 (1962).

28. K. Kodaira, On stability of compact submanifolds of complex manifolds, *Am. J. Math.* **85**, 79–94 (1963).

29. K. Kodaira and D. C. Spencer, On deformations of complex analytic structures I, II, *Ann. Math.* **67**, 328–401, 403–466 (1958).

30. C. R. LeBrun, Twistors, ambitwistors, and conformal gravity, in *Twistors in Mathematical Physics*, LMS Lec. note ser. 156, eds. T. N. Bailey and R. J. Baston (Cambridge Univ. Press, Cambridge, 1990).

31. C. R. LeBrun, On four-dimensional Einstein manifolds, in *The Geometric Universe; Science, Geometry, and the Work of Roger Penrose*, eds. S. A. Huggett, L. J. Mason, K. P. Tod, S. T. Tsou and N. M. J. Woodhouse (Oxford Univ. Press, Oxford, 1998).

32. S. Mac Lane, *Categories for the Working Mathematician* (Springer-Verlag, Berlin, 1988).

33. L. J. Mason and N. M. J. Woodhouse, *Integrability, Self-Duality, and Twistor Theory* (Oxford University Press, Oxford, 1996).

34. J. P. Moussouris, Quantum models of space-time based on recoupling theory, Oxford D.Phil. thesis, (1983) unpublished.

35. V. Nair, A current algebra for some gauge theory amplitudes, *Phys. Lett.* **B214**, 215 (1988).

36. E. T. Newman, Heaven and its properties, *Gen. Rel. Grav.* **7**, 107–11 (1976).

37. E. T. Newman, Deformed twistor space and $\mathcal{H}$-space, in *Complex Manifold Techniques in Theoretical Physics*, eds. D. E. Lerner and P. D. Sommers (Pitman, San Francisco, 1979), pp. 154–65.

38. S. Parke and T. Taylor, An amplitude for N gluon scatterings, *Phys. Rev. Lett.* **56**, 2459 (1986).

39. R. Penrose, The apparent shape of a relativistically moving sphere, *Proc. Camb. Phil. Soc.* **55**, 137–9 (1959).

40. R. Penrose, Twistor quantization and curved space–time, *Int. J. Theor. Phys.*
     1, 61–99 (1968).
41. R. Penrose, Solutions of the zero rest-mass equations, *J. Math. Phys.* **10**,
     38–9 (1969).
42. R. Penrose, Angular momentum: an approach to combinatorial space-time,
     in *Quantum theory and Beyond*, ed. Ted Bastin (Cambridge University Press,
     Cambridge, 1971).
43. R. Penrose, Twistor theory: its aims and achievements, in *Quantum Grav-
     ity, an Oxford Symposium*, eds. C. J. Isham, R. Penrose and D. W. Sciama
     (Oxford University Press, Oxford, 1975).
44. R. Penrose, Non-linear gravitons and curved twistor theory, *Gen. Rel. Grav.*
     7, 31–52 (1976).
45. Penrose, R. (1987) On the origins of twistor theory, in *Gravitation and Geom-
     etry: a volume in honour of I. Robinson*, eds. W. Rindler and A. Trautman
     (Bibliopolis, Naples). http://users.ox.ac.uk/~tweb/00001/index.shtml
46. R. Penrose, Holomorphic linking and twistors. Holomorphic linking:
     postscript, *Twistor Newsletter* **27**, 1–4 (1988).
47. R. Penrose, $\mathcal{H}$-space and Twistors, in *Recent Advances in General Relativity*,
     (Einstein Studies, Vol. 4) eds. Allen I. Janis and John R. Porter (Birkhäuser,
     Boston, 1992) 6–25.
48. R. Penrose, On gravity's role in quantum state reduction, *Gen. Rel. Grav.*
     28, 581–600 (1996).
49. R. Penrose, Wavefunction collapse as a real gravitational effect, in *Mathe-
     matical Physics 2000*, eds. A. Fokas, T. W. B. Kibble, A. Grigouriou, and B.
     Zegarlinski (Imperial College Press, London, 2000), pp. 266–282.
50. R. Penrose, Towards a twistor description of general space-times; introduc-
     tory comments. In *Further Advances in Twistor Theory, Vol.III: Curved
     Twistor Spaces*, Chapman & Hall/CRC Research Notes in Mathematics 424,
     eds. L. J. Mason, L. P. Hughston, P. Z. Kobak and K. Pulverer (Chapman
     & Hall/CRC, London, 2001a), pp. 239–255, ISBN 1-58488-047-3.
51. R. Penrose, Physical left-right symmetry and googlies, in *Further Advances
     in Twistor Theory, Vol. III: Curved Twistor Spaces*, Chapman & Hall/CRC
     Research Notes in Mathematics 424, eds. L. J. Mason, L. P. Hughston, P. Z.
     Kobak and K. Pulverer (Chapman & Hall/CRC, London, 2001b), 274–280,
     ISBN 1-58488-047-3.
52. R. Penrose, *The Road to Reality: A complete guide to the Laws of the Universe*
     (Jonathan Cape, London, 2004).
53. R. Penrose and W. Rindler, *Spinors and Space-Time, Vol. 1: Two-Spinor
     Calculus and Relativistic Fields* (Cambridge University Press, Cambridge,
     1984).
54. R. Penrose and W. Rindler, *Spinors and Space-Time, Vol. 2: Spinor and
     Twistor Methods in Space-Time Geometry* (Cambridge University Press,
     Cambridge, 1986).
55. J. Terrell, Invisibility of the Lorentz contraction, *Phys. Rev.* **116**, 1041–5
     (1959).

56. W. Tittel, J. Brendel, H. Zbinden and N. Gisin, Violations of Bell inequalities by photons more than 10 km apart, *Phys. Rev. Lett.* **81**, 3563 (1998).

57. R. S. Ward, On self-dual gauge fields, *Phys. Lett.* **61A**, 81–2 (1977).

58. R. S. Ward and R. O. Wells, Jr., *Twistor Geometry and Field Theory* (Cambridge University Press, Cambridge, 1989).

59. R. O. Wells, Jr., *Differential Analysis on Complex Manifolds* (Prentice-Hall, Englewood Cliffs, New Jersey, 1991).

60. E. T. Whittaker, On the partial differential equations of mathematical physics, *Math. Ann.* **57**, 333–55 (1903).

61. E. Witten, Topological quantum field theory, *Comm. Math. Phys.* **118**, 411 (1988).

62. E. Witten, Perturbative gauge theory as a string theory in twistor space, *Commun. Math. Phys.* **252**, 189-258 (2003).

INDEX

e United States
zs

Printed in the United States
By Bookmasters